PLANETARY NEBULAE

INTERNATIONAL ASTRONOMICAL UNION
UNION ASTRONOMIQUE INTERNATIONALE

SYMPOSIUM No. 103

HELD AT UNIVERSITY COLLEGE, LONDON, U.K.
AUGUST 9–13, 1982

PLANETARY NEBULAE

EDITED BY

D. R. FLOWER
Physics Department, University of Durham, U.K.

D. REIDEL PUBLISHING COMPANY
DORDRECHT : HOLLAND / BOSTON : U.S.A. / LONDON : ENGLAND

Library of Congress Cataloging in Publication Data
Main entry under title:

Planetary Nebulae

At head of title: International Astronomical Union, Union Astronomique Internationale.
Includes index.
1. Planetary nebulae–Congresses. I. Flower, D. R. (David Roger), 1945- II. International Astronomical Union.
QB855.5.P54 1983 523.1′135 83-3010
ISBN 90-277-1557-2
ISBN 90-277-1558-0 (pbk.)

Published on behalf of
the International Astronomical Union
by
D. Reidel Publishing Company, P. O. Box 17, 3300 AA Dordrecht, Holland

Sold and distributed in the U.S.A. and Canada
by Kluwer Boston, Inc.,
190 Old Derby Street, Hingham, MA 02043, U.S.A.

In all other countries, sold and distributed
by Kluwer Academic Publishers Group,
P. O. Box 322, 3300 AH Dordrecht, Holland

D. Reidel Publishing Company is a member of the Kluwer Group

Printed in The Netherlands

TABLE OF CONTENTS

INVITED PAPERS

SECTION I: OBSERVATIONS OF PLANETARY NEBULAE

SECTION II: PHYSICAL PROCESSES IN PLANETARY NEBULAE

ABSTRACTS OF CONTRIBUTED PAPERS

SCIENTIFIC ORGANISING COMMITTEE

L. H. Aller
D. R. Flower
D. E. Osterbrock
B. Paczyński
N. Panagia
M. Peimbert (Chairman)
D. Péquignot
S. R. Pottasch
M. J. Seaton
I. S. Shklovsky
Y. Terzian
V. Weidemann

LOCAL ORGANISING COMMITTEE

M. J. Barlow
R. E. S. Clegg
D. R. Flower
J. Meaburn
M. M. Morris
N. K. Reay
I. W. Roxburgh
M. J. Seaton (Chairman)

PREFACE

IAU Symposium 103 was held at University College London, August 9-13 1982. This volume contains the proceedings of the meeting - invited papers, abstracts of contributed papers, and discussion. As is now customary with the proceedings of IAU Symposia, the manuscript was compiled from camera-ready copy. The Editor was responsible for the preparation of the abstracts of the contributed papers and the discussion, the authors of the invited papers for the preparation of their own reviews. The discussion at the meeting was lively and informative, and the Editor hopes that a reasonably faithful and readable record of the discussion is to be found in these proceedings.

In accordance with the wish of the Scientific Organising Committee, an object index has been compiled and appended. It is to be hoped that the index will augment the usefulness of the volume. The Editor is greatly indebted to M. J. Barlow for his help in preparing the index.

Financial assistance for the meeting was provided by the IAU and University College London. The hospitality received during the excursion to the Old Royal Observatory and National Maritime Museum, Greenwich, is gratefully acknowledged.

The task of editing these proceedings has been greatly facilitated by the excellent secretarial assistance of V. A. Kerr.

David Flower
Durham, October 1982

D. R. Flower (ed.), Planetary Nebulae, xiii.

The excursion to the Old Royal Observatory and National Maritime Museum, Greenwich.

The excursion to the Old Royal Observatory and National Maritime Museum, Greenwich.

The excursion to the Old Royal Observatory and National Maritime Museum, Greenwich.

LIST OF PARTICIPANTS

A. ACKER, Observatoire, 11 rue de l'Universite, 67000 - Strasbourg, France.

J. ADAM, Institut für Theoretische Astrophys., Im Neuenheimer Feld 294, 6900 Heidelberg, W. Germany.

S. ADAMS, Physics Department, University College London, Gower Street, London, WC1E 6BT.

D. AITKEN, Department of Physics and Astronomy, University College London, Gower Street, London, WC1E 6BT.

S. M. V. ALDROVANDI, Instituto Astronomico e Geofisico, Av. Miguel Stefano, 4200, 04301 - S. Paulo, Brazil.

L. H. ALLER, Department of Astronomy, University of California, Los Angeles, Ca. 90024, U.S.A.

D. BAADE, European Southern Observatory, Karl-Schwarzchild-Str.2, D8046 Garching, W. Germany.

T. BARKER, Wheaton College, Norton, Mass. 02766, U.S.A.

M. J. BARLOW, Physics Department, University College London, Gower Street, London, WC1E 6BT.

P. BARR, Physics Department, University College London, Gower Street, London, WC1E 6BT.

B. BAUD, Space Sciences, P.O. Box 800, 9700 Av Groningen, The Netherlands.

A. BAYES, Physics Department, University College London, Gower Street, London, WC1E 6BT.

S. C. BECK, Astronomy Department, Space Sciences Building, Cornell University, Ithaca, NY 14853, U.S.A.

M. BESSELL, Mt Stromlo Observatory, Private Bag, Woden PO, Act 2606, Australia.

C. BIGNELL, NRAO, VLA, P.O. Box O, Socorro, N.M. 87801, U.S.A.

J. BLACK, Centre for Astrophysics, 60 Garden Street, Cambridge, Ma. 02138, U.S.A.

B. BOHANNAN, Sommers-Bausch Observatory, University of Colorado Box 391, Boulder, Colorado 80302, U.S.A.

K. BUTLER, Physics Department, University College London, Gower Street, London WC1E 6BT.

C. CACCIARI, ESA Satellite Tracking Station, Apartado 54065, Madrid, Spain.

V. CALOI, Istituto de Astrofisica Spaziale, C.P. 67, 00044 Frascati, Italy.

E. R. CAPRIOTTI, Ohio State University, Astronomy Department, 174 W. 18th Avenue, Columbus, Ohio, 43210, U.S.A.

U. CARSENTY, Max-Planck Inst. fur Astronomy, 69 Heidelberg, Konigstuhl, W. Germany.

M. CERRUTI-SOLA, Osservatorio Astrofisico di Arcetri, Largo Enrico Fermi, 5, 50125 Florence, Italy.

A. CHE, Hamburger Sternwarte, Gojenbergsweg 112, 2050 Hamburg 80, W. Germany.

R. CLEGG, Physics Department, University College London, Gower Street, London, WC1E 6BT.

M. COHEN, Radioastronomy Laboratory, 601 Campbell Hall, University of California, Berkeley, Ca. 94720, U.S.A.

R. COSTERO, Instituto de Astronomia, Apartado Postal 70-264, Mexico, D.F. 04510, Mexico.

H. L. DINERSTEIN, NASA Ames Research Center, MS 245-6, Moffett Field, Ca. 94035, U.S.A.

M. A. DOPITA, Mt Stromlo Observatory, Private Bag, Woden PO, Act 2606, Australia.

J. E. DREW, Physics Department, University College London, Gower Street, London, WC1E 6BT.

W. EISSNER, Daresbury Laboratory, Nr. Warrington, Cheshire.

H. FORD, Department of Astronomy, University of California, Los Angeles, Ca. 90024, U.S.A.

H. B. FRENCH, Department of Physics and Astronomy, University of Oklahoma, Norman, Oklahoma 73019, U.S.A.

R. H. GARSTANG, Joint Institute for Laboratory Astrophysics, University of Colorado, Boulder, Colorado 80309, U.S.A.

R. GATHIER, Kapteyn Astronomical Institute, Postbus 800, 9700 Av. Groningen, The Netherlands.

J. R. GIDDINGS, Physics Department, University College London, Gower Street, London, WC1E 6BT.

F. GIESEKING, Observatorium Hoher List, D-5568 Daun, W. Germany.

A. E. GLASSGOLD, Physics Department, New York University, New York, N.Y. 10003, U.S.A.

F. GLEIZES, Universite des Sciences et Techniques du Languedoc, 34060 Montpellier Cedex, France.

A. A. GOHARJI, Physics Department, Durham University, Durham, DH1 3LE.

F. A. GOLDSWORTHY, School of Mathematics, University of Leeds, Leeds, LS2 9JT.

T. R. GULL, 9275 Brush Run, Columbia, MD. 21045, U.S.A.

J. P. HARRINGTON, Astronomy Program, University of Maryland, College Park, Md. 20742, U.S.A.

S. R. HEAP, Code 681, Goddard Space Flight Center, Greenbelt, Md. 20771, U.S.A.

K. G. HENIZE, Code CB, Johnson Space Centre, Houston, Texas, 77058, U.S.A.

J. HERMAN, Sterrewacht, Huygens Laboratorium, P.O. Box 9513, NL-2300 RA, Leiden, The Netherlands.

J. R. HOUCK, 220 Space Science Building, Cornell University, Ithaca, New York 14853.

I. HOWARTH, Physics Department, University College London, Gower Street, London, WC1E 6BT.

C. T. HUA, Laboratoire d'Astronomie Spatiale, CNRS, Les Trois Lucs, 13012 Marseille, France.

D. G. HUMMER, Joint Institute for Laboratory Astrophysics, University of Colorado, Boulder, Colorado 80309, U.S.A.

D. HUTSEMEKERS, Institute of Astrophysics, Universite de Liege, B-4200 Cointe Ougree, Belgium.

I. IBEN, Department of Astronomy, University of Illinois, 1011 West Springfield, Urbana, Illinois 61801, U.S.A.

R. ISAACMAN, UKIRT, 900 Leilani St., Hilo, Hawaii 96720, U.S.A.

G. H. JACOBY, KPNO, 950 North Cherry Avenue, P.O. Box 26732, Tucson, Arizona 85726, U.S.A.

F. D. KAHN, Astronomy Department, The University, Manchester, M13 9PL.

J. B. KALER, 341 Astronomy Building, 1011 West Springfield, Urbana, IL. 61801, U.S.A.

H. U. KAUFL, Max-Planck Institut fur Physik und Astrophysik, Institut fur Extraterr Physik, D-8046 Garching, W. Germany.

C. D. KEYES, Astronomy Department, University of California, Los Angeles, CA 90024, U.S.A.

C. KINDL, Institute of Astronomy, ETH-Zentrum, CH-8092 Zurich, Switzerland.

L. KOHOUTEK, Hamburg Observatory, Gojenbergsweg 112, D-2050 Hamburg 80, W. Germany.

J. KOPPEN, Inst. fur Theoretische Astrophysik, Im Neuenheimer Feld 294, D6900 Heidelberg 1, W. Germany.

K. B. KWITTER, Department of Physics and Astronomy, Williams College, Williamstown, Ma. 01267, U.S.A.

S. KWOK, National Research Council Canada, Ottawa, Canada, K1A OR6.

D. G. LAWRIE, Department of Astronomy, Ohio State University, 174 W. 18th Avenue, Columbus, Ohio 43210, U.S.A.

D. A. E. LEAVER, 88 St. James's Avenue, Hampton, Middlesex, TW12 1HN.

D. J. LENNON, Flat B, 38 Windsor Park, Belfast, Northern Ireland.

J. A. LOPEZ, Astronomy Department, The University, Manchester, M13 9PL.

J. LUTZ, Associate Professor of Astronomy, Washington State University, 428 French Administration Building, Pullman, Washington 99164, U.S.A.

T. E. LUTZ, Washington State University, 428 French Administration Building, Pullman, Washington 99164, U.S.A.

A. E. LYNAS-GRAY, Physics Department, University College London, Gower Street, London, WC1E 6BT.

R. McCARROLL, Laboratoire d'Astrophysique, University de Bordeaux, 33405 Talence, France.

W. J. MACIEL, Instituto Astronomico e Geofisico da USP, Caixa Postal 30,627, 01000 Sao Paulo, Sp - Brazil

D. C. V. MALLIK, Indian Institute of Astrophysics, Bangalore 560034, India.

W. MARTIN, Max-Planck-Institut fur Astronomie, D-6900 Heidelberg, 1, Konigstuhl, W. Germany.

J. MATHIS, Department of Astronomy, University of Wisconsin 475 N. Charles Street, Madison, WI 53705, U.S.A.

R. H. MENDEZ, Instituto de Astronomia y Fisica del Espacio, C.C.67 - Suc 28, 1428 Beunos Aires, Argentina.

C. MENDOZA, Physics Department, University College London, Gower Street, London, WC1E 6BT.

D. L. MOORES, Physics Department, University College London, Gower Street, London, WC1E 6BT.

A. NATTA, Osservatorio di Arcetri, Largo Enrico Fermi 5, 50125 Florence, Italy.

H. NUSSBAUMER, Institute of Astronomy, ETH Zentrum, 8092 Zurich, Switzerland.

D. E. OSTERBROCK, Lick Observatory, University of California - Santa Cruz, Santa Cruz, California 95064, U.S.A.

K. PANG, Jet Propulsion Lab, MS 183-901, 4800 Oak Grove Drive, Pasadena, CA91109, U.S.A.

K. PAPP, Department of Physics, University of Waterloo, Waterloo, Ontario, Canada, N2L 3G1.

P. PATRIARCHI, Villafranca Satellite Tracking Station, Apartado 54065, Madrid, Spain.

M. PEIMBERT, Instituto de Astronomia, UNAM, Apartado Postal 70-264, Mexico, 04510, Mexico.

C. PENN, Department of Physics, Durham University, South Road, Durham, DH1 3LE.

D. PEQUIGNOT, Observatoire de Meudon, 92190 Meudon, France.

A. PERAIAH, Indian Institute of Astrophysics, Bangalore 560034, India.

M. PERINOTTO, Osservatorio Astrofisico Arcetri, Largo E. Fermi 5, 50125, Florence, Italy.

M. PERL, Racah Institute of Physics, Hebrew University, 91904 Jerusalem, Israel.

J. P. PHILLIPS, Astronomy Division, Space Science Department, European Space Agency, Estec, Noordwijk, The Netherlands.

S. R. POTTASCH, Postbus 800, 9700 Av. Groningen, The Netherlands.

A. PREITE-MARTINEZ, Istituto di Astrofisica Spaziale, C.P. 67, I-00044 Frascati, Italy.

C. R. PURTON, Dominion Radio Astrophysical Observatory, Box 248, Penticton, British Columbia, Canada, V2A 6K3.

N. K. REAY, Astronomy Group, Blackett Laboratory, Imperial College, South Kensington, London, SW7 2BZ.

E. RECILLAS-CRUZ, Instituto de Astronomia, Apartado Postal 70-264, Unam, Mexico.

A. RENZINI, Osservatorio Astronomico, CP 596, 40100 Bologna, Italy.

E. RIBAK, Wise Observatory, Tel-Aviv University, Ramat Aviv, Tel Aviv 69978, Israel.

R. R. ROBBINS, Astronomy Department, University of Texas, Austin, Texas 78712, U.S.A.

P. F. ROCHE, Physics Department, University College London, Gower Street, London, WC1E 6BT.

L. F. RODRIGUEZ, Instituto de Astronomia, Unam, Apartado Postal 70-264, Mexico 04510, Mexico.

I. W. ROXBURGH, Department of Applied Maths., Queen Mary's College, Mile End Road, London, E1 4NS.

F. SABBADIN, Asiago Astrophysical Observatory, 36012 Asiago (VI), Italy.

A. W. SABTI, 31 Wilton Road, Salford, Lancashire

W. SCHMUTZ, Institute of Astronomy, ETH-Zentrum, CH - 8092, Zurich, Switzerland.

S. E. SCHNEIDER, 208 Space Sciences Building, Cornell University, Ithaca, New York 14853, U.S.A.

D. SCHONBERNER, Department of Physics and Astronomy, Louisiana State University, Baton Rouge, LA70803, U.S.A.

P. F. SCOTT, Mullard Radio Astronomy Observatory, Cavendish Laboratory, Cambridge.

M. J. SEATON, Physics Department, University College London, Gower Street, London, WC1E 6BT.

A. SERRANO, Instituto de Astronomia, Apartado Postal 70-264, Mexico 04510, Mexico.

R. A. SHAW, 341 Astronomy, 1011 W. Springfield Avenue, Urbana, Illinois 61801, U.S.A.
G. A. SHIELDS, Department of Astronomy, University of Texas at Austin, Austin, Texas 78712, U.S.A.
B. M. SHUSTOV, Astronomical Council, 48 Piatnitskaya Street, Moscow 109017, USSR.
M. A. SNIJDERS, Physics Department, University College London, Gower Street, London, WC1E 6BT.
G. STASINSKA, DAF Observatoire de Meudon, 92190 Meudon, France.
T. P. STECHER, Code 680, Goddard Space Flight Center, Greenbelt, MD 20771, U.S.A.
P. J. STOREY, Physics Department, University College London, Gower Street, London, WC1E 6BT.
J. SURDEJ, Institut d'Astrophysique, 5 Avenue de Cointe, Cointe-Ougree B-4200, Belgium.
S. TAMURA, Astronomical Institute, Tohoku University, Aobayama, Sendai, Japan 980.
C. B. TARTER, Lawrence Livermore Laboratory L-71, University of California, Livermore, California 94550, U.S.A.
M. E. TAYLOR, Department of Astronomy, University of Manchester, Brunswick Street, Manchester, M13 9PL.
Y. TERZIAN, 109 Brandywine Drive, Ithaca, New York 14850, U.S.A.
A. C. THEOKAS, Department of Pure and Applied Maths, Washington State University, Pullman, Washington 99164, U.S.A.
S. TORRES-PEIMBERT, Instituto de Astronomia, UNAM, Apartado Postal 70-264, Mexico 04510, Mexico.
Y. TUCHMAN, Theoretical Physics, Hebrew University, Jerusalem, Israel.
R. TYLENDA, Lab. of Astrophysics, Copernicus Astronomical Center, Chopina 12/18, 87-100 Torun, Poland.
G. A. VICTOR, Harvard-Smithsonian Centre for Astrophysics, 60 Garden Street, Cambridge, MA 02138, U.S.A.
R. A. WADE, Institute of Astronomy, Madingley Road, Cambridge, CB3 0HA.
J. WALSH, Department of Astronomy, University of Manchester, Oxford Road, Manchester, M13 9PL.
P. G. WANNIER, Caltech Code 102-24, Pasadena, California 91125, U.S.A.
R. WEHRSE, Institut fur Theoretische Astrophysik, Im Neuenheimer Feld 294, D6900 Heidelberg, W. Germany.
V. WEIDEMANN, Institut fur Theoretische Physik, U. Sternwarte, Univ 23, Kiel, W. Germany.
R. WEINBERGER, Institut fur Astronomie, Universitatsstrasse 4, A-6020 Innsbruck, Austria.
A. J. WILLIS, Physics Department, University College London, Gower Street, London, WC1E 6BT.
S. L. WRIGHT, Physics Department, University College London, Gower Street, London, WC1E 6BT.
C. J. ZEIPPEN, Observatoire de Paris, Section d'Astrophysique, 92190 Meudon, France.
B. ZUCKERMAN, Astronomy Program, University of Maryland, College Park, Md. 20740, U.S.A.

Planetary Nebulae: An Introductory Review

Lawrence H. Aller
Astronomy Department
University of California
Los Angeles, California USA

Planetary nebulae constitute a popular field of investigation because they offer a unique opportunity to study the final stages of stellar evolution before a star dies. They supply challenges not only to the stellar evolution expert but also to the spectroscopist, to the student of galactic structure, and to all observers, whether one works in the optical, ultraviolet (UV), infrared (IR) or radio-frequency (r.f.) ranges. They offer insights into properties of dusty, low-density plasmas which may help unravel problems of the interstellar medium (ISM) and quasars. Finally, many are aesthetically beautiful art creations.

Technological strides make possible great advances in studies of planetary nebulae (PN); the International Ultraviolet Explorer (IUE) has revolutionized PN researches in the UV, as will become abundantly clear in this conference. At the extreme other end of the spectrum, the Very Large Array (VLA) has been employed by Johnson, Balick, and Thompson (1979) to study compact PN, but it can also be used to obtain high-resolution images of more extended nebulae, and even investigate mass loss rates of planetary nebulae nuclei (PNN) (Thompson and Sinha 1980). The flying Kuiper Infrared Observatory has opened a new window on important IR atomic and molecular transitions in a dusty spectral domain. These techniques have improved our understanding of PN, but the most exciting advances may lie ahead. The astrometric satellite can supply trigonomeric parallaxes of bright planetary nebulae nuclei PNN. Alas, most PNN are too faint. High spatial resolution, now partly available for NGC 6853 and 7293, will be attained for many PN, and when we combine these data with correspondingly detailed kinematical information, we can substantially advance our understanding of structures of PN.

Certain basic statistics of PN are not well fixed. A reliable distance scale is fundamental. Among recent efforts are those of Acker (1978), Khromov (1979), Maciel and Pottasch (1980), and of Milne (1982). A popular idea is that we can start with an evolutionary scenario giving relations between the absolute magnitude M_s and temperature T_s of the central star, the radius R_n and surface brightness S_n of the nebula and lo presto we get a distance scale.

D. R. Flower (ed.), Planetary Nebulae, 1–13.

These parameters must have an intrinsic spread and it may be a while before we get distances better than 10%. Our basic statistics on stellar PN may be fairly good to a faint limiting magnitude, but data on low surface brightness objects are badly affected by selection effects, especially interstellar extinction (ISE). Isaacman used the distribution of galactic center PN to get a galactic bulge mass model. His best fit gives 1.1×10^{10} m(sun) for the inner kpc and a mass luminosity (M/L) ratio ~ 5. He finds 300 PN within R = 0.3 kpc.

As for spatial distributions and motions of PN, although Paralta (1978) found low latitude PN moved in orbits with $\varepsilon \sim 0.2$, and Khromov (1979) concluded PN belong to an old flat Population I system, PN orbits are distinctly elliptical (Khromov; Purgathofer and Perinotto 1980). Other statistics important to PN as a group include the number per kpc^3, total number in our galaxy and other galaxies, masses of ejected shells, masses of progenitor stars (and spatial distribution of the same), the birth rate and luminosity function, see, e.g., Maciel's (1981) summary. Total masses of ejected shells are popularly taken ~ 0.2 to 0.4 m(sun); halo PN masses $\lesssim$ 0.1 m(sun), with the ionized mass only a small fraction of the total (Pottasch 1980). Jacoby (1980) finds luminosity functions in local group galaxies to be remarkably similar, fitting the idealized Henize-Westerlund (1963) model in which PN are regarded as uniformly expanding, homogeneous, gaseous spheres ionized by nonevolving PNN.

Approximate coincidences between PN birth rates and death rates of main sequence (MS) stars suggest that most in the range of 1 to 5 m(sun) eventually produce PN and return much of their mass to the ISM. Identification of immediate giant precursors of PN is difficult. Carbon stars, Mira variables, and "symbiotic" variables have all been invoked. At the risk of being vague, we can sketch reasonable scenarios. At the peak of its asymptotic giant branch (AGB) evolution, a star loses its outer envelope, maybe by pulsational instabilities or a "superwind." In the Kwok-Purton-FitzGerald KPF (1978) model, for example, the outflowing gas in a gentle wind that had been blowing during the AGB phase is suddently blasted by a stellar wind of ~ 1000 km/sec from the exposed hot core of the expiring star. Although this scenario has much to recommend it, many details need examination. Fragmentation of the ejected outer envelope and subsequent motions can be complex. Many investigations dating to the 1950's attest to the inhomogeneities in supergiant envelopes such as those of ζ Aurigae or 31 Cygni. It should not surprise us that PN start their lives as broken shells, rings, etc., with many condensations or blobs. Some of the proto-PN must be objects such as the bipolar nebulae GL 618 and M 2 - 9 (cf. Schmidt and Cohen 1981). Often, the character of the blobs can be inferred only indirectly from plasma diagnostics, but in some objects, e.g., NGC 6853 or NGC 7293, they can be studied directly and their relationship to the overall nebular structure established.

Dust seems to play an important role throughout the entire lifetime of a PN, although effects may be more dramatic in earlier stages. Dust masses seem similar from one PN to another, T(dust) ~ 100°K, and dust emission seems to favor H^+ regions (Moseley 1980). Despite its small contribution to a PN mass, dust can have important effects on gas phase fractions of refractory elements, on spectral line intensities, and on PN evolution. Iron (Shields 1978), magnesium (Pequinot and Stasinska 1980; Harrington and Marionni 1981), calcium (Aller 1978), and aluminum are examples of metals depleted by being locked in solid grains. In most PN, the solid particles are probably carbon; in Abell 30, Greenstein (1981) found that the extinction could be explained by soot! Cohen and Barlow (1980) separate PN from very low excitation (VLE) H II regions by the silicate emission in the latter. Aitken et al. (1979) find that some PN, e.g., Hb 12 and Swst 1 show silicate emission as do the Trapezium stars. Dust and gas can affect T_s determinations by Zanstra's method (Helfer et al. 1981).

Molecules can supply data on physical conditions in PN condensations, shielded zones, etc. We would expect large abundances of H_2, H_2^+, OH, and CH^+ in transition zones of ionized nebulae (Black 1978). In NGC 6720, Beckwith et al. (1978) find H_2 emission correlated with [O I], not H II zones. H_2 emission can be produced in clumps heated by an expanding H II region (Smith et al. 1981).

Turning to spectroscopy, note the great leap forward given by opening up of the UV, where fall precious transitions of He, C, N, O, Ne, and Si. Although most lines here are excited by collisions, processes such as dielectronic recombination become important for some. These UV lines offer many advantages for diagnostics and abundances. Much information is buried in many weak permitted transitions of abundant elements in the optical region. This spectroscopic source has been inadequately tapped since the pioneering work of Wyse (1942). Nikitin et al. (1981) used permitted lines of N III, C III, and C IV to get abundances of C and N in high-excitation PN, and to assess the role of spectral lines corresponding to two-electron excited states. Wilkes et al. (1981) used weak permitted ionic lines to get N abundance in NGC 3242.

Infrared fine structure transitions involving ions such as Ne II, S IV, and Ar III, which include some ionization stages not otherwise observed, offer important help for PN diagnostics and compositions (Grasdalen 1979; Aitken et al. 1979; Dinerstein 1980; Beck et al. 1981). We hope these data can be supplemented by observations from satellites.

Of fundamental importance to all spectroscopic studies and plasma diagnostics are accurate atomic parameters: A-values, Ω-values, recombination coefficients, and charge-exchange cross sections. Much progress has been made since Osterbrock's (1974) compilation. C.D. Keyes is preparing a summary of recent data. We are heavily indebted here, particularly to Seaton, Nussbaumer, Dalgarno, Czyzak, and their respective associates, and to the Belfast group.

Accurate chemical compositions of PN should assist in investigations of stellar evolution, nucleogenesis, and the history of the ISM. Some elements in all PN and perhaps all elements in some PN have not been transmuted in the nuclear furnace. What counts is not just the processing of H into He, and He subsequently into C, N, Ne, and N production by the CN cycle, but the mixing of these products into the outer layers which are eventually ejected as a PN. He, C, N, and possibly O, may be affected by nucleogenesis but elements heavier than

O certainly will not be, for m(*) < 5 m(sun). To what extent are observed abundance fluctuations real and how much do they reflect errors in analysis? Of course, the abundances of C and N which are important in the nucleogenesis scenario, are among the hardest to get, at least from optical data. N is represented only by [N I] and [N II], while most of its atoms are usually in higher ionization stages. Fortunately, the persevering efforts of Barker and Kaler have shown that $N/O = N^+/O^+$ to within a factor of 2 for many PN.

Given the observed line intensities and diagnostics, i.e., (T_ε, N_ε), we can get reasonably good estimates of $N(X_i)$, the number of ions of element X in the ith ionization stage. The big problem is always to get N(X), the elemental abundance, expressed in terms of N(H), the hydrogen abundance. Three possibilities seem promising: a) simple ratios such as $(O^+ + O^{++})/H^+ = O/H$, $O/O^{++} \sim Ne/Ne^{++}$ work well in PN of moderate excitation but not for ions of S, Cℓ, Ar, etc.; b) we can use theoretical models embracing all relevant physics of ionization and recombination, charge exchange, radiative transfer, and include complications introduced by condensations, dust, etc. The beautiful model by Harrington *et al.* (1982) for NGC 7662 shows an approach that is rigorous, elegant, and well suited to symmetrical PN for which detailed observational data exist. I believe it is the method of the future; c) we can use theoretical models as interpolation devices. We fit certain line ratios as closely as possible, notably 4686/4471, He II/He I, 3727/4959 + 5007 [O II]/[O III], 3426/3868 [Ne V]/[Ne III] and then adjust abundances to represent line intensities as well as we can, thus obtaining "model" abundances. Also, we can use the models to get ionization correction factors (ICF's) and apply them to get "extrapolated" abundances. The two techniques give log N(X)'s differing normally by about 0.1. Our team applied this procedure to study some forty-odd PN. For most of them, application of the Harrington *et al.* procedure would have exceeded our computing budget. Limitations imposed by geometrical irregularities and big density fluctuations are assuredly real, but we believe that by starting with the best physics we can, and later introducing modifications imposed by dust and blobs, we can approach accurate results asymptotically.

Table 1 shows some mean results, obtained primarily by models. These data refer mostly to relatively bright, nearby PN (Aller and Czyzak 1982). The assignment of population types was taken from Kaler (1978). High-excitation objects are those for which λ4686 He II of nebular origin is certainly present. Nitrogen-rich objects are defined as those with log N(N) > 8.0; carbon-rich objects are those with C/O > 1. The He/H ratio tends to be larger in N-rich PN, as has been noted by Peimbert (1978) and by Kaler (1978). Elements heavier than He are less abundant in Population Type II PN than in those of Population I. Ne, S, Cℓ, and Ar seem to have similar abundances in Population I, high-excitation, C-rich, and N-rich objects. C and N show considerable fluctuations as has been noted by many workers (e.g., Peimbert, Kaler). Comparisons with solar abundances show some close similarities but also conspicuous differences. In contrast to Kaler's (1981) conclusion, we find the C/O ratio in PN to be generally higher than in the sun. Although there is a tendency for the mean value of the C/O ratio found from $\lambda 4267$ to exceed that from the UV lines, log <C/O> [4267] - log <C/O> [UV] ~ 0.16, there seems to be a large intrinsic spread (>> 0.16 dex) in the C/O ratio, no matter whether we use recombination or collisionally-excited C lines. Oxygen appears to be less abundant in PN than in the sun. Peimbert and Serrano's (1980) value, log N(O) = 8.83, was obtained with a temperature fluctuation parameter $\Delta\tau^2 = 0.035$.

Table 1. Mean Logarithmic Abundances (Mostly From Models)

	Pop. I	Pop. II	High Excitation	N-rich	C-rich	Solar
He			11.04	11.3.	11.03	11.0:
C	8.82	8.72	8.89	8.50	8.99	8.65
N	8.12	7.94	8.39	8.88	8.15	7.96
O	8.68	8.58 K	8.66	8.71	8.69	8.87
Ne	8.08	7.88	8.02	8.05	8.05	8.05
S	6.96	6.88 K	7.03	6.98	7.09	7.23
Cℓ	5.28	5.13	5.27	5.4	5.27	5.5
Ar	6.42	6.22	6.48	6.65	6.46	6.57

See also Peimbert and Serrano (1980); Kaler (1978, 1980); Dinerstein (1980); Beck et al. (1981); French (1980). Solar values are those presented at a 1980 Santa Cruz workshop, these come from Lambert (1978) and Ross and Aller (1976). In high-excitation planetaries, the following logarithmic mean abundances were found: F (4.6), Na (6.18), K (5.0), and Ca (5.0). K denotes values from Kaler (1980).

Distinctive composition anomalies are well known in halo PN. The He/H ratio is essentially cosmic, but big differences occur in other elements, the most dramatic effect being in Ar which Barker (1980) finds to be about two orders of magnitude less abundant than in general field PN. Galactic bulge PN merit more study. Do they have "normal" compositions (Webster 1976) or do many have high O/H ratios as Price (1981) found for H 1-55? Chemical composition gradients in disks of galaxies are well-known from studies of H II regions. Many attempts have been made to use PN for this purpose in our galaxy (D'Odorico et al. 1976; Aller 1976; Barker 1978; Peimbert and Torres-Peimbert (PTP) 1977; Kaler 1978; Peimbert and Serrano 1980). The derived gradients depend on the selection of objects and the size of the temperature fluctuations that are assumed. C and N are manufactured in the PN themselves and cannot be used to obtain galactic composition gradients unless one can clearly separate those PN whose progenitors have manufactured C and N in their interior from those that did not. More massive stars, which presumably evolve into the brighter PN, certainly produce C and N. Peimbert and Serrano (1980) suggested that N-rich stars that enrich the ISM with He and H constitute 20% of the total. In classifications by Acker (1980) and by Kaler (1980), two distinct groups are discussed. One includes relatively young PN (age $\leqslant 10^9$ years), many with high Ne and N enrichment, and presumably more massive progenitor stars; they seem to be essentially Population I objects moving close to the galactic plane in orbits of low eccentricity. The other group includes old PN (ages up to 10^{10} years), small He or N enrichments; these are old disk PN plus some halo objects (in Acker's group). They move in more elliptical orbits and often show lower excitation (Kaler 1980).

Theories of the late stage of stellar evolution (Paczyński 1970; Renzini and Voli 1981; Iben and Renzini 1982) predict the enrichment of certain chemical elements in outer stellar envlopes that are destined to be ejected as PN shells. As the star evolves up the AGB, the mass M_e above the H-burning shell decreases because H is converted to He at its base and sinks into the core, while the upper part is lost in wind. At some point this outer envelope is ejected. It may form a dense, dusty shell that temporarily hides the core which eventually settles down to become a white dwarf. Paczyński's and subsequent theories have predicted that the more massive the core, the faster it fades. Calculations of advanced stages of AGB evolution as a function of progenitor star mass give L, T_{eff}, and dM/dt and also surface abundance ratios, He/H, C/O, and N/O as functions of time. Detailed results depend on the initial He/H ratio and on the ratio of mixing length to scale height.

Figures 1, 2, and 3 show the predicted and observed relationships; C/O versus He/H, N/O versus He/H, and C/O versus N/O. The He/H ratio should be good to $\sim$ 0.01; the log C/O ratios may have uncertainties of $\sim$ 0.2 to 0.3, while the N/O ratios should be more accurate for most objects. The detailed results differ slightly from those of Peimbert (1981) since different data are employed here. He concluded that the ejected shell had a smaller mass than the predicted one, that the He enrichment could be explained if the PN originated from stars in the 1 to 5 solar mass range, and that N enrichment exceeded predicted values. Note that the theoretical curves can be shifted leftward in Figs. 1 and 2 by changing the initial He content of the progenitor stars. Masses of progenitor stars are indicated by numbers in [ ] brackets and tick marks on the curves. Although the ordering of successive masses is correct, one cannot guarantee that exact locations are accurate; the whole system of tick marks may have to be displaced up or down along the curves. Most of the PN would appear to have come from progenitors of less than 3 solar masses; Hu 1 - 2 and NGC 6778 may have come from more massive stars. The N-rich object NGC 6302 (Aller *et al.* 1981) has a higher He/H and a lower C/O ratio than theory suggests. C and perhaps O were processed to N and excess He was produced in a vigorous CN cycle operation. Also, C may have been preferentially converted to N in the progenitors of Hu 1 - 2, NGC 2440, 6741, and 6778. Note that IC 351, 4593, and 4634 seem N weak. The N-rich object in NGC 6822 (Dufour and Talent 1980) may be similar to NGC 6302, or even a more extreme example.

Although much of the scatter in Figs. 1, 2, and 3 may come from observational uncertainties, some deviation seems real. Some PN do not follow the prescribed scenario. Improved stellar evolution calculations and abundances are both needed. In Schönberner's 1981 model, where m(residual core) = 0.58 for all PN, it appears difficult to find dredge-up mechanisms that can give such a variety of chemical compositions. Who knows? PN composition investigations may yet help dredge-up theory.

Additional insights are offered by PN in other galaxies, although observations can be difficult. Particularly noteworthy are the measurements by H.C. Ford and associates of O/H gradients from PN in M 31, and PN detection in the Virgo group. So far, a relatively small number of objects has been studied in the Magellanic Clouds. Comparing their results for seven PN in the SMC with analyses of H II regions, Aller et al. (1981) found He/H and N/H to be enhanced by factors of 1.2 and 5, respectively, although the abundances of O, Ne, S, and Ar seemed consistent with H II region data.

Table 2 compares data for eight PN observed in the LMC from Cerro Tololo. Except for P40, simple approximations have been used to get N(X) from $N(X_i)$. For Ar we observe most of the ionization stages but for S, I have employed the approximation $S/O = (S^+ + S^{++})/O^+$; the resultant, highly uncertain numbers are placed in [ ]. More accurate values must await proper models, but results for He, N, O, Ne, and Ar ought to be good to a factor of about two and enable us to draw some conclusions. For comparison, log N_{el} for LMC H II regions from models by Dufour et al. (1982) and Aller et al. (1979) are He: 10.93; N: 6.98; O: 8.41; Ne: 7.73, S: 6.96; and Ar: 6.24. P07, P08, and P38 are He rich (Webster (1976); N is enhanced in P07 and P08 by about a factor of 36 over LMC H II regions. For the other PN, the factor is about 5. To within the accuracy of the analysis, log $\langle N(PN)\rangle \sim$ log $\langle N(H\ II)\rangle$ for O, Ne, and Ar. Notice the depletion of O in P7 and P9 where N is greatly enhanced, suggesting that O may have been converted to N. P25 shows prominent lines of [Ar V], [Ne IV], λ4724, 26 [Ca V], [Fe V] λ4227 and probably [Fe VI] and [Fe VII] as well; O is depleted with respect to the local ISM, but N is enhanced. Is P25 a PN where Fe isn't locked in grains? We do not know the proper mix of N super-rich and N less-rich objects, so we cannot yet estimate how important PN are as suppliers of N, but there can be no doubt that they are copious sources of C. IUE observations by Maran et al. (1982) show the C enhancements to be spectacular. Carbon is about 40 times more abundant in the SMC PN than in SMC H II regions, with a corresponding ratio of about six for the LMC. In fact, the C abundances are comparable with those found in galactic PN, suggesting that the process of C synthesis and subsequent dredge-up are the same in PN progenitors, whether they be in the galaxy, the SMC or the LMC. Exploring yet further, the Fornax PN is also found to have C/O > 1 (Stecher et al. 1982). I want to emphasize that in each instance we are observing intrinsically bright PN; reaching fainter objects will be difficult.

At the 1977 conference, Terzian enumerated some unsolved PN problems. Many remain unresolved but progress occurs. We can now attack Minkowski's apparent paradox of the fast PNN wind as contrasted

Table 2. Logarithmic Abundances in Planetaries in the LMC

Element	P02	P07	P08	P09	P25	P33	P38	P40
He	11.06	11.20	11.03	11.23	11.04	11.08	11.176	11.02
N	7.83	8.53	----	8.55	7.24	7.68	7.74	7.5
O	8.26	8.04	8.46	8.02	8.07	8.43	8.40	8.33
Ne	7.65	7.61	7.67	7.54	7.56:	7.69	7.62	7.62
S	[7.35:]	[7.26:]	[7.9:]	[7.66:]	[7.11:]	[7.35:]	[7.51:]	6.42
Ar	6.06	6.23	>5.8	6.32	6.11	~6.4	6.84:	6.19

with gentle PN expansion rates. Low mass loss rates, $10^{-8} < \dot{M}\ 10^{-7}$ m(sun)/year Perinotto et al. 1982; Castor et al. 1981) can't explain the detachment of PN shells, although 1500 to 2000 km/sec winds might affect the excitation of some spectral lines in PN.

Great advances should be coming in PNN studies, e.g., checking Paczyński's relation between mass and fading time for progenitor stars. UV observations can help us check theoretical models of stellar fluxes, derive $T_{eff}(^{*})$ and log g. Pottasch (1981) suggested large masses and high T_s's for a number of PNN. The notorious discordance between $T_{Zanstra}$ and T_s inferred from dark-line spectra of some PNN warns us against simplistic stellar atmospheric models. Planckian fluxes are outmoded; we must allow for stellar coronae and chromospheres. Another clue to T_s is given by flux distribution needed for excitation of an observed nebular spectrum, e.g., for the PNN in NGC 7662, $T_s \sim 95{,}000°K$. To apply this method extensively, we need a fine-grid network of stellar models.

To improve our statistics, we need Hβ fluxes, spectra, A_V, and PNN data for many more objects. More information is needed even for traditional PN, e.g., accurate isophotic contours are necessary to interpret PN spectra. Most of the discordances in line intensity measurements arise not from a lack of precision in the same, but because different measurements, photoelectric, ptg., and ITS refer to different places in a PN image. The effect is especially bad in PN like NGC 6720 or 7009, where large excitation differences occur on a small spatial scale, and for comparing IUE and optical data.

The space telescope can give us monochromatic images for tantalizing PN like NGC 7027 or 2392, resolve the structure of PN in the Magellanic Clouds, and yield statistics on PN in the Virgo cluster.

Theoretical problems run a gamut from atomic physics (Ω's, A's, charge exchange cross sections), molecular structure and "dirty" chemistry (including grain formation and binding of Ca, Al, Fe, etc., in solid structures), the complete hydrodynamical structure and history of a typical PN, to the details of the last active chapter of stellar evolution. Our work is laid out for us!

Illustrations

1) Comparison of Log n(C)/n(O) With the He/H Ratio. The solid curve gives Renzini and Voli's prediction (1981) for their parameter $\alpha = 2$, the dotted curve for $\alpha = 1.5$. The numbers in brackets indicate the initial masses of the progenitor stars, 1.0 to 5.0 M(sun). The more accurate determinations are indicated by solid dots, the less accurate ones by open circles.
2) Comparison of Log n(N)/n(O) With the He/H Ratio. Notation and symbols are as in Figure 2.
3) Comparison of Log n(C)/n(O) With Log n(N)/n(O). The notation and symbols are as in Figure 2. Compare Peimbert (1981).

REFERENCES

Acker, A. 1978, Astron. Astrophys. Suppl. Ser., 33, 367.
________ 1980, Astron. Astrophys., 89, 33.
Aitken, D.K., Roche, P., Spencer, P., Jones, B. 1979, Ap. J., 233, 925.
Aller, L.H. 1976, Publ. Astron. Soc. Pac., 88, 574.
________ 1978, I.A.U. Symposium No. 76, p. 225 (ed. Y. Terzian).
________ 1982, Astrophys. Space Sci., 83, 225.
Aller, L.H., and Czyzak, S.J. 1982, Ap. J. Suppl., in press.
Aller, L.H., Keyes, C.D., and Czyzak, S.J. 1979, Proc. Natl. Acad. Sci. USA, 76, 1525.
Aller, L.H., Keyes, C.D., Ross, J.E., and O'Mara, B.J. 1981, M.N.R.A.S., 194, 613 (SMC); 197, 95 (NGC 6302).
Barker, T. 1978, Ap. J., 220, 193.
________ 1980, Ap. J., 237, 482.
Beck, S.C., Lacy, J.H., Townes, C.H., Aller, L.H., and Baas, F. 1981, Ap. J., 249, 592.
Beckwith, S., Persson, S.E., Gatley, I. 1978, Ap. J., 219, L33.
Black, J.H. 1978, Ap. J., 222, 125.
Castor, J.L., Lutz, J., and Seaton, M.J. 1981, M.N.R.A.S., 194, 574.
Cohen, M., and Barlow, M.J. 1980, Ap. J., 238, 585.
Dinerstein, H. 1980, Ap. J., 237, 486.
D'Odorico, S., Peimbert, M., and Sabbadin, F. 1976, Astron. Astrophys., 47, 341.
Dufour, R.J., Shields, G.A., and Talbot, R.J. 1982, Ap. J., 252, 461.
Ford, H.C., and Jacoby, G.K. 1978, Ap. J. Suppl., 38, 351.
French, H. 1980, Bull. Amer. Astron. Soc., 12, 842.
Grasdalen, G. 1979, Ap. J., 229, 587.
Greenstein, J.L. 1981, Ap. J., 225, 124.
Harrington, J.P., Seaton, M.J., Adams, S., and Lutz, J.H. 1982, M.N.R.A.S., 199, 517.
Harrington, J.P., and Marionni, P.A. 1981, First Two Years of IUE, NASA Conference Publ. CP-2171, pp. 623, 633.
Helfer, H.L., Herter, T., Lacasse, M.G., Savedoff, M.P., and Van Horn, H.M. 1981, Astron. Astrophys., 94, 109.
Henize, K.G., and Westerlund, B.E. 1963, Ap. J., 137, 747.
Iben, I., and Renzini, 1982, Ann. Rev. Astron. Astrophys., 20, in press.

REFERENCES (Continued)

Isaacman, R. 1981, Astron. Astrophys., 95, 46; Suppl., 43, 405.
Jacoby, G.H. 1980, Ap. J., Suppl., 42, 1.
Johnson, H.M., Balick, B., and Thompson, A.R. 1979, Ap. J., 233, 919.
Kaler, J.B., 1978a, Ap. J., 225, 527.
__________ 1978b, Ap. J., 226, 947.
__________ 1981, Ap. J., 249, 201.
Khromov, G. 1979, Astrofizika, 15, 269.
Kwok, S., Purton, C.R., and FitzGerald, M.P. 1978, Ap. J., 219, L125.
Lambert, D.L. 1978, M.N.R.A.S., 182, 249.
Maciel, W.J., 1982, Astron. Astrophys., 98, 406.
Maciel, W.J., and Pottasch, S.R. 1980, Astron. Astrophys., 88, 1.
Maran, S.P., Aller, L.H., Gull, T.R., and Stecher, T.P. 1982, Ap. J., 253, L43.
Milne, D.K. 1982, in preparation.
Moseley, H. 1980, Ap. J., 238, 892.
Nikitin, A., Sapar, A., Keklistova, T.Kh., Kohltygin, A.F. 1981, Soviet Astronomy, 25, 1.
Osterbrock, D.E. 1974, Astrophysics of Gaseous Nebulae, San Francisco, W.H. Freeman Co.
Paczyński, B. 1970, Acta Astron., 20, 47, 287.
Paralta, J.O. 1978, Astron. Astrophys., 64, 127.
Peimbert, M. 1978, I.A.U. Symposium No. 76, p. 215 (ed. Y. Terzian).
__________ 1981, Physical Processes in Red Giants, ed. I. Iben and A. Renzini, Dordrecht, Holland, Reidel, p. 409.
Peimbert, M., and Serrano, A. 1980, Rev. Mex. Astron. y Astrofis., 5, 9.
Peimbert, M., and Torres-Peimbert, S. 1977, Rev. Mex. Astron. y Astrofis., 2, 181.
Pequinot, D., and Stasinska, G. 1980, Astron. Astrophys., 81, 121.
Perinotto, M., Benvenuti, P., and Cerruti-Sola, M. 1982, Astron. Astrophys., 108, 314.
Pottasch, S.R. 1980, Astron. Astrophys., 89, 336.
__________ 1981, Astron. Astrophys., 94, L13.
Price, C.M. 1981, Ap. J., 247, 540.
Purgathofer, A., and Perinotto, M. 1980, Astron. Astrophys., 81, 215.
Renzini, A., and Voli, M. 1981, Astron. Astrophys., 94, 125.
Ross, J.E., and Aller, L.H. 1976, Science, 191, 1223.
Schmidt, G.D., and Cohen, M. 1981, Ap. J., 246, 444.
Schönberner, D. 1981, Astron. Astrophys., 103, 119.
Shields, G. 1978, Ap. J., 219, 559.
__________ 1980, Publ. Astron. Soc. Pac., 92, 418.
Smith, H.A., Larson, H.P., and Fink, U. Ap. J., 244, 835.
Thompson, A.R., and Sinha, R.P. 1980, A.J., 85, 1240.
Webster, B.L. 1976, M.N.R.A.S., 174, 513.
__________ 1978, I.A.U. Symposium No. 76, 11 (ed. Y. Terzian).
Wilkes, R.J., Ferland, G.J., Hanes, D., and Truran, J.W. 1981, M.N.R.A.S., 197, 1.
Wyse, A.B. 1942, Ap. J., 95, 356.

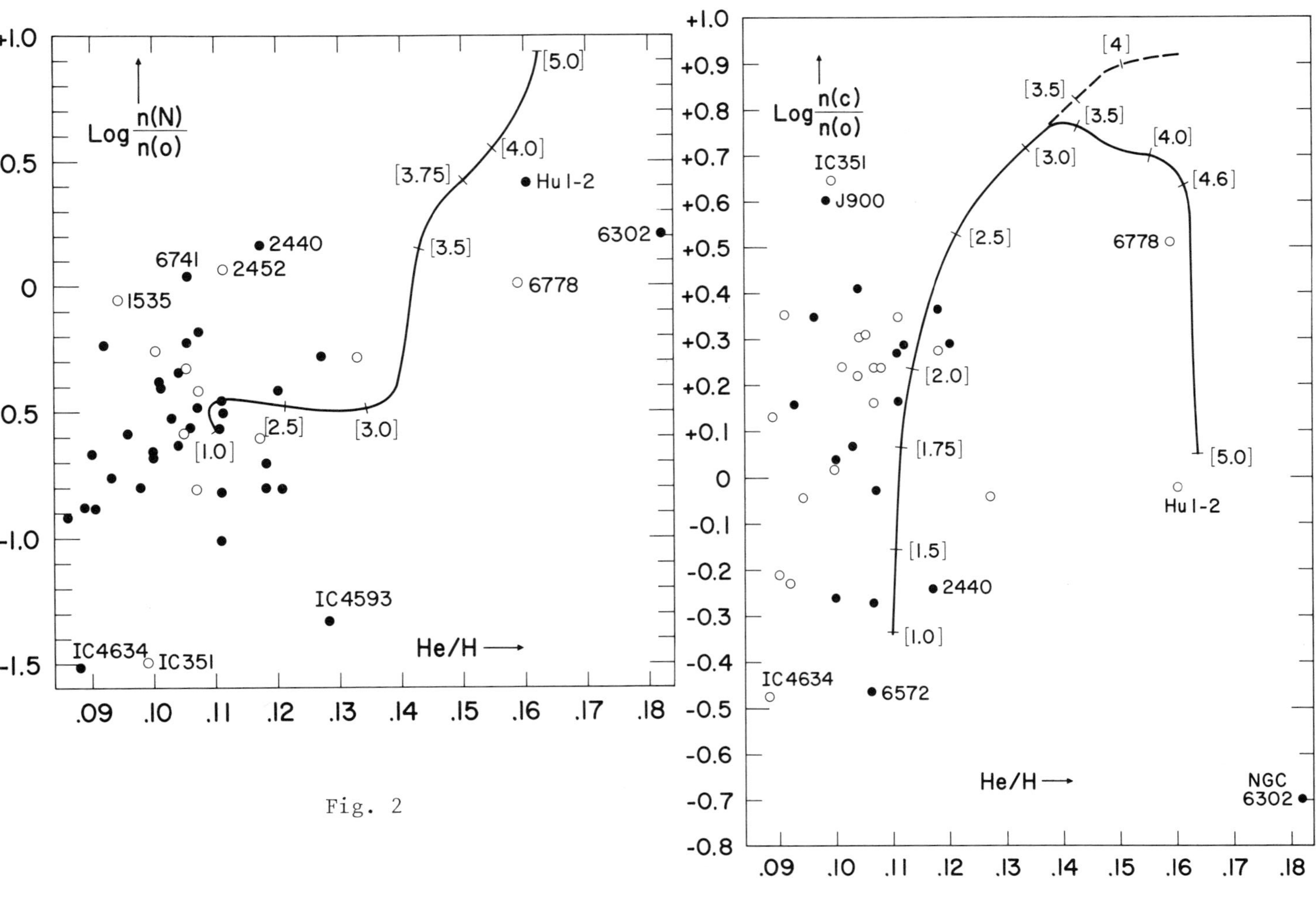
Log n(N)/n(o)
[5.0]
[4.0]
[3.75]
Hu I-2
[3.5]
6302
2440
6741
2452
6778
1535
[2.5]
[3.0]
[1.0]
IC4593
He/H
IC4634
IC351
Log n(c)/n(o)
[4]
[3.5]
[3.5]
[4.0]
[3.0]
[4.6]
IC351
J900
[2.5]
6778
[2.0]
[1.75]
[5.0]
Hu I-2
[1.5]
2440
[1.0]
IC4634
6572
He/H
NGC 6302

Fig. 2

Fig. 1

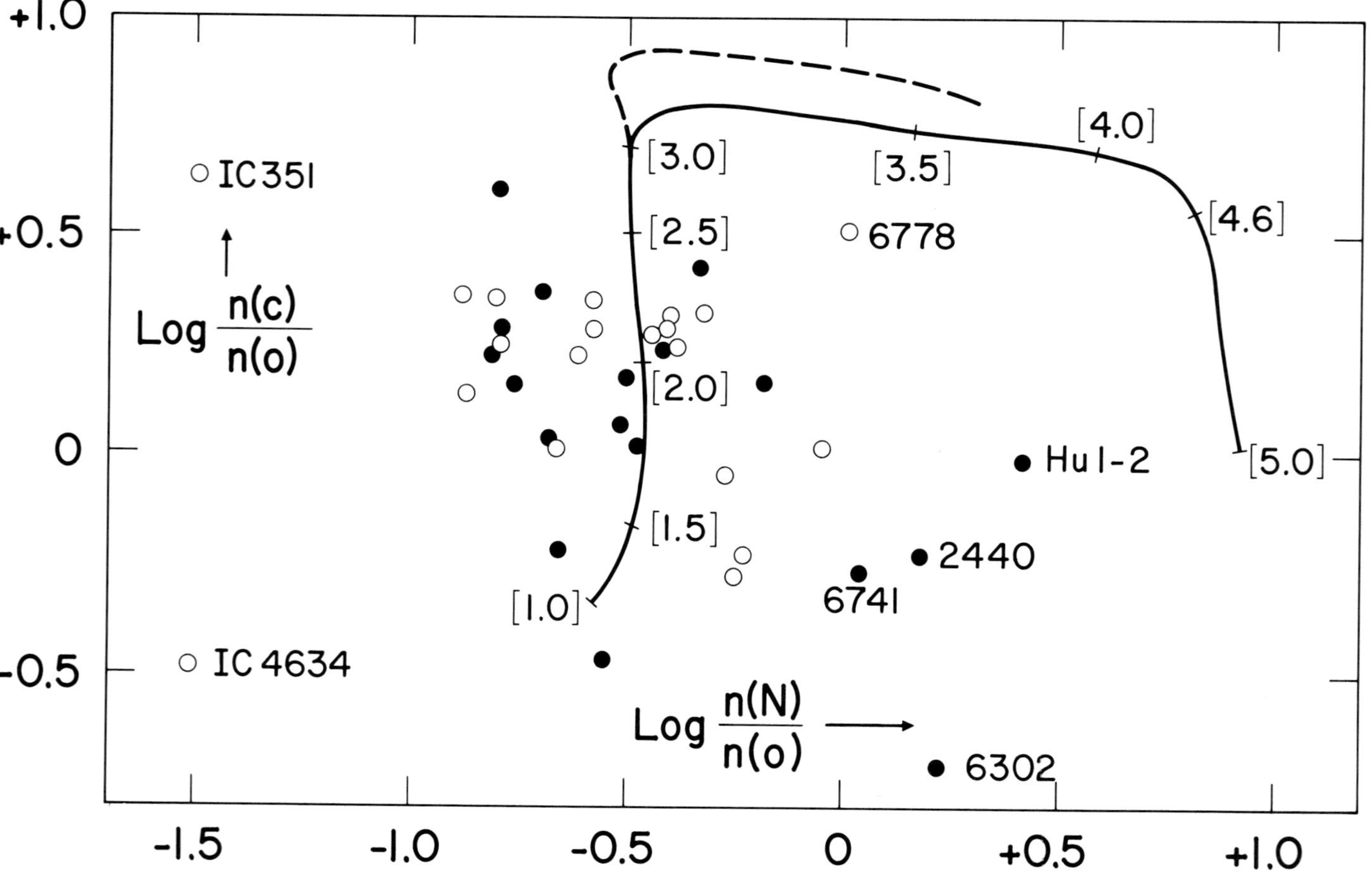

+1.0
+0.5
0
-0.5
IC351
IC 4634
Log n(c)/n(o)
[3.0]
[2.5]
[2.0]
[1.5]
[1.0]
[3.5]
[4.0]
[4.6]
[5.0]
6778
Hu1-2
2440
6741
6302
Log n(N)/n(o)
-1.5
-1.0
-0.5
0
+0.5
+1.0

Fig. 3

OSTERBROCK: Are any elements with higher Z than Fe observed in any PN shells?

ALLER: Beyond the iron peak, the abundance curve declines rapidly so it is not surprising that lines of these elements have not been seen. There is also a tendency for the iron group elements to be tied-up in grains, as Shields has pointed out.

SECTION I

OBSERVATIONS OF PLANETARY NEBULAE

NEW AND MISCLASSIFIED PLANETARY NEBULAE

L. Kohoutek
Hamburg Observatory
Hamburg-Bergedorf, W. Germany

ABSTRACT: Since 1978 85 new objects have been classified as planetary nebulae. They are listed in Table 1 which gives the designations, names, coordinates and the references to the discovery. In the list of misclassified planetary nebulae (Table 2) 33 objects have been included. Principal criteria have been summarized which can help to distinguish planetaries from other objects. They refer to observable properties and are valid only in a statistical sense.

The second supplementary list to the "Catalogue of Galactic Planetary Nebulae" (CGPN - Perek, Kohoutek, 1967) of new planetary nebulae is presented containing 85 discoveries which were published between 1978 and 1981. As in the case of the first supplementary list (Kohoutek, 1978 - discoveries between 1966 and 1977) the designations, names, coordinates and references to the discovery are given in Table 1. An asterisk affixed to the galactic number means an uncertain classification (suspected, possible or probable planetary nebula).

The considerable number of discoveries of PN since the CGPN indicates the large activity in this field, as shown in the following short statistics:

Period:	Number of discoveries:
CGPN (- 1965)	1036
1966 - 1969	17
1970 - 1973	78
1974 - 1977	132
1978 - 1981	83

The majority of the new planetaries reported in the last period are objects of very low surface brightness and

D. R. Flower (ed.), Planetary Nebulae, 17–30.

TABLE 1 NEW PLANETARY NEBULAE (1978-1981)

Design.	Name	R.A. (1950)	Decl.		Discovery	Rem.
120 -5.1	Sh 2-176	$0^h28^m.9$	$+57^\circ01'$		Sabbadin, etal. 1977	
124 -7.1	WeSb 1	0 57.93	+54 47.5	F	Weinberger, Sabbadin 1981	
181 +0.1	Pu 1	5 49.65	+28 05.8	F	Purgathofer 1978	
228-22.1	DeHt 1	5 53.02	-22 54.4	F	Dengel, etal. 1980	R
175 +6.1	Pu 2	5 59.20	+36 07.7	F	Purgathofer 1980	
183 +5.1	WeSb 2	6 13.02	+28 23.2	F	Weinberger, Sabbadin 1981	
158+17.1	PuWe 1	6 15.38	+55 38.0	F	Purgathofer,Weinberger 1980	
249-22.1*	ESO-308-08	6 24.35	-41 09.6		Holmberg, etal. 1978	R
239-18.1*	ESO-426-13	6 25.63	-31 09.8		Holmberg, etal. 1978	R
239-12.1*	ESO-427-19	6 53.23	-29 03.5		Holmberg, etal. 1978	R
217 -0.1*	MaC 1-1	6 56.24	-3 37.0	F	MacConnell 1978	
217 +2.1	Sp 3-1	7 04.3	-3 00	F	Stephenson 1978	
248-12.1*	ESO-367-03	7 08.60	-37 14.2		Holmberg, etal. 1978	R
242 -3.1*	ESO-429-04	7 33.47	-28 02.5		Holmberg, etal. 1978	R
245 -3.1*	ESO-429-17	7 42.93	-30 10.8		Holmberg, etal. 1978	
241 +0.1*	ESO-493-13	7 46.02	-25 06.4		Holmberg, etal. 1978	R
251 -4.1*	ESO-369-01	7 50.47	-36 18.8		Holmberg, etal. 1978	
254 -6.1*	ESO-311-18	7 52.22	-39 41.6		Holmberg, etal. 1978	R
263 -8.1*	ESO-209-15	8 03.68	-48 14.8		Holmberg, etal. 1978	R
262 -4.1*	ESO-259-06	8 22.25	-45 21.4		Holmberg, etal. 1978	R
265 -4.1*	ESO-259-10	8 32.50	-47 06.3		Holmberg, etal. 1978	R
265 +5.1*	ESO-314-12	9 17.23	-40 58.5		Holmberg, etal. 1978	R
283 +9.1*	ESO-215-04	10 52.48	-48 30.9		Holmberg, etal. 1978	
291+19.1	ESO-320-28	11 49.97	-42 00.9		Holmberg, etal. 1978	R
298+34.1	CTIO 1230-275	12 30.6	-27 32	F	Smith, etal. 1976	R
339+88.1	LoTr 5	12 53.13	+26 09.7	F	Longmore, Tritton 1980	
312+25.1	LoTr 6	13 31.20	-36 35.1	F	Longmore, Tritton 1980	

Design.	Name	R.A. (1950)	Decl.		Discovery	Rem.
310 -5.1	LoTr 7	$14^h11^m.35$	$-67^\circ18'.0$	F	Longmore, Tritton 1980	
315 +5.2	LoTr 8	14 18.52	-54 48.6	F	Longmore, Tritton 1980	
315 -1.1	LoTr 9	14 37.43	-61 07.1	F	Longmore, Tritton 1980	
316 -1.1	LoTr 10	14 42.47	-61 01.2	F	Longmore, Tritton 1980	
327+14.1*	ESO-328-04	14 58.15	-41 43.1		Holmberg, etal. 1978	R
329+12.1*	ESO-328-40	15 14.32	-42 26.4		Holmberg, etal. 1978	R
313-12.1	LoTr 11	15 15.97	-72 03.3	F	Longmore, Tritton 1980	
341+17.1*	ESO-450-16	15 45.58	-31 58.1		Holmberg, etal. 1978	R
325 -1.1	VB 2	15 47.3	-56 12	F	van den Bergh 1979	
326 -1.2	VB 3	15 49.0	-56 15	F	van den Bergh 1979	
343+16.1*	ESO-451-03	15 55.92	-31 03.3		Holmberg, etal. 1980	R
346+19.1*	ESO-515-19	15 56.92	-27 05.9		Lauberts, etal. 1981	R
345+10.1*	ESO-390-05	16 21.13	-34 08.2		Holmberg, etal. 1978	R
347 +7.1*	ESO-391-02	16 38.50	-34 11.6		Holmberg, etal. 1978	
343 +3.1	SuWt 3	16 40.95	-39 57.8	F	West, Schuster 1980	
339 -0.1*	VB 1	16 41.8	-46 03	F	van den Bergh 1978	R
339 -3.1*	MaC 1-3	16 57.72	-47 41.2	F	MacConnell 1978	
7+10.1*	MaC 1-4	17 23.74	-16 46.0	F	MacConnell 1978	
358 +3.9	Ae 2-B	17 24.63	-28 08.6	F	Allen 1979	
357 +2.8	Ae 2-D	17 26.27	-29 44.8	F	Allen 1979	
359 +3.4	Ae 2-E	17 27.10	-27 28.1	F	Allen 1979	
358 +2.4	Ae 2-F	17 27.34	-28 33.7	F	Allen 1979	
1 +4.1*	MaC 1-5	17 28.98	-24 42.6	F	MacConnell 1978	
359 +2.5	Ae 2-G	17 29.22	-28 12.4	F	Allen 1979	
357 +1.2	Ae 2-H	17 30.07	-30 24.4	F	Allen 1979	
359 +2.6	Ae 2-I	17 31.08	-27 54.0	F	Allen 1979	
0 +2.1	Ae 2-J	17 32.45	-27 22.2	F	Allen 1979	
357 +1.3	TrBr 4	17 32.52	-30 19.6	F	Terzan, etal. 1978	
359 +2.7	Ae 2-K	17 33.09	-27 58.9	F	Allen 1979	

Design.	Name	R.A. (1950)	Decl.		Discovery	Rem.
359 +1.3	19 W 32	$17^h35^m.87$	$-28^\circ55'.0$		Wouterloot, Dekker 1979	R
0 +2.2*	ESO-520-13	17 36.25	-27 14.3		Holmberg, etal. 1978	
27+16.1*	DeHt 2	17 39.18	+3 08.4	F	Dengel, etal. 1980	
357 -2.1	Ae 2-M	17 44.52	-32 19.8	F	Allen 1979	
358 -2.3*	MaC 1-7	17 48.15	-31 13.0	F	MacConnell 1978	
358 -2.4	Ae 2-O	17 48.50	-32 02.4	F	Allen 1979	
0 -1.7	Ae 2-Q	17 50.23	-29 16.5	F	Allen 1979	
358 -2.5	Ae 2-R	17 50.38	-31 25.0	F	Allen 1979	
358 -3.3*	MaC 1-8	17 52.79	-31 38.0	F	MacConnell 1978	
13 +5.2*	MaC 1-9	17 53.03	-14 06.4	F	MacConnell 1978	
28+10.1	WeSb 3	18 03.46	+0 22.3	F	Weinberger, Sabbadin 1981	
5 -2.2*	MaC 1-10	18 06.12	-25 05.2	F	MacConnell 1978	
3-4.10*	ESO-456-64	18 07.40	-27 51.5		Holmberg, etal. 1978	R
0 -6.1*	ESO-456-73	18 09.37	-32 06.8		Holmberg, etal. 1978	
8 -2.1*	MaC 1-11	18 11.82	-22 44.9	F	MacConnell 1978	
21 +2.1*	MaC 1-12	18 18.62	-8 33.2	F	MacConnell 1978	
22 +1.1*	MaC 1-13	18 25.85	-8 45.4	F	MacConnell 1978	
9 -6.1*	ESO-522-29	18 26.00	-23 53.3		Lauberts, etal. 1981	R
44+10.1*	We 3-1	18 31.77	+14 46.9	F	Weinberger 1978	
20 -3.1*	MaC 1-14	18 38.40	-13 14.6	F	MacConnell 1978	
31 -0.1	WeSb 4	18 48.09	-0 06.8	F	Weinberger, Sabbadin 1981	
19 -8.1*	MaC 1-15	18 54.37	-15 33.3	F	MacConnell 1978	
23 -7.1*	MaC 1-16	18 58.56	-12 02.7	F	MacConnell 1978	
30 -7.1*	MaC 1-17	19 10.29	-5 26.5	F	MacConnell 1978	
19-13.1*	DeHt 3	19 14.17	-18 07.0	F	Dengel, etal. 1980	
48 -1.1*	DeHt 4	19 24.12	+13 13.6	F	Dengel, etal. 1980	
58 -5.1	WeSb 5	19 59.48	+19 46.3	F	Weinberger, Sabbadin 1981	
111+11.1*	DeHt 5	22 18.36	+70 40.9	F	Dengel, etal. 1980	
110 -1.1	WeSb 6	23 10.89	+59 01.5	F	Weinberger, Sabbadin 1981	

REMARKS

3-4.10	Nebulous oval.
9 -6.1	Faint ring around stellar centre.
228-22.1	Discovered indep. by Longmore, Tritton (1980), F.
239-12.1	PN or galaxy, starlike centre.
239-18.1	PN or galaxy, starlike centre.
241 +0.1	Starlike centre.
242 -3.1	PN or galaxy.
248-12.1	PN or galaxy, starlike centre.
249-22.1	Starlike centre.
254 -6.1	PN or galaxy.
262 -4.1	PN or galaxy. LoTr 2, F.
263 -8.1	PN or galaxy.
265 +5.1	PN or galaxy.
265 -4.1	LoTr 3, F.
291+19.1	LoTr 4, confirmed by Longmore,Tritton (1980), F.
298+34.1	See also Hawley (1981).
327+14.1	Starlike centre.
329+12.1	Starlike centre.
339 -0.1	G 339.2-0.4, orig. classif. by Clark, etal. (1973) as radio SNR.
341+17.1	PN or galaxy.
343+16.1	PN or galaxy, starlike centre.
345+10.1	PN or galaxy.
346+19.1	B star in ring.
359 +1.3	Confirmed by Isaacman, etal. (1980), F.

* Possible planetary nebula. F Finding chart. R Remarks.

REFERENCES TO TABLE 1

Ae 2 Allen D.A., 1979, Obs. 99, 83.

-- Clark D.H., Caswell J.L., Green A.J., 1973, Nature 246, 28.

DeHt Dengel J., Hartl H., Weinberger R., 1980, Astron. Astrophys. 85, 356.

-- Hawley S.A., 1981, Publ.Astron.Soc.Pacific 93, 93.

ESO Holmberg E.B., Lauberts A., Schuster H.-E., West R.M., 1978, Astron.Astrophys.Suppl. 31, 15.

ESO Holmberg E.B., Lauberts A., Schuster H.-E., West R.M., 1978, Astron.Astrophys.Suppl. 34, 285.

ESO Holmberg E.B., Lauberts A., Schuster H.-E., West R.M., 1980, Astron.Astrophys.Suppl. 39, 173.

-- Isaacman R., Wouterloot J.G.A., Habing H.J., 1980, Astron.Astrophys. 86, 254.

ESO Lauberts A., Holmberg E.B., Schuster H.-E., West R.M., 1981, Astron.Astrophys.Suppl. 43, 307.

LoTr Longmore A.J., Tritton S.B., 1980, Monthly Notices 193, 521.

MaC 1 MacConnell D.J., 1978, Astron.Astrophys.Suppl. 33, 219.
Pu Purgathofer A., 1978, Astron.Astrophys. 70, 589.
Pu Purgathofer A., 1980, Astron.Astrophys. 88, 275.
PuWe Purgathofer A., Weinberger R., 1980, Astron.Astrophys. 87, L5.
Sh 2 Sabbadin F., Minello S., Bianchini A., 1977, Astron. Astrophys. 60, 147.
CTIO Smith M.G., Aguirre C., Zemelman M., 1976, Astrophys.Journal Suppl. 32, 217.
Sp 3 Stephenson C.B., 1978, Publ.Astron.Soc.Pacific 90, 396.
TrBr Terzan A., Bernard A., Ju K.H., 1978, C.R.Acad.Sci. Paris, t.287, Serie B, 235.
VB van den Bergh S., 1978, Astrophys.Journ.Suppl. 38, 119.
VB van den Bergh S., 1979, Astrophys.Journ. 227, 497.
We 3 Weinberger R., 1978, Obs. 98, 137.
WeSb Weinberger R., Sabbadin F., 1981, Astron.Astrophys. 100, 66.
SuWt West R.M., Schuster H.-E., 1980, Astron.Astrophys. 88, 350.
19 W 32 Wouterloot J.G.A., Dekker E., 1979, Astron.Astrophys.Suppl. 36, 323.

TABLE 2 MISCLASSIFIED PLANETARY NEBULAE

Design.	Name	Remarks and references
0 -0.1	Bl 3-4	H_α only, in RCW 141 but perhaps not related (Sanduleak, 1976) No emission (Allen, 1979) Not a PN (Kohoutek, 1982)
0 -0.2	Bl 3-22	Faint H_α , RCW 139? (Sanduleak, 1976) No emission (Allen, 1979) Not a PN (Kohoutek, 1982)
0 -1.3	Bl O	H_α only (Sanduleak, 1976) No emission (Allen, 1979)
0 -1.4	Bl 3-14	Unresolved, H_α only (Webster, 1975) Symbiotic star (Allen, 1979) Not a PN (Kohoutek, 1982)
1 +0.1	Bl 3-2	Only H_α in emission and cont. (Sanduleak, 1976) No emission (Allen, 1979)
1 -0.1	Bl 3-11	Not a PN (Vorontsov-Velyaminov, etal. 1973) Be pec (Sanduleak, 1976; Allen, 1979)

Design.	Name	Remarks and references
1 -0.2	Bl 3-3	No emission (Allen, 1979) Not a PN (Kohoutek, 1982)
1 -0.3	Bl 3-19	Be star (Allen, 1979) Not a PN (Kohoutek, 1982)
1 -1.1	Bl M	H_α only (Sanduleak, 1976) No emission (Allen, 1979)
65-27.2	CiPg	No [O III] emission in M 15 detected except for Ps 1 (Aurière, etal. 1978) Peterson's identification of enhanced H_α core emission with a PN unlikely (Phillips, etal. 1978)
74 +1.1	M 1-76	IR spectrum quite unusual,may be highly reddened P Cygni-type star (Bidelman, Krumenaker, 1972) Not a PN but a BQ[ ] star (Sabbadin, Bianchini, 1979)
176 -0.1	NGC 1985	Not a emission nebula (Purgathofer, Perinotto, 1980) Reflection nebula around a F1(V) star (Sabbadin, Hamzaoglu, 1981)
248 +8.1	He 2-10	Dwarf em. galaxy (Allen, etal. 1976) Emission line dwarf galaxy, d = 10 Mpc (D'Odorico, Rosa, 1981)
328-17.1	He 2-269	Dwarf emission galaxy (IC 4662) (Pastoriza, 1970)
356 +1.1	Th 3-21	No emission (Allen, 1979) Not a PN (Kohoutek, 1982)
356 -2.2	M 1-27	VLE? (Sanduleak, 1976) Be star with [N II], VLE (Allen, 1979)
357 +3.3	Th 3-17	H_α only (Sanduleak, 1976) Symbiotic star (Allen, 1979)
357 +3.5	Th 3-16	H_α only (Sanduleak, 1976) Be star (Allen, 1979)
357 +2.1	Ap 1-1	No emission (Allen, 1979)
357 +2.3	Th 3-20	Only faint H_α (Sanduleak, 1976) Symbiotic star (Allen, 1979)
357 -3.1	He 2-294	Faint H_α only (Sanduleak, 1976) Symbiotic star (Allen, 1979)
358 +3.5	Th 3-18	Faint H_α only (Sanduleak, 1976) Symbiotic star (Allen, 1979)

Design.	Name	Remarks and references
358 +2.1	Ap 1-3	No emission (Allen, 1979) Not a PN (Kohoutek, 1982)
358 +2.3	Th 3-29	H_α only (Sanduleak, 1976) Little or no [O III] (Allen, 1979) Not a PN (Kohoutek, 1982)
358 +1.2	Th 3-31	Faint H_α only (Sanduleak, 1976) Symbiotic star (Allen, 1979) Not a PN (Kohoutek, 1982)
358 -0.1	Bl 3-5	Early M star (Allen, 1979) Not a PN (Kohoutek, 1982)
358 -1.2	Sa 3-80	H_α only (Sanduleak, 1976) Symbiotic star (Allen, 1979)
358 -2.2	Bl 3-6	H_α only (Sanduleak, 1976) No emission (Allen, 1979) Not a PN (Kohoutek, 1982)
359 +2.1	Th 3-30	H_α only (Sanduleak, 1976) Symbiotic star (Allen, 1979) Not a PN (Kohoutek, 1982)
359 +2.2	Ap 1-5	No emission (Allen, 1979) Not a PN (Kohoutek, 1982)
359 +2.4	Th 3-32	Be star with [N II] (Allen, 1979) Not a PN (Kohoutek, 1982)
359 +1.2	Th 3-36	M star (Allen, 1979) Not a PN (Kohoutek, 1982)
359 -2.1	Bl L	H_α only (Webster, 1975) H_α only (Sanduleak, 1976) Symbiotic star (Allen, 1979)

REFERENCES TO TABLE 2

Allen D.A., 1979, Obs. 99, 83.
Aurière M., Laques P., Leroy J.L., 1978, Astron.Astrophys. 63, 341.
Allen D.A., Wright A.E., Goss W.M., 1976, Monthly Notices 177, 91.
Bidelman W.P., Krumenaker L.E., 1972, Publ.Astron.Soc.Pacific 84, 685.
D'Odorico S., Rosa M., 1981, ESO Preprint No. 180.
Kohoutek L., 1982, in preparation.
Pastoriza M., 1970, Bol.Asoc.Arg.Astron. No.15, 1.
Phillips J.P., Reay N.K., Worswick S.P., 1978, Astron. Astrophys. 70, 625.

Purgathofer A., Perinotto M., 1980, Astron.Astrophys. 81,215.
Sabbadin F., Bianchini A., 1979, Publ.Astron.Soc.Pacific 91, 278.
Sabbadin F., Hamzaoglu E., 1981, Astron.Astrophys. 94, 25.
Sanduleak N., 1976, Publ.Warner Swasey Obs. 2, No.3, 55.
Vorontsov-Velyaminov B.A., Kostyakova E.B., Dokuchaeva O.D., Arhipova V.P., 1973, Mem.Soc.Roy.Liège, 6e Serie, t.V, 79.
Webster B.L., 1975, Monthly Notices 173, 437.

---

of large angular dimensions which it was still possible to find on the Palomar Observatory Sky Atlas (15 objects), or discovered in the ESO/SRC Southern Sky Survey (34 objects). They increase substantially the statistics of near-by planetaries, which is important for determining the space density, the total number, and the birth-rate of planetaries in our Galaxy.

It is recommended that 33 objects be removed from the CGPN, the classification of which as emission-line stars of various types, symbiotic stars, reflection nebulae or emission galaxies seems now to be more or less guaranteed. The present list of misclassified planetary nebulae (Table 2) includes only a small fraction of all doubtful objects which still can be found in various lists of planetaries and in the CGPN. This is not surprising as - as we should remember - one of the main aims of the CGPN was to provoke further and more detailed observations just in order to enable a reliable classification of PN. The list of confirmed PN is indeed very desirable, but it can only result from sufficient observational data as well as from a necessary theory which would interpret the observations and give a correct picture of a PN in any evolutionary stage.

Planetary nebulae are generally discovered and classified according to their morphology and emission spectrum. They can possibly be mistaken for objects showing some morphological or spectroscopical similarities,but having physical properties very different from those of typical PN. The main criteria which can help to distinguish PN from other objects are summarized below (Table 3); they refer to observable properties (and not to absolute or derived parameters, like mass or luminosity), and one should be careful in applying them because they are valid only in a statistical sense.

The correct classification of extended PN does not in general bring large problems if sufficient observational material is available. On the contrary, the unresolved young planetaries still cannot be reliably recognized due to the incomplete theory of the origin of PN. For such objects the

TABLE 3a

PN mistaken for:	Distinguishing criteria:
Emission-line galaxies	Morphology: non-stellar nucleus unresolved, very bright nucleus (Sy gal.) High galactic latitude (generally) High radial velocity Continuum and em. lines have the same spatial extension Broad em. lines (Sy gal.); H_β comparable to [OIII] 5007 for narrow-emission-gal. and for Sy 1 gal., H_β/[OIII] 5007 $\sim$ 0.1 for Sy 2 gal. (Atlas of Seyfert gal.: Khachikian, Weedman, 1974)
Supernova remnants	Morphology: predominantly curved or crisp filaments, fragmentary nebulosities or part-shell structure (see van den Bergh, etal. 1973; van den Bergh, 1978) High expansion velocity (except for very old SNR) Emission-line ratios H_α/[NII], H_α/[SII], [SII] 6717/6731 (see Sabbadin, D'Odorico, 1976) Great strength of the forbidden lines compared with the H lines Non-thermal radio spectrum; X-ray source
Reflection nebulae	Morphology: mostly amorphous Continuous spectrum; the nebula is generally bluer than the star; frequently associated with dark nebulae, H II regions, young star clusters
H II regions (dense, classical, giant)	Associated with molecular clouds Spectra of lower excitation than PN: [OIII] 5007 + 4959/H_β <7, HeII 4686 absent (see Chopinet, L.-Zuckermann, 1976; for a typical spectrum see Terzian, Balick 1974) Emission-line ratios H_α/[NII], H_α/[SII], [SII] 6717/6731 (see Sabbadin, D'Odorico, 1976) High reddening, strong local obscuration For the IR and radio criteria see Panagia (1978)

TABLE 3b

PN mistaken for:	Distinguishing criteria:
Nebulae associated with WR stars	Morphology: amorphous or shell-structured H II regions, clumpy appearance, thin sheets of gas, filaments, bubbles (for classification see Chu, 1981) WR star (Population I) located at a preferred position inside the nebula
H II regions (compact)	Associated with molecular clouds Spectra of lower excitation than PN: [OIII] 5007 + 4959/H_β <7, He II 4686 absent (see Chopinet, L.-Zuckermann, 1976) High reddening, strong local obscuration Classification, criteria for IR and radio observations: see Habing, Israel (1979)
Type I Symbiotic stars (Z And, AG Peg, CI Cyg)	Composite spectrum (G or later + B + nebular) Absorption lines of a late component visible (CaI, CaII, TiO) Emission lines of HeII, [OIII], [FeIII] or higher ionized atoms present (the width $\lesssim$ 100 km/s) Variability possible (amplitude up to 3 mag) IR photometry: S - type (see Duerbeck, Seitter, 1982)
Slow novae (RR Tel, RT Ser, AS 239?)	Outburst similarities with classical and recurrent novae (RS Oph); a single nova-like eruption typically of 2-7 mag amplitude; possible mira-like variability TiO absorption in most of the slow novae IR photometry: bright continua of M stars, additional presence of circumstellar dust in some objects (see Allen, 1980)

TABLE 3c

PN mistaken for:	Distinguishing criteria:
Be and related stars	Spectrum late O to early A, high luminosity (V to III) Visual spectrum: emission lines of H, sometimes of HeI, FeII, MgII - not HeII, [OIII], [NeIII], [OII] Brightness variations up to about 1 mag frequent IR excess reported in about 50% Be stars No detectable radio emission in normal Be stars (see Seitter, Duerbeck, 1982)
Ke, Me, Ce, Se stars	Mostly long period variables (Mira stars), semiregular, irregular, flare stars Emission lines of H, CaII, HeI, FeI, FeII, TiII, [SII], sometimes [FeII] - not HeII, [OIII], [NeIII]
Pre-main sequence emission objects	Generally very strong association with interstellar clouds or star formation regions; strong IR sources (see Strom, etal., 1975)
T Tauri stars	Irregular optical variability (range 1-2 mag, FU Ori stars as large as 5 mag) Spectral types Fe to Me; UV excess; IR excess Emission lines: H, CaII, FeI, FeII, TiII, [SII] - but not HeII, [OIII]; P Cygni profiles
Herbig Ae, Be stars	Spectral types B1e to F8e Optical spectrum ranging from those with H only through the rich emission spectrum (V 380 Ori); similar to T Tauri stars
Herbig-Haro objects	Variable "semistellar" patches of nebulosities, sometimes complex structure associated with a single IR source Very high reddening Generally low excitation emission spectrum dominated by the B. lines and by [SII] and [OII]; faint continuum

morphological criterium is naturally not applicable, and their spectra may be unlike those of conventional PN. The symbiotic stars of Type II (also called BQ[] stars, e.g. V 1016 Cyg, HM Sge, HBV 475) are frequently regarded as PN in an early evolutionary stage, although some spectral similarities between very young PN and slow novae also exist. More detailed UV, visual and IR spectroscopy, as well as radio observations will be necessary in order to establish definitively the relationship between young PN and various emission line objects.

I would like to encourage the observers operating with appropriate telescopes and measuring facilities to occupy themselves with doubtful PN.We have started a comprehensive programme of a spectroscopic verification of suspected PN, but due to the large number of such objects any help or collaboration would be appreciated.

REFERENCES

Allen D.A., 1980, Monthly Notices Roy.Astron.Soc. 192, 521

Chopinet M., Lortet-Zuckermann M.C., 1976, Astron.Astrophys. Suppl. 25, 179.

Chu Y.-H., 1981, Astrophys.Journ. 249, 195.

Duerbeck H.W., Seitter W.C., 1982, Variable Stars, in Landolt-Börnstein, Band 2 (ed.K.Schaifers, H.H.Voigt), Springer-Verlag, Berlin-Heidelberg-New York, p.197.

Habing H.J., Israel F.P., 1979, Ann.Rev.Astron.Astrophys. 17, 345.

Khachikian E.Ye., Weedman D.W., 1974, Astrophys.J. 192, 581.

Kohoutek L., 1978, IAU Symp.No.76 (ed.Y.Terzian), D.Reidel Publ.Comp., Dordrecht,Boston, p.47.

Panagia N., 1978, IAU Symp.No.76 (ed.Y.Terzian), D.Reidel Publ.Comp., Dordrecht,Boston, p.315.

Perek L., Kohoutek L., 1967, Catalogue of Galactic Planetary Nebulae, Academia Praha.

Sabbadin F., D'Odorico S., 1976, Astron.Astrophys. 49, 119.

Seitter W.C., Duerbeck H.W., 1982, Peculiar Stars, in Landolt-Börnstein, Band 2 (ed. K.Schaifers, H.H.Voigt), Springer-Verlag, Berlin-Heidelberg-New York, p.269.

Strom S.E., Strom K.M., Grasdalen G.L., 1975, Ann. Rev. Astron.Astrophys. 13, 187.

Terzian Y., Balick B., 1974, Fundamentals of Cosmic. Phys. 1, 301.

van den Bergh S., 1978, Astrophys.Journ.Suppl. 38, 119.

van den Bergh S., Marscher A.P., Terzian Y., 1973, Astrophys.Journ.Suppl. 26, 19.

SEATON: Aller mentioned NGC 6302 several times. Is it a PN or a slow nova?

KOHOUTEK: The classification of this peculiar nebula is very uncertain, but it is still included in the PN.

TERZIAN: Do you think that NGC 7027 is a PN?

KOHOUTEK: This nebula really is rather strange but is, I believe, the most investigated object among PN. At present, it would not be advisable to change the classification of NGC 7027. Even more observations have to be awaited, especially of its central star.

TERZIAN: Do you know if the Southern Sky Atlas has been completely surveyed for PN?

KOHOUTEK: Longmore (1977) and Longmore and Tritton (1980) published two lists of new planetary nebulae found in the Southern Sky Atlas. I have also discovered some new objects in it, but I am sure that this Atlas still contains many PN which have not yet been found.

REAY: The lack of new discoveries in the southern sky is probably because no-one has conducted a systematic search for PN on the scale of the searches on the POSS plates.

FORD: What is known about the emission line object discovered by Peterson near the globular cluster NGC 6401?

PEIMBERT: Recillas-Cruz and I have obtained spectra of this object in two different observing seasons and we find that it is a symbiotic star that shows H and He lines in emission and an underlying M-type spectrum with Ti O bands in absorption.

KOHOUTEK: G. Schnur and myself took spectra of this object too and we can confirm Peimbert's result. For this reason, this symbiotic star has not been included in the list of new PN.

WALSH: Are there plans for a second edition of the Perek and Kohoutek catalogue?

KOHOUTEK: The second edition of this catalogue is in preparation but not in the same form as the first edition. The new edition will probably contain only data relating to the discovery and the identification of the individual objects, mainly their positions and identification charts.

MORPHOLOGY AND KINEMATICS OF PLANETARY NEBULAE

N. K. REAY,
Blackett Laboratory,
Imperial College,
London, SW7 2BZ,
England.

The application of monochromatic images and internal kinematic data to the construction of 3-dimensional models for the HII regions of planetary nebulae is discussed, and the role of models in investigating envelope evolution is commented upon.

The quality and usefullness of available data is critically reviewed, and examples of 'seeing limited' kinematic and isophotometric data obtained with a new imaging Fabry-Perot system are presented.

INTRODUCTION

The remarkable variation in the appearance of different planetary nebulae has led to interpretations in terms of spherical, elliptical, cylindrical, toroidal and helical density distributions, often with double and sometimes triple shells comprising combinations of one or more of the above 3-dimensional shapes. In addition, knots, filaments, ansae and stratification effects contribute to produce a very large variety of nebular forms.

Since planetary nebulae appear to constitute a physically related class of astronomical object, however, their spatial forms must be governed by similar dynamical forces, and they should be expected to have similar or related intrinsic forms. The diversity of observed shapes must be the result of a range of conditions at the point of nebular ejection and/or evolution of the nebula after ejection, coupled with stratification effects and projection effects onto the plane of the sky.

If we accept this statement to be true, then in principle geometric modelling of a sufficiently large sample of planetary envelopes will give a clearer view of the range of intrinsic 3-dimensional structures, and when specific geometries are used in conjunction with ionisation models, a more complete understanding of the physical processes at work in the shell.

D. R. Flower (ed.), Planetary Nebulae, 31–43.

Modelling will also enable us to investigate evolutionary links between different nebular geometries and, possibly, to comment on conditions prevailing on the surface of the progenitor at the time of envelope ejection.

To make progress in this direction it will be necessary to model a large number of nebulae (60 or more), chosen as representative examples of known visual forms, and for each nebula we will require a uniform data set comprising

(a) Photometrically calibrated monochromatic images in Hα or Hβ and a number of other lines covering a range of ionisation potentials e.g. [OI] 6300Å, [OII] 3727/29Å, [OIII] 4363Å and 5007Å, [NII] 6584Å, [SII] 6717/31Å, [NeV] 3426Å and HeII 4686Å.

(b) Velocity field information at 'seeing limited' spatial resolution over the envelope, and at a spectral resolution of 50,000 ($\sim$ 6 km s^{-1}).

In the following I review briefly the data currently available, commenting on its scope and limitations and the way it has been used to deduce the 3-dimensional structure of planetary envelopes. Finally I will discuss methods currently employed by my colleagues and I to obtain data in catagories (a) and (b) above, in order to establish a uniform morphological and kinematic data base for a large sample of planetary nebulae.

OBSERVATIONS AND MODELLING

Studies of the distribution of emission intensities in planetary nebulae, usually photographically, are too numerous to cite individually. References to past literature are contained in the proceedings of the two planetary nebulae symposia IAU 34 (1968) and IAU 76 (1978). Much, although not all, of this early work was broad-band, and whilst its value in establishing the overall structure of nebular envelopes should not be underestimated (cf Curtis 1918, Aller 1956), it is not in general a source of quantitative data useful for the construction of detailed geometric or ionisation structure models.

Broad-band photography is however unsurpassed in detecting faint extended nebulosity. This is best illustrated by Figure 1 which shows a beautiful exposure of NGC 7293 obtained by David Malin at the prime focus of the Anglo-Australian telescope. Faint hitherto undetected extensions, sweeping out to a radius of 18 arcmin, are revealed on this high contrast print.

In the last few years Goad and Chaisson (1973), Feibelman (1970) and others have published detailed monochromatic photographs or contour maps which show often quite dramatic changes in intensity distribution with ionisation potential. Figure 2 compares Hα and [NII]6584Å contour maps of NGC 6543 published by Phillips, Reay and

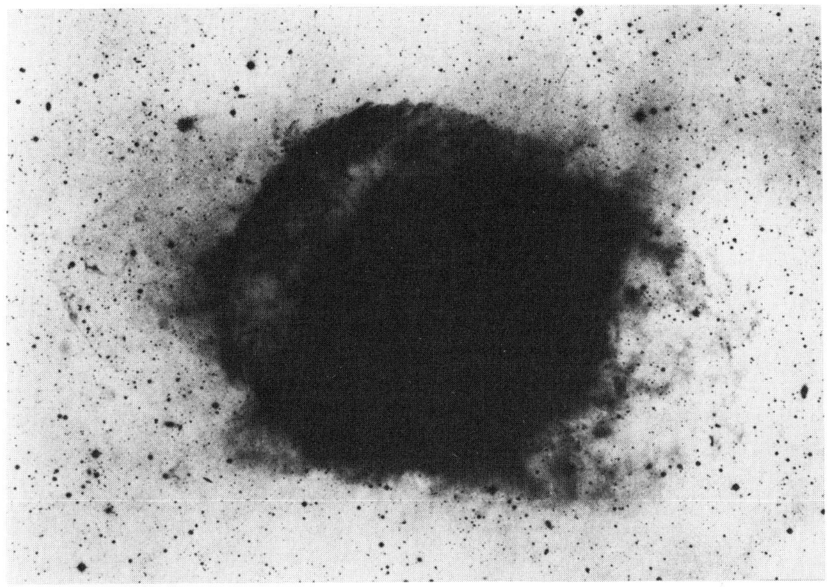

FIGURE 1. NGC 7293

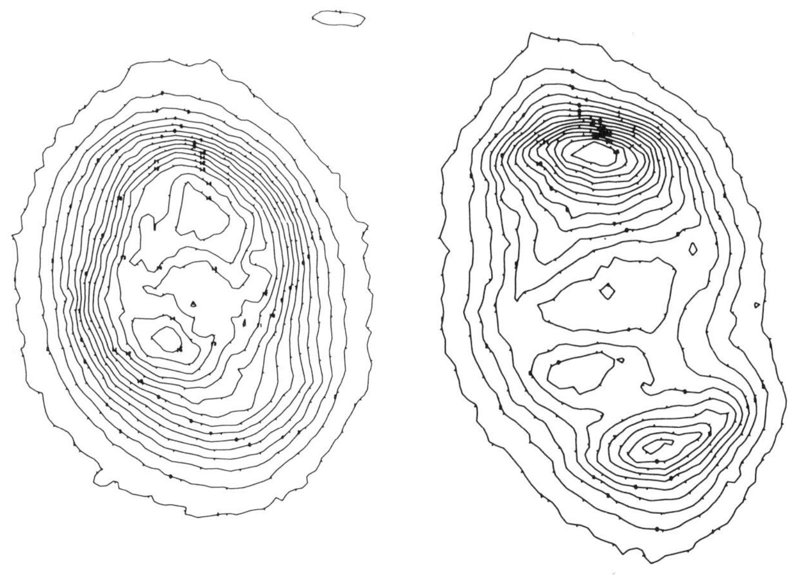

FIGURE 2. NGC 6543. (a) Hα 6563Å (b) [NII] 6584Å
North is at the top and east to the left.

Worswick (1977) which illustrated well this point and demonstrates that considerable caution should be exercised in using morphological data for which the observational parameters are not very well known.

Internal velocities in planetary nebulae were first observed by Campbell and Moore (1918), but it was Wilson (1948, 1950) who systematically measured line splitting, interpreting his data in terms of envelope expansion, typically 20 km s^{-1}, and showed that stratification effects within the envelopes often led to ions of low ionisation potential expanding more rapidly than those with higher ionisation potential. Osterbrock et al (1966) interpreted the shapes of high dispersion emission line profiles, obtained at the centre of a number of bright planetary nebulae, in terms of a velocity distribution of emitting ions of typically 30 km s^{-1} in the line of sight through the shell. Weedman (1968) took the technique one step further by obtaining slit spectroscopy at points across 10 planetary nebulae, including NGC 7009, 3242 and IC418 and interpreted his results in terms of 3-dimension prolate spheroidal shells. He also deduced the distribution of electron density within the shells, and showed that expansion velocities were in general proportional to distance from the central star.

Theoretical work on the overall shapes of shells expanding from rotating stars supports to a large extent the general observational conclusion that planetary envelopes are spheroidal in shape, either prolate or oblate. Kirkpatrick (1976) has considered the development of shells under the influence of a continuous accelerative process (of the kind produced by ablation from the inner surface) and concluded that prolate forms would develop from an initially oblate spherical shell ejected by a rotating giant star. Louise (1973) discussed the development of gas shells ejected from a spherical rotating star. He showed that nebulae acquired prolate forms, although omitted to account for gravitational retardation during the ejection process. Phillips and Reay (1977) modelled the effect of gravitational braking, differential stellar rotation and radiation pressure on the development of ejected gas shells. They concluded that both oblate and prolate forms could develop, and that the principle forms observed in planetary nebulae can be explained reasonably well in terms of these factors alone.

To facilitate comparison with theoretical models my group at Imperial College have, over the last seven years used electronography to obtain monochromatic images of 50 or more planetary nebulae in a number of emission lines of widely differing ionisation potentials. The electronographic technique has the advantage over photography that it is linear over a wide density range, and does not suffer from reciprocity failure (Worswick 1975). It is suitable therefore for recording both the bright inner and faint outer regions of planetary nebulae on a single exposure. Figure 3 shows, for example the faint halo surrounding NGC 7027 recorded electronographically by Atherton et al (1979). The halo is about one thousand times fainter than the

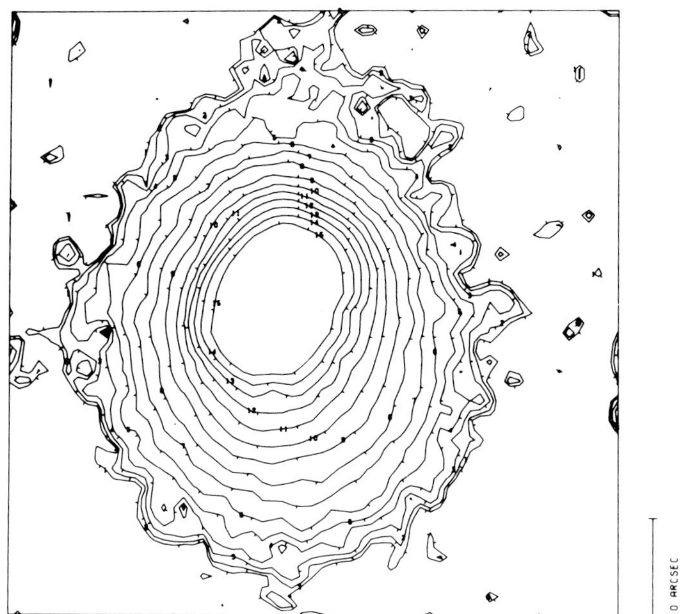

FIGURE 3. NGC 7027. Hα 6563Å map with contour levels set at logarithmic intervals.

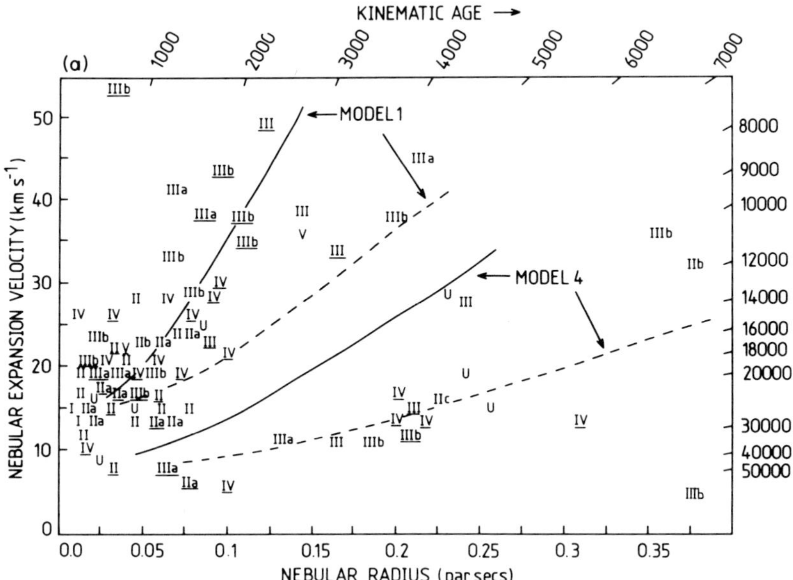

FIGURE 4. [OIII] expansion velocities versus radius for nebular envelopes. Points are shown as morphological classes (Verontsov-Velyaminov 1948). Full lines are model curves from Ferch and Salpeter (1975). Broken lines are same models but taking radius as being two times radius of maximum density.

brightest part of the NGC 7027 image.

In general, because of its large dynamic range electronographic data is more capable than is photography of detecting underlying geometric symmetry which may otherwise be masked by large differences in surface brightness. Monochromatic electronography of NGC 5189 (Phillips and Reay 1982) and NGC 2440 (Phillips, Reay and Worswick 1980), for example, show these apparently irregular, chaotic objects to have a considerable degree of symmetry when viewed in sufficient detail. Both, in fact, show evidence for pairs of condensations symmetrically disposed about their respective central stars.

Reay and Worswick (1982) have used the linearity of electronography in order to ratio [OIII] 5007Å with 4363Å to map the electron temperature variation across 5 nebulae. Phillips, Reay & White (1982) have used the ratio technique to search for cool condensations in NGC 6302 and Atherton et al (1979) to locate the central star in NGC 7027 and estimate its brightness as $m_V = 19.4 \pm 1.0$.

Taylor (1977, 1979), and Hicks, Phillips and Reay (1976) have used Fabry-Perot interferometers to study velocity fields in planetary envelopes, and Atherton et al (1978) have combined monochromatic electronography with Fabry-Perot velocity field measurements to construct a 3-dimensional model for the HII region of NGC 6720 which shows it to have a closed spheroidal structure and not, as sometimes suggested necessary, a toroidal structure. Phillips, Reay and Worswick (1977) have shown using monochromatic electronography and re-analysing existing kinematic data, that the intrinsic form of NGC 6543 is spheroidal and not, as previously suggested, helical. The significance of this result is, more than anything else, in showing that an apparently complex nebula can be accommodated within a simple spheroidal model scheme.

EXPANSION OF THE HII REGION

Valuable insight into the evolution of planetary envelopes, and into the mechanism which drives or accelerates them, can be obtained by studying in a statistical way the relationship between expansion velocity and nebula radius.

In a comprehensive study of the dynamical effect of dust within nebular envelopes, Ferch and Salpeter (1975) have computed the accelerative effect of radiation pressure on the dust for a range of central star and nebular parameters. In general they predict an expanding envelope, with expansion velocity increasing with radius and a central 'hole' swept clear of gas by radiation pressure.

Bohuski and Smith (1974), Robinson, Reay and Atherton (1982) and Sabbadin and Hamzaoglu (1982) have made observational studies of the expansion velocity-radius relationship. Figure 4 (from Robinson et al, 1982) summarises their data and compares it with two of the

Ferch and Salpeter models. The data points include data from Bohuski and Smith (1974), Johnson (1977), Wilson (1950) and others, but not the new data from Sabbadin and Hamzaoglu which appeared in print in June 1982. The agreement between theory and experiment is good, and Robinson et al conclude that there may be evidence in the data for two distinct velocity-radius sequences, consistent with Ferch and Salpeter (1975) Models 1 and 4. Sabbadin and Hamzaoglu confirm the general trend of increasing expansion velocity with radius up to 0.2 parsec beyond which they see a fall in expansion velocity which they attribute to possible interaction of the envelope with the interstellar medium.

Thus we appear to have some quantitative agreement between theory and observation which may well be improved upon by deriving the envelope expansion velocity from a 3-dimensional model of the nebula rather than from a single point measurement. The importance of the agreement lies in the ability it gives to attribute a relative age to a nebula and so to study evolution of the 3-dimensional forms.

NEW METHODS

Returning to the methods by which kinematic and monochromatic isophotometric data is acquired, it is worth noting the limitations in current technique and asking what modern instruments and electronic detectors enable us to achieve.

Photographic photometry is limited by dynamic range and calibration difficulties (although not everyone would agree with that!)

Electronography overcomes these difficulties, but has its own limitations to do with difficulties in flat fielding caused by poor quality electronographic emulsion. This conspires to limit the achievable photometric accuracy obtainable with an electronographic device to approximately 2%.

Kinematic studies are limited to simultaneously measuring a number of points along a spectrograph slit at a spatial resolution determined by the slit width, or to measuring points sequentially with more luminous Fabry-Perot interferometers. Spatial resolutions in this latter case are usually 5 arcsec or worse.

Multislit spectrometers and Fabry-Perot interferometers working in non-classical mode (see for example Meaburn, 1976) have been used to achieve better spectral and spatial coverage, but none are completely satisfactory.

The advent of imaging electronic detectors has however shown the way to an instrument which my colleague and I at Imperial College and at the Anglo-Australian Observatory now use to obtain simultaneous seeing limited photometry and kinematic information across the envelopes

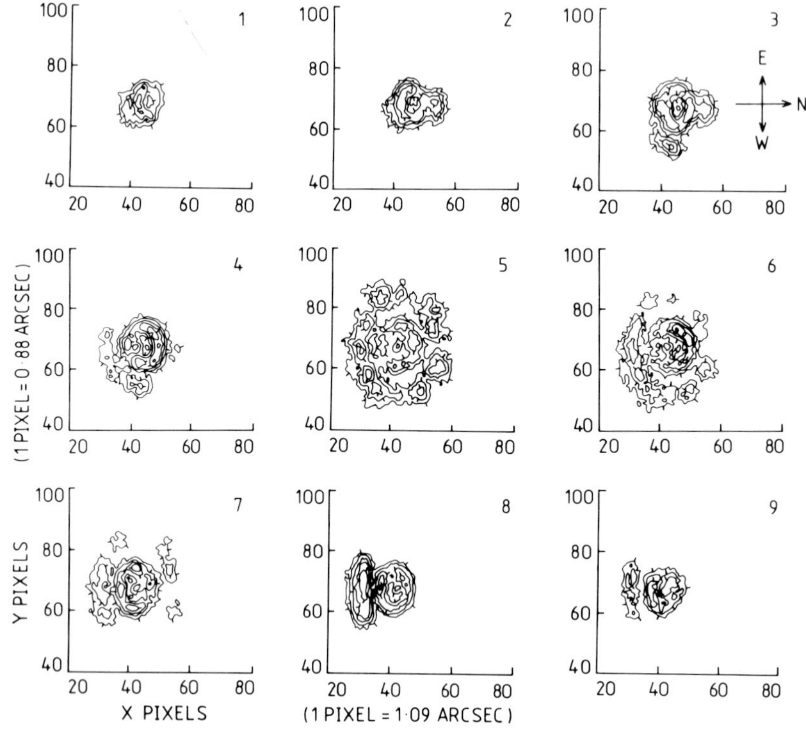

FIGURE 5. NGC 2392. A sequence of monochromatic slices through the [NII] 6584Å line. Resolution 0.33Å, spacing 0.66Å. Wavelength increases with frame number.

of planetary nebulae. The instrument is described fully by Atherton et al (1982), suffice to say it is a large aperture servo-controlled scanning Fabry-Perot interferometer which uses as a detector the Image Photon Counting System (Boksenberg 1972). A range of etalons is available to give spectral coverage in the green and red with a resolving power of up to 50,000. The instrument is scanned under microprocessor control and simultaneously records an array of up to 500 by 500 spectra over a field of up to 9 arcminute in diameter.

The data can be presented either as an array of spectra or a series of monochromatic slices. A comprehensive data reduction package is available on the SERC STARLINK image processing system to extract from the data emission line velocities and intensities, monochromatic maps, simulated slit spectra and a host of other parameters.

The beauty of this technique is that we acquire kinematic and photometric data simultaneously. The data is recorded at 'seeing limited' spatial resolution and is stored digitally, so making it

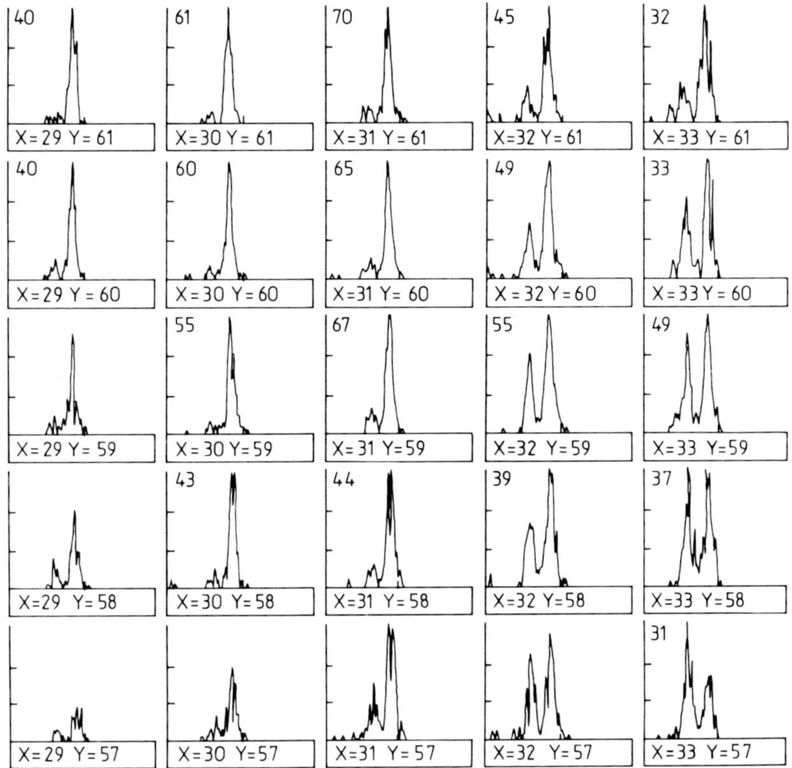

FIGURE 6. NGC 2392. Sample spectra from the [NII] array. Wavelength scans from red to blue. When maximum count exceeds 30, its value is indicated in top left corner

easier to extract, for example, electron density and temperature maps from forbidden line ratios and to attempt on-line 3-dimensional model fitting.

Figure 5 shows an example of [NII] 6584Å data obtained on NGC 2392, Figure 6 gives a 5 x 5 sample from the grid of 128 x 128 spectra we have obtained across the envelope, and in Figure 7 we show an image of NGC 5189 with synthetic spectra reconstructed from our data array. The inner regions of NGC 5189 can be interpreted on the basis of this data as a ring of condensations expanding from the central star at 23 km s^{-1} and lying in a plane, the normal to which is tilted an an angle of 80° to the line of sight.

CONCLUDING REMARKS

A full size TAURUS data 'cube' consists of 500 x 500 pixels with 100 or more interferometer steps and 8 bit numbers - that is

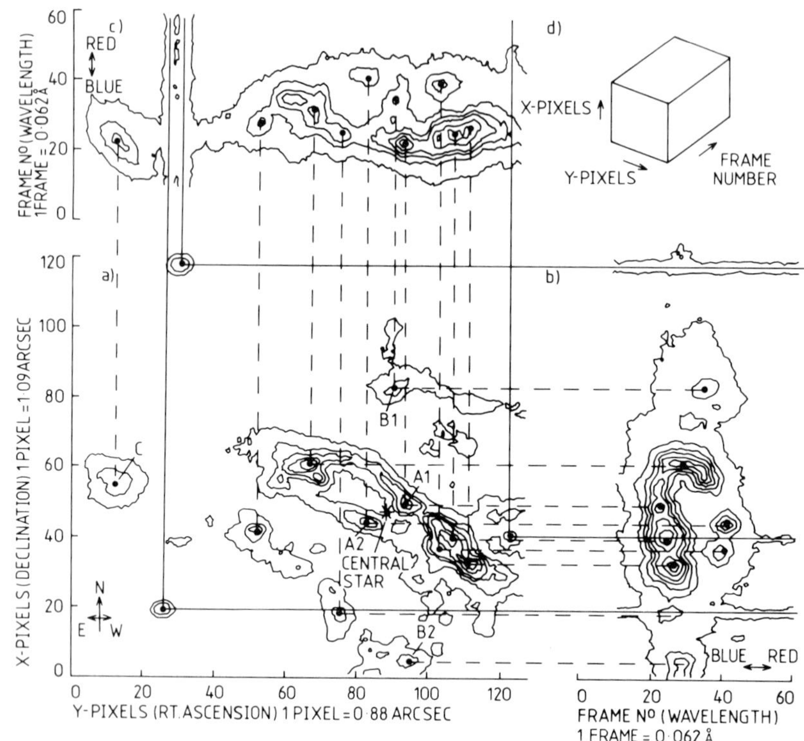

FIGURE 7. NGC 5189. (a) [NII] 6584Å contour map. (b) Synthetic spectral line obtained by adding X PIXEL-WAVELENGTH slices from the data array (c) Synthetic spectral line obtained by adding Y PIXEL-WAVELENGTH slices. (d) Schematic data cube.

a staggering 200 Mbits of data.

In practice we use smaller data arrays, perhaps 150 x 150 pixels, but even so produce 18 Mbit data sets for a typical one hour observation of say the [SII] 6717Å line in NGC 2440.

The CFH Telescope and the new UK observatory in the Canary Islands will be equipped with similar wide field instruments. In the next ten years I anticipate that imaging Fabry-Perot or Michelson interferometers will be standard equipment at all major observatories, and with infrared array detectors they will be capable of working at wavelengths to 10 μm and beyond.

The problem will then not be one of how to acquire sufficient data, but how to cope with the data we have acquired.

ACKNOWLEDGMENTS

I thank David Malin of the Anglo-Australian Observatory for permission to use the photograph of NGC 7293 (Figure 1).

REFERENCES

Aller, L.H., 1956. Gaseous Nebulae, Chapman Hall, London.
Atherton, P.D., Hicks, T.R., Reay, N.K., Robinson, G.J., Worswick, S.P., 1979. Ap.J. 232, p.786.
Atherton, P.D., Hicks, T.R., Reay, N.K., Worswick, S.P., Hayden Smith, W., 1978. Planetary Nebulae, IAU Symposium No.76, ed. Y. Terzian.
Atherton, P.D., Taylor, K., Pike, C.D., Harmer, C.W.F., Parker, N., Hook, R.N., Mon. Not. R. astr. Soc. In Press (1982).
Bohuski, T.J., Smith, M.G. 1974. Ap.J. 193, p.197.
Boksenberg, A. 1972. Auxiliary Instrumentation for Large Telescopes p.205. Proc. ESO/CERN Conference, Geneva.
Campbell, W.W. and Moore, J.H. 1918. Pub. Lick Obs. 3, p.161.
Curtis, H.D. 1918. Pub. Lick Obs. No.13.
Feibelman, W.A. 1970. J. R. Astron. Soc. Canada, 64, p.305.
Ferch, B.L., Salpeter, E.E. 1975. Ap.J. 202, p. 195.
Goad, L.E., Chaisson, E.J. 1973. Mem. Soc. R. des Sci de Liege. No.5.
Hicks, T.R., Phillips, J.P., Reay, N.K. 1976. Mon. Not.R. astr. Soc., 147, p.339.
IAU Symposium No. 34 1968. eds. D. E. Osterbrock and C.R. O'Dell. D. Reidel publishing Co.
IAU Symposium No. 76 1978. ed. Y. Terzian. D. Reidel publishing Co.
Johnson, H.M. 1977. Ap.J. 208, p.127.
Kirkpatrick, R.C. 1976. Astrophys. Lett. 17, p.7.
Louise, R. 1973. Mem. Soc. Roy. Liege, 5, p.465.
Meaburn, J. 1976. Detection and Spectrophotometry of Faint Light. Reidel-Holland.
Osterbrock, D.E., Miller, J.S. Weedman, D.W. 1966. Ap.J., 145, p.697.
Phillips. J.P., Reay, N.K., 1977. Astron. Astrophys. 59, p.91.
Phillips. J.P., Reay, N.K. Mon. Not. R. astr. Soc. - In Press (1982).
Phillips, J.P., Reay, N.K., White, G.J. Astron. Astrophys - In Press (1982).
Phillips, J.P., Reay, N.K., Worswick, S.P., 1977. Astron. Astrophys. 61, p.695.
Phillips, J.P., Reay, N.K., Worswick, S.P., 1980. Mon. Not. R. astr. Soc., 193, p.231.
Reay, N.K., Worswick, S.P. 1982. Mon. Not. R. astr. Soc., 199, p.581.
Robinson, G.J., Reay, N.K., Atherton, P.D., 1982. Mon. Not. R. astr. Soc., 199, p.649.
Sabbadin, F., Hamzaoglu, E., 1982. Astron. Astrophys. 110, p.105.
Taylor, K., 1977. Mon. Not. R. astr. Soc., 181, p.475.
Taylor, K., 1979. Mon. Not. R. astr. Soc., 189, p.511.
Vorontsov-Velyaminov, B.A. 1948. Gasnebel und Neue Sterne, Verlagkulter und Fortschritt, Berlin.
Weedman, D.W., 1968. Ap.J. 153, p.49.
Wilson, O.C., 1948. Ap.J. 108, p.201.

Wilson, O.C., 1950. Ap.J. 111, p.279.
Worswick, S.P., 1975. Ph.D. Thesis, University of London.

KAHN: Could you give us an indication of the likely age of the PN (with ansae) that you have just been talking about?

REAY: The outer envelope of A 30 has an expansion velocity of 40 km s^{-1} and a kinematic age of 10^4y. The system of inner ansae has a radial velocity $\simeq$ 25 km s^{-1} and an actual expansion velocity of 35 km s^{-1} if we assume a projection angle of 45°. The kinematic age of the ansae is, therefore, about 1.5 x 10^3y. A distance of 1.4 kpc has been assumed.

SURDEJ: How many PN are known to show double structure in their envelopes? What are the radial velocities of these structures? Has any mechanism been proposed to explain the formation of such structures?

REAY: From my own observations, I have found strong evidence for pairs of ansae in NGC 2440, NGC 5189 and NGC 6826. In the first two objects, there is some evidence for multiple pairs at different position angles. A30 has two pairs of ansae and recent evidence suggests that NGC 7026 also has ansae. The prototype object is, of course, NGC 7009.

Radial velocities of ansae are typically 20 km s^{-1}, similar to the expansion velocities of nebulae. I know of no mechanism proposed for the ejection of multiple pairs of ansae. Could it be symmetric ejection from a precessing central star? I would be interested in hearing from anyone with ideas on this topic.

ALLER: It is extremely important to have monochromatic isophotes for PN. The discordances between different sets of line intensity measurements arise often not from errors in the data but from differences in what was actually observed. Photoelectric measurements often give the integrated light from the entire nebula, whilst IDS or IPCS observations are necessarily restricted to small strips or areas. With isophotic contours, we can at least correct some lines for this effect. Such observations are especially important to relate IUE and optical observations.

KALER: Are the Greig and Khromov and Kohoutek classification schemes still valid in the light of the new work?

REAY: We have not yet observed a sufficiently large number of nebulae to enable me to answer your question. My initial impression is that the current classification schemes are valid but inadequate.

MALLIK: Which distances did you use in the plot of V against R?

REAY: All distances are from Acker (1978, Astron. Astrophys. Suppl. 33, 367).

OSTERBROCK: The monochromatic images are a very important step forward in your work. In the case of NGC 6543, can you explain how the two "closed shell" models for Hα and [N II] combine to give the helix appearance as their sum? Are the two models tipped? Do other helix nebulae - for instance NGC 7293 - have similar morphology?

REAY: Electronographic observations of NGC 6543 were made through narrow band (≈ 10 Å) filters centred on [O I], [S II], [N II], [O III], [O II], Hβ and other lines. The low ionisation structures are similar but quite different from [O III] and Hβ. None of the lines show strong evidence for a helical structure. My suggestion was that the helical structure may be an illusion produced by adding high and low excitation images. Even if real, the structure can be accommodated in a simple, three-dimensional, kinematic model. However, new 20 cm VLA data to be presented at this conference by Bignell suggest that the helical structure may, indeed, be real.

NGC 7293 is too large for me to observe, but like NGC 6720 the overall structure can be accommodated within a three-dimensional closed shell model.

KEYES: Have you used your observations of other emission lines to obtain three-dimensional maps of electron density (from [S II] or [O II]) or electron temperature ([O III]) in any objects? If so, what are the ranges of N_e and T_e?

REAY: I have obtained data in the [S II] λλ 6717, 6731 lines for NGC 2440, NGC 5189 and a number of other PN which I shall use to construct three-dimensional electron density maps. The data have not yet been reduced.

My colleagues and I have published electron temperature maps of a number of PN using two-dimensional electronographic data ([O III] λ 5007 and λ 4363). We find temperature changes of typically a few hundred degrees over the [O III] - emitting region (1982, Mon. Not. Roy. Astron. Soc. 199, 581).

CARSENTY: Leaving the small-scale structure aside and considering the general structure, do you find any cases of special symmetry (cylindrical, prolate or oblate spheroids)?

REAY: It is difficult to leave aside the condensations and look at the underlying structure. In the cases of NGC 2440 and NGC 5189, the condensations appear to dominate the structure to such an extent that their removal leaves very little behind! The inner condensations in NGC 5189 do, however, lie in a ring centred on the central star, in a plane the normal to which is tilted at 80° to the line of sight, and are expanding at about 25 km s^{-1} from the star. In a sense, this is the underlying structure; it is very similar to that of NGC 650-1.

RECENT WORK ON BIPOLAR NEBULAE

Martin Cohen
Radio Astronomy Laboratory, University of California, Berkeley
and NASA-Ames Research Center

Recent results obtained from studies of bipolar nebulae with a variety of techniques are described. Nebular polarization maps and spectropolarimetry, near-infrared spectroscopy, far-infrared photometry, radio maser and continuum work all have contributed to our knowledge of this heterogeneous class of objects. Some are certainly pre-main-sequence; others are likely to represent the rapid transition from red giant to planetary nebula. At least one dust-shrouded carbon star (CIT 6) and one visual binary with an O-star primary (MWC 349) have bipolar structure. Equatorial dusty disks must be common occurrences at different phases of stellar evolution.

Bipolar nebulae (BPNs) represent a class of objects defined purely morphologically. As such, they constitute a very heterogeneous group of nebulae in evolutionary terms (Calvet and Cohen 1978). Fig. 1 indicates the locations of 19 reasonably well-studied BPNs in the Hertzsprung-Russell diagram (HRD). Some nebulae are associated with pre-main-sequence stars; others seem to represent the transition from red giants to bona fide planetary nebulae; still others are well on their way to becoming white dwarfs. (It should be noted that the most problematic aspect of constructing such an HRD is still the determination of distances to individual nebulae; indeed, M2-9 is plotted twice - for 900 pc and again for 50 pc distance (the open circle in Fig. 1; see Kohoutek and Surdej 1980)).

A number of BPNs have been discovered in the past few years. It does seem that, to first order, pre-main-sequence objects can be recognised by their ragged and amorphous outer structures (e.g. see the Centaurus BPN, Wegner and Glass 1979), while highly evolved nebulae appear more highly symmetrical (e.g. GL 2688). Several techniques have been fruitfully applied to the study of BPNs and it is valuable to examine the results of this work. The principal techniques are polarimetric mapping, spectroscopic mapping, spectropolarimetry, infrared spectroscopy, far-infrared photometry, the study of radio molecular masers and continuum monitoring.

D. R. Flower (ed.), Planetary Nebulae, 45–55.

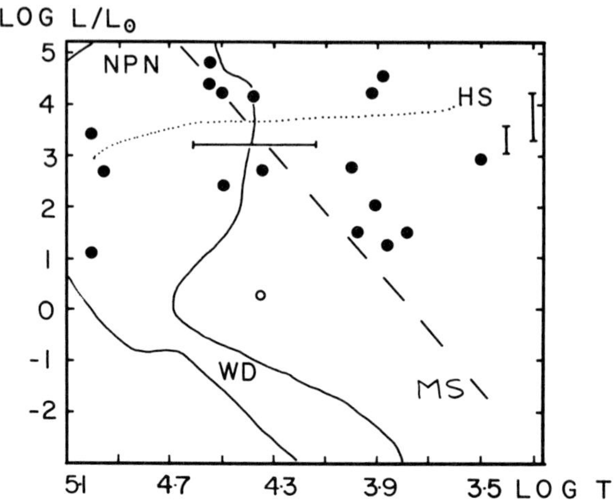

Fig. 1: HRD for 19 BPN. Indicated are the domains of the nuclei of planetary nebulae (NPN) and of the white dwarfs (WD), the main sequence (MS), and a typical evolutionary track for a red giant from Harm and Schwarzschild (1975). Where the temperature of the central star is uncertain, a horizontal bar is plotted; likewise for ranges in luminosity vertical bars are shown. M2-9 is shown at 50 pc (the open circle) and at 900 pc.

All investigators interpret their observations using the same model, namely a central star/infrared object, perhaps optically invisible, surrounded by an equatorial disk of gas and dust, embedded in a more extensive, bipolar nebula that contains small grains that scatter the light of the star in our direction. The photographic structure of some BPNs strongly suggests that that their disks are discernible; for example, M1-92, where the fainter nebular lobe appears truncated by an overlying circular mass. Nebular polarimetric mapping vindicates this view by revealing perturbations of the usual centrosymmetric pattern of polarization vectors in the vicinity of the central object. The centrosymmetric maps are explained as scattering of central starlight by small dust grains. The central disturbances of these patterns, resulting in vectors parallel to the minor axes, are thought to arise by viewing the stars straight through the disks in which are found aligned grains (e.g. Taylor and Scarrott 1980; Perkins, King and Scarrott 1981a; Perkins, King and Scarrott 1981b). Maps of near-infrared polarization reveal an identical situation (e.g. Allen _et al_. 1980a; Staude et al. 1982).

The BPN GL 2688 (Ney _et al_. 1975) is rich in molecules, both in the optical and microwave. High-spatial-resolution spectroscopy optically reveals the distribution of molecular emission of C_2, C_3 and SiC_2 across the nebular lobes. These features appear to arise through resonance fluorescence in the stellar radiation field, as for comets in sunlight

(Cohen and Kuhi 1980).

Some BPNs are characterised by very rich atomic emission-line spectra. For these, spectropolarimetry has proved extremely fruitful (Schmidt and Cohen 1981). For the brighter lobe of GL 618 (Westbrook et al. 1975) one sees greatly reduced degrees of polarization at the locations of the emission lines, with rotations in position angle for the very strong forbidden lines (Fig. 2). The permitted lines exhibit the same position angle as the highly polarized continuum so some fraction of this emission arises close to the star and is scattered to us by grains. The degree of polarization, however, is less than that of the continuum, implying that a portion of the permitted line flux is emitted in the lobes themselves. GL 618 has a spectrum that greatly favours the

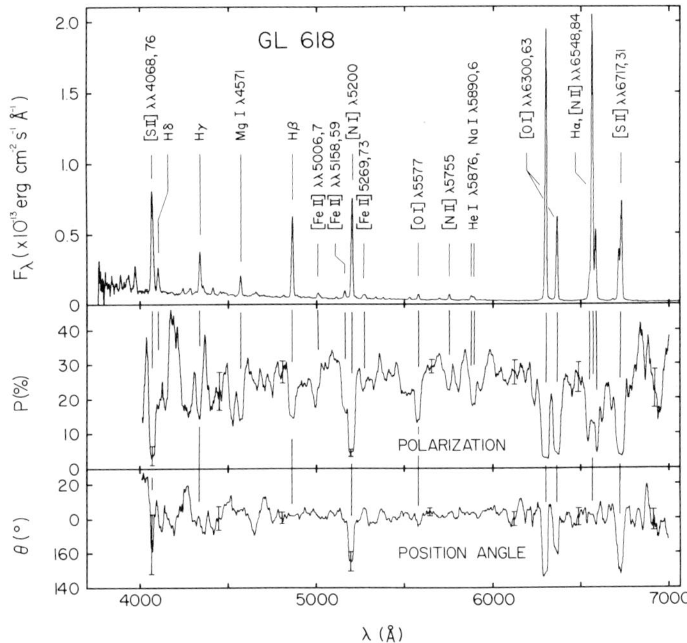

Fig. 2: Spectropolarimetry of the brighter lobes of the BPN GL 618 showing the different characters of polarization for continuum, permitted and strong forbidden lines.

cooling lines of heavy elements. An analysis of the line intensities intrinsic to the lobes enables a model to be constructed of the nebulae in which two very different regimes are found. Ionized, hot (18000K), high-density ($N_e > 20000\ cm^{-3}$) filaments occur but neutral (10000K), moderate-density ($N_e \sim 1000\ cm^{-3}$) gas is the primary constituent. These zones are identified with unshadowed and shadowed material, respectively (Fig. 3). From both observational and theoretical viewpoints, this BPN fits naturally into present conceptions of the initial development of planetary nebulae where condensations are important. A similar picture

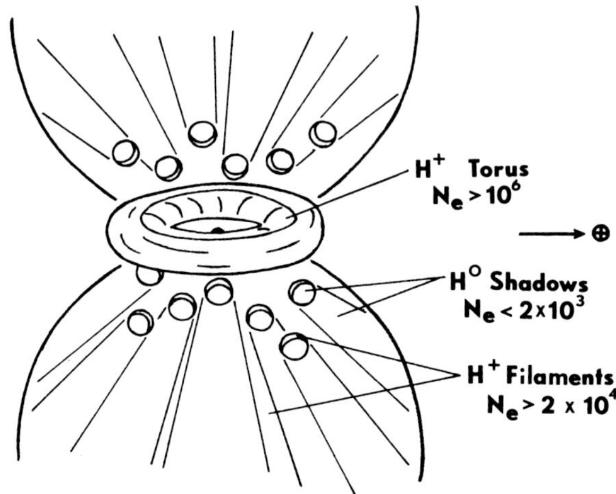

Fig. 3: Model for GL 618 showing the shadowed (neutral) and unshadowed (ionized) zones in the lobes of this BPN.

showing the presence of condensations emerges from spectropolarimetry of M2-9 (Schmidt and Cohen 1981) and nebular polarimetric mapping (King et al. 1981).

The near-infrared spectrum of GL 618 is also of great interest as it shows a number of vibration-rotation lines of molecular hydrogen (Thronson 1981). These seem to arise in a shocked zone of $T \sim 2000K$, and velocity ~ 10 km/s, consistent with little ionization of hydrogen. Radial velocity differences between atomic lines in the two lobes have enabled the identification of the blue component of Hα in the brighter lobe as locally ionized matter, whereas the red one is scattered from the central small HII region (Carsenty and Solf 1982). The kinematic analysis suggests an extremely young age for GL 618, perhaps only 600 yr, comparable with the time scale for expansion of the CO molecular shell (1500 yr). Strong support for the view that BPNs like GL 618 are now in rapid evolution comes from the increase in radio continuum flux in only 2 yr (Kwok and Feldman 1981), and the steady brightening of GL 2688 over some 20 yr (Gottlieb and Liller 1976), with some short-term fluctuations (Franz 1980). Short-term variability in the structure of M2-9 may also indicate rapid evolution although Kohoutek and Surdej (1980) favour rotation. Incidentally, these authors present a beautiful picture of M2-9 that reveals an outer very faint loop structure that may illuminate the processes of formation of BPNs.

Substantial infrared flux characterises all BPNs. Airborne far-infrared photometry indicates a grain emissivity like $\lambda^{-1.5}$ for the cool emitting grains, and appreciable long wavelength fluxes speak for circum-stellar disks that are greatly extended in temperature range (Kleinmann et al. 1978). Near-infrared polarimetry (Jones and Dyck 1978) of the BPNs shows a degree of polarization that diminishes with increasing wavelength beyond 1 μm, arguing that the scattering grains are less than about 0.3 μm in radius.

Radio masers are now known to be associated with several BPNs, principally lines of OH. OH0739-14 (Morris and Bowers 1980) is a remarkable OH object, possessing weak emission over a broad velocity range with a prominent narrow spike that moves within this plateau. Although it had not previously been bright enough for optical spectroscopy, one lobe of this BPN was visible beyond 6000Å at a time of strong OH emission. The reflected spectrum (Fig. 4) is clearly that of an extremely cool star, an M9III, the coolest star known within a BPN (Cohen 1981). In fact, OH maser activity extends across the HRD for BPNs from spectral type M9III (OH0739-14), through A2I (Roberts 22: Allen, Hyland and Caswell 1980), to M1-92 (B1V) where it is weakly observed.

GL 2789 has excited considerable controversy as to its distance (Cohen 1977; Humphreys, Merrill and Black 1980). It was first recognised as bipolar from nebular spectropolarimetry (Cohen 1977); it is associated with a CO cloud (Harvey and Lada 1980), and an H_2O maser coincides with its brightest portion (Lada et al. 1981). The exciting star is of type late O. Another late O-star that excites and illuminates a BPN is found in Sh2-106. This very complicated region, with several infrared sources, has been extensively mapped in the mid-infrared (Gehrz et al. 1982), studied through optical and near-infrared polarimetry (Lacasse et al. 1981; Tokunaga, Lebofsky and Rieke 1981), and by optical spectropolarimetry (Staude et al.1982). All results are consistent with a single exciting star that drives a bipolar molecular flow away from its equatorial dust disk.

The optical spectrum of "The Red Rectangle", illuminated by HD44179 (Cohen et al. 1975), has been found to contain a wealth of so far unidentified apparently molecular emission features (Schmidt, Cohen and Margon 1980). In the nebular lobes (Fig. 5) the spectrum is dominated by a broad diffuse peak with several groups of sharp emissions, some resolved. None of these structures is polarized, indicating that their source(s) are found within the lobes and that their emission merely dilutes the scattered starlight. It is tempting to identify the sharper features as due to gaseous species, and the diffuse peak as due to solid-state resonances when these gaseous molecules are bound into grains. As yet, however, no satisfactory matches have been found with known molecular spectra.

A feature common to several BPN is the presence of nebular spikes; for example, those of the Rectangle, of GL 2688, and of Parsamyan 22. Nebular polarization maps in the visible suggest that, for HD44179, these arise because of enhanced dust density on a biconical surface (Perkins et al. 1981c). On rather general theoretical grounds (Icke 1981) it has been argued that mass loss from an object embedded within an accretion disk would result in biconical outflow, independent of the driving mechanism. Stellar winds have been favored as producing some BPN by means of expanding circumstellar shells (M2-9: Walsh 1981), or intermittent outflows of

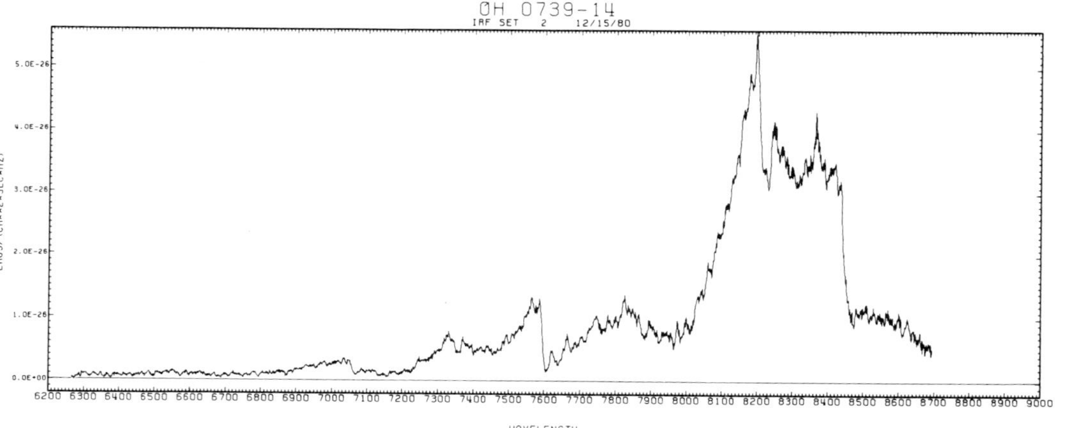

Fig. 4: Red spectrum of the embedded star of the BPN OH0739-14 as reflected from its southern lobe; the spectral type is M9III.

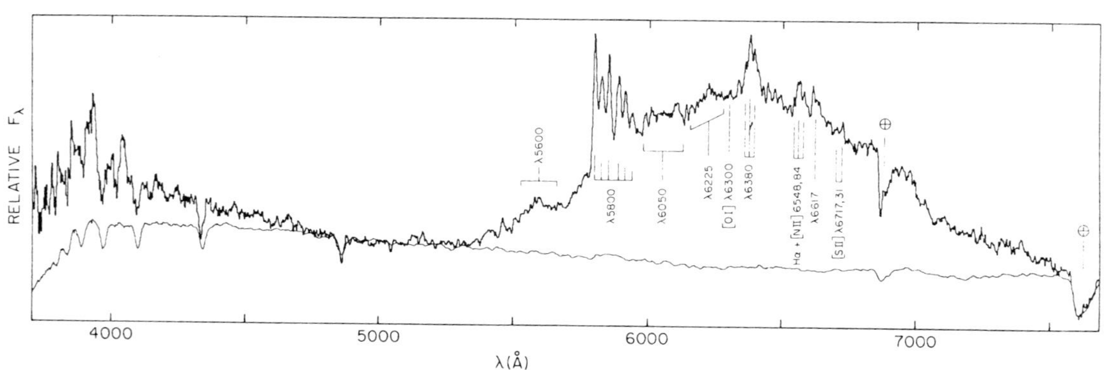

Fig. 5: Optical spectrum of the nebulosity associated with HD44179. Lower spectrum is that of the central star; heavy line refers to the nebular spectrum.

high mass loss rate from very hot central stars (Mz-3: López and Meaburn 1982). Even the complex "poly-polar" nebula NGC 6302 has been interpreted in this light (Barral *et al.* 1982). In order to produce adequate ionization in Roberts 22, with an A2I central star, appeal was made to sudden deceleration of a rapid outflow at the inner edge of the torus, producing a shock (Allen *et al.* 1980b). Several independent models for different BPNs require mass loss rates of order 10^{-4} to 10^{-5} $M_{\odot}$/yr, a magnitude suggestive of a short-lived phenomenon.

Another mechanism advanced in the context of stars that have driven out bipolar CO flows is that of supersonic expansion through a de Laval nozzle, in the direction of a decreasing density gradient (Königl 1982). This picture unifies conical/fan and biconical/bipolar cometary nebulae (Cohen 1982) since one nebular lobe is often deeply embedded in a dust cloud. Within this framework, the displacement between fan-tip and associated infrared object (e.g. for PV Cep, Cohen *et al.* 1981; for R Mon, Cohen and Schwartz 1982) is explained - the fan-tip apparently representing the nozzle above the star/infrared source. The association between ambient cloud and BPN has been probed by a 2.6-mm CO occultation of Lk Hα-208 (Good *et al.* 1981). No compact molecular source (e.g. a disk) was resolved.

Stimulated by OH0739-14, Morris (1981) proposed that in a binary with primary evolving up the red giant branch, and achieving its tidal radius before corotation with the secondary, the primary envelope could be ejected and would sink to the equatorial plane. Such a model would predict that BPNs would evolve into planetaries whose nuclei would be binaries that ultimately become cataclysmic variables. In this context, it may be of interest to note the 16-day period for the spectroscopic binary nucleus of NGC 2346 (e.g. Méndez and Niemela 1981).

It is remarkable that so many diverse objects are found to possess a bipolar structure. Let me conclude with two very recently discovered examples of this diversity. GL 1403 (CIT-6) is an extreme carbon star, embedded in a thick circumstellar dust shell, whose spectropolarimetry presents an intriguing phenomenon (Cohen and Schmidt 1982), namely abrupt rotation of position angle through 90 degrees, accompanied by almost a nulling of polarization at the same wavelength (Fig. 6). GL 1403 also exhibits a strange, smooth, blue continuum, not typical of extreme carbon stars, and due to thermal emission by hot (2000K) grains viewed over the poles of the star. The red continuum is starlight scattered through an equatorial dust disk (in whose plane we lie) and hence polarized orthogonally to the blue spectrum (Fig. 7).

MWC349 is a highly intriguing late O star with a curiously slow wind and associated with radio continuum emission. Very recently, deep VLA maps (Cohen *et al.* 1982) have been made at 6 cm that clearly demonstrate MWC349 to be a binary, not merely an optical double star. Complex gas streaming is seen (Fig. 8). However, the radio contours that surround the primary are conspicuously square, with corners "pinched out" along the diagonals. This morphology is strikingly reminiscent of the optical

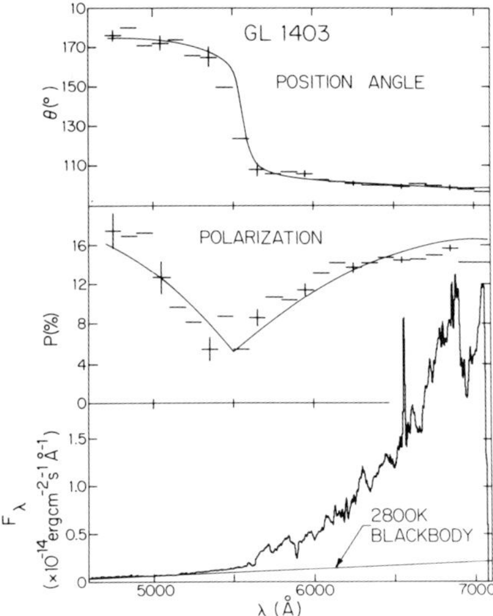

Fig. 6: Spectropolarimetry of the extreme carbon star GL 1403 (CIT-6).

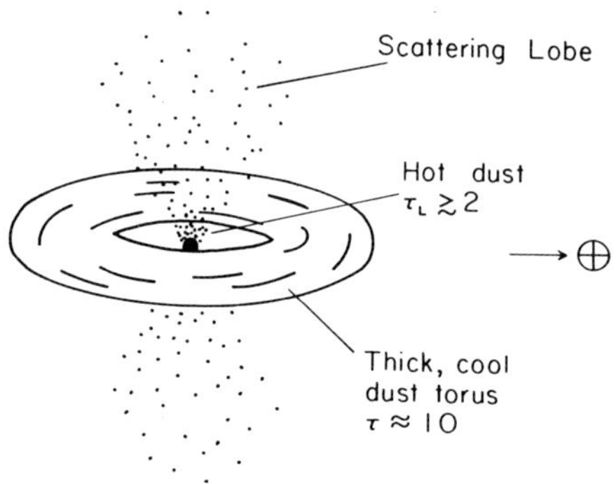

Fig. 7: The bipolar model proposed for GL 1403.

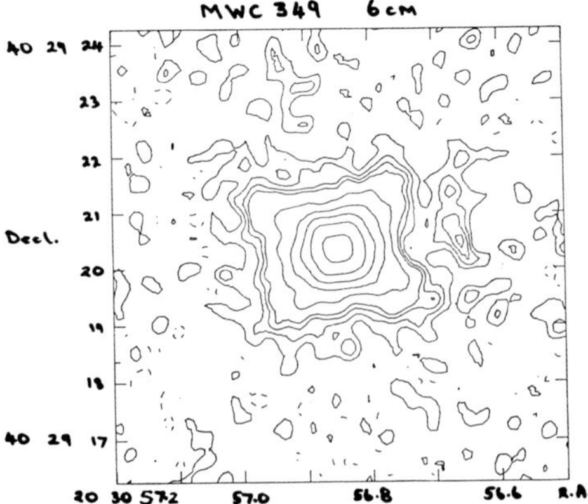

Fig. 8: VLA 6-cm map of the binary MWC349 showing the squaring of the contours around the primary which may represent a biconical surface flow akin to that in The Red Rectangle.

structure of The Red Rectangle. This serves to indicate that equatorial dusty tori are common occurrences at very different phases of stellar evolution. Consequently, bipolar nebulae are also abundant, although apparently representative of phenomena of very short duration.

REFERENCES

Allen, D.A., Barton, J.R., Gillingham, P.R., and Phillips, B.A.: 1980a, Mon. Not. R. Astr. Soc. 190, p. 531.

Allen, D.A., Hyland, A.R., and Caswell, J.L. : 1980b, Mon. Not. R. Astr. Soc. 192, p. 505.

Barral, J.F., Cantó, J., Meaburn, J., and Walsh, J.R. : 1982, Mon. Not. R. Astr. Soc. 199, p. 817.

Calvet, N. and Cohen, M. : 1978, Mon. Not. R. Astr. Soc. 182, p. 687.

Carsenty, U. and Solf, J.: 1982, Astron. Astrophys. 106, p. 307.

Cohen, M.: 1977, Astrophys. J. 215, p. 533.

Cohen, M.: 1981, Publ. Astr. Soc. Pacific 93, p. 288.

Cohen, M.: 1982, Publ. Astr. Soc. Pacific 94, p. 266.

Cohen, M. and Kuhi, L.V.: 1980, Publ. Astr. Soc. Pacific 92, p. 736.

Cohen, M. and Schmidt, G.D.: 1982, Astrophys. J., in press.

Cohen, M. and Schwartz, R.D.: 1982, Astrophys. J., in press.

Cohen, M., Kuhi, L.V., Harlan, E.A., and Spinrad, H.: 1981, Astrophys. J. 245, p. 920.

Cohen, M. et al.: 1975, Astrophys. J. 196, p. 179.

Cohen, M., Bieging, J., Dreher, J., and Welch, W.J.: 1982, in preparation

Franz, O.G.: 1980, Bull. Amer. Astr. Soc. 12, p. 445.

Gehrz, R.D., Grasdalen, G.L., Castelaz, M., Gullixson, C., Mozurkewich, D., and Hackwell, J.A.: 1982, Astrophys. J. 254, p. 550.

Good, J., Scoville, N., Schloerb, F.P., and Bally, J.: 1981, Astron. J. 86, p. 892.

Gottlieb, E.W. and Liller, W.: 1976, Astrophys. J. Letters 207, p. L135.

Harm, R. and Schwarzschild, M.: 1975, Astrophys. J. 200, p. 324.

Harvey, P.M. and Lada, C.J.: 1980, Astrophys. J. 237, p. 61.

Humphreys, R.M., Merrill, K.M., and Black, J.H.: 1980, Astrophys. J. Letters 237, p. L17.

Icke, V.: 1981, Astrophys. J. 247, p. 152.

Jones, T.J. and Dyck, H.M.: 1978, Astrophys. J. 220, p. 159.

King, D.J., Perkins, H.G., Scarrott, S.M., and Taylor, K.N.R.: 1981, Mon. Not. R. Astr. Soc. 196, p. 45.

Kleinmann, S.G., Sargent, D.G., Moseley, H., Harper, D.A., Loewenstein, R.F., Telesco, C.M., and Thronson, H.A.: 1978, Astron. Astrophys. 65, p. 139.

Kohoutek, L. and Surdej, J.: 1980, Astron. Astrophys. 85, p. 161.

Königl, A.: 1982, Astrophys. J., in press.

Kwok, S. and Feldman, P.A.: 1981, Astrophys. J. Letters 247, p. L67.

Lacasse, M.G., Boyle, D., Levreault, R., Pipher, J.L., and Sharpless, S.: 1981, Astron. Astrophys. 104, p. 57.

Lada, C.J., Blitz, L., Reid, M.J., and Moran, J.M.: 1981, Astrophys. J. 243, p. 769.

López, J.A. and Meaburn, J.: 1982, Mon. Not. R. Astr. Soc., in press.

Mendez, R.H. and Niemela, V.S.: 1981, Astrophys. J. 250, p. 240.
Morris, M.: 1981, Astrophys. J. 249, p. 572.
Morris, M. and Bowers, P.R.: 1980, Astron. J. 85, p. 724.
Ney, E.P., Merrill, K.M., Becklin, E.E., Neugebauer, G., and Wynn-Williams, C.G.: 1975, Astrophys. J. Letters 198, p. L129.
Perkins, H.G., King, D.J., and Scarrott, S.M.: 1981a, Mon. Not. R. Astr. Soc. 196, p. 7P.
Perkins, H.G., King, D.J., and Scarrott, S.M.: 1981b, Mon. Not. R. Astr. Soc. 196, p. 403.
Perkins, H.G., Scarrott, S.M., Murdin, P., and Bingham, R.G.: 1981c, Mon. Not. R. Astr. Soc. 196, p. 635.
Schmidt, G.D. and Cohen, M.: 1981, Astrophys. J. 246, p. 444.
Schmidt, G.D., Cohen, M., and Margon, B.: 1980, Astrophys. J. Letters 239, p. L133.
Staude, H.J., Lenzen, R., Dyck, H.M., and Schmidt, G.D.: 1982, Astrophys. J. 255, p. 95.
Taylor, K.N.R. and Scarrott, S.M.: 1980, Mon. Not. R. Astr. Soc. 193, p. 321.
Thronson, H. A.: 1981, Astrophys. J. 248, p. 984.
Tokunaga, A. T., Lebofsky, M. J., and Rieke, G.H.: 1981, Astron. Astrophys. 99, p. 108.
Walsh, J.R.: 1981, Mon. Not. R. Astr. Soc. 194, p. 903.
Wegner, G. and Glass, I.S.: 1979, Mon. Not. R. Astr. Soc. 188, p. 327.
Westbrook, W.E., Becklin, E.E., Merrill, K.M., Neugebauer, G., Schmidt, M., Willner, S.P., and Wynn-Williams, C.G.: 1975, Astrophys. J. 202, p. 407.

EDITOR'S NOTE: Figs. 2,3,5,6 and 7 are reproduced by courtesy of the Astrophys. J., Fig. 4 by courtesy of the Pub. Astron. Soc. Pacific.

DOPITA: Some of the bipolar nebulae show spectra like those of Herbig-Haro objects whose spectra can be well reproduced by low velocity shocks. The same idea might help in understanding bipolar nebulae spectra.

COHEN: For nebulae such as GL 618, this might be fruitful, but many bipolar nebulae show just reflection spectra.

KALER: In GL 618, are the emission lines unpolarized and what are the errors?

COHEN: The permitted lines are polarized at a level of about 10% ± 1%. The forbidden lines are about 3% polarized, but this is attributable to the interstellar medium.

SURDEJ: Do you think that the outer loops of bipolar nebulae are produced by magnetic fields?

COHEN: It is quite possible that this is the case.

CARSENTY: We have recently found that the motions in M2-9 cannot be attributable to rotation for the N and S lobes would be rotating in opposite senses.

COHEN: Good! Then I can remove the open circle from Fig. 1 representing M2-9 at a distance of 50 pc.

BLACK: Is it true that for the bipolar nebulae which you claim to be excited by O-stars, the spectral type rests solely on counting ultraviolet photons?

COHEN: It really depends on the distance: if you have a credible distance, then you have an estimate of the bolometric luminosity. For some nebulae, you have to rely on photon-counting. Of course, even if you do see absorption lines, as in GL 2789, you may be looking only at a shell spectrum.

NUSSBAUMER: Could the shapes of bipolar nebulae arise from magnetic confinement? Material may be carried away from the stars along open magnetic field lines as in the case of the Sun.

COHEN: Yes; in fact, it may be possible to interpret bipolar nebulae in terms of supersonic expansion through a de Laval nozzle under magnetic confinement.

LEAVER: Are the axes of bipolar nebulae aligned with the Galactic magnetic field direction?

COHEN: I believe that it has been claimed that the orientations of nebular disks or axes are correlated with the Galactic plane or the direction of the local Galactic magnetic field. However, I am not convinced that this is true of the entire sample of bipolar nebulae.

THE STRUCTURE OF THE BINARY STAR NEBULA NGC 2346

J.R. Walsh
Department of Astronomy, Manchester University, Manchester, England

NGC 2346 is a high excitation nebula which has a bipolar appearance and whose obvious central star is of too late a spectral type (A5) to account for the photo-ionization of the nebula. Spectral observations at high and low resolution in the visible, and low dispersion IUE observations in the UV are combined to explain the structure of the object. The radial velocity structure indicates a cylindrical geometry, or possibly two cavities, with the ionized material in the form of thin sheets. The variation of extinction has been investigated at many positions over the nebula; the obscuration is irregular although a ring of low extinction around the central star is apparent. The central star has previously been found to be a single-lined spectroscopic binary, and its zero velocity lies midway between the positive and negative radial velocity components of the ionized material. Visible spectrophotometry of the central star is combined with long and short wavelength IUE spectra in order to determine the parameters of the cool central star and the hot binary companion. The evolutionary status of this peculiar nebula is briefly discussed.

IUE OBSERVATIONS OF THE BIPOLAR PLANETARY NEBULA NGC 2346

W.A. Feibelman
Laboratory for Astronomy and Solar Physics, NASA-GSFC

L.H. Aller
Astronomy Department, University of California, Los Angeles, USA

The nucleus of NGC 2346 (=PK 215 + 3°1; $\alpha = 7^h06^m49.^s7$, $\delta = -0^o43'29''$, 1950) was recently discovered by Kohoutek (1982a,b) to be an eclipsing binary with a deep ($\sim 2.^m2$) minimum in the visible and an orbital period of 17.2 ± 0.4 days. We observed NGC 2346 with the IUE on 4 occasions between 1982, Feb. 24 and May 13 and obtained 6 SWP and 4 LWR low dispersion spectrograms. These were taken at different phases, using the large entrance aperture centered on the nucleus.

The radiation background was high on two occasions. A set of Vilspa observations for 1981, Feb. 6 was obtained from the National Space Science Data Center and was incorporated in our analysis.

For our 1982 observations, the IUE FES magnitude varied between m_v = 11.7 and 13.0. Emission line fluxes of NV, C IV, He II, O III], N III], Si III] and C III] were measured from the SWP spectra, and C II + O III] and [Ne IV] from the LWR range. At some phases, a relatively strong emission line near λ 1700 and another, weaker one at λ 1856 is seen. In addition, at certain phases the continuum level rises from about λ 1600 to λ 1950 and is believed to be the Rayleigh-Jeans tail of the A-star seen in the visible. As was already pointed out by Mendez (1978), this star may be merely a chance coincidence with the nebula and may not be physically associated with it. The eclipsing binary nucleus may contain gas streams (cf. Kohoutek, 1982b) or possibly an accretion disk, as suggested by the asymmetric UV line profiles.

Our magnitude estimates are only accurate to ± 0.1 m and probably contain some nebular contribution, although the IUE large entrance aperture of 10 x 23 arcsec is roughly equivalent to the circular apertures used by Kohoutek (15, 21, 30 and 30 arcsec). However, we have difficulty fitting the IUE magnitudes to the 17.2 days partial light curve by Kohoutek, based in primary minimum at JD 2445010.85. (Paper to be submitted to Ap. J.).

Kohoutek, L.: 1982a, IAU Circular No. 3667.
Kohoutek, L.: 1982b, Info. Bull. Var. Stars No. 2113, March 19.
Mendez, R.H.: 1978, Mon. Not. Roy. Astr. Soc., 185, 647.

THE UNPRECEDENTED LIGHT VARIATIONS OF NGC 2346*

R.H. Méndez
Instituto de Astronomia y Fisica del Espacio, Buenos Aires, Argentina; Visiting Astronomer, Cerro Tololo Inter-American Observatory, operated by the Association of Universities for Research in Astronomy, Inc., under contract with the National Science Foundation; Member of the Carrera del Investigador Cientifico, CONICET, Argentina

R. Gathier
Kapteyn Laboratorium, Groningen, The Netherlands

V. S. Niemela
Instituto de Astronomia y Fisica del Espacio, Buenos Aires, Argentina; also Instituto Argentino de Radioastronomia, Villa Elisa, Buenos Aires, Argentina; Visiting Astronomer, Cerro Tololo Inter-American Observatory, operated by the Association of Universities for Research in Astronomy, Inc., under contract with the National Science Foundation; Member of the Carrera del Investigador Cientifico, CIC, Provincia de Buenos Aires, Argentina

The central star of the planetary nebula NGC 2346 is now well confirmed as a single-lined spectroscopic binary, with P = 16d (Méndez and Niemela 1981, Ap. J., 250, 240). Unexpected photometric variations were recently reported by L. Kohoutek (1982, IAU Circular 3667). From additional photoelectric measurements and visual estimates we have found that these variations are periodic, with the same period as the orbital motion of the A-type primary component. From previous observations we can ascertain that such variations did not exist before, and must have started in 1981. The light minimum occurs at phase 0.75, that is to say when the A-type component is moving towards us. Radial velocities measured on spectrograms obtained during the light minimum are more positive, by about 40 km s^{-1}, than expected from the orbital motion; while the radial velocities corresponding to the light maximum agree with what is expected from the orbital motion. The spectral type of the A-type star does not change significantly as a function of brightness.

Although additional observations are clearly needed, we can advance the suggestion that these phenomena are due to the progressive occultation of the binary system by a dense dust cloud external to the system, which may or may not be associated with the planetary nebula. The proposed cloud probably has stellar dimensions and a very small mass. The full paper will be submitted to Astron. Astrophys.

* Based partly on observations made at the European Southern Observatory.

COHEN: With which star, the visible A star or the hot star seen in the ultraviolet, is this dust cloud associated?

MENDEZ: Probably with neither: if our simple assumption of a single, isolated, spherical cloud is correct - and it might well be just a small part of something more complex - then the high degree of dust concentration towards the centre implies a very low temperature. The cloud could not then be very near the binary system because the hot star would heat the dust.

POTTASCH: How well is the spectral type of the A star known? Somewhat different values are given in the literature.

MENDEZ: The spectral type, as determined from our spectrogrammes, is A5 V ± one spectral subtype. Julie Lutz has determined A2 V. All other spectral classifications in the literature were derived from photometry and cannot be given the same weight.

LUTZ: My spectral classification is less accurate than that of Mendez and collaborators because my classification is based on lower dispersion spectra.

KEYES: A technique used by Mick Plavec and his collaborators at UCLA on Algol eclipsing binaries and by myself on symbiotics may be of some use in more accurately determining the spectral type of the A star. Simultaneous ground-based optical spectrophotometric scans and IUE ultraviolet images are obtained in the eclipses and at quadrature. By subtraction of the appropriate observations, it is possible to obtain the spectrum of the cool (A) component uncontaminated by nebular continuum or hot component light.

KALER: There seems to be an interesting parallel between NGC 2346 and R Aqr, in which there may be an eclipse by a dust cloud. Furthermore, the R Aqr nebula has a vaguely similar bipolar structure.

KOHOUTEK: Our photoelectric observations of NGC 2346 in the period January - April 1982 show a light curve similar to that which has already been published (I.B.V.S. Budapest No. 2113) but with rather a high scatter. The scatter might be due to instrumental effects (subtraction of the contribution of the nebula), to the uncertain period (we used 16.2 d), or to real changes in the light curve.

BAADE: Dr. Walsh mentioned possible similarities between NGC 2346 and NGC 6302. We have used TAURUS data (imaging Fabry-Perot observations, cf. our poster, this volume) for NGC 6302 and constructed a velocity map, i.e. a two-dimensional image with the pixel values presenting radial velocities instead of intensities. The velocity map reproduces the biconical structure of direct images quite well. Any contribution of rotation to the observed radial velocities seems, therefore, to be small in NGC 6302.

RADIO OBSERVATIONS OF PLANETARY NEBULAE

P.F. Scott
Mullard Radio Astronomy Observatory,
Cambridge, England.

1. INTRODUCTION

A discussion of radio observations for the period up to 1979 forms part of the general review given at the 1979 IAU General Assembly by Terzian (1980) and I shall be concerned here mainly with later results. With the advent of improved radio telescopes and techniques, in particular the coming into full operation of the Very Large Array (VLA), there has been an increasing interest in compact and proto-planetary nebulae and I shall discuss these observations in a separate section of this review.

2. RADIO SURVEYS

The Parkes radiotelescope has been used to observe a further 167 nebulae at 5 GHz (Milne 1979) and 98 nebulae at 2.7 GHz (Milne & Webster 1979). Combined with earlier observations, reliable data on a total of 332 sources at 5 GHz and 144 sources at 2.7 GHz are available and have been used to derive extinction coefficients and distances for the majority of these objects. Calabratta (1982) has used the Molonglo telescope to observe 43 planetary nebulae at a frequency of 408 MHz, achieving a detection rate of 22%. He found no evidence for non-thermal emission. Mross, Weinberger and Hartl (1981), using the 100m Effelsberg telescope at 5 GHz, detected 19 out of 39 objects selected from the list of new planetary nebulae compiles by Weinberger (1977). Much of the motivation for radio surveys relates to their use in deriving radio distances. Maciel and Pottasch (1980) have discussed the errors in distance determinations resulting from an assumed constant ionised mass for nebulae which may in fact be ionisation bounded; they found that the use of an empirical mass-radius relation leads to more consistent distances and distributions of intrinsic size. Subsequently, Milne (1982) has shown that ionisation-bound nebulae might be expected to follow an $M \propto R^{3/2}$ mass-radius relation. As applied to the radio data, nebulae can be expected to fall into the optically-thick or optically-thin (to L_α) regime according as their radio surface brightness is greater or less than a certain critical value which, at a frequency

D. R. Flower (ed.), Planetary Nebulae, 61–68.

of 5 GHz, corresponds to a brightness temperature of 22K. Disagreement concerning distance scales remains, however, Maciel & Pottasch deriving distances some 50% smaller than those quoted by Milne. The Westerbork telescope has been used in an extensive programme of observations of planetary nebulae near the Galactic centre (Wenterloop & Dekker 1979; Isaacman 1980a/b, 1981; Isaacman, Wenterloop & Habing 1980) as part of a study of the mass distribution in the central parts of the Galaxy. These data are presented elsewhere in this Symposium and will not be discussed further here.

3. RADIO RECOMBINATION LINES

Walmsley, Churchwell & Terzian (1981) have reported measurements of radio recombination lines for six planetary nebulae and upper limits for a further three, extending earlier data to include low-frequency (2.37 GHz) and high-frequency (14.7 GHz) transitions. A non-LTE analysis produces electron temperatures substantially in agreement with those derived optically. The frequency dependence of the observed line/continuum ratio for NGC 7027 is best fitted by a model having an electron density of 5×10^4 cm^{-3}, although some discrepancies with the observations remain. Viner, Vallée & Hughes (1979) have fitted the radio continuum and recombination line emission from NGC 7027 with a series of models which incorporate power-law radial density variations and allow for nebular expansion. Although the continuum data can be fitted by a range of models, including one having uniform electron density, the recombination line data appear to require an $n_e \propto R^{-2}$ density gradient. It should be noted however that such a variation is unlikely to be compatible with the sharp outer boundary seen on radio continuum maps; it is also inconsistent with the theoretical model considered by Guiliani (1981) which predicts an <u>increasing</u> radial density gradient. Walmsley et al. (1980) have noted that the line intensities used by Viner et al. may be in error by up to 50% and further modelling should probably incorporate the results of recent radio continuum mapping.

4. RADIO MAPPING

Felli & Perinotto (1979) have used 5 GHz observations made with the Westerbork telescope (WSRT) to compare the radio and optical brightness distributions of 8 planetary nebulae. None showed any sign of the patchy extinction apparent in NGC 7027. The VLA has been used to map a number of 'classical' planetary nebulae and some of these results will be presented in the next contribution. Radio mapping at longer wavelengths is beyond the capability of current instruments due to the low surface brightness of the sources, although some lunar occultation observations have been made at 327 MHz (Gopal Krishna 1978). With the planned improvements in the sensitivity of existing telescopes such as the WSRT and the prospect of further measurements from the VLA it can be expected that more high quality radio maps will be appearing.

5. COMPACT AND PROTO-PLANETARY NEBULAE

Johnson, Balick & Thompson (1979), using part of the VLA at 4.9 GHz, successfully detected 8 out of 13 stellar planetary nebulae which were observed; these sources were resolved by the VLA and appeared to be moderately compact but comparatively distant nebulae. Kwok, Purton & Keenan (1981) list 40 planetary nebulae observed at the Algonquin Radio

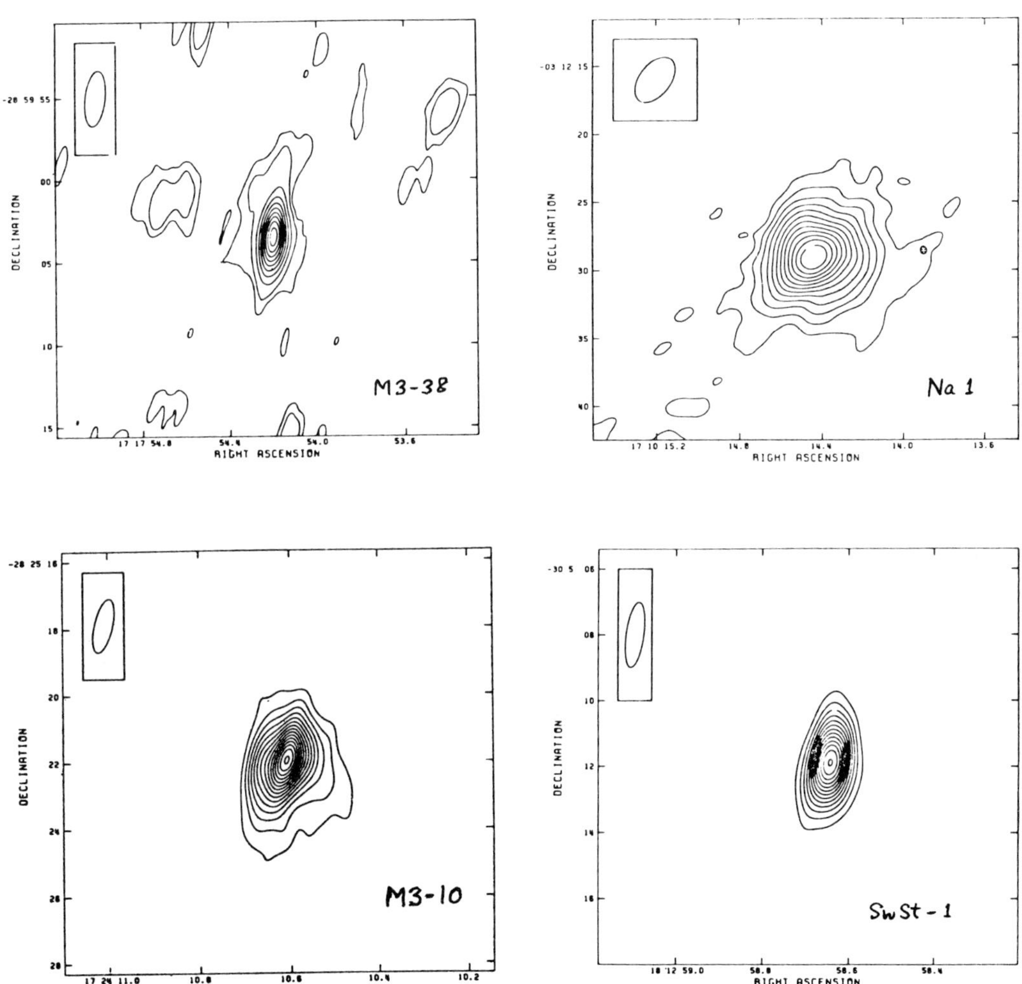

Figure 1. Maps of four compact planetary nebulae obtained with the VLA at 5 GHz. [From Kwok, Purton & Keenan (1981)]

Observatory (ARO) at a frequency of 10.6 GHz. The majority were found to be optically thin above 5 GHz but six of the objects appeared to have a higher turn-over frequency and were subsequently observed with the VLA.

Maps of the four of these sources which were detected by the VLA are shown in Figure 1. Two of the objects, M3-38 and SwSt-1, were found to have the high emission measures ($\gtrsim 10^8$ cm^{-6} pc) appropriate to young planetary nebulae. Purton et al. (1982) have recently reported a comprehensive series of observations of mainly emission-line objects but including a number of planetary and proto-planetary nebulae. 44 of the 325 objects studied were detected as radio sources, the probability of detection correlating strongly with the presence (optical) forbidden lines and, less strongly, with the existence of dust emission. Thirteen of the objects exhibited low frequency spectra approximating an $S \propto \nu$ law, consistent with mass outflow: this category included the suggested proto-planetaries HM Sge, V1016 and Hb 12. On the colliding stellar wind model proposed by Kwok, Purton & Fitzgerald (1978) this emission represents the ionisation of the original red giant stellar wind; the application of this model to the case of HM Sge has been discussed in detail by Kwok & Purton (1979) who find the model can explain satisfactorily the existing optical, infrared and radio data.

A particular property of some proto-planetary nebulae is the existence of radio variability on time scales of years or less. Much of this information derives from measurements carried out over a period of years at the Algonquin Observatory. The results for four sources are shown in Figure 2. In the case of GL 618, for which spectral information is also available, the changes are consistent with the progressive ionisation of an optically thick HII region (Kwok & Feldman 1981). It may be remarked in passing that the accurate flux measurements of weak sources can present observational uncertainties as is illustrated by the example of K648, below.

Frequency GHz	Flux density mJy	Reference
2.7	4.4)	Johnson 1976
8.1	3.3)	
4.9	16 ± 4	Johnson, Balick & Thompson 1979
10.6	16 ± 5	Kwok, Purton & Keenan 1981
5.0	9 ± 4	Mross, Weinberger & Hartl 1981
5.0	4.0 ± 1.7	Birkinshaw, Downes & Pooley 1981

Flux density measurements of the planetary nebula K648

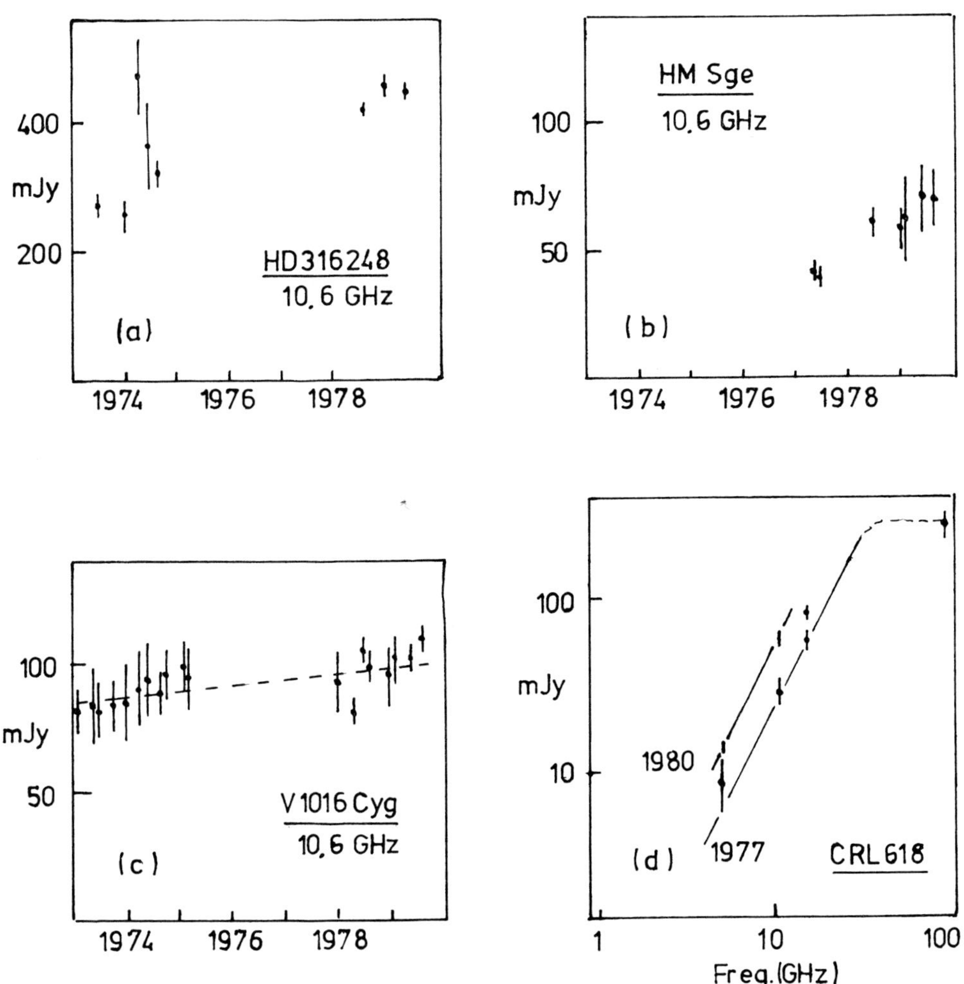

Figure 2. Flux density variations. [(a) (b) and (c) from Purton et al. 1982; (d) from Kwok & Feldman 1981]

6. CONCLUDING REMARKS

As mentioned earlier, much of the recent progress has been concerned with proto-planetary objects and it is clear that a distinct class of object has been identified. The link between these objects and the 'classical' planetary nebulae is perhaps less well proven. Infrared studies (Aitken, Roche, Spenser & Jones 1979) have found silicate emission features in proto-planetary objects but not in other planetaries, such as the compact nebulae NGC 7027 and BD $30^{o}3639$. There is considerable evidence supporting a binary interpretation for

V1016 and HM Sge (Pazcynski & Rudak 1980; Thronson & Harvey 1981) whereas a relatively small proportion of planetary nebulae are known to comprise binary systems. An interesting test of the colliding stellar wind model (Kwok et al. 1978) will be provided by the evolution of the radio spectrum of HM Sge, where, as noted by Kwok & Purton (1979), one would expect to see the emergence of an $S \propto \nu^2$ spectrum. The spectral dependence of flux variability in this and other sources will clearly provide useful data for models of nebular evolution.

Although not mentioned previously in this review, current advances in millimetre and sub-millimetre telescopes and receivers should allow the detection of neutral material associated with planetary nebulae to be extended to the more compact nebulae and help to elucidate their dynamics and origin. The first detection of HI associated with a planetary nebula, reported in a poster contribution to this Symposium (Rodriguez & Moran) illustrates one of the possibilities opened up by the availability of improved angular resolution. As in other areas of astrophysics, future advances are likely to result from a synthesis of observational data from many different parts of the spectrum and the improvement in radio techniques should ensure that radio observations continue to play an important part.

REFERENCES

Aitken, D.K., Roche, P.F., Spenser, P.M. & Jones, B., 1979. Astrophys. J., 233, 925.

Birkinshaw, M., Downes, A.J.B. & Pooley, G.G., 1981. The Observatory, 101, 120.

Calabretta, M.R., 1982. Mon. Not. R. astr. Soc., 199, 141.

Felli, M. & Perinotto, M., 1979. Astr. Astrophys., 76, 69.

Gopal Krishna, 1978. Mon. Not. R. astr. Soc., 182, 723.

Guiliani, J.L., 1981. Astrophys. J., 245, 903.

Isaacman, R., 1980a. Astron. Astrophys., 81, 359.

Isaacman, R., 1980b. Astron. Astrophys. Suppl., 43, 405.

Isaacman, R., 1981. Astron. Astrophys., 95, 46.

Isaacman, R., Wouterloot, J. & Habing, H.J., 1980. Astron. Astrophys., 86, 254.

Johnson, H.M., 1976. Astrophys. J., 208, 706.

Johnson, H.M., Balick, B. & Thompson, A.R., 1979. Astrophys. J., 233, 919.

Kwok, S. & Feldman, P.A., 1981. Astrophys. J., 247, L67.

Kwok, S. & Purton, C.R., 1979. Astrophys. J., 229, 187.

Kwok, S., Purton, C.R. & Fitzgerald, P.M., 1978. Astrophys. J., 219, L125.

Kwok, S., Purton, C.R. & Keenan, D.W., 1981. Astrophys. J., 250, 230.

Maciel, W.J. & Pottasch, S.R., 1980. Astron. Astrophys., 88, 1.

Milne, D.K., 1979. Astron. Astrophys. Suppl., 36, 227.

Milne, D.K., 1982. Mon. Not. R. astr. Soc., 200, 51P.

Milne, D.K. & Webster, B.L., 1979. Astron. Astrophys. Suppl., 36, 169.

Mross, R., Weinberger, R. & Hartl, H., 1981. Astron. Astrophys. Suppl., 43, 75.

Paczynski, B. & Rudak, B., 1980. Astron. Astrophys., 82, 349.
Purton, C.R., Feldman, P.A., Marsh, K.A., Allen, D.A. & Wright, A.E., 1982. Mon. Not. R. astr. Soc., 198, 321.
Terzian, Y., 1980. Q.J. R. astr. Soc., 21, 82.
Thronson, H.A. & Harvey, P.M., 1981. Astrophys. J., 248, 584.
Viner, M.R., Vallée, J.P. & Hughes, V.A., 1979. Astrophys. J. Suppl. Ser. 39, 405.
Walmsley, C.M., Churchwell, E. & Terzian, Y., 1981. Astron. Astrophys., 96, 278.
Weinberger, R., 1977. Astron. Astrophys. Suppl., 30, 335.
Wouterloot, J. & Dekker, E., 1979. Astron. Astrophys. Suppl., 36, 323.

OSTERBROCK: Is it possible, from radio measurements, to fairly unambiguously identify "new" (previously uncatalogued) PN? Would it be possible, by a radio survey, to find all planetary nebulae within, say, 2 kpc of the Sun, eliminating extinction effects of interstellar dust?

SCOTT: This would require high resolution surveys at two or three frequencies to pick out the thermal sources. The problem would be to distinguish PN from compact H II regions (W3 (OH) and NGC 7027 are very similar in their radio properties).

ISAACMAN: The Westerbork search for Galactic centre PN was at λ = 21 cm and 6 cm and yielded only one unambiguous and one possible PN. Unless one has sufficient spatial resolution to resolve the shell, some kind of optical or infrared spectral information is required to distinguish PN from compact H II regions.

TERZIAN: Together with K. Turner at the Arecibo Observatory, I have just completed a radio interferometric survey at λ = 12 cm of compact PN. Analysis is in progress.

WADE: Which (the radio or the optical) values of electron temperature and density have changed in order to bring about the present "agreement" between radio and optical studies of PN?

TERZIAN: The radio values have changed through non-LTE analysis of the data. However, the optical results also change, depending on the observer! It is too early to say that there is reasonable agreement between radio and optical determinations of electron temperature and density.

SEATON: With regard to optical and radio observations, the radio recombination lines involve a transfer problem and hence probe in depth in a way which cannot be achieved by optical observations.

MALLIK: Could radio recombination line observations tell us something about electron temperature fluctuations in PN?

TERZIAN: Not as yet: we need radio recombination line measurements at high angular resolution. Perhaps the VLA will be used for such work.

FLOWER: IUE and optical observations of Sw St 1 and Hb 12, mentioned in Scott's talk, indicate electron densities of 10^5 cm^{-3} and 10^6 cm^{-3}, respectively. In the case of Sw St 1, the density is determined directly from high dispersion observations which resolve the C III $\lambda\lambda$ 1907, 1909 doublet.

HIGH RESOLUTION MAPS WITH THE VLA

R.C. Bignell
National Radio Astronomy Observatory

The Very Large Array is a radio picture making instrument operated by the National Radio Astronomy Observatory in Socorro, New Mexico. The array, which has been in full time operation for more than 2½ years, operates at four main wavelengths, 1.3, 2.0, 6.0 and 20 cm with achievable resolutions of .05, .08, .25 and 0.8 arc seconds respectively.

One of the first objects to be mapped (May 1977) with the VLA was the planetary nebula, NGC 40 (Hjellming, Bignell, Balick 1978). Since then there have been numerous VLA observations of different planetary nebulae (PN).

The classical planetaries are being mapped with the intention of learning more about the physical processes of the later stage evolution whereas the early evolution is being studied from high resolution and high sensitivity VLA maps of young planetaries. The latter class of objects is a sub-class of active Galactic objects consisting of symbiotic sources, cataclysmic binaries, novae and young planetary nebulae which are being studied extensively with the VLA. More fundamental problems such as distances to planetary nebulae are also being tackled. The current review is not aimed at presenting a complete list of all the available VLA observations of PN, but will be slanted towards summarizing some of the more interesting sources and recent results.

"CLASSICAL" PLANETARY NEBULAE

These PN are typically ≥0.5 arc minute (large) in diameter and have the optical appearance of a (modified) doughnut, shell or other peculiar structure (such as loops, etc). The radio spectrum of these objects is generally optically thin at wavelengths shorter than 20 cm.

D. R. Flower (ed.), Planetary Nebulae, 69–78.

NGC 6853

The Dumbbell nebula is a well known relatively close PN approximately 8 arc minutes in size. The nebular emission exhibits a moderate range of excitation with some stratification (Goudis et al 1978). One interesting feature is the enhancemnt of [NII] emission relative to Hα in the outer parts of the nebula (Hua and Louise 1981). There is also a secondary shell of optical emission from a region of about 15 arc minutes centered about the main nebula (Perek and Kohoutek 1967, Millikan 1974). The VLA radio maps (Bignell 1982) shown in Figure 1, indicate two characteristics. Firstly, the general radio features most closely resemble the Hα optical emission distribution with significant differences from the low and high excitation emissions, both of which differ significantly themselves. Secondly, there is no indication of radio emission from the secondary shell down to a low emission level.

NGC 6543

The PN, NGC 6543, has a striking optical appearance in that most photographs show a structure of two overlapping elliptical rings. This source has been mapped both in the radio from VLA observations and in selected optical lines using the Kitt Peak 4 and 2 meter telescopes by Balick et al 1982. The 20 and 6 cm data are presented in Figure 1. Although the gross features of the radio appear similar to those of most optical pictures, a comparison of the radio maps with optical line maps of different excitations reveals that the distribution in the lines of Hα, Hβ and the high excitation lines of [OIII] appear to resemble the radio emission approximately, wheras the distribution in low excitation lines of [NII] and [SII] differs substantially from the radio and shows very unusual structures some of which resemble a "string of pearls".

NGC 40

This is a moderately close PN approximately 40 arc seconds in size. Optical photographs show a bubble type morphology. The VLA maps (Balick et al 1982) at 6 and 20 cm show similar gross features to optical pictures. It is worth noting that the [OIII] line differs significantly from the radio whereas the low excitation line maps more closely resemble the radio emission. This trend is opposite to that found in NGC 6543. Both NGC 40 and NGC 6543 show different radio and optical correlations than NGC 6853! Since these nebulae are probably about the same age it would be interesting to know why the excitation structures relative to the radio distributions differ so much. Do these differences result from different evolving processes, environments, parent objects or some combination of these factors?

Further VLA observations of both NGC 6543 and NGC 40 will allow one to look for evidence of a circumstellar wind and any narrow region

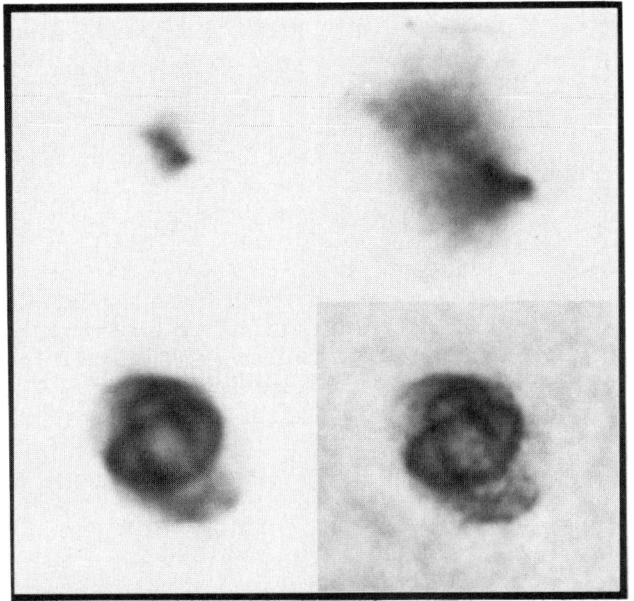

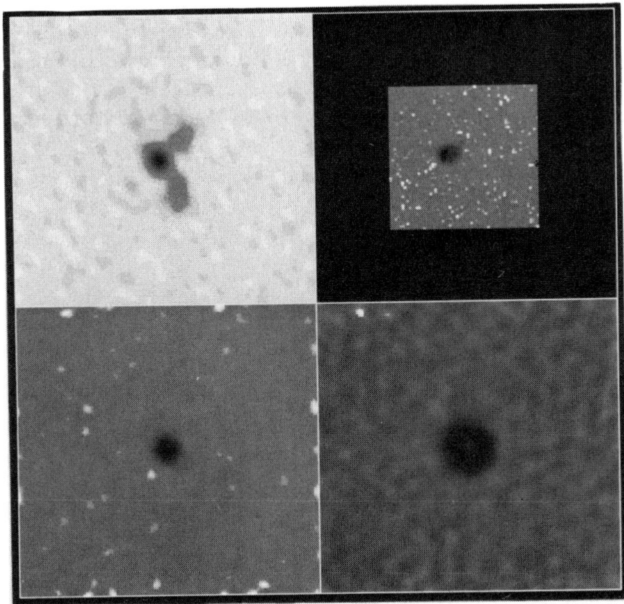

Figure 1. Top row: VLA images of NGC6853 at 20 cm (left) and 6 cm (right) from observations in D array. Second row: VLA images of NGC 6543 at 20 cm (left) and 6 cm (right) from observations in A array. Third row: VLA images of Hubble 12 at 6 cm (left) and 1.3 cm (right) from observations in the B and A arrays respectively. Bottom row: VLA images of VY2-2 at 6 cm (left) and 2 cm (right).

at or near the nebular boundary which may show signs of anomolous spectral lines.

IC 3568

The "thereotician's" nebula is a very symmetric object about 8 arc seconds in diameter and of medium excitation. The VLA 6 cm map (Balick et al 1982) shows a very symmetric doughnut shape. Harrington and Feibelman (1982) have combined recent IUE and VLA observations with optical data to propose a detailed comprehensive model for this source. The low resolution VLA data was used to obtain the gas density distribution throughout the nebula and they find that the total mass of the nebula ~.005 Mo. Their UV data show strong P Cygni line profiles indicating a stellar wind with a significant mass loss. If this is the case there probably should be radio emission from the central star. The current VLA observations (A configuration) shows no strong evidence of radio emission from the central star, however, the observations are noise limited and the additions of the most recent B configuration should prove more informative.

NGC 7027

This is one of the strongest radio and IR emitting planetary nebulae. The VLA 6 cm map (Balick et al) shows that the nebula is about 12 arc seconds in size and exhibits a doughnut structure with two bright peaks in the ring which are consistent with a tilted cylinder geometry. Although there is strong differential extinction in the direction of NGC7027, deep optical pictures are consistent with the radio maps (Atherton et al 1979).

YOUNG PLANETARIES

The set of objects considered as young PN have observational properties strongly indicating significant mass loss. The sources tend to have strong IR emission and their radio spectra tend to be optically thick at wavelengths longer than 1 cm with spectral indices consistent with mass loss. Many of these sources are small in angular extent.

VY2-2

This is classified as a PN based on its optical spectrum. It is a strong IR source undergoing mass loss as indicated by both the IR and optically thick radio emissions. It is one of the few PN to exhibit molecular radio emission (Davis et al 1979). The OH emission properties of VY2-2 resemble some type II OH/IR stars and show a single asymmetric shape suggestive of an expanding shell in which emission from the shell on the reverse side of the nebula is absorbed by the PN. The 2 cm VLA map of Seaquist and Davis (1982) shows that the source is approximately 0.4 arc seconds in size and has the doughnut shape characteristic of many PN. It is interesting to note that the peak of

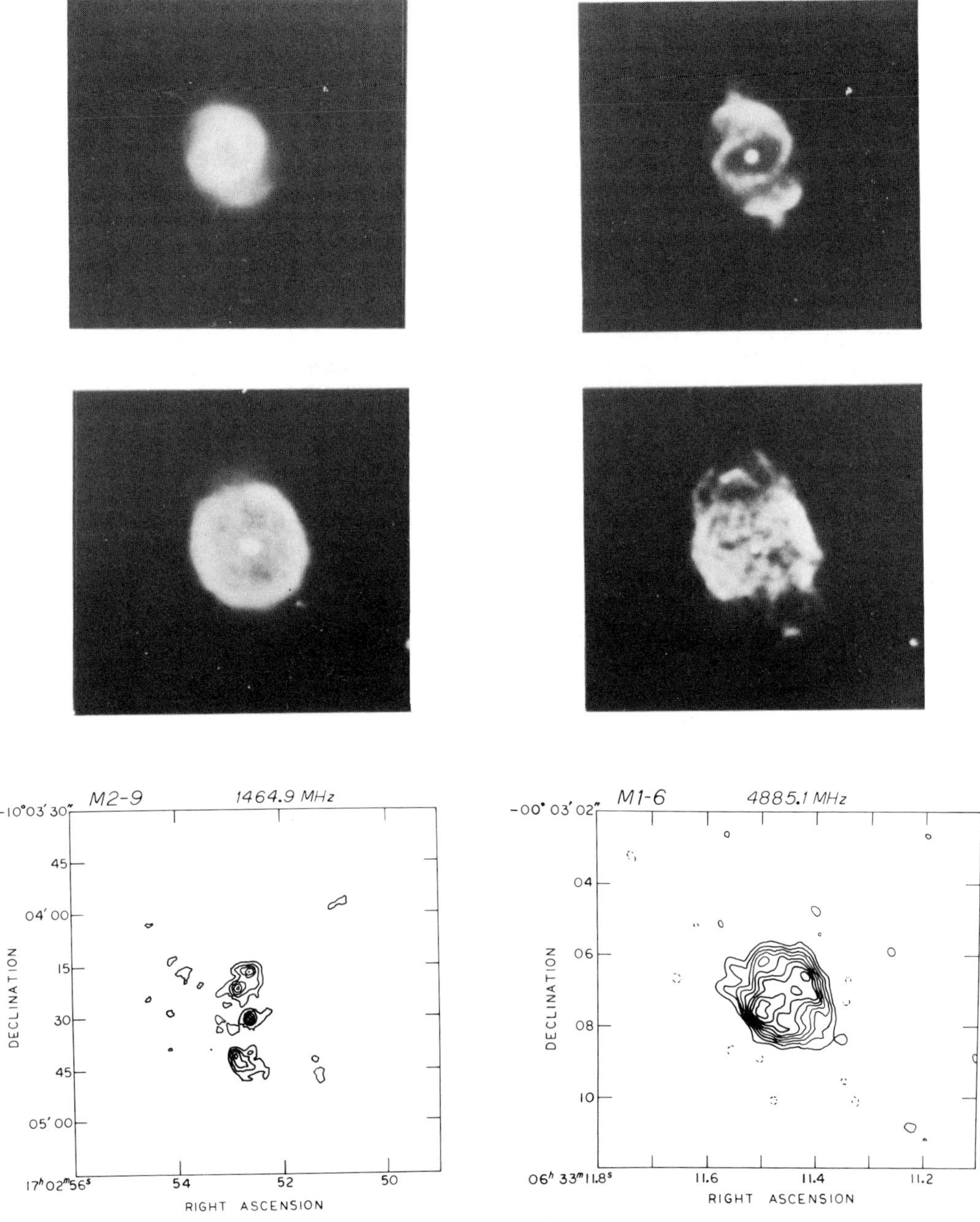

Figure 2. Top row: KPNO pictures of NGC 6543 in the λ5007Å [OIII] line (left) and λ6731Å [SII] line (right). Second row: KPNO pictures of NGC 40 in the λ5007Å [OIII] line (left) and λ6584Å [NII] line (right). Bottom row: VLA contour maps of M2-9 at 20 cm and M1-6 at 6 cm.

the OH emission lies in the shell (Seaquist and Davis 1982) and not at the center of nebula! In addition to the ring there appears to be faint emission extending to the north which is most evident at 6 cm. What are the relationships of these observational features?

M2-9

This pretty bipolar nebular is about 40 arc seconds in size with a 13 magnitude central condensation (Calvet and Cohen 1982). The optical spectra suggest ionization stratification in the core with high electron densities and a different environment in the wings where [OI] densities greatly exceed the [NII] densities. The 20 cm and 6 cm VLA maps (Purton and Kwok 1982) show extended structure both north and south overlapping the bipolar structure seen optically. The radio core exhibits an optically thick spectrum at wavelengths longer than 1 cm whereas the wings are optically thin over most of the radio spectrum. The core is unresolved (<.1 arc second) at 1.3 cm. The radio properties of this nebula are consistent with the contact binary model proposed by Morris (1981) for the origin of bipolar nebulae (Purton and Kwok 1982).

Hubble 12

This nebula was classified by Perek and Kohoutek (1967) as a PN smaller than 10 arc seconds. The finding chart shows a stellar image. Its spectrum shows both forbidden and permitted lines. It is also a strong IR source. The VLA maps at 20 and 6 cm show a central unresolved object with wings extending outwards in the shape of a "V" with an overall extent of about 10 arc seconds (Newell 1981). The shape of these lobes is difficult to understand but may be related to the biconical flow model for bipolar nebula and motion through the interstellar medium. The high resolution VLA map at 1.3 cm (Newell and Hjellming 1982) indicate very interesting structure in the core. The central source has a scale size of several tenths of an arc second in the shape of a disc with very bright emission to one side. It should be pointed out that the radio wings are similar to those of the symbiotic star V1016 Cygni and the small scale structure of the core is similar to the symbiotic star RY Scuti. What gives rise to this core structure and what is its relation to the larger extended wings? The radio spectrum of the core is optically thick below 1 cm and the wings are optically thin over most of the spectrum.

M1-6

It has been suggested that this low excitation PN (Kondrat'eva 1979) is a young PN. The optical image in the PK catalogue (Perek and Kohoutek 1967) is stellar. The VLA map at 6 cm shows a "typical" looking planetary nebula in the shape of a 2 arc second ring with a bright spot on one side of the ring. Unlike VY2-2, M2-9 and Hubble 12 the decimeter radio spectrum of this object is optically thin. Although

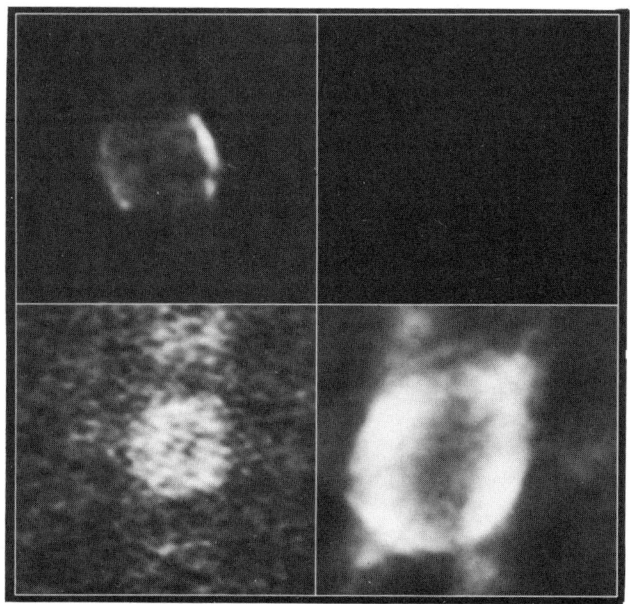

Figure 3. VLA images of NGC 40 at 20 cm (top Left), IC3568 at 6 cm (bottom left) and NGC7027 at 6 cm(bottom right). All observations we taken in A configuration.

this PN maybe young there is no evidence in the radio spectrum of mass loss.

AFGL 618

This is a bipolar object with interesting IR, optical and radio properties. The radio source lies between the optical nebulosities and has CO emission lines resembling those found in circumstellar envelopes of early type stars. The radio spectrum, like VY2-2, M2-9 and Hubble 12, is optically thick below 1 cm but unlike these sources the spectral index is about 2 (cf .6 for the others) and is consistent with a black body spectrum (Kwok and Feldman 1981). The radio spectrum has shown variations over a three year period consistent with an expanding mass of ionized gas. The VLA 2 cm map shows a small source ≤ .3" in size coincident with the IR source and a 3" extension to the south-east. VLA observations at 1.3 cm indicate that the core consists of three components aligned approximately east-west and roughly coincident with the bipolar axis (Newell and Hjellming 1982). It has been suggested that this object may be undergoing the brief transition from a red giant to the young PN stage typified perhaps by M2-9.

RADIO EMISSION FROM CENTRAL STARS OF PLANETARY NEBULAE

The use of the VLA to map the classical planetary nebulae to very low brightness levels will not only help understand the evolution of the older PN but may also lead to the study of the link or connection between these and the young PN. Although early VLA observations with a small number of antennas (Thompson and Sinha 1980) of three classical PN showed no evidence of radio emission from the central stars, the more recent observations will extend the sensitivity of this study substantially and will include a larger number of more likely candidates.

SUMMARY

The above sample of sources, although not complete, does typify the recent research work carried with the VLA. The high resolution and high sensitivity of the VLA is currently being utilized to probe the cores of active objects such as symbiotic stars, novae and young planetary nebulae and through these studies more will be learned about the mechanisms responsible for the mass loss and early formation of the nebulae. Perhaps radio observations of this kind will also help lead to an understanding of the relationship or commonality, if any, between the different mass loss objects such as novae, red giant stars, young planetary nebulae, etc.

Finally, the VLA has and will continue to be used to study the radio spectrum and structure of small PN (Kwok et al. 1981, Issacman et al 1980, Johnson et al 1979) in an attempt to give more information on the physics of these objects, to identify PN and tackle more fundamental questions, such as envelope masses and distances to PN.

REFERENCES

Atherton, P.D., Hicks, T.R., Reay, N.K., Robinson, G.J., Worswick S.P., Astrophys. J., 1979, 232, pp786

Bignell, R.C., 1982, in preparation

Balick, B., Hjellming, R.M., Bignell, R.C., Goad, L., 1982, inpreparation

Calvet, N., Cohen, M., Mon. Not. R. A. S., 1978, 182,pp687

Davis, L.E., Seaquist, C.R., Purton, C.R., 1979, 230,pp434

Goudis, C., McMullan, D., Meaburn, N.J., Tebbutt, N.J., Terrett, D.L., Mon. Not. R. A. S.., 1978, 182, pp13

Harrington, J.P., Feibelman, W.A., 1982, preprint

Hjellming, R.M., Bignell, R.C., Balick, B., Sky and Telescope, 1978, 56, pp199

Hua, C.T., Louise, R., Astron. Astrophys., 1981, 98, pp397

Isaacman, R., Wouterloot, J.G.A., Habing, H.J., Astron. Astrophys., 1980, 86, pp254

Johnson, H.M., Balick, G., Thompson, A.R., Astrophys. J. 1979, 233, pp919

Kondrat eva, L.M., Sov. Astron., 1979, 23, pp193

Kwok, S., Feldman, P.A., Astrophys. J., 1981, 247, ppL67

Kwok, S., Purton, C.R., Keenan, D.W., Astrophys.J., 1981, 250, pp232

Kwok, S., Purton, C.R., 1982, private communication

Millikan, A.G., Astron. J., 1974, 79, pp1259

Morris, M., Astrophys. J., 1981, 249, pp572

Newell, R., Ph. D. Thesis, 1981

Perek, L., Kohoutek, L., Academia Publishing House of the Czechoslovak Academy of Sciences, 1967

Purton, C.R., Kwok, S., 1982, private communication

Seaquist, E.R., Davis. L., 1982, private communication

Thompson, A.R., Sinha, R.P., Astron. J., 1980, 85, pp1240

SEATON: I think that one of the best ways of getting distances is from measured angular expansion rates. Perhaps ten or twenty years are required between first and second epoch "plates". What is being done about the first epoch?

BIGNELL: A VLA proposal has been submitted to obtain the first epoch maps of some dozen close PN.

REAY: Measurements of rates of angular expansion of PN must be supplemented by a good measurement of the expansion velocity in order to obtain a distance. The expansion velocity measured spectroscopically at the centre of the nebula may not be a good measure of the tangential velocity, which governs the angular expansion rate. Models for NGC 2392, for example, suggest that the tangential expansion velocity is 2 to 3 times less than the central (maximum) expansion velocity.

BIGNELL: Care will have to be exercised when choosing the radial velocity to be used in the determination of the distance. It may be necessary to map the radial velocity across the nebula.

SEATON: Overall expansion should be deducible from auto-correlation of first and second epoch observations. Comparison with Doppler velocities should give good distances.

CLEGG: Have you any preliminary results from your attempt to detect the central star of NGC 40?

BIGNELL: The data are only one week old and we do not yet have even a preliminary estimate.

INFRARED EMISSION LINES IN PLANETARY NEBULAE

Harriet L. Dinerstein
NASA Ames Research Center
Moffett Field, California 94035

ABSTRACT

Advances in infrared spectroscopy have led to the detection of many forbidden fine-structure emission lines in planetary nebulae. Measurements of these lines offer sensitive probes of the physical conditions and ionization structure, and lead to improved abundance determinations. This paper reviews recent observations and discusses the ways in which infrared line measurements can contribute to our knowledge of planetary nebulae.

I. INTRODUCTION

Infrared fine-structure emission lines have unique properties which make them useful tools for studying the physical conditions, ionization structure, and chemical abundances in planetary nebulae. These forbidden transitions fall into two categories: (1) lines arising from the single ground-state transition of p^1 and p^5 ions and (2) lines arising from the triplet ground term of p^2 and p^4 ions (see Fig. 1). Lines of the first category offer the opportunity to sample ions that are usually unobservable in other spectral regions. They can be used to study the ionization structure of nebulae and, in combination with other ions of the same elements, to derive better total abundances. As indicated in Figure 1, p^2 and p^4 ions have optical or UV transitions connecting the ground term and higher terms of the same electron configuration, as well as infrared transitions among levels of the ground term. The fine-structure line emissivities have very different dependences on physical conditions than do the optical lines of the same ion. As a result, observations of lines in the second category make it possible to analyze in detail the electron temperature and density, and variations in these parameters within a nebula. Such studies are important not only in understanding the density and ionization structure, but also strongly influence abundance determinations, which depend on the assumed physical conditions.

D. R. Flower (ed.), Planetary Nebulae, 79–88.

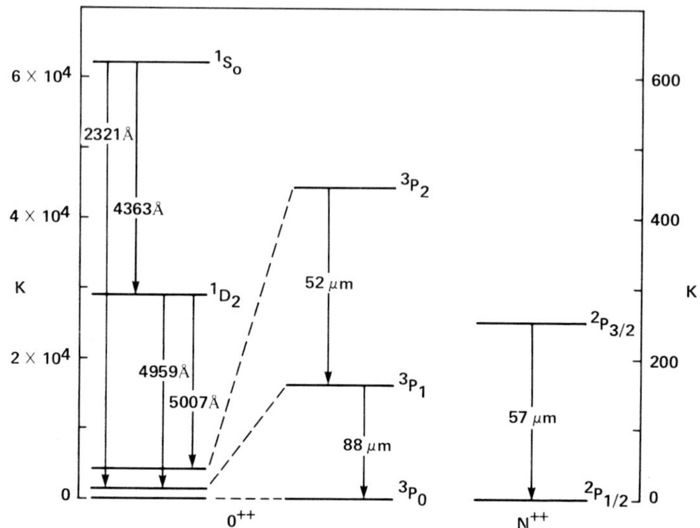

Figure 1. Energy-level diagrams are shown for the ground electron configurations of O III and N III. The vertical axis is the excitation energy in units of degrees Kelvin, hν/k. The O III ion is representative of the p^2 configuration, and N III is an example of the two-level p^1 and p^5 configurations.

The theory describing the emission of forbidden lines has been presented by many authors (e.g., Osterbrock 1974), the infrared fine-structure lines in particular have been discussed by Simpson (1975). Many of the recent observations described in this review have been made possible by the development of moderate-to-high spectral resolution infrared spectrometers and their use on the Kuiper Airborne Observatory, as well as on ground-based telescopes. Descriptions of some of these instruments can be found in Soifer and Pipher (1978) and in references cited below.

II. DETECTION OF NEW LINES AND IONS

The 8-14 μm window, observable from the ground, offers three major fine-structure lines: Ar III 9.0 μm, S IV 10.5 μm, and Ne II 12.8 μm. These lines were first detected in planetary nebulae about 12 years ago (see Rank 1978), and recent observations of them will be discussed below in Section IV. Since then, the availability of a low water-vapor platform, the Kuiper Airborne Observatory, has opened up the spectral regions 5-8 μm and 16-100 μm. This has enabled the detection of a variety of ions both of very high and very low ionization potential, many of these measured first in the bright, complex planetary NGC 7027. Some of these lines have made it possible to observe important ions for the first time.

The low-resolution 4-8 μm spectrum of NGC 7027 taken by Russell et al. (1977) showed, in addition to broad, dust-related features, several emission peaks which they attributed to ionic lines. Recent higher spectral resolution observations by Beckwith et al. (1982) have confirmed the presence of the Mg IV 4.5 μm and Mg V 5.6 μm lines (see Table 1), which indicate a large amount of magnesium in these very highly ionized species. Two other high-ionization lines, O IV 25.9 μm and Ne V 24.3 μm, have been measured by Forrest et al. (1980). They also find that large fractions of the respective elements are in the form of these highly ionized species. The relative intensities of Ne V 24.3 μm and 3426 Å are consistent with $T_e = 12{,}500$ K and $n_e = 2.5 \times 10^5\ cm^{-3}$ in the Ne V-emitting zone, within the uncertainty in the extinction correction in the blue. For these conditions, Forrest et al. derive $Ne^{+4}/H^+ \sim 10^{-4}$ and $O^{+3}/H^+ \sim 4 \times 10^{-4}$.

TABLE 1

INFRARED LINES MEASURED IN PLANETARY NEBULAE

Ion	λ (μm)	Transition	I.P. Range (eV)
Mg IV	4.49	$^2P_{3/2} - {^2P_{1/2}}$	80.1 - 109.3
Mg V	5.61	$^3P_2 - {^3P_1}$	109.3 - 141.3
Ni II	6.62	$^2P_{1/2} - {^2P_{3/2}}$	7.6 - 18.2
Ar II	6.98	$^2P_{1/2} - {^2P_{3/2}}$	15.8 - 27.6
Ar III	8.99	$^3P_1 - {^3P_2}$	27.6 - 40.7
S IV	10.52	$^2P_{3/2} - {^2P_{1/2}}$	34.8 - 47.3
Cl IV	11.76	$^3P_1 - {^3P_2}$	39.6 - 53.5
Ne II	12.81	$^2P_{1/2} - {^2P_{3/2}}$	21.6 - 41.0
S III	18.71	$^3P_2 - {^3P_1}$	23.3 - 34.8
Ne V	24.28	$^3P_1 - {^3P_0}$	97.1 - 126.2
O IV	25.87	$^2P_{3/2} - {^2P_{1/2}}$	54.9 - 77.4
O III	51.81	$^3P_2 - {^3P_1}$	35.1 - 54.9
	88.36	$^3P_1 - {^3P_0}$	
N III	57.33	$^3P_{3/2} - {^3P_{1/2}}$	29.6 - 47.5
O I	63.17	$^3P_1 - {^3P_2}$	0 - 13.6

Other p^1 and p^5 ions, observable only by means of their fine-structure lines, are dominant species in nebulae ionized by very cool stars. Ne II, mentioned above, is one example. More recently, Ar II 6.98 μm has been measured in IC 418 (Willner et al. 1979) and BD+30°3639 (Dinerstein et al. 1982). Since the Ar II ion has an ionization potential similar to that of neutral helium (which is not directly observable), measurements of Ar II may prove to be useful for understanding the ionization and abundance of helium in gaseous nebulae, as well as in determining argon abundances.

Emission lines from ions with ionization potentials lower than that of hydrogen (13.6 eV) may be expected to arise from neutral gas outside the ionized region or at the interface between these regions. The O I 63 μm line has been detected in NGC 7027 by Melnick et al. (1981). The flux they measure is substantially higher than expected on the basis of models, so the 63 μm line may be produced by a different process, possibly arising from the extensive neutral cloud known to surround NGC 7027. A recent spectrum of NGC 7027 taken by Bregman et al. (1981) identifies an emission peak at 6.62 μm with the ground-state transition of Ni II, probably also arising from gas outside the fully ionized region.

The ongoing development of high-resolution infrared spectrometers will inevitably lead to the measurement of other previously undetected lines. Lists of potentially observable lines can be found in Petrosian (1970), Olthof and Pottasch (1975), Simpson (1975), Garstang et al. (1978), and Watson and Storey (1980). These include the fine-structure lines of other p^2 ions, such as N II (at 122 and 204 μm); low-ionization species, such as CII (157 μm) and Fe II (5.34 μm); and ions of less abundant elements, such as Na, Ca, and P. Some of these lines are observable from ground-based telescopes; for example, the 11.8 μm line of Cl IV, searched for unsuccessfully in various H II regions, has only recently been detected in a planetary nebula (Roche 1982). Many of the lines lying in the far-infrared ($\lambda > 14$ μm) can be observed from the Kuiper Airborne Observatory (KAO). However, certain lines, e.g., Ne III 15.4 μm, have wavelengths that fall in terrestrial absorption bands that are opaque even at the altitude of the KAO. Observations of these lines will await the development of infrared spectroscopy from space platforms.

III. DIAGNOSTICS OF PHYSICAL CONDITIONS

Fine-structure lines of p^2 and p^4 ions, which have five levels in the ground configuration, offer a unique opportunity to probe the physical conditions in nebulae. The optical transitions yield sensitive indicators of the electron temperature T_e via the intensity ratios of transitions from the highest and second-highest energy levels (e.g., O III 4363 Å/5007 Å). The infrared lines, on the other hand, are insensitive to T_e, but sensitive to the electron density n_e. Different transitions within a triplet, such as O III 52 μm and 88 μm, generally have different critical densities and therefore their ratio is a good indicator of n_e. The combination of infrared and optical lines makes possible a simultaneous and independent determination of n_e and T_e in the O III zone, which comprises a large fraction of the gas in most planetary nebulae.

Measurements of the 88 μm O III line have been made for four planetary nebulae by this author, and the 52 μm line has been observed in two nebulae (Watson et al. 1981; Moseley 1982). Figure 2 demonstrates how such measurements can be combined with optical lines to derive n_e and T_e. The ratio I(5007)/I(88 μm) is used as a density indicator, and I(4363)/I(5007) yields the temperature. The grid lines indicate the

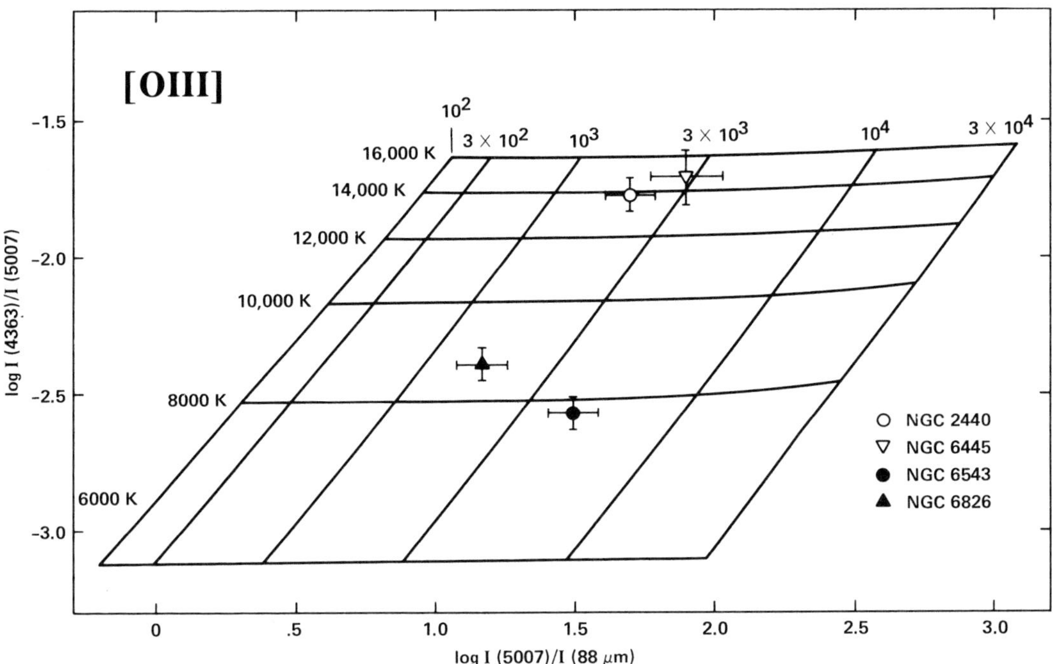

Figure 2. This figure demonstrates the use of observed line-intensity ratios of p^2 ions such as O III as diagnostics of the physical conditions. The vertical axis, 4363 Å/5007 Å, measures the electron temperature; the ratio 5007 Å/88 μm (horizontal axis) is a density indicator. The grid of intensity ratios was computed by solving the five-level equilibrium (see text).

loci of the line ratios for a variety of n_e and T_e values, from a five-level equilibrium solution. Lines of constant n_e and T_e are not orthogonal in this diagram because I(5007)/I(88 μm) is sensitive to temperature as well as density. The calculation assumed the transition probabilities of Garstang (1968) and the collision strengths of Eissner and Seaton (1974) for transitions between terms and the values of Saraph et al. (1969) for the fine-structure lines. The 88 μm line measurements in Figure 2 are fluxes measured by the author from the KAO using the spectrometer described by Storey et al. (1980) with a large beam, integrating over the nebula. Integrated 5007 Å fluxes were measured in the same beam (45") at Lick Observatory for NGC 6445, NGC 6543, and NGC 6826. The 4363 Å/5007 Å ratios for NGC 6543 and NGC 6826 are the large-beam (35") values of Bohuski et al. (1974). Fluxes of the optical lines for NGC 2440 are from Shields et al. (1981), and for NGC 6445, from Torres-Peimbert and Peimbert (1977). The O III densities derived for these objects are similar to values derived from O II in the same objects (Aller and Epps 1976). This result is important because it shows that these particular nebulae are quite homogeneous in their density structure.

The optical O III line emissivities depend strongly on the electron temperature, and the accuracy of ionic abundance determinations relative to hydrogen is mainly limited by the accuracy with which T_e is estimated. In addition, values of T_e determined from high-lying levels are weighted toward regions of highest T_e and highest line emissivity, so that ionic abundances will be underestimated if the effects of temperature variations within a nebula are neglected (Peimbert 1967; Rubin 1969). The magnitude of the variation in T_e is parameterized by the quantity t^2, the rms temperature fluctuation. Correction for this effect, using t^2 values derived empirically from line observations (e.g., Peimbert and Torres-Peimbert 1971), can lead to increases by a factor of about 2 in the derived ionic aboundances. The appropriate value of t^2 has been a matter of some controversy (Barker 1979). Probably the best way to detect the effects of a spread in T_e is to examine two temperature-sensitive line ratios of a single ion. For O III these could be I(4363)/I(5007) and I(5007)/I(88 μm), and the electron density n_e can be independently determined from the ratio I(52 μm)/I(88 μm). Thus, measurements of both infrared and optical lines would yield not only n_e and T_e, but also the magnitude of inhomogeneities in these quantities within the O III zone. Density fluctuations may also lead to an apparent positive value for t^2, but these effects are probably less important than those of temperature variations since the line ratios depend on a higher power of temperature than of density.

A similar analysis can be carried out for other p^2 ions such as Ar III and S III. The 18.7 μm S III line has been measured in several planetaries: NGC 7027 and BD+30°3639 (Greenberg et al. 1977), and NGC 6543 (McCarthy 1979). Combination of these observations with measurements of the $^1D-^3P$ lines near 9000 Å (Barker 1978b; Dinerstein 1980a,b) and of 6312 Å will yield n_e and T_e in the S III region, as was done above for O III. Other ions which could potentially be studied in this way include N II, Ne III, and Ar III, but these studies await the detection of infrared lines that have not yet been observed.

IV. ABUNDANCE DETERMINATIONS

As discussed above, infrared lines make it possible to observe ions that are difficult or impossible to measure in other spectral regions, and can therefore greatly improve abundance determinations for certain elements. Two of the best examples are Ne II and S IV, which have only been measured in the infrared. In the last few years, these lines have been measured sprectroscopically in a large number of planetary nebulae: Bregman (1978), Grasdalen (1979), Aitken et al. (1979), Dinerstein (1980a,b), and Beck et al. (1981).

The S IV ion in particular usually contains most of the sulfur in typical planetaries. Previous sulfur abundance determinations, even those making use of observations of S III as well as S II, showed systematic effects indicating that ionization correction formulae based on ionization potential coincidences with oxygen were not adequate (Barker 1978b, Pagel 1978). Measurement of accurate abundances for individual nebulae, and selection of the best empirical correction scheme (e.g., Natta et al. 1980) or of reliable models, requires direct measurements of the S IV ion. Table 2 presents results from the two infrared surveys in which total abundances were derived (Dinerstein 1980b and Beck et al. 1981). Average values for S/H are compared with values from Barker (1978b) based on the original ionization correction formula of Peimbert and Costero (1969), and with results from model-fitting reported by Aller (1978). Table 2 also compares Ne/H values from the infrared studies with values from Barker (1978a), Kaler (1978, 1980), and Aller (1978) for similar groups of nebulae. Most of the nebulae observed in the infrared lines so far have been relatively bright, compact objects belonging to a fairly young population (groups 1 and 2 of Peimbert 1978). Observations of a few high-velocity nebulae (previous references) show

TABLE 2

ABUNDANCES OF PLANETARY NEBULAE

	[S/H][a]	[Ne/H][a]	[Ar/H][a]
Infrared Studies			
Dinerstein (1980b)	7.02	8.11	---
Beck et al. (1981)	7.08	8.03	6.44
Optical Studies			
Barker (1978a,b)[b]	7.60	7.8	---
Kaler (1978, 1980)[c]	---	8.0	6.5
Aller (1978)	7.35	8.1	6.4

[a] Logarithmic abundance, [X/H] = log N(X)/N(H) + 12.

[b] Values for intermediate population objects (Group II).

[c] Assumes O/H = 4.5×10^{-4}, the value found by Kaler (1980) for intermediate disk population planetaries, and Ne/O = 0.225 (Kaler 1978).

no evidence of lower S/H in kinematically older nebulae, in contrast to the conclusion of Kaler (1980) that these nebulae have somewhat lower O/H. This question will be examined as part of an extensive study of S/H and Ne/H currently in progress (Dinerstein and Rank 1982).

Infrared observations can contribute even to measurements of such well-studied elements as oxygen. As discussed in Section III comparison of the infrared and optical O III lines will lead to better determinations of the electron temperature and its variations within a nebula. Through the dependence of the derived O^{++}/H^{+} abundances on the assumed values of T_e and t^2, this will yield more accurate abundance determinations. Observations of O IV will clarify the ionization equilibrium between O III and O IV and improve total abundance determinations in very highly ionized nebulae.

Another important p^1 ion, with an infrared transition at 57 μm, is N III. Most N/O determinations in nebulae are based on the optical N II/O II ratio (Kaler 1979); however, N III, like O III, is generally the most populous ion in planetaries. The ratio of N/O is particularly interesting because it can have an elevated value owing to processing within the star, and because planetaries are thought to be a significant source of nitrogen for enriching the interstellar medium. Recent UV observations (reviewed elsewhere in this volume) have led to better N/O determinations from N III and higher ions. Such determinations depend on the assumed extinction curve and electron temperatures. An alternative method is to use the infrared N III and O III lines, which are insensitive to T_e and sensitive to n_e, but from which the density can be measured, as discussed above. This approach has been applied to H II regions and can be used for some planetaries as well. Determination of N/O from N III/O III should make it possible to check the large overabundances of nitrogen derived for certain planetary nebulae.

I wish to thank D. Lester, D. M. Rank, J. P. Simpson, and R. Rubin for their helpful comments, and R. Genzel, D. Lester, J. Storey, C. Townes D. M. Watson, and M. Werner for their collaboration in obtaining the 88 μm measurements. I acknowledge the support of a National Academy of Sciences/National Research Council Research Associateship at Ames Research Center.

REFERENCES

Aitken, D. K., Roche, P. F., Spenser, P. M., and Jones, B.: 1979, Ap. J. 233, 925.
Aller, L. H.: 1978, in "Planetary Nebulae," IAU Symposium No. 76, Y. Terzian (ed.), Dordrecht: Reidel, pp. 225-234.
Aller, L. H. and Epps, H. W.: 1976, Ap. J. 204, 445.
Barker, T.: 1978a, Ap. J. 220, 193.
Barker, T.: 1978b, Ap. J. 221, 145.
Barker, T.: 1979, Ap. J. 227, 863.
Beck, S. C., Lacy, J. H., Townes, C. H., Aller, L. H., Geballe, T. R., and Baas, F.: 1981, Ap. J. 249, 592.
Beckwith, S., Evans, N. J. II, Natta, A., Russell, R. W., and Wyant, J.: 1982, in preparation.

Bohuski, T. J., Dufour, R. J., and Osterbrock, D. E.: 1974, Ap. J. 188, 529.
Bregman, J. D.: 1978, Pub. Astron. Soc. Pac. 90, 548.
Bregman, J. D., Dinerstein, H. L., Goebel, J. H., Lester, D. F., Witteborn, F. C., and Rank, D. M.: 1981, Bull. Am. Astron. Soc. 13, 852.
Dinerstein, H. L.: 1980a, Ap. J. 237, 486.
Dinerstein, H. L.: 1980b, Ph.D. thesis, University of California, Santa Cruz.
Dinerstein, H. L. and Rank, D. M.: 1982, in preparation.
Dinerstein, H. L., Lester, D. F., Bregman, J. D., Witteborn, F. C., and Rank, D. M.: 1982, in preparation.
Eissner, W. and Seaton, M. J.: 1974, J. Phys. B: Atom. Molec. Phys. 7, 2533.
Forrest, W. J., McCarthy, J. F., and Houck, J. R.: 1980, Ap. J. Lett. 240, L37.
Garstang, R. H.: 1968, in "Planetary Nebulae," IAU Symposium No. 34, D. E. Osterbrock and C. R. O'Dell (eds.), Dordrecht: Reidel, pp. 143-152.
Garstang, R. H., Robb, W. D., and Rountree, S. P.: 1978, Ap. J. 222, 384.
Grasdalen, G.: 1979, Ap. J. 229, 587.
Greenberg, L. T., Dyal, P., and Geballe, T. R.: 1977, Ap. J. Lett. 213, L71.
Kaler, J. B.: 1978, Ap. J. 225, 527.
Kaler, J. B.: 1979, Ap. J. 228, 163.
Kaler, J. B.: 1980, Ap. J. 239, 78.
McCarthy, J. F.: 1979, Ph.D. thesis, Cornell University.
Melnick, G., Russell, R. W., Gull, G. E., and Harwit, M.: 1981, Ap. J. 243, 170.
Moseley, H.: 1982, private communication.
Natta, A., Panagia, N., and Preite-Martinez, A.: 1980, Ap. J. 242, 596.
Olthof, H. and Pottasch, S. R.: 1975, Astr. Ap. 43, 291.
Osterbrock, D. E.: 1974, "Astrophysics of Gaseous Nebulae," San Francisco: Freeman.
Pagel, B. E. J.: 1978, Mon. Not. Roy. Astron. Soc. 183, 1 p.
Peimbert, M.: 1967, Ap. J. 150, 825.
Peimbert, M.: 1978, in "Planetary Nebulae," IAU Symposium No. 76, Y. Terzian (ed.), Dordrecht: Reidel, pp. 215-224.
Peimbert, M. and Costero, R.: 1969, Bol. Obs. Tonant. y Tacu. 5, 3.
Peimbert, M. and Torres-Peimbert, S.: 1971, Ap. J. 168, 413.
Petrosian, V.: 1970, Ap. J. 159, 833.
Rank, D. M.: 1978, in "Planetary Nebulae," IAU Symposium No. 76, Y. Terzian (ed.), Dordrecht: Reidel, pp. 103-110.
Roche, P. F.: 1982, Ph.D. thesis, University College, London.
Rubin, R.: 1969, Ap. J. 155, 841.
Russell, R. W., Soifer, B. T., and Willner, S. P.: 1977, Ap. J. Lett. 217, L149.
Saraph, H. E., Seaton, M. J., and Shemming, J.: 1969, Phil. Trans. Roy. Soc. London, Ser. A 264, 77.
Shields, G. A., Aller, L. H., Keyes, C. D., and Czyzak, S. J.: 1981, Ap. J. 248, 569.

Simpson, J. P.: 1975, Astron. Ap. 39, 43.
Soifer, B. T. and Pipher, J. L.: 1978, Ann. Rev. Astron. Ap. 16, 335.
Storey, J. W. V., Watson, D. M., and Townes, C. H.: 1980, Int. J. Infr. Millim. Waves 1, 15.
Torres-Peimbert, S. and Peimbert, M.: 1977, Rev. Mex. Astron. Ap. 2, 181.
Watson, D. M. and Storey, J. W. V.: 1980, Int. J. Infr. Millim. Waves 1, 609.
Watson, D. M., Storey, J. W. V., Townes, C. H., and Haller, E. E.: 1981, Ap. J. 250, 605.
Willner, S. P., Jones, B., Puetter, R. C., Russell, R. W., and Soifer, B. T.: 1979, Ap. J. 234, 496.

ALLER: This splendid paper points very clearly to the importance of infrared data in the study of PN. It is absolutely essential to observe as many ionization stages as possible in order to minimize reliance on ionization correction factors, whether obtained from theoretical models or by less elaborate procedures.

Also, the importance of atomic parameters needs to be emphasized. In Ar III, for example, the nebular-type line λ 7135 tended to give different abundances from the infrared fine structure line.

ISAACMAN: Do the temperatures and densities derived from the [O III] lines agree with those inferred from the radio data?

DINERSTEIN: The temperatures which I derive in Fig. 2 are essentially the same as previous optical determinations, since they are based on the λ 4363/λ 5007 radio. The densities are similar to the values derived from the optical lines of [O II], and so the comparison with radio results is unchanged. I might add that I have just made a measurement of the intensity of the [O III] 52 μm line in NGC 6543. Although the analysis is in a preliminary stage, it appears that the observed intensity is roughly consistent with the values of density and temperature in Fig. 2.

HOUCK: We have seen [O IV] 25.87 μm in a total of eight PN and [S III] 33.46 μm has been detected in NGC 6543.

10 μm SPECTRAL OBSERVATIONS OF MODERATELY EXTENDED PLANETARY NEBULAE

P.F. Roche, D.K. Aitken, B.R. Whitmore
University College London, UK

We have obtained 8-13 μm spectra of a sample of eight moderately extended planetary nebulae at a resolution of 0.22 μm using a 20" circular aperture.

Most compact planetaries which have been studied in this way show generally a strong continuum due to thermal emission from dust, together with fine structure line emission. In contrast the extended objects have weak or undetected continuum emission and are dominated by fine structure line emission, especially by [SIV] which in some cases accounts for most of the broad band 10 μm flux.

MATHIS: Do oxygen-rich PN always show a "silicate" feature at 10 μm? Do carbon-rich PN ever show it?

ROCHE: There are not many objects on which we can base comparisons, but for the nebulae with accurate abundances derived from optical and ultraviolet work the agreement with the infrared spectra, in terms of oxygen- or carbon-richness, is 100 per cent. The only oddity is Sw St 1, which has an oxygen-rich nebula but apparently a WC central star.

MOLECULES IN PLANETARY NEBULAE

John H. Black
Harvard-Smithsonian Center for Astrophysics
60 Garden Street, Cambridge, MA 02138, USA

1. INTRODUCTION

The study of molecules in planetary nebulae is still in its infancy; therefore, it is appropriate both to review existing knowledge and to anticipate developments that will arise as the subject matures. Briefly stated, this subject consists of observations of hydrogen, H_2, and carbon monoxide, CO, molecules in a very few nebulae; the tentative identification of the hydrogen molecular ion, H_2^+; unsuccessful searches for various other species; some theoretical work on molecular processes relevant to planetary nebulae; and the study of molecules in possible precursors of planetary nebulae. Such a facile summary might suggest that molecules are mere curiosities of peripheral interest; on the contrary, it can be asserted that molecular studies of planetary nebulae will yield important new information about their origins, structures, and evolution.

Some of the reasons for studying molecules in nebulae derive from their spectroscopic advantages over atomic species. Molecules are versatile diagnostic probes: they can be coupled to ultraviolet radiation fields through electronic transitions, their vibrational transitions can be excited in shock-heated neutral gas at temperatures $T \simeq 1000$ K, and they can reveal the coldest ($T \simeq 10$-100 K) neutral regions through rotational lines. Molecules are ideally suited for the determination of isotope abundances because a small difference in the mass of a molecule produces a large, and easily observable, difference in the rotational and vibrational structure of its spectrum. With the exception of the microwave line of $^3He^+$ (Rood et al. 1979), atomic spectra are unsuitable for isotope studies of nebulae. Isotope abundances may someday prove as valuable as element abundances in establishing the evolutionary role of planetary nebulae and the properties of the stars that form them. Molecules are particularly useful for studying the neutral gas associated with nebulae: in some cases, this otherwise unobservable material accounts for a large fraction of the total mass of a nebula. The interpretation of the spectra of planetary nebulae has traditionally provided problems of basic interest in atomic physics. There are many molecular processes important in

D. R. Flower (ed.), Planetary Nebulae, 91–102.

nebulae, which may stimulate a similar symbiotic relationship with basic molecular physics. Molecules in the solid phase (i.e. dust grains) are of interest, but are beyond the scope of this brief review.

2. OBSERVATIONS AND INTERPRETATION

Mufson et al. (1975) observed CO rotational line emission (J=1-0) associated with NGC 7027 and reported marginal detections in IC 418 and NGC 6543. Knapp et al. (1982) observed the J=2-1 line of CO in the first two objects, but failed to detect it in the third. Zuckerman et al. (1977) searched unsuccessfully for CO in several other nebulae. Thronson (1982a) has made improved measurements of CO J=1-0 in NGC 7027. The CO line profiles in NGC 7027 are broad (Δv=32 km s^{-1} at the base), with steep wings, suggestive of an expansion velocity of 16 km s^{-1} for the molecular shell. The expansion velocity of the ionized nebula is 5-8 km s^{-1} larger. Thronson observed the isotopic species ^{13}CO and inferred an abundance ratio $n(^{12}CO)/n(^{13}CO)=36\pm6$, somewhat larger than the value suggested earlier by Mufson et al. This ratio must be regarded as a lower limit if the stronger ^{12}CO line is optically thick. Thronson argues that if the molecular isotope ratio reflects the overall nuclear abundances in the atmosphere of the star that formed the nebula, and if the carbon/oxygen ratio in the nebula can be used to place an upper limit on the mass of the star (cf. Iben 1981), then the nebula was formed during the asymptotic giant branch phase of a star which began life with approximately 3 $M_\odot$. Such arguments are not definitive, but it seems clear that careful studies of isotope abundances in planetary nebulae will provide useful tests of theories of stellar evolution (cf. Finzi and Yahel 1978). Such abundance determinations will also help elucidate the contribution of planetary nebulae to the enrichment of heavy elements in the general interstellar medium. The angular extent of the CO emission in NGC 7027 is 40-50 arcsec (Mufson et al. 1975, Knapp et al. 1982), which is approximately 3 times larger than the mean diameter of the ionized nebula (Becklin et al. 1973). There are several ways to estimate the total neutral mass from CO emission, and all of them require questionable assumptions; however, it seems reasonable to accept Thronson's lower limit $M \geq 0.05\ d^2\ M_\odot$, where d is the distance in kpc. The most recent re-examination of the distance to NGC 7027 (Pottasch et al. 1982) suggests that $1 \leq d \leq 1.5$ kpc. Far infrared emission due to cool dust in NGC 7027 ($T_d \simeq 90$ K) is confined to a region smaller than 20 arc-sec, and any more extended dust shell must be colder than 35 K (Moseley 1980). Both a search for very cold dust (perhaps at 350 μm as suggested by Moseley) and a better measurement of the CO excitation temperature would help determine the conditions in the extended molecular cloud.

The first high-resolution infrared spectrum of NGC 7027 revealed v=1-0 vibration-rotation emission lines of H_2 near 2 μm (Treffers et al. 1976). By now, 7 lines in the v=1-0 band have been measured (Smith et al. 1981) and the spatial distribution of the v=1-0 S(1) line emission has been mapped (Beckwith et al. 1980). The excitation energies of the v=1-0 lines are $\Delta E/k \simeq 6800$ K, and the hydrogen molecules thus observed represent a small, but highly-excited, fraction of the molecular gas. The relative strengths of the v=1-0 lines arising in different rotational levels (v=1,J) can be described well by thermal populations at an excitation temperature T_{ex}=1100 K in NGC 7027 (Smith et al. 1981). For fully thermalized level populations and an emitting source that fills a 7x7 arcsec beam, the total column density of H_2 at this temperature is $N(H_2)=8.9\times10^{19}$ cm^{-2}. In ionization-bounded nebulae, a shock front will often precede the ionization front, and produce a compressed layer of neutral gas at an elevated temperature ($T\gtrsim1000$ K). Additional molecule formation processes that are not effective in cold gas become important in shock-heated regions (cf. Dalgarno 1981), and vibrational line emission in H_2 can be excited readily (Hollenbach and Shull 1977, Kwan 1977, London et al. 1977). There exist other processes by which nebular molecules can be excited to radiate: in H_2, both formation processes and ultraviolet pumping (Black et al. 1981) produce vibrational and rotational lines, and molecules like CH^+ and C_2^- can be excited by resonance fluorescence involving nebular emission lines (Gahm et al. 1977). Beckwith et al. (1980) have used a shock model with a steep radial density gradient to account for the excitation of the hot H_2 and to estimate the total mass of cooler H_2 contained in a volume of the same size as the extended CO emission: the model-dependent result is $M\simeq0.9\ d^2\ M_\odot$. The central star of NGC 7027 is a copious source of ultraviolet photons which can excite vibrational lines by absorption and fluorescence (UV pumping). If UV pumping dominated the excitation, then the v=2-1 lines would be nearly as strong as the well-observed v=1-0 lines (cf. Black and Dalgarno 1976). Smith et al. (1981) placed upper limits on v=2-1 lines and Beckwith et al. (1980) measured the v=2-1 S(1) line at a level much below that expected from pure UV pumping: thus the rates of radiative excitation must be suppressed relative to the rates of collisional excitation. These observations permit a limit to be placed upon the ratio of the UV flux at 1000 Å to the density in the H_2 emitting region. The UV flux at the boundary of NGC 7027 can be estimated from the known size and emission measure of the nebula, and it is too large to be consistent with the weak v=2-1 lines, unless there is a significant amount of internal extinction, corresponding to $A_V\simeq0.6$-1.5 mag, between the central star and the emitting region. The observed line strengths in NGC 7027 can be explained without a shocked neutral zone by the combined effects of UV pumping and electron-impact excitation on H_2 inside the transition zone of the nebula, provided that the transition zone is thick enough (Black et al. 1981). The spatial distribution of H_2 emission, both in NGC 7027 and in NGC 6720 (Beckwith et al. 1978) differs from that of the bulk of the ionized gas, and is consistent with a relatively narrow emitting zone near the boundary. Infrared line emission from H_2 has also been observed in the nebulae NGC 6720, BD+30,3639, and Hb 12 (Beckwith et al. 1978). Additional sensitive searches for hot H_2 in planetary nebulae are needed.

Heap and Stecher (1981) observed anomalous flux distributions in the ultraviolet spectra of central stars of several planetary nebulae, and attributed the depressed fluxes shortward of 1500 Å to absorption by nebular H_2^+. Feibelman et al. (1981) presented similar data on other nebulae, and Seaton (1980) noted a flux anomaly shortward of 1500 Å in the central star of NGC 1514, which can probably also be attributed to H_2^+. The relative absorption, $\exp\{-N(H_2^+)\ \sigma(\lambda)\}$, using the cross sections $\sigma(\lambda)$ of Dunn (1968a,b), provides an excellent match of the observed flux deficiency if the H_2^+ is predominantly in its lowest vibrational state v=0. In NGC 6210, a column density $N(H_2^+)=8\times10^{16}\ cm^{-2}$ is implied. We will argue below on theoretical grounds that the identification of nebular H_2^+ in its ground state is plausible.

Despite careful searches for the 18 cm lines of OH (Silverglate et al. 1979) only one detection of 1612 MHz maser emission, in Vy 2-2, has been reported (Davis et al. 1979). OH emission attributed to NGC 2438 (Turner 1971) was later shown to be due to an unrelated source 6.5 arcmin away (Hardebeck 1972, Goss et al. 1973).

An ionization-bounded nebula will possess a narrow transition zone at its boundary in which the hydrogen goes from being fully ionized to being almost completely neutral. Here, where the concentrations of electrons, protons, and neutral hydrogen atoms are comparable, the conditions are optimal for the formation of H_2, H_2^+, and HeH^+ by gas phase processes. The abundances of such simple molecules inside a steady state nebula have been discussed (Black 1978), and the variety of processes that control the abundance and excitation of HeH^+ have been studied (Flower and Roueff 1979, Roberge and Dalgarno 1982). Although HeH^+ was suggested as the source of the unidentified emission feature at 3.3 μm in NGC 7027 and other objects (Dabrowski and Herzberg 1977), subsequent observations have indicated that some other species is probably responsible for this broad, intense feature (Scrimger et al. 1978, Tokunaga and Young 1980).

Thronson and Lada (1982) have searched for the J=2-1 rotational line of SiO in seven planetary nebulae without success. While SiO appears to be underabundant in the gas phase relative to CO in NGC 7027, its absence from other nebulae may result from inadequate excitation conditions. Fairly sensitive searches failed to reveal either CH or HCN in NGC 7027 (Sume and Irvine 1977, Zuckerman et al. 1977). Nitrogen is overabundant in some objects, and careful searches for nitrogen-bearing molecules, e.g. CN, HCN, HC_3N, and NH_3, would be valuable.

Certain evolved stars that possess thick, dusty, molecular shrouds and that show evidence of high rates of mass loss have been identified as "proto-planetary nebulae" (Zuckerman 1978, 1980). Characteristics other than thick molecular envelopes have led to the identification of some emission line objects as the precursors of planetary nebulae (Kwok et al. 1978, Kwok and Purton 1979). These may represent a different evolutionary stage or an entirely distinct population of objects from the proto-planetary nebulae described by Zuckerman. The proto-planetary nebulae are usually distinguished by large infrared fluxes and by strong, broad CO emission lines indicative of expanding clouds (Zuckerman et al. 1976,1977,1978; Lo and Bechis 1976; Knapp et al. 1982). Two proto-planetary nebulae, AFGL 618 and AFGL 2688, exhibit H_2 line emission similar to that from NGC 7027 (Beckwith et al. 1978; Thronson 1981, 1982b), and in AFGL 618 the intensity has increased significantly over two years (Beck and Beckwith 1983). In order to understand fully the properties of the precursor stars and the processes by which they form planetary nebulae, it is necessary to determine the total amounts of mass involved and the element abundances. Element abundances (including isotope ratios) can be used to infer the processes of nucleosynthesis at work up to the time of rapid mass loss, and hence to help characterize the evolutionary state of a star when it forms a nebula. Moreover, the C/O abundance ratio is crucial for inferring masses and mass loss rates from observations of CO lines (Zuckerman 1980). In this connection, it is important to recognize that the overall abundances of elements like C and O may be disguised by the manner in which the atoms have been divided up between the gas and the dust. Whether dust grains are enriched in carbon, for example, can sometimes be determined from the 8-13 μm spectra of planetary nebulae (Aitken and Roche 1982 and references therein), and the broad infrared feature at λ>24 μm that is common to carbon stars and the nebulae IC 418 and NGC 6572 (Forrest et al. 1981). As indicated by the small number of planetary nebulae observed in CO lines, it is fairly rare for a *bona fide* planetary nebula to possess an extensive molecular envelope, while such envelopes are characteristic of the objects identified as proto-planetary nebulae. Evidently molecular processes are most important during the early evolution of the nebula.

In the atmospheres of stars that form planetary nebulae, high densities enable three-body processes to control molecular abundances; hence the abundances of many species approach their thermodynamic equilibrium values. These abundances may be frozen into the gas at some stage as the atmosphere expands. Relevant theoretical studies include those of Goldreich and Scoville (1976), Scalo and Slavsky (1980), McCabe et al. (1979), Clegg and Wootten (1980), Lafont et al. (1982), and Huggins and Glassgold (1982). Later, as the nebula expands and the optical depth of the molecular envelope decreases, the molecules become exposed to the destructive effects of the ambient interstellar radiation from outside and the intense ultraviolet radiation of the central star from inside. If the typical molecular lifetime becomes smaller than the nebular lifetime, then the molecular abundances will be determined by a statistical equilibrium of two-body processes.

Previous theoretical work has concerned the processes by which molecules form inside nebulae and models to explain their excitation. The photochemistry of a dense molecular shell close to a very hot central star still requires careful study: preliminary results of such an investigation are discussed in the following section. There is also a great need for detailed theoretical models of transition zones of nebulae in which charge transfer processes and the effects of dust have been fully incorporated.

3. DISCUSSION

Some problems concerning the abundance and excitation of molecules in planetary nebulae can be discussed in terms of simple models of the ionized nebulae and their surrounding molecular envelopes. Consider an idealized model in which the ionization front expands more rapidly than the molecular envelope. In the reference frame of the ionization front, molecules of initial density n_o cm^{-3} flow with relative velocity v_o across the front into an ionized region of uniform electron density $n(e)$ and temperature $T_e \simeq 10^4$ K. Outside the ionization boundary, the molecules are completely shielded from H-ionizing photons ($h\nu > 13.6$ eV). The abundant species of particular interest, H_2 and CO, are special in that dissociation by less energetic photons ($h\nu < 13.6$ eV) occurs through line absorptions. Dissociation is rapid only into depths where these lines remain unsaturated; at greater depths the destruction rate decreases greatly and H_2 and CO molecules effectively shield themselves. Once they enter the nebula, the molecules will be exposed to large ionizing fluxes and to hot positive ions, and their lifetimes against destruction will be short, $\tau \simeq 10^5 - 10^6$ s. The principal reactions that affect the abundances of H_2, CO, and their ions, H_2^+ and CO^+, inside the nebula are summarized in the table, together with their rates. Photoionization

	Reaction	Rate
1.	$H_2 + h\nu \rightarrow H_2^+ + e$	$k_{pi} = 1.4 \times 10^{-6}$ s^{-1}
	$CO + h\nu \rightarrow CO^+ + e$	$k_{pi} = 8.6 \times 10^{-6}$ s^{-1}
2.	$H_2 + H^+ \rightarrow H_2^+ + H$	$k_{ct} = 7.7 \times 10^{-11}$ cm^3 s^{-1}
	$CO + H^+ \rightarrow CO^+ + H$	$k_{ct} = 8.0 \times 10^{-11}$ cm^3 s^{-1}
3.	$H_2^+ + e \rightarrow H + H$	$k_{dr} = 8.9 \times 10^{-9}$ cm^3 s^{-1}
	$CO^+ + e \rightarrow C + O$	$k_{dr} = 1.1 \times 10^{-7}$ cm^3 s^{-1}

rates assume a blackbody stellar radiation field of temperature $T_* = 3.29 \times 10^5$ K and a ratio of nebular radius to stellar radius $R_o/R_* = 5.7 \times 10^7$. The rates of charge transfer of H^+ with ground-state H_2 and CO have been estimated from the measured rates of the reverse processes (Karpas et al. 1979). Theory suggests that the low energy cross sections for dissociative recombination of H_2^+ (reaction 3) increase by an order of magnitude as the vibrational state of H_2^+ goes from v=0 to v=1 and from v=1 to v=2 (Zhdanov 1980, Derkits et al. 1979). The adopted rate coefficient, k_{dr}, represents an average over the states v=0,1,2 (Auerbach et al. 1977), and it is almost certain to be excessively large for ground-state H_2^+(v=0). The rate k_{dr} for CO^+ is from Mentzoni and Donohoe (1969), and all collisional rates in the table are evaluated at $T_e = 10^4$ K. The destructions of H_2 and CO by reactions 1 and 2 are the sources of the ions H_2^+ and CO^+. The rates of processes like photodissociation and e-impact dissociation, which form neutral products, are factors of at least 10 smaller than those of reactions 1 and 2 at the temperatures and radiation fields of interest (cf. Black 1978). The ions have even shorter lifetimes than their parent molecules. For a uniform spherical nebular boundary of radius R_o, the expressions governing the densities of these molecules as functions of distance r from the central star are just those used to describe density distributions in the coma of an idealized comet (Haser 1957). The distance scale must be reversed, however, because the nebular molecules flow inward from the boundary, rather than outward from the nucleus. In the cases of interest, the molecules are confined to a very thin ($\Delta r \simeq \tau v_o < 10^{13}$ cm) zone near the boundary; therefore, $y \equiv R_o - r << R_o$, and the number densities of molecule X (=H_2 or CO) and its ion X^+ are given approximately by

$$n(X) = n_o(X) \exp(-\beta(X)y) \quad (1)$$

$$n(X^+) = n_o(X)\ \beta(X)\ \{\exp(-\beta(X)y) - \exp(-\beta(X^+)y)\}/\{\beta(X^+)-\beta(X)\}\ . \quad (2)$$

The factors β are given by $\beta(X) \approx (k_{ct}\ n(e) + k_{pi})/v_o$ cm^{-1} and $\beta(X^+) = k_{dr}\ n(e)/v_o$ cm^{-1}, when it is assumed that the systematic inflow velocities of the initial and product species are equal and independent of y. This approximation permits useful simplifications, but is not strictly valid. The initial density will be $n_o(H_2) \simeq 200\ n(e)$ for a molecular shell at T=100 K in approximate pressure balance with an ionized nebula at $T_e=10^4$ K. The corresponding initial density of CO can be at least as large as $n_o(CO) \simeq 4\times10^{-4}\ n_o(H_2)$, even if half the carbon is in grains. We adopt v_o=10 km s^{-1}. The resulting column densities, $N(X) = \int_0^\infty n(X)\,dy$ cm^{-2}, computed from equations (1) and (2) are 2×10^{18}, 2×10^{16}, 4×10^{14}, and 7×10^{11} cm^{-2}, for H_2, H_2^+, CO, and CO^+, respectively. The column densities are quite insensitive to the value of n(e) in the range 10^3-10^5 cm^{-3}, because a decrease in $n_o(H_2)$ is largely compensated by a decrease in β in the integration over path length y.

The observable properties of these molecules depend upon their degree of vibrational and rotational excitation. The homonuclear species H_2 and H_2^+ lack permanent dipole moments and thus have long radiative lifetimes in excited states (cf. Black and Dalgarno 1976, Posen et al. 1982). H_2 will be substantially excited in nebulae. H_2^+, however, will exist primarily in v=0 because it is formed preferentially in that state by charge transfer and because excited molecules are removed rapidly by dissociative recombination. Indeed, if realistic, v-dependent rates of dissociative recombination are used, a ratio of populations n(v=0)/n(v=1) >25 is expected, and the total column density will be increased to $N(H_2^+) \simeq 10^{17}$ cm^{-2}, in harmony with the value suggested by the observations of Heap and Stecher (1981). Both CO and CO^+ will exist mostly in v=0, and the rotational populations of the latter should be subthermal because of its large dipole moment and very rapid dissociative recombination.

We conclude that the identification of H_2^+(v=0) in nebulae is plausible. If the identification is correct, H_2^+ must be accompanied by observable concentrations of H_2, CO, and CO^+. The predicted column density of nebular H_2 is 40 times smaller than that ascribed to a shocked neutral layer in NGC 7027 (Smith et al. 1981, Beckwith et al. 1980) and could have escaped detection through infrared emission lines. The nebular H_2 will, however, produce ultraviolet absorption lines in the spectra of central stars: e.g., $N(H_2)=2\times10^{18}$ cm^{-2} at $T=10^4$ K will yield an equivalent width W=0.2 Å (including curve of growth effects) in the $B^1\Sigma_u^+ - X^1\Sigma_g^+$ 0-3 R(7) line at 1298.2 Å. Ultraviolet absorption lines of CO have been sought, but not detected, in the spectrum of the central star of IC 418 (Clavel and Flower 1980). The quoted upper limit, N(CO) $\lesssim 3\times10^{14}$ cm^{-2}, is similar to the model prediction; therefore, a more sensitive search for CO in nebulae suspected of harboring H_2^+ might prove fruitful. Because of its subthermal excitation, the small amount of CO^+ might be detectable, either by the comet tail ($A^2\Pi - X^2\Sigma^+$) lines in absorption against the central star, or by millimeter wavelength rotational lines. It is interesting that rotational lines of CO^+ have been observed towards the Orion Nebula (Erickson et al. 1981), and that CO^+ has been suggested as the source of possible absorption bands in the spectra of several stars in and near the Orion Trapezium (Tamura and Ishii 1977). The flow of molecules from the Orion Molecular Cloud into the nebula is expected to produce nebular molecules in the manner outlined above for planetary nebulae: perhaps the CO^+ rotational lines in Orion can be explained in this way also. Searches for CO^+ in planetary nebulae would be interesting.

A simplified photochemical model has been constructed for predicting the abundances of H, H_2, C, C^+, CO, OH, and other simple species as functions of distance through the molecular envelope (details will be published elsewhere). The distances beyond the ionization front at which hydrogen becomes mostly H_2 and carbon becomes mostly CO are sensitive to the total density of the envelope. These models can be compared with observational data on the envelope around NGC 7027. Pottasch et al. (1982) observed no H 21 cm line and derived a strong limit $N(H)<1.2\times10^{20}$ cm^{-2} and a weak (model-dependent) limit $N(H)\lesssim(2-4)\times10^{19}$ cm^{-2}. The CO column density is $N(CO)\gtrsim2\times10^{17}$ cm^{-2} (Thronson 1982a). These constraints are satisfied by the model only if $n=n(H)+2n(H_2)>4\times10^5$ cm^{-3} (strong H limit) or $n>10^6$ cm^{-3} (weak H limit) near the boundary. Beyond the C/CO transition distance, the density can decrease without serious effects on the abundances until the molecules become exposed to unattenuated galactic background radiation that enters from outside. An additional constraint is supplied by an upper limit on the intensity of the C 76α radio recombination line (Bignell 1974). At low temperatures, $T\simeq20$ K, implied by the excitation temperature of CO (Thronson 1982a) and the relative sizes of the CO envelope and far IR emitting region (Moseley 1980), the limit $\int n(e)\, n(C^+)\, dr < 3.2$ cm^{-6} pc requires a model with $n>2\times10^6$ cm^{-3} near the boundary. At densities this high, the important chemical time-scales are short compared with a nebular lifetime, and steady-state abundances are realistic. At lower densities, more elaborate time-dependent models will be required.

In summary, even though only a few molecules have been observed in a small number of planetary nebulae, the existing observations augur well for future molecular studies of planetary nebulae and their precursors.

This research has been supported by the U.S. National Science Foundation (Grant AST-81-14718). I am very grateful to H.A. Thronson and S.C. Beck for providing useful information prior to publication, and to A. Dalgarno, T.P. Stecher, and E.F. van Dishoeck for helpful comments.

REFERENCES

Aitken, D.K., and Roche, P.F.: 1982, Monthly Notices Roy. Astron. Soc. 200, 217.
Auerbach, D., Cacak, R., Caudano, R., Gaily, T.D., Keyser, C.J., McGowan, J.W., Mitchell, J.B.A., and Wilk, S.F.J.: 1977, J. Phys. B 10, 3797.
Beck, S.C. and Beckwith, S.V.W.: 1983, this volume, p. 103.
Becklin, E.E., Neugebauer, G., and Wynn-Williams, C.G.: 1973, Astrophys. Letters 15, 87.
Beckwith, S., Persson, S.E., and Gatley, I.: 1978, Astrophys. J. Letters 219, L33.
Beckwith, S., Neugebauer, G., Becklin, E.E., Matthews, K., and Persson, S.E.: 1980, Astron. J. 85, 886.
Bignell, R.C.: 1974, Astrophys. J. 193, 687.
Black, J.H.: 1978, Astrophys. J. 222, 125.
Black, J.H. and Dalgarno, A.: 1976, Astrophys. J. 203, 132.
Black, J.H., Porter, A., and Dalgarno, A.: 1981, Astrophys. J. 249, 138.
Clavel, J. and Flower, D.R.: 1980, Monthly Notices Roy. Astron. Soc. 190, 1P.
Clegg, R.E.S. and Wootten, H.A.: 1980, Astrophys. J. 240, 828.
Dabrowski, I. and Herzberg, G.: 1977, Trans. N. Y. Acad. Sci. 38, 14.
Dalgarno, A.: 1981, Phil. Trans. Roy. Soc. London A 303, 513.
Davis, L.E., Seaquist, E.R., and Purton, C.R.: 1979, Astrophys. J. 230, 434.
Derkits, C., Bardsley, J.N., and Wadehra, J.M.: 1979, J. Phys. B 12, L529.
Dunn, G.H.: 1968a, Phys. Rev. 172, 1.
Dunn, G.H.: 1968b, Joint Inst. for Lab. Astrophys. Rept. No. 92.
Erickson, N.R., Snell, R.L., Loren R.B., Mundy, L., and Plambeck, R.L.: 1981, Astrophys. J. Letters 245, L83.
Feibelman, W.A., Boggess, A., McCracken, C.W., and Hobbs, R.W.: 1981, Astron. J. 86, 881.
Finzi, A. and Yahel, R.: 1978, Astron. Astrophys. 68, 173.
Flower, D.R. and Roueff, E.: 1979, Astron. Astrophys. 72, 361.
Forrest, W.J., Houck, J.R., and McCarthy, J.F.: 1981, Astrophys. J. 248, 195.
Gahm, G.F., Lindgren, B., and Lindroos, K.P.: 1977, Astron. Astrophys. Suppl. 27, 277.
Goldreich, P. and Scoville, N.: 1976, Astrophys. J 205, 144.
Goss, W.M., N.-Q.-Rieu, Winnberg, A.: 1973, Astron. Astrophys. 29, 435.
Hardebeck, E.G.: 1972, Astrophys. J. 172, 583.
Haser, L.: 1957, Bull. Acad. Roy. Belgique 43, 740.
Heap, S.R. and Stecher, T.P.: 1981, in R.D. Chapman (ed.), *The Universe at Ultraviolet Wavelengths*, NASA Conference Publ. 2171, p. 657.
Hollenbach, D.J. and Shull, J.M.: 1977, Astrophys. J. 216, 419.
Huggins, P.J. and Glassgold, A.E.: 1982, Astrophys. J. 252, 201.

Iben, I.: 1981, Astrophys. J. 246, 278.
Karpas, Z., Anicich, V., and Huntress, W.T., Jr.: 1979, J. Chem. Phys. 70, 2877.
Knapp, G.R., Phillips, T.G., Leighton, R.B., Lo, K.Y., Wannier, P.G., Wootten, H.A., and Huggins, P.J.: 1982, Astrophys. J. 252, 616.
Kwan, J.: 1977, Astrophys. J. 216, 713.
Kwok, S., Purton, C.R., and FitzGerald, M.P.: 1978, Astrophys. J. Letters 219, L125.
Kwok, S. and Purton, C.R.: 1979, Astrophys. J. 229, 187.
Lafont, S., Lucas, R., and Omont, A.: 1982, Astron. Astrophys. 106, 201.
Lo, K.Y. and Bechis, K.P.: 1976, Astrophys. J. Letters 205, L21.
London, R., McCray, R., and Chu, S.I.: 1977, Astrophys. J. 217, 442.
McCabe, E.M., Smith, R.C., and Clegg, R.E.S.: 1979, Nature 281, 263.
Mentzoni, M.H. and Donohoe, J.: 1969, Can. J. Phys. 47, 1789.
Moseley, H.: 1980, Astrophys. J. 238, 892.
Mufson, S.L., Lyon, J., and Marionni, P.A.: 1975, Astrophys. J. Letters 201, L85.
Posen, A.G., Dalgarno, A., and Peek, J.M.: 1982, Atomic Data Nucl. Data Tables,(in press).
Pottasch, S.R., Goss, W.M., Arnal, E.M., and Gathier, R.: 1982, Astron. Astrophys. 106, 229.
Roberge, W.G. and Dalgarno, A.: 1982, Astrophys. J. 255, 489.
Rood, R.T., Wilson, T.L., and Steigman, G.: 1979, Astrophys. J. Letters 227, L97.
Scalo, J.M. and Slavsky, D.B.: 1980, Astrophys. J. Letters 239, L73.
Scrimger, J.N., Lowe, R.P., Moorhead, J.M., and Wehlau, W.H.: 1978, Publ. Astron. Soc. Pacific 90, 257.
Seaton, M.J.: 1980, Quart. J. Roy. Astron. Soc. 21, 229.
Silverglate, P., Zuckerman, B., Terzian, Y., and Wolff, M.: 1979, Astron. J. 84, 345.
Smith, H.A., Larson, H.P., and Fink, U.: 1981, Astrophys. J. 244, 835.
Sume, A. and Irvine, W.M.: 1977, Astron. Astrophys. 60, 345.
Tamura, S. and Ishii, H.: 1977, 21st Coll. Astrophys. Liège, p. 66.
Thronson, H.A.: 1981, Astrophys. J. 248, 984.
Thronson, H.A.: 1982a, Astrophys. J. (in press).
Thronson, H.A.: 1982b, Astron. J. (in press).
Thronson, H.A. and Lada, C.J.: 1982, Publ. Astron. Soc. Pacific 94, 226.
Tokunaga, A.T. and Young, E.T.: 1980, Astrophys. J. Letters 237, L93.
Treffers, R.R., Fink, U., Larson, H.P., and Gautier, T.N.: 1976, Astrophys. J. 209, 793.
Turner, B.E.: 1971, Astrophys. Letters 8, 73.
Zhdanov, V.P.: 1980, J. Phys. B 13, L311.
Zuckerman, B.: 1978, in Y. Terzian (ed.), *Planetary Nebulae, Observations and Theory*, D. Reidel, Dordrecht, p. 305.
Zuckerman, B.: 1980, Ann. Rev. Astron. Astrophys. 18, 263.
Zuckerman, B., Gilra, D.P., Turner, B.E., Morris, M., and Palmer, P.: 1976, Astrophys. J. Letters 205, L15.
Zuckerman, B., Palmer, P., Morris, M., Turner, B.E., Gilra, D.P., Bowers, P.F., and Gilmore, W.: 1977, Astrophys. J. Letters 211, L97.
Zuckerman, B., Palmer, P., Gilra, D.P., Turner, B.E., and Morris, M.: 1978, Astrophys. J. Letters 220, L53.

OSTERBROCK: What is the evidence that the H_2 v = 1 → 0 emission is concentrated near the boundary of NGC 7027? What is meant by the boundary - that of the ionized region or that of the ionized, high density region?

BLACK: The spatial distribution of the S(1) line intensity was mapped by Beckwith et al. (1980) with resolution adequate to show a marked contrast with the distribution of most of the atomic line emission but inadequate to define precisely the H_2 emitting region. The observations are consistent with a narrow emitting zone near the boundary of the ionized region.

REAY: Using the U.K. Infrared Flux Collector (Hawaii), my colleagues and I have detected H_2 S(1) v = 1 → 0 emission from NGC 6302 at one position only, near the centre where the (so far undetected) central star would be expected to be. Flux levels $\simeq 5 \times 10^{-18}$ W cm^{-2} μm^{-1}.

BLACK: This is extremely interesting, particularly in view of the detection of H I 21 cm line absorption in the same direction (L.F. Rodriguez and J.M. Moran, contributed poster paper, this volume). Further study of the distribution of neutral gas associated with NGC 6302 may help to partly explain the complicated dynamical structure of the nebula.

CLEGG: At UCL, we have just completed a survey of the ultraviolet fluxes from about twenty PN central stars with the IUE satellite. However, we were not able to confirm the detection of H_2^+ absorption by Heap, Stecher and others. In particular, many stars cooler that about 50 000 K show line blanketing between 1200 Å and 1500 Å.

HEAP: Yes, I agree that the effect of line blanketing can be important, although this should be diminished in a differential analysis, which we did. Another source of uncertainty in the identification of H_2^+ in NGC 6210 is the extinction by dust internal to the nebula.

ZUCKERMAN: I have been using the 20 m radiotelescope at Onsala (Sweden) with Olofsson and Johansson (Onsala), Rieu (Meudon) and Sopka (Maryland) to study HCN in Red Giant stars and a few PN. We have detected weak J = 1 → 0 HCN emission from NGC 7027. This is the first time that a polyatomic molecule has been seen in a PN environment. There has been some controversy in the literature as to whether NGC 7027 is oxygen-rich (O/C > 1) or carbon-rich (C/O > 1). The detection of HCN shows reasonably conclusively that the envelope is carbon-rich or S-type (C/O ≈ 1) since this molecule is expected to have an extremely small abundance in an oxygen-rich environment. Observation of HCN in Red Giant stars is consistent with this expectation.

BLACK: This is very encouraging news, and I hope that more such observations will be made. The absence of thermal Si O emission (Thronson and Lada, 1982) would also seem to support C/O ≳ 1 in the envelope of NGC 7027.

KALER: There is evidence that C → N conversion takes place on the AGB which affects the $^{12}C/^{13}C$ ratio. It is very important to observe this ratio in high N-abundance nebulae such as NGC 2440. However, the observed CO may relate to an earlier stellar wind, and the measured $^{12}C/^{13}C$ ratio may not reflect that in the nebular proper. Have you any comments?

BLACK: I agree that the molecular material may often have been ejected earlier than most of the nebular gas. The significance of the molecular isotopic abundance ratios cannot be established without more and better observations. Attempts to measure isotope abundances of other elements, such as N, although very difficult, should be made.

MATHIS: You showed processes for producing H_2 by gas-phase reactions. Do these dominate the production of H_2 in PN? Do you still believe that H_2 is produced almost entirely on grains in the general interstellar medium?

BLACK: Although it is uncertain, inside nebulae the dust may be too hot for H_2 to form on dust at a sufficiently high rate to compete with gas-phase processes. In cool, extended molecular envelopes, on the other hand, formation of H_2 on dust grains probably dominates, although the formation time scale can be long compared with a nebular lifetime if the density is low. Gas-phase processes are totally inadequate to account for the large amounts of H_2 observed in the general interstellar medium, but are probably responsible for most other molecules.

INFRARED SPECTROSCOPY OF THE TRANSITION OBJECTS CRL 618 AND CRL 2688

S.C. Beck, S.V.W. Beckwith
Astronomy Department, Cornell University

We have observed CRL 618, an infrared object believed to be in transition between the red giant and planetary nebula stages, in seven lines of vibrationally excited H_2 and the Brackett α and γ lines of ionized hydrogen. The H_2 observations were made in 1979 and repeated in 1982, extending over more than four years monitoring of H_2 emission from this source. The intensity of H_2 emission approximately tripled between August 1977 and November 1979 and may have increased by another 15-20 per cent between November 1979 and February 1982. This increase in emission occurred during the same time period that the radio free-free flux roughly doubled (Kwok and Feldman, 1981). The object may now be in a more quiescent phase following a strong flare. The H_2 lines appear to be in thermal equilibrium at ≈ 2000 K and to suffer only small extinction. The Brackett α and γ lines are more heavily obscured, with ≈ 2 mag. of 2.17 μm extinction. These results support the model that the compact H II region is behind and expanding into the dense molecular material where the H_2 is found. The mass of hot H_2 is ≈ 4×10^{-4} $M_\odot$ and of total ejecta ≈ .1 $M_\odot$, so the mass-loss rate is probably > 5×10^{-5} $M_\odot$ y^{-1}.

CRL 2688 is apparently at an earlier stage than CRL 618 and has no observed H II region. The source has been mapped in the near-infrared continuum and in the 1-0 S(1) line of H_2, and the spatial distributions of intensity show clearly how the H_2 is excited in an outflow from a central disk of dense material. Continued monitoring of these sources may provide further detail on the expansion and brightening of the H II regions and neutral envelopes as the central stars evolve rapidly into planetary nebulae.

Kwok, S. and Feldman, P.A.: 1981, Ap. J. Lett., 247, L67.

COHEN: Could the failure to detect H_2 in the fainter lobe of CRL 618 be due to the extra extinction of that lobe?

BECK: The extra extinction to that lobe is very small (A_v = 2 mag, equivalent to about 0.2 mag at our wavelength). We detected H_2 in the bright lobe with a signal/noise ratio of over 20, so the extra extinction to the faint lobe will not explain our failure to detect H_2.

KWOK: Dr. Bignell and I have been monitoring the radio flux of GL 618 at the VLA during the last two years, and the radio flux increase has indeed slowed down.

BECK: Will you continue to monitor it?

KWOK: Yes.

WANNIER: In CRL 618 and CRL 2688 (as well as NGC 7027, discussed by the previous speaker), we have mapped the CO $J = 1 \rightarrow 0$ line using the 20 m telescope at Onsala. We have found that the molecular clouds have sizes of 40-50 arcsec in all three sources, with no indications of asymmetry. These results confirm previous tentative observations of extended CO emission. This work was carried out in collaboration with H. Olofsson and L.-A. Nyman (Onsala) and with R. Sahai (Caltech).

OBSERVATIONS OF DUST IN PLANETARY NEBULAE

M. J. Barlow

Dept. of Physics and Astronomy, University College London,
Gower St., London, WC1E 6BT, England

1. INTRODUCTION

This review summarises those observations which relate to the existence of dust in planetary nebulae. The most direct evidence for dust, namely infrared emission by grains, is considered in Section 2. Section 3 describes a variety of optical and ultraviolet observations which bear more indirectly on the properties of the dust, while Section 4 discusses the power sources of the observed infrared emission. The peculiar nebulae A 30 and A 78 are considered in Section 5, and Section 6 concludes with a discussion of the various infrared features which have been observed in emission from planetary nebulae.

2. INFRARED PHOTOMETRIC STUDIES OF PLANETARY NEBULAE

Since the discovery by Gillett, Low and Stein (1967) of 10 μm emission from NGC 7027, which was greatly in excess of the expected free-free and line fluxes, the study of dust in planetary nebulae has virtually been a study of dust in NGC 7027. Although a variety of observational data are now available for other nebulae, they still lack the wealth of detail afforded by the brightness of NGC 7027 in the infrared. Table 1 summarises the published ground-based photometry of planetary nebulae.

Krishna Swamy and O'Dell (1968) interpreted the infrared continuum from NGC 7027 in terms of emission from small graphite particles heated by resonantly trapped Lyman-α photons. They were not able to distinguish whether the grains were inside the ionized zone or in a surrounding neutral shell. They suggested that the grains could have originated in the envelope of a precursor red giant star, which is now the accepted explanation. Terzian and Sanders (1972) calculated the expected IR flux distributions under the assumption that the grains resided in a neutral shell and predicted mean grain temperatures of ∿85 K, with fluxes peaking at ∿35 μm. The 10-18 μm observations of Cohen and Barlow (1974) did not seem to support the prediction,

D. R. Flower (ed.), Planetary Nebulae, 105–128.

TABLE 1: GROUND-BASED INFRARED PHOTOMETRY OF PLANETARY NEBULAE

OBSERVERS	REFERENCE	NUMBER OF PN DETECTED AT EACH WAVELENGTH (μm)						OBJECT NAME(S)
		1.6	2.2	3.5	10	11	20	
Gillett, Low, Stein	1967,Astrophys.J.,149,L97				1	1		NGC 7027
Woolf	1969,Astrophys.J.,157,L37					5		
Gillett, Stein	1970,Astrophys.J.,159,817			2		2		NGC 6572, BD+30°3639
Neugebauer, Garmire	1970,Astrophys.J.,161,L91			1	1			NGC 7027
Gillett, Knacke, Stein	1971,Astrophys.J.,163,L57			3		3		IC 4997, VV8, FG Sge, Sh 2-71
Gillett, Merrill, Stein	1972,Astrophys.J.,172,367					15	3	
Willner, Becklin, Visvanathan	1972,Astrophys.J.,175,699	11	15	8				
Persson, Frogel	1973,Astrophys.J.,182,503	19	24	11				
Danziger, Frogel, Persson	1973,Astrophys.J.,184,L29	1	1	1	1		1	NGC 6302
Allen	1973,M.N.R.A.S.,161,145	5	5					
Becklin, Neugebauer, Wynn-Williams	1973,Astrophys.Lett.,15,87				1		1	NGC 7027
Khromov	1974,Sov.Astr.,18,195	15	5					
Webster, Glass	1974,M.N.R.A.S.,166,491	2	2	2				He 2-113, CPD-56°8032
Allen, Glass	1974,M.N.R.A.S.,167,337	63	100	28				
Allen	1974,M.N.R.A.S.,168,1	7	15	8	2	2	2	
Cohen, Barlow	1974,Astrophys.J.,193,401		10	16	51	15	40	
Cohen, Barlow	1975,Astrophys.Lett.,16,165	1	1	1	1			NGC 2346
Cohen	1975,M.N.R.A.S.,173,489			1	1	1	1	He 2-113
Dyck, Simon	1976,P.A.S.P.,88,738						2	NGC 7027, BD+30°3639
Cohen, Hudson, O'Dell, Stein	1977,M.N.R.A.S.,181,233	2	2	2				A 30, A 78
Cohen, FitzGerald, Kunkel, Lasker, Osmer	1978,Astrophys.J.,221,151	1	1	1	1	1	1	Mz 3
Cohen, Barlow	1980,Astrophys.J.,238,585				23	18	20	

yielding mean colour temperatures of ∿190 K. It later became clear that broadband ($\Delta\lambda$ ∿ 5 μm) observations of planetary nebulae at 10 μm can significantly overestimate true continuum levels due to the presence of numerous emission lines and features in this region, thus yielding too high 10-18 μm colour temperatures. More importantly though, airborne far-infrared observations by Telesco and Harper (1977), McCarthy, Forrest and Houck (1978) and Moseley (1980) have shown that (1) the IR energy distributions of most planetaries do in fact peak near 30 μm, (2) the flux distributions shortward of the peak cannot be fitted by single temperature grain models and therefore (3) ground-based 10-20 μm observations sample a hotter dust component than 30-100 μm observations.

Telesco and Harper (1977) found that their far-infrared data for NGC 7027 could be fitted by a 95 K blackbody with a λ^{-2} emissivity. Classical Mie theory calculations predict such an emissivity law for small graphite particles. McCarthy et al (1978) obtained 15-40 μm spectrophotometry of NGC 7027 and found that a 90 K blackbody with λ^{-2} emissivity yielded the best fit. Shortward of 20 μm, the 90 K spectrum left an excess of smooth continuum, establishing the presence of hotter dust. The 90 K spectrum, which makes only a small contribution at 10 μm, accounts for 68% of the total infrared flux. Of the remaining 32%, 3% is emitted in the unidentified features between 3.3 and 11.3 μm and 29% is emitted by hotter dust. Kwok (1980) has fitted this excess short wavelength continuum with a λ^{-2} emissivity, 230 K blackbody spectrum.

Moseley (1980) has observed thirteen planetaries at four wavelengths between 37 and 108 μm. Most of the flux distributions were similar to that of NGC 7027, despite the fact that the range of nebular densities and ages was large. The mean colour temperature was 80 K (for λ^{-2} emissivity). Figure 1 shows the IR flux distributions of some of the planetary nebulae observed by Moseley. IC 418 was found to have a more sharply peaked flux distribution than the other nebulae. Recent spectrophotometric observations by Forrest, Houck and McCarthy (1981) have shown that this was due to the presence of a strong emission feature in Moseley's 28-52 μm passband.

Many attempts have been made to derive the masses of emitting dust in planetary nebulae. As pointed out by Panagia (1975) and Mathis (1978), to do this accurately requires a knowledge of the composition and size distribution of the emitting particles, since differently sized grains will have different emissivities and temperatures in the same radiation field. Masses derived assuming a single grain temperature and size are thus susceptible to large errors, especially at shorter wavelengths. The errors should decrease at longer wavelengths where emissivities approach the Rayleigh limit and the cool grain component dominates. Telesco and Harper (1977) have shown that for grains with a ν^n emissivity, the total mass of IR emitting dust can be derived from the relation:

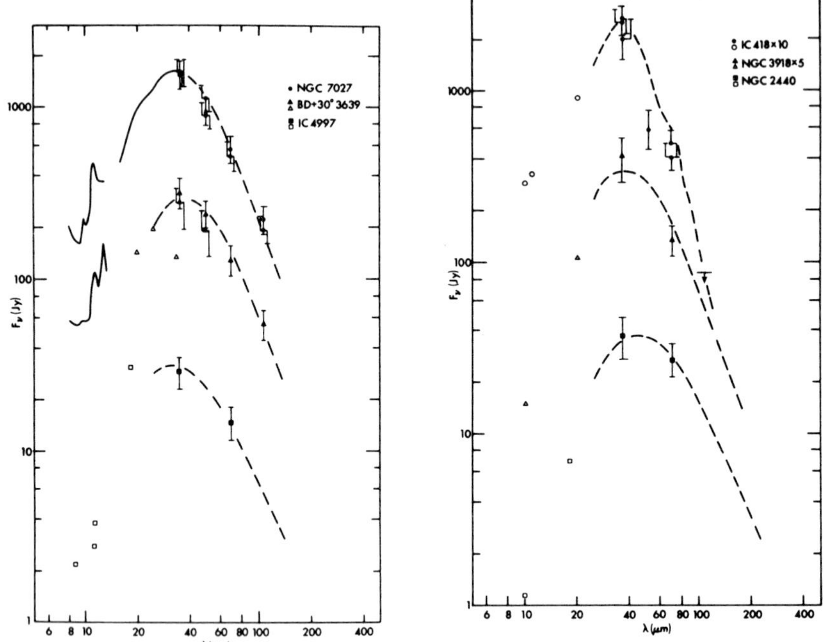

Figure 1. Infrared flux distributions of six planetary nebulae in the form of F_ν versus λ. Dashed lines indicate fits to the far-infrared photometric points (Figure from Moseley 1980).

$$M_d = \frac{a\rho_d}{Q(\nu_{max})} \cdot \frac{\nu^n_{max} L_{IR}(\lambda > 17\mu m)}{K(n) T_d^{4+n}} \quad (1)$$

where a and ρ_d are the radius and mean density of the emitting grains of temperature T_d which have an emissivity $Q(\nu_{max})$ at the peak of the infrared energy distribution. K(n) is a numerical constant for each value of n and L_{IR} is the luminosity of the dominant grain component. Moseley (1980), by observing out to 108 μm, was able to make use of the expression derived by Hildebrand et al (1977) in which grains are assumed to emit in the Rayleigh-Jeans limit at long enough wavelengths:

$$M_d = \frac{a\rho_d}{Q(\nu)} \cdot \frac{4D^2F_\nu}{3B(\nu, T_d)} \propto \frac{a\rho_d}{Q(\nu)} \cdot \frac{L_\nu}{T_d} \quad (2)$$

where D is the distance of the nebula. This method is less sensitive to the grain temperature, but an estimate of the emissivity is still required.

The observed form of the infrared spectrum of NGC 7027 suggests an n=2 grain emissivity, for which graphite seems appropriate. However, Forrest, McCarthy and Houck (1980) have noted that the recent determination of the far-infrared constants of graphite by Philipp (1977) would,

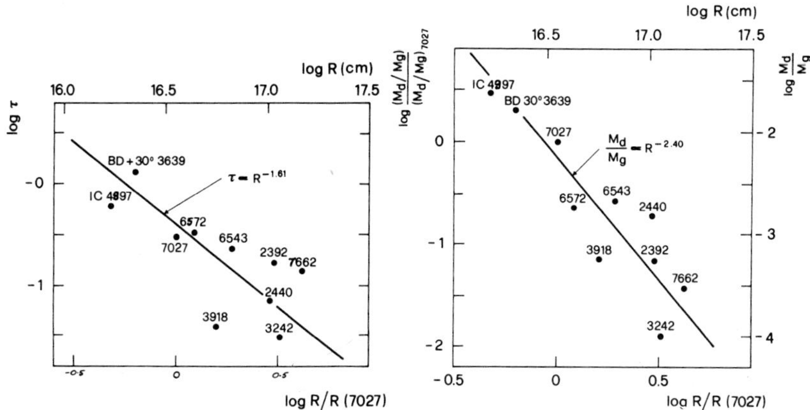

Figure 2. Left hand figure: the UV optical depth of internal dust in ten planetary nebulae versus nebular radius. Right hand figure: the dust-to-gas ratio for the ten planetary nebulae versus nebular radius (Figures from Natta and Panagia 1981).

if appropriate for small grains, predict an n=3 emissivity beyond 20 μm. On the other hand Koike, Hasegawa and Manabe (1980) have experimentally found an n=1 emissivity for small amorphous carbon grains.

In order to overcome the uncertainty associated with the absolute values of far-infrared grain parameters, Natta and Panagia (1981) have carried out a differential analysis of the planetary nebulae observed by Moseley. The analysis, based on the methods they had already applied to HII regions, assumed the dust to have an internal UV extinction, τ_{int}, in the ionized zone. Their results are shown in Figure 2. They found that the dust column density, τ_{int}, decreased with nebular radius as $R^{-1.6}$ (Figure 2). The gas column density, on the other hand, decreased as $R^{-0.46}$, similar to the $R^{-0.5}$ law expected for ionization bounded nebulae, thus providing evidence for increasing dust depletion with increasing nebular age. The absolute values of τ_{int} for NGC 7027 and NGC 7662 were 0.3 and 0.14, similar to the values derived from UV observations of CIV λ1549 (see section 3.4). The dust-to-gas ratio was found to decrease as $R^{-2.4}$ (Figure 2). Evidence was also found for a decrease in the mean grain radius with increasing nebular radius. Natta and Panagia noted that the low values of τ_{int} which they derived for most nebulae implied that dust does not affect the ionization structure of planetary nebulae significantly. They also argued that the low derived values of M_d/M_{gas} ($\leq 3 \times 10^{-4}$) for the older nebulae implied that they could not enrich the dust content of the interstellar medium, but would dilute it instead.

3. OTHER OBSERVABLE EFFECTS OF DUST IN PLANETARY NEBULAE

3.1 Red/blue line profile asymmetries

Osterbrock (1974) pointed out that if significant internal dust existed in the ionized region of a planetary nebula, it would produce asymmetric emission line profiles, since the red wing, originating from the far side of the nebula, would suffer more extinction than the blue wing. Since an asymmetric distribution of gas could give rise to spurious results if only a single profile was analysed, Osterbrock instead compared the profiles of three different Balmer lines. For normal extinction laws, the effect of internal extinction would be most pronounced on the profile with the shortest rest wavelength. However the Hα, Hβ and Hγ lines of NGC 7027 showed no noticeable differences in their profiles and Osterbrock set a conservative upper limit of $\tau(H\beta) < 0.6$.

Hicks, Phillips and Reay (1976) analysed the less thermally broadened [OIII] λ5007 profile at various points on NGC 7027. A red/blue asymmetry was found, consistent with a non-uniform gas distribution or $\tau_{int}(\lambda=5007) \sim 0.34$. However, Hicks et al also pointed out that the effects of internal extinction could be mimicked by a non-uniform distribution of external dust across the face of an inclined spheroidal nebula, preferentially extinguishing the emission from one side (and velocity). NGC 7027 is known to have just such an asymmetric distribution of external dust since the 5 GHz radio contours of Scott (1973) are very symmetric, whereas the optical image is cut off on the southern side. The spatial distribution of extinction across NGC 7027 has been mapped by Atherton et al (1979). The external dust appears to be associated with the dense molecular CO envelope discovered by Mufson, Lyon and Marionni (1975).

Doughty and Kaler (1982) have analysed the extensive coudé plate collection for seven planetaries observed by O. C. Wilson. They compared both short and long wavelength lines in order to eliminate the effects of asymmetric gas distributions. For NGC 7027 they derived $\tau(H\beta) = 0.15$, under the assumption that the dust is internal.

3.2 Scattered light continua

Galactic Population I HII regions have long been known to exhibit optical continua dominated by dust-scattered starlight (e.g. O'Dell and Hubbard 1965) and in the UV the observed ratio of scattered light to nebular continuum is even larger (e.g. Perinotto and Patriarchi (1980) for the Orion Nebula). To my knowledge, no continuum in excess of that attributable to atomic processes has ever been observed from the ionized region of a planetary nebula.

Page (1936) studied the continuous and Balmer line spectrum of NGC 7662 on plates taken with the DAO 72" reflector. He found that the nebular continuum in the 3900-4800 Å region was much stronger than the predicted Paschen continuum and suggested that starlight scattered by dust particles within the nebula was responsible. This represents one of the first suggestions of the existence of dust in planetary nebulae. However, allowance for the existence of a hydrogen two-photon continuum later removed the discrepancy (Seaton 1955). The line to continuum ratios derived by Page agree to within 20% with the ratios predicted using the nebular parameters of Harrington et al (1982).

Persson and Frogel (1973) observed the Hβ line to continuum ratio in BD+30°3639, one of the intrinsically strongest infrared emitters amongst planetary nebulae. Their initial results indicated the detection of a dust-scattered continuum (corresponding to τ_s(Hβ) $\sim$ 0.07) but later analysis, published as an erratum, weakened the evidence.

Telesco and Harper (1977) noted the existence of a faint Hβ halo in electronographic maps of NGC 7027 made by Coleman, Reay and Worsick (1975) and suggested that the halo was due to light scattered by dust in the neutral shell surrounding the ionized zone. A deeper exposure in Hα by Atherton et al (1979) has shown the reflection nebula to have a major axis diameter of $\sim$50 arcsec (versus $\sim$13 arcsec in the radio), which is comparable to the diameter of the CO emitting region mapped by Mufson et al (1975). To test this interpretation, optical polarisation maps would be very useful. It would also be interesting to obtain polarisation measurements of the giant haloes surrounding many planetary nebulae, in order to see whether the haloes emit intrinsically or are due only to the scattering by dust of light from the nebular cores. Such dust could have been produced during the mass loss phase of a progenitor red giant star.

3.3 Abnormal extinction laws

The infrared continuum emission from planetary nebulae has often been attributed to graphite grains, which have also been proposed to be responsible for the interstellar 2200 Å extinction feature. Pottasch et al (1977) investigated the strength of the 2200 Å feature towards 30 planetary nebulae using ANS photometric data. The extinctions that they derived from the 2200 Å feature, assuming a standard UV extinction law, were in very good agreement with the extinctions derived from radio-Hβ fluxes, with no evidence for an abnormal component. They concluded that the infrared emitting dust inside planetary nebulae did not have large UV optical depths. Adams and Barlow (these Proceedings) have measured the strength of the 2200 Å feature towards four planetary nebulae which show 10 μm silicate emission features. They found the 2200 Å strengths to be weaker than predicted by the radio-Hβ extinctions, consistent with some of the reddening being due to local silicate dust without a corresponding graphite component.

On the basis of a radio map published by Sistla and Kaftan-Kassim (1976) and its comparison with a Hβ map by Phillips, Reay and Worsick (1979), it has been suggested that the region around the central star of IC 3568 is more highly reddened than the nebula as a whole. However,

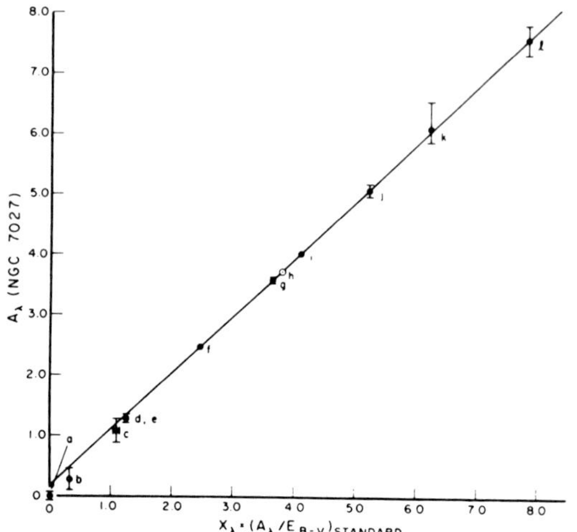

Figure 3. Extinction, A_λ, of NGC 7027 in magnitudes, deduced from radio continuum and HI and HeII recombination line measurements, plotted against standard interstellar extinction at the same wavelengths (Figure from Seaton 1979).

---

from an analysis of UV, optical and radio data, Harrington and Feibelman (1982) have found no evidence for a difference in the reddening suffered by the central star compared with the nebular lines.

Seaton (1979) has analysed the relative intensities of HI and HeII recombination lines emitted by NGC 7027 in the infrared, optical and ultraviolet. He found remarkable agreement between the standard Galactic extinction law and the law derived for NGC 7027 (Figure 3). The straight line in Figure 3 does not pass through the origin and Seaton showed that this could be explained by heavier extinction over a fraction of the nebula, with the same standard extinction law. Since the ionized nebula is believed to be carbon-rich (Section 6), as by extension are the surrounding neutral shell and dust, it might seem surprising that the extinction law towards NGC 7027 appears to be normal, because interstellar dust is thought to consist of a mixture of both graphite and silicate grains (e.g. Mathis, Rumpl and Nordsieck, 1977). However, Kwok (1980) has pointed out that if, prior to PN envelope ejection, a red giant star goes from an oxygen-rich mass loss phase producing silicate grains to a carbon-rich phase producing graphite grains, the silicate grains would be too far out and too cool to emit appreciably in the infrared during the planetary nebula phase. Since red giant stars are thought to be the main source of interstellar dust, it might not, therefore, be surprising to find a normal reddening law through the envelope surrounding NGC 7027.

3.4 Depletion of UV resonance lines by internal dust

Bohlin, Marionni and Stecher (1975) obtained the first UV spectrum of NGC 7027 and found that the CIV λ1549 resonance doublet was a factor of three weaker, relative to CIII] λ1909, than predicted by model nebula calculations. They ascribed the discrepancy to dust absorption of the resonantly scattered CIV photons and derived a radial dust optical depth within the ionized zone of $\tau_d(1549) \sim 0.2$, using the resonance line transfer results of Panagia and Ranieri (1973). Their nebular model had an ionized hydrogen column density of $N_H = 1.9 \times 10^{21}$ cm^{-2} and they noted that this column density, combined with a normal interstellar reddening law and colour excess ratio, $N_H/E_{B-V} = 5.4 \times 10^{21}$

cm^{-2} mag^{-1}, would predict $\tau_d(1549) = 2.6$, so that the dust within the ionized zone must be depleted by a factor of ten if it has normal reddening characteristics.

Harrington et al (1982) have carried out a very extensive analysis of the optical and UV spectrum of NGC 7662. By comparing their model predictions with the relative strengths of CIII λ2297 and CIV λ1549, they inferred that CIV λ1549 was depleted by a factor of three by internal dust (see Seaton, these Proceedings). Using the resonance line transfer results of Hummer and Kunasz (1980), along with the CIV line optical depths from their own model, they derived an internal dust optical depth of $\tau_d(1549) = 0.08$-0.13. These results may imply that a $\tau_d(1549)$ of only 0.1 is required to explain the depletion of CIV in NGC 7027, but a more extensive nebular analysis is required. The hydrogen column density in Harrington et al's model of NGC 7662 was 6×10^{20} cm^{-2} which, with a normal interstellar colour excess ratio and reddening law, would predict $\tau_d(1549) = 0.8$. Thus the dust in the ionized zone of NGC 7662 must also be severely depleted compared to interstellar ratios if it has normal reddening characteristics.

Clavel, Flower and Seaton (1981) have derived $\tau_d(1335) = 0.08$ for the ionized zone of IC 418, from an analysis of the relative intensities of the CII λ1335 and λ4267 lines (see Seaton, these Proceedings).

4. THE ENERGETICS OF DUST EMISSION FROM PLANETARY NEBULAE

Following the initial suggestion by Krishna Swamy and O'Dell (1968), the absorption by dust of resonantly trapped Lyman-α photons has often been suggested to be the power source for the infrared emission from planetary nebulae. For example, Cohen and Barlow (1974, 1980) proposed that Lyman-α absorption dominated the infrared energetics of the medium and low density planetaries in their survey.

Becklin, Neugebauer and Wynn-Williams (1973) mapped NGC 7027 at 10 μm (broadband) with a 1.8 arcsec beam and found that the infrared contours closely followed those of the 5 GHz map made by Scott (1973). They inferred that the emitting dust was uniformly mixed with the ionized gas. Since their derived ratio of total infrared to Ly-α flux, $L_{IR}/L_{Ly\text{-}\alpha}$, was much greater than unity, they argued that the dust was heated by direct absorption of stellar continuum photons shortward of the Lyman limit, with $\tau_d(UV) \sim 1$. This is much larger than the value of τ_d implied by the depletion of CIV λ1549.

Telesco and Harper (1977) measured the total IR flux of NGC 7027 to be 2.4×10^{-7} ergs cm^{-2} s^{-1}. With $S_\nu = 6.3$ Jy at 5 GHz and $T_e = 1.35 \times 10^4$ K, the relation of Rubin (1968) implies $L_{IR}/L_{Ly\text{-}\alpha} = 3.4$. Telesco and Harper noted that this ratio could be achieved if dust absorbed half of the stellar continuum photons plus all the Ly-α photons. They noted that, alternatively, the IR emission could be powered by the diffuse nebular radiation emitted longward of 912 Å, on the basis of the flux predictions of nebular models. With the availability of IUE data, it is now possible to investigate this case more fully. Table 2 presents the energy and photon budget of NGC 7027 between 912 Å and 5100 Å.

TABLE 2: ENERGY AND PHOTON BUDGET OF NGC 7027

Spectral Component	λ (Å)	$I_{Dered}-I_{Obs}$ ($\times 10^{-9}$ ergs cm^{-2} s^{-1})	$N_{Photons}$ (cm^{-2} s^{-1})
HI Ly-α	1216	70	4300
CIV	1548,1550	28.2 × 3	6600
CIII]	1909	19.3	1900
[OIII]	4959,5007	32.3	8100
Other lines	1200-5000	35.7	4600
Nebular continuum	1200-5000	20.8	2600
	Total:	262.7	28100

The observed UV and optical line fluxes have been taken from Perinotto, Panagia and Benvenuti (1980) and Kaler et al (1976). They have been dereddened by C(Hβ) = 1.43 (Seaton 1979), corresponding to A_V = 3.1, and the observed fluxes subtracted to give the total flux available for heating dust (column 3). The flux of CIV λ1549 was multiplied by three to allow for resonance line depletion. The nebular continuum was calculated using $n_e = 5.9 \times 10^4$ cm^{-3} and $T_e = 1.35 \times 10^4$ K. The Hβ flux and He^+ and He^{2+} abundances were taken from Miller and Mathews (1972). Column 4 gives the corresponding photon fluxes. The available dereddened nebular flux of 2.6×10^{-7} ergs cm^{-2} s^{-1} can comfortably supply the 2.4×10^{-7} ergs cm^{-2} s^{-1} observed in the infrared. One can therefore envisage a two-component model, whereby warm dust in the ionized region, with τ_d(UV) ∿ 0.1-0.2, is heated by Ly-α and CIV resonance photons, while colder dust in the surrounding neutral shell is heated by the remaining nebular photons. In this model the cool component could be identified with the $\lambda^{-2} B_\lambda$(90 K) dust, which supplies 70% of the total IR luminosity, while the warm dust in the ionized zone could be identified with the $\lambda^{-2} B_\lambda$(230 K) component, which has been shown by Kwok (1980) to fit the remaining continuum. Since the 90 K dust will make only a very small contribution at 10 μm, the dust continuum at 10 μm would be expected to have the same spatial distribution as the radio emission, in agreement with the observations of Becklin et al (1973). If the flux of Ly-α and CIV photons were depleted by a factor of 2.3 in the ionized region, instead of by the factor of 3 assumed in Table 2, the total energy available for powering dust emission would be 2.4×10^{-7} ergs cm^{-2} s^{-1}, with 30% absorbed in the HII region, exactly in keeping with the proportions derived observationally.

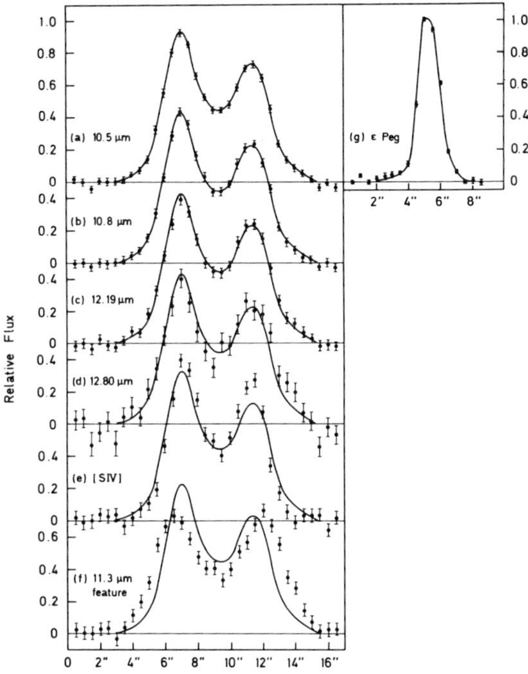

Figure 4. Spatial profiles along the minor axis of NGC 7027 at 4 continuum wavelengths (a - d), in the [SIV] line (e), and the 11.3 μm emission feature (f). The system response to a stellar source is shown in (g). Figure taken from Aitken and Roche (1982b).

Natta and Panagia (1981) have constructed a model in which the total infrared luminosity of NGC 7027 is explained by an optical depth $\tau_d(UV) = 0.3$ within the ionized region alone. In their model, the far infrared emission arises from large, cool grains in the ionized zone, while the 10 μm emission comes from smaller and hotter grains in the same region. One way to discriminate between the two alternative models might be to make an accurate estimate of $\tau_d(1549)$, while comparing the total infrared luminosity with the integrated stellar luminosity deduced from a Stoy-type analysis. Another would be to compare the spatial distribution of dust emission at 10 μm and 35 μm, since the cool component should originate outside the ionized zone according to the two-zone model.

Aitken and Roche (1982b) have used a grating spectrometer to make spatial scans along the minor axis of NGC 7027 simultaneously in 30 wavelength channels between 10 and 13 μm, with a FWHM beamwidth of 1.5 arcsec. They also mapped the nebula in the 10.5 μm [SIV] line and in the adjacent continuum. The spatial profiles which they obtained at 6 wavelengths are shown in Figure 4. The top 4 profiles of the continuum, at wavelengths between 10.5 and 12.8 μm, are identical within the errors, showing that there is no pronounced temperature gradient within the emitting dust at these wavelengths. Aitken and Roche argued that this supported heating by resonantly trapped photons, rather than by direct absorption of starlight. Their 10.25 μm continuum map had the same FWHM as the 5 GHz radio map of Scott (1973), while the [SIV] map FWHM was 1 arcsec smaller. On the other hand, the unidentified 11.3 μm dust feature (see Section 6) has a FWHM 1.5 arcsec larger than that of the continuum (Figure 4f) and clearly originates in the neutral shell surrounding the ionized zone. Can this narrow dust shell be identified with the dominant 90 K component? Kwok (1980) has shown that if the dust in the surrounding neutral shell has an r^{-2} density distribution and a λ^{-2} emissivity, then the 20-50 μm emission will be sharply peaked just outside the boundary of the ionized zone. To test this model, high spatial resolution ground-based observations are needed, with simultaneous mapping at 10, 20 and 35 μm. Non-simultaneous maps are of little use since an increase in size from 10 to 35 μm of a few arcseconds could not be reliably proven or disproven.

Since NGC 7027 is unusual in respect of the amount of material surrounding it, one might ask if a similar two-component model is appropriate for explaining the infrared luminosities of other planetary nebulae. Moseley (1980) has measured the total IR fluxes of eleven other nebulae. As some of the values of $L_{Ly-\alpha}$ used by Moseley were incorrect, corrected $L_{IR}/L_{Ly-\alpha}$ ratios are listed in Table 3. With the exception of NGC 7662 (Harrington et al 1982) the values of $L_{Ly-\alpha}$ used in Table 3 have been calculated using the relation of Rubin (1968) and are not adjusted for hydrogen recombinations which result in two-photon emission. For the lower density nebulae ($n_e < 10^4$ cm^{-3}) this correction would increase $L_{IR}/L_{Ly-\alpha}$ by a factor of ∿1.4.

TABLE 3: IR DATA FROM MOSELEY (1980)

Planetary Nebula	$L_{IR}/L_{Ly-\alpha}$	$T_d(n=2)$
BD+30°3639	7.9	80
IC 4997	4.3	80
NGC 7027	3.4	90
NGC 7662	2.7	80
NGC 6543	2.1	75
NGC 6572	2.1	90
NGC 2392	1.5	70
IC 418	1.4	
NGC 2440	1.0	65
NGC 3918	0.9	85
NGC 3242	0.4	85

The highest value of $L_{IR}/L_{Ly-\alpha}$ (= 7.9) belongs to BD+30°3639, whose central star has a spectral type, WC9, for which $T_{eff} \simeq 3 \times 10^4$ K. At this effective temperature 80% of the stellar radiation is emitted longwards of 912 Å, while 12% of the stellar flux will be reradiated as nebular Ly-α photons, in the absence of direct absorption of ionizing starlight by dust. Therefore, if all of the stellar flux longwards of 912 Å was absorbed by dust in a neutral shell, accompanied by the absorption of Ly-α photons in both the ionized and neutral zones, the ratio $L_{IR}/L_{Ly-\alpha}$ would be 7.8.

Moseley finds IC 418 to have a total infrared flux of 3×10^{-8} ergs cm^{-2} s^{-1}. The total (dereddened - observed) flux of IC 418 between 912 Å and 2000 Å is 5×10^{-8} ergs cm^{-2} s^{-1}, with 42% in Ly-α and 58% in the stellar continuum. Here, the power requirments can be met without requiring a large fraction of the overall extinction to be local.

All of the planetaries mentioned above are fairly high density ($n_e \gtrsim 10^4$ cm^{-3}), ionization bounded nebulae, for which it is reasonable to expect a surrounding neutral shell. However, at lower nebular densities planetary nebulae will become optically thin in the Lyman continuum and neutral shells will not exist. NGC 7662 is probably such a nebula, for which Moseley found a total infrared flux of 1.2×10^{-8}

ergs cm^{-2} s^{-1}. The (dereddened - observed) flux between 912 Å and 5000 Å is 1.17×10^{-8} ergs cm^{-2} s^{-1} and so could in principle power the infrared emission (Ly-α 34%, CIV 5%, other lines 22%, continuum 39%). However, this would require virtually all of the reddening towards NGC 7662 to be local and would conflict with the analysis of Harrington et al (1982) who found the hydrogen Lyman continuum to be optically thin. Harrington et al showed instead that the optical depth of dust inside the ionized zone, $\tau_d(1549) = 0.1$, could explain the observed infrared emission by absorbing two-thirds of the nebular Ly-α and CIV λ1549 photons, along with 10% of the overall stellar continuum, yielding a flux of 1.3×10^{-8} ergs cm^{-2} s^{-1}(Ly-α 20%, CIV 13%, stellar continuum 67%). Although there is a problem in that 95% of the nebular Hβ flux from NGC 7662 is emitted in a 27 arcsec aperture versus only 50% of the total infrared flux, this may be due to inaccuracies in the infrared data. It seems clear that the far-infrared emission from the older density bounded nebulae originates from dust inside the observed ionized zones.

It is suggested in Section 6, for other reasons, that carbon grains are present in the neutral zone around carbon-rich planetary nebulae (C/O > 1) but are destroyed while passing through the advancing ionization-dissociation fronts, leaving only less abundant grain materials, such as iron, present in the ionized zones. This would explain the decreasing dust-to-gas ratios found by Natta and Panagia (1981) in the progression from young ionization bounded nebulae to older density bounded nebulae and would also explain the severe dust depletions in the ionized zones of both types of nebulae implied by the low derived values of $\tau_d(1549)$ (Section 3.4). According to this scenario, in young, dense, carbon-rich nebulae the far-infrared continuum would be emitted by carbon grains in the neutral shell, while the 10 μm continuum would come from iron, or other grains in the ionized region. In older, density-bounded nebulae these grains within the ionized zone would provide all the infrared emission. Grains in the ionized zone, with a λ^{-2} emissivity, heated solely by trapped resonance line radiation whose energy density is proportional to n_e^2, will have $T_d^6 \propto n_e^2$. Thus the T_d = 230 K component in the ionized zone of NGC 7027 would have T_d = 85 K if the density was reduced from $n_e = 6 \times 10^4$ cm^{-3} to $n_e = 3 \times 10^3$ cm^{-3}, a value typical of the less dense planetary nebulae in Moseley's sample.

Shields (1978) has shown that the observed depletion of iron by a factor of 25 over cosmic values in NGC 7027 would, if the iron was tied up in grains, give $\tau_d(1549) \sim 0.1$, which is sufficient to explain the reduction in CIV λ1549 intensity over computed values and to give agreement with the depleted dust-to-gas ratio deduced by Bohlin et al (1975). Similarly, Harrington et al (1982) noted that their derived $\tau_d(1549)$ for NGC 7662 could be explained by a cosmic abundance of iron locked up in grains. Scaling the ionized hydrogen column density in their model of NGC 7662 up to the column density in the Shields (1978) model of NGC 7027 gives $\tau_d(1549)$ = 0.12-0.20 for NGC 7027.

For a constant dus-to-ionized gas ratio, n_d/n_H, the dust optical depth $\tau_d(II)$ within the ionized radius R_{II} of a linearly expanding spherical nebula will be given by $\tau_d(II) \propto n_d R_{II} \propto R_{II}^{-2}$, if the nebula is density bounded. In an ionization bounded nebula, recombinations balance ionizations and $n_H^2 R_{II}^3$ is constant, so $\tau_d(II) \propto R_{II}^{-1/2} \propto n_H^{1/3}$. Since NGC 7662, with $n_H = 3 \times 10^3$ cm^{-3}, is only just density bounded (Harrington et al 1982), we can extrapolate the latter relation backwards to predict that only at densities in excess of a few times 10^5 cm^{-3} should dust compete significantly with gas for ionizing stellar photons.

5. A 30 AND A 78

A 30 and A 78 are a pair of large, low surface brightness planetary nebulae with OVI central stars. Cohen and Barlow (1974) found that both nebulae exhibited strong 10 and 20 μm dust emission from their central zones with colour temperatures of ∿140 K. A 30 was found to have a FWHM at 10 and 18 μm of ∿25 arcsec in both NS and EW directions, compared to an optical diameter of 130 arcsec. Since there were no known optical nebulosities on existing photographs of these nebulae which corresponded to the extent of the dust emission, Cohen and Barlow suggested that the dust had condensed in the outflowing stellar winds from the OVI central stars.

Cohen et al (1977) made multiaperture observations of A 30 and A78 between 1.6 and 3.5 μm and found significant emission from hot dust grains (T_d ∿ 1000 K). The angular diameters at 2.3 and 3.5 μm were similar to those found by Cohen and Barlow at 10 and 18 μm, but the infrared spatial distribution was not in agreement with that predicted by an R^{-2} stellar wind density distribution. The excess fluxes were instead linearly proportional to aperture area.

Moseley (1980) found A 30 to have an angular diameter of ∿25 arcsec at 37 μm, with the dominant dust component having a colour temperature of 60 K for a λ^{-2} emissivity. Moseley noted that the dust temperature and total infrared luminosity of A 30 was similar to those of the high surface brightness planetaries in his sample, implying that the amount of dust in the central 15 arcsec of A 30 was comparable to the total amount of dust in a typical high surface brightness nebula.

A first step towards an understanding of the origin of the centrally concentrated dust in A 30 and A 78 came with the independent discovery by Jacoby (1979) and by Hazard et al (1980) of centrally condensed knots of strong [OIII] emission in both nebulae. The major axis diameters of the central nebulosities, at 24 and 23 arcsec in A 30 and A 78 respectively (Jacoby 1979), are coextensive with the infrared dust emission. Optical spectroscopy of the central knots in A 30 by Hazard et al revealed no detectable hydrogen lines and a He/H ratio of at least 20 was derived.

Greenstein (1981) obtained optical and ultraviolet spectrophotometry of the central star of A 30. He found that the ultraviolet extinction curve deviated from the normal interstellar extinction law, exhibiting a broad peak at 2470 Å rather than at the normal wavelength of 2200 Å. Greenstein attributed the abnormal reddening to the dust inside the nebula which is responsible for the infrared emission. The (dereddened-observed) stellar flux longwards of 912 Å is insufficient to power the observed infrared emission. Hence, direct absorption of photons below the Lyman limit by dust must occur. However, with no hydrogen in the inner nebula, it is not clear that the dust affects the ionization structure. Greenstein noted that an extinction curve for carbon smoke measured by Stephens (1980) gave a peak at the observed wavelength of 2470 Å and provided a better fit than a graphite grain model.

Iben et al (1982) have recently provided a detailed theoretical interpretation of the characteristics of the nebulosities and central stars in A 30 and A 78. According to their interpretation, the nuclei represent stars which have passed along the normal hydrogen-shell-burning central star evolutionary track, but then have experienced a final thermal pulse just after achieving a white dwarf configuration. As a result, most of the remaining hydrogen was completely burned following incorporation into the helium-burning convective shell and the stars swelled to red giant dimensions once more. The stars are now burning helium on a long timescale and retracing their path on the H-R diagram. The precise means of ejection of the inner helium-rich knots in A 30 and A 78 was not established by Iben et al, but ejection during the renewed red giant phase would provide a natural explanation for the existence of large quantities of dust in the knots.

6. INFRARED EMISSION FEATURES

Gillett, Forrest and Merrill (1973) discovered two emission features at 8.6 μm and 11.3 μm in the 8-13 μm spectra of NGC 7027 and BD+30°3639. Their initial suggestion that the 11.3 μm feature was due to $MgCO_3$ was not supported by the subsequent failure to detect other expected $MgCO_3$ features at 7 μm and 24 μm (Russell, Soifer and Willner 1977, McCarthy et al 1978). Further ground-based observations of NGC 7027 revealed an emission feature at 3.3 μm, with a weaker feature at 3.4 μm (Merrill, Soifer and Russell 1975, Grasdalen and Joyce 1976). Airborne 4-8 μm spectrophotometry of NGC 7027 by Russell, Soifer and Willner (1977) revealed two more features at 6.2 μm and 7.7 μm. The latter is the strongest of the six unidentified features. The six features have now been observed in other classes of objects besides planetary nebulae and their relative intensities have been found to vary from object to object. The 2-13 μm spectrum of HD 44179, the Red Rectangle, is shown in Figure 5 in order to illustrate the features uncontaminated by nebular emission lines. This bipolar reflection nebula, whose central star is of late B - early A spectral type, may be a planetary nebula precursor (see Cohen, these proceedings).

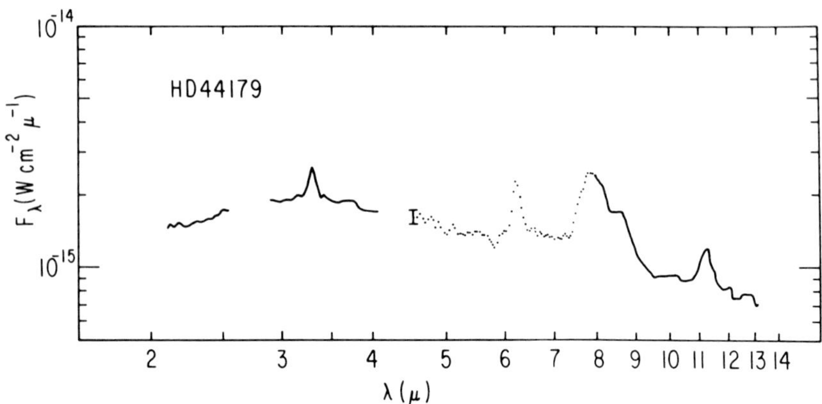

Figure 5. The 2-13 μm spectrum of HD 44179, showing the unidentified features at 3.3, 3.4, 6.2, 7.7, 8.6 and 11.3 μm (Russell et al 1978).

High spectral resolution observations of the features at 11.3 μm (Bregman and Rank 1975), 3.3 μm (Tokunaga and Young 1980) and 7.7 μm (Russell et al 1982) have shown that they are broad and not due to unresolved lines. A solid state origin is therefore implied. Although the six features are unidentified at present, an identification of at least some of them with hydrocarbons seems very probable (see below).

Spectrophotometry of the 3.3 and 3.4 μm features, in planetary nebulae other than NGC 7027, has been obtained by Russell, Soifer and Merrill (1977; IC 418 and BD+30°3639), Willner et al (1979; IC 418 and NGC 6572), Allen et al (1982; M 1-11, IC 2501, He 2-113 and CPD-56°8032) and Martin (these Proceedings). This is an insufficient sample from which to draw statistical conclusions and observations of more nebulae are needed. Similarly, 4-8 μm spectrophotometry has been published for only NGC 7027 (see above) and IC 418 (Willner et al 1979). By contrast, 8-13 μm spectrophotometry has been published for 26 different planetary nebulae: by Gillett et al (1973; 3 PN), Grasdalen (1979; 2 PN), Willner et al (1979; 2 PN), Aitken et al (1979, 1980; 5 and 2 PN) and Aitken and Roche (1982a; 20 PN). The existing sample is biased towards compact nebulae. A first step towards obtaining a less biased sample has been made by Roche, Aitken and Whitmore (these Proceedings).

Of the ten planetary nebulae so far found to show strong 11.3 μm and 8.6 μm features, five are high excitation (HeII λ4686/Hβ > 0.2), while the other five, of very low to intermediate excitation, have WC-type central stars.

Willner et al (1979) found a broad emission feature between 10.5 μm and 13 μm in the 8-13 μm spectra of IC 418 and NGC 6572. They identified the feature with SiC grains. Such grains had already been identified in the 8-13 μm spectra of many carbon stars. The SiC feature has since been seen in the 8-13 μm spectra of NGC 6790 (Aitken et al

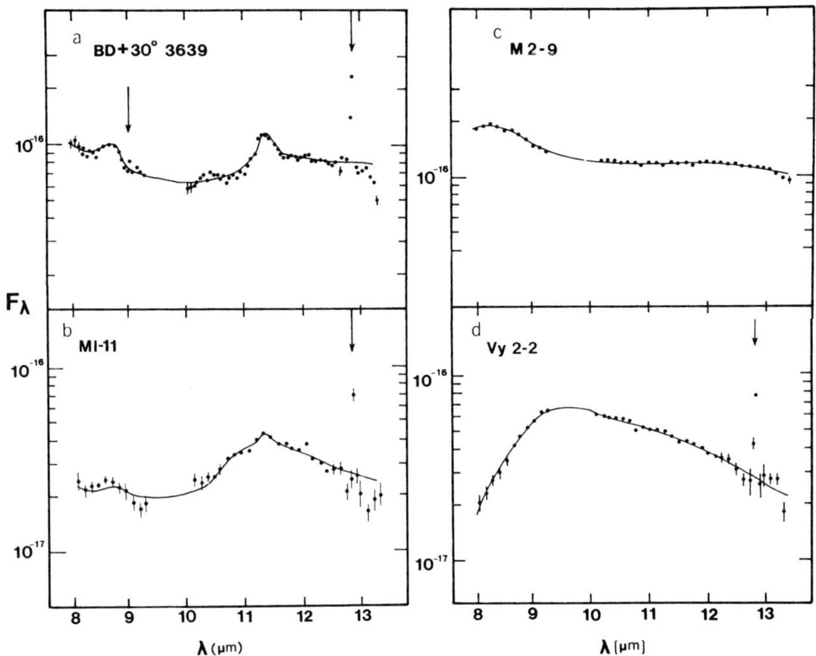

Figure 6. 8-13 μm spectrophotometry of four planetary nebulae, illustrating the various features encountered. The solid lines are best fits to the continua utilising varying proportions of featureless continuum, plus the SiC, 8.6 μm and 11.3 μm features (Figures 6a, b); and continuum plus the silicate feature in emission (6d), or in emission plus absorption (6c). Arrows indicate the positions of nebular emission lines. All figures are taken from Aitken and Roche (1982a).

1979) and IC 2501 and M 1-11 (Aitken and Roche 1982a). The 8-13 μm spectrum of M 1-11 is shown in Figure 6b, where a weak 11.3 μm feature can also be seen. The 11.3 μm feature is always weak in the spectra of planetary nebulae with strong SiC features. Conversely, Aitken and Roche (1982a) found that small amounts of the SiC component improved the fits to the spectra of planetary nebulae with strong 11.3 μm features. Unlike the 11.3 μm feature, the 3.3 μm feature shows good contrast in those planetary nebulae with strong SiC features. Three of the planetary nebulae with strong SiC features are of intermediate excitation (HeII/Hβ ∿ 0, [OIII]/Hβ ∿ 10), while IC 418 and M 1-11 have [OIII]/Hβ < 2.

Aitken et al (1979) discovered the well-known silicate feature in emission, in the 8-13 μm spectra of M 1-26, SwSt 1 and Hb 12. Since then, Aitken and Roche (1982a) have also found silicate emission in the spectra of IC 4997, He 2-47, Vy 2-2 and He 2-131. Four of this set of seven nebulae are of very low excitation ([OIII]/Hβ < 1), while Hb 12, IC 4997 and Vy 2-2 have [OIII]/Hβ $\gtrsim$ 5. Figure 6d shows the 8-13 μm spectrum of Vy 2-2.

Aitken and Roche (1982a) found that the 8-13 μm spectra of M 2-9, Mz 3, He 2-90 and M 2-56 required a combination of silicate emission with overlying absorption by colder silicate grains to fit them. The required silicate absorption optical depths implied visual extinctions $A_V \sim 15$ mag to the infrared cores, which is much larger than the interstellar extinction to these objects. These nebulae can all be loosely classified as Type I bipolar nebulae (see the reviews by Cohen and by Peimbert, these Proceedings). The 8-13 μm spectrum of M 2-9 is shown in Figure 6c.

Forrest, Houck and McCarthy (1981) found a strong emission feature in the 16-30 μm spectra of IC 418, NGC 6572 and four carbon stars. The feature extended from 24 μm to at least 30 μm. All these objects also have strong 10.5-13 μm SiC features, although some carbon stars with SiC features did not show the λ>24 μm feature. Forrest et al suggested that the grain material responsible for the feature was either Fe_3C or carbyne, the latter a long-chain allotrope of carbon which had been proposed to exist in meteorites. Recent work (e.g. Smith 1981) now suggests carbyne does not exist. Goebel (1980) proposed MgS grains.

The standard picture of dust formation in late-type stellar envelopes is that the C/O ratio determines the type of grain formed. In oxygen-rich envelopes, CO molecules should lock up all carbon atoms, leaving only oxygen-rich grains such as silicates to form, while in carbon-rich envelopes the locking up of all available oxygen atoms in CO would allow carbon-rich grains such as silicon carbide and graphite to condense. Available measurements of C/O ratios in planetary nebulae confirm this picture (Table 4, below). The nebulae with silicate emission all have C/O < 0.5 (Table 4; Flower and Penn and Cohen, Flower and Goharji, this volume). The nebulae with strong SiC emission all have C/O very close to unity and are all consistent with C/O > 1 when the small contribution from C locked up in SiC is allowed for. The nebulae with strong unidentified features all appear to have C/O > 2. The carbon abundance determinations are particularly difficult for this

TABLE 4: INFRARED EMISSION FEATURES AND NEBULAR C/O ABUNDANCE RATIOS

NEBULA	FEATURE	C/O	REFERENCE
IC 4997	silicate	0.4	see Table 1, Seaton, this volume
He 2-131	"	0.3	" " "
M 1-26	"	0.5	" " "
IC 418	SiC	1.3	" " "
NGC 6572	"	1.1	" " "
IC 2501	"	0.9	" " "
NGC 5315	11.3 μm	2.5	Torres-Peimbert and Pena (1981)
BD+30°3639	"	≥2.8	" " "
NGC 7027	"	3.5	{ Shields (1978)
"	"	3.2	{ Perinotto et al (1980)

group of nebulae. Allowance must be made for the attenuation of CIV λ1549 by internal dust in the high-excitation nebulae, while contamination of the nebular carbon lines by stellar features is a problem in those with WC type central stars. However, the evidence seems to suggest that the planetary nebulae with strong SiC features have lower C/O ratios than the planetary nebulae with strong 11.3 μm features.

The general progression, from mainly low-excitation planetaries showing silicate emission, through intermediate-excitation nebulae with strong SiC features, to high-excitation nebulae showing strong unidentified features, suggests that a correlation between initial stellar mass and C/O ratio is involved. If the high-excitation, high luminosity nebulae are identified with the more massive progenitors, this would imply that the nebulae with silicate emission have originated from the least massive progenitors. This would be consistent with the models of Renzini and Voli (1981), who predict C/O < 1 for low initial masses and C/O > 1 for higher masses. Figure 9 of Renzini and Voli also predicts that C/O can fall below unity again for the most massive progenitors ($M \geq 5\ M_\odot$), and these nebulae should also have He/H > 0.15. It might be possible to identify such objects with bipolar nebulae such as Mz 3 and M 2-9 which have been found to show silicate features by Aitken and Roche (1982a).

I conclude this section with a brief discussion of the origin of the unidentified emission features, along with a personal interpretation of the observations.

Allamandola and Norman (1978) proposed that the 3.3 μm and 7.7 μm features were due to C-H stretching and H-C-H bending modes of solid methane, in which the normal rotational structure of gaseous CH_4 is suppressed. Duley and Williams (1981) proposed that the features were due to radicals bound to the surface of graphite or amorphous carbon grains. They showed that such surface functional groups would give rise to spectral features characteristic of aromatic molecules. For example, an H atom bound to a carbon atom in a hexagonal graphitic platelet would correspond to an aromatic -CH complex. The identifications proposed by Duley and Williams included the CH stretch and H wag of aromatic -CH, at 3.3 μm and 11.3 μm respectively, and the asymmetric stretch of aromatic $-CH_3$, at 3.4 μm.

Barlow and Silk (1977) proposed that graphite grains would react with atomic or ionized hydrogen in the vicinity of HII regions, if their temperatures exceeded $T_d \gtrsim 70$ K, leading to the destruction of the grains on a short timescale. Draine (1979) criticised the reaction model of Barlow and Silk on the grounds that the predicted chemical sputtering yields did not agree with low temperature laboratory data. However, Barlow (1982) has analysed all the existing experimental data and shown that H_2 recombination dominates at laboratory densities, whereas chemical sputtering will dominate at densities typical of the interstellar medium and planetary nebulae. Chemical sputtering by hydrogen occurs both for graphite and amorphous carbon. The analysis yielded a revised critical grain temperature, T_d(react) $\sim$ 110 K, above which chemical sputtering proceeds rapidly. The timescale for the destruction of a carbon grain of radius 10^{-6} cm was found to be $\sim 10^7/n_H$ years for a 100 K HI region and $\sim 10^6/n_H$ years for a 10^4 K HII region.

It is proposed here that surface hydrocarbon complexes formed during the chemical sputtering of carbon grains are responsible for the observed infrared emission features. The main product of chemical sputtering is the abstraction of a carbon atom to yield CH_4. Excitation of the CH_4 on the surface could give the 3.3 μm and 7.7 μm features. The steps along the path to the creation of CH_4 give -CH, -CH_2 and -CH_3. These aromatic complexes could give the 3.3 μm and 11.3 μm features, as discussed by Duley and Williams. C_2H_4 and similar hydrocarbons are also formed during the chemical sputtering of graphite and could be responsible for the other features.

The observations of the spatial distribution of the infrared features in NGC 7027 are consistent with this model. Jones et al (1980) found that the 7.7, 8.6 and 11.3 μm features showed greatest contrast in the outer regions. The high spatial resolution observations of Aitken and Roche (1982b) confirm this and show conclusively that the 11.3 μm feature originates from a shell outside the ionized zone (see Section 4 and Figure 4). Although molecular H_2 in a neutral zone will not react with graphite, a dissociation front will propagate ahead of the advancing ionization front, yielding H atoms which can then react rapidly with the grains.

Two excitation mechanisms have been suggested for the infrared features. Allamandola and Norman (1978) proposed excitation by ultra-violet fluorescence, in which absorption of short wavelength photons by surface molecules leads to the excitation of vibrational transitions. Aitken and Roche (1982b) have calculated the implied fluorescence yields in NGC 7027 for this mechanism. The observed flux of photons in the infrared features is 2.9×10^4 cm^{-2} s^{-1}. The available flux of nebular photons between 1200 and 5007 Å is 2.8×10^4 cm^{-2} s^{-1} (see Table 2), implying a yield of unity if this mechanism operates. We thus have the remarkable situation that the number and luminosity of nebular photons between 1200 and 5007 Å is equal to the number of infrared feature photons and the continuum infrared luminosity, respectively.

An alternative to UV fluorescence, namely thermal excitation has been suggested by Dwek et al (1980), who propose that the emitting material is thermally excited on warm grains. The most severe constraint on this model is provided by the 3.3 μm feature, which requires significantly hotter grain temperatures for excitation than the other features. However, it is interesting that the brightness temperature for the 3.3 μm feature in NGC 7027, quoted by Aitken and Roche (1982b), is equal to the temperature of the $\lambda^{-2}B_\lambda$(230 K) continuum which seems to originate from the ionized zone (Section 4). A comparison of the spatial distribution of the 3.3 μm and 11.3 μm features in NGC 7027 could be very useful.

The association of the infrared features with carbon-rich planetary nebulae can be understood if they originate from carbon grains. Objects with C/O ratios only slightly larger than unity will have most of their excess carbon locked up in SiC grains which condense before graphite grains. This would explain the weakness of the 11.3 μm feature in planetary nebulae with the SiC feature and C/O $\sim$ 1. For C/O > 2, SiC grains will not use up a significant fraction of the excess carbon, allowing large quantities of carbon grains to form. This would be followed by the production of strong infrared features when the grains are destroyed.

The progressive destruction of carbon grains in the envelopes of carbon-rich planetary nebulae would explain the decrease of the dust-to-gas ratio with increasing nebular age, which was found by Natta and Panagia (1981). The confinement of carbon grains to the neutral shells and ionization fronts of optically thick planetary nebulae would be consistent with the depleted dust-to-gas ratios found in the ionized zones of both optically thick and thin planetary nebulae (Section 3.4). The anomalously large quantities of dust in the inner regions of A 30 and A 78, along with the carbon composition of the dust in A 30, suggested by its extinction curve (Section 5), also find an explanation. Since no hydrogen exists in the inner regions of these nebulae, carbon grains formed during the revisited red giant phase would not have been destroyed.

Silicate grains in oxygen-rich planetary nebulae will not be destroyed chemically and so there should be no decrease of the dust-to-gas ratio in such nebulae with increasing age. Although carbon grains can be destroyed in planetary nebulae envelopes, this does not necessarily imply that carbon-rich progenitor stars do not enrich the interstellar medium carbon grain content. The differences between derived progenitor star masses and current central star masses are often much larger than the mass of the planetary nebula envelopes (see the reviews by Renzini and by Schönberner and Weidemann in this volume). Most of the missing mass may have been lost as a late-type stellar wind prior to PN envelope ejection. Carbon grains in these winds will not react with the ambient H_2 during the carbon star phase and will be too distant to be heated sufficiently for reaction with atomic or ionized hydrogen during the planetary nebula phase. Thus, providing planetary nebulae envelopes represent a small fraction of the total mass lost, carbon star progenitors can enrich the grain content of the interstellar medium.

I would like to thank J. Drew, D. Aitken and P. Roche for their help in the preparation of this review.

REFERENCES

Aitken, D.K., Roche, P.F., Spenser, P.M., Jones, B., 1979, Astrophys.J., 233, 925.

Aitken, D.K., Barlow, M.J., Roche, P.F., Spenser, P.M., 1980, MNRAS, 192, 679.
Aitken, D.K., Roche, P.F., 1982a, MNRAS, 200, 217.
Aitken, D.K., Roche, P.F., 1982b, MNRAS, in press.
Allamandola, L.J., Norman, C.A., 1978, Astr.Astrophys., 63, L23.
Allen, D.A., Baines, D.W.T., Blades, J.C., Whittet, D.C.B., 1982, MNRAS, 199, 1017.
Atherton, P.D., Hicks, T.R., Reay, N.K., Robinson, G.J., Worswick, S.P., Phillips, J.P., 1979, Astrophys.J., 232, 786.
Barlow, M.J., 1982, in preparation.
Barlow, M.J., Silk, J., 1977, Astrophys.J., 215, 800.
Becklin, E.E., Neugebauer, G., Wynn-Williams, C.G., 1973, Astrophys.Lett., 15, 87.
Bohlin, R.C., Marionni, P.A., Stecher, T.P., 1975, Astrophys.J., 202, 415.
Bregman, J.D., Rank, D.M., 1975, Astrophys.J., 195, L125.
Clavel, J., Flower, D.R., Seaton, M.J., 1981, MNRAS, 197, 301.
Cohen, M., Barlow, M.J., 1974, Astrophys.J., 193, 401.
Cohen, M., Barlow, M.J., 1980, Astrophys.J., 238, 585.
Cohen, M., Hudson, H.S., O'Dell, S.L., Stein, W.A., 1977, MNRAS, 181, 233.
Coleman, C.L., Reay, N.K., Worswick, S.P., 1975, MNRAS, 171, 415.
Doughty, J.R., Kaler, J.B., 1982, PASP, 94, 43.
Draine, B.T., 1979, Astrophys.J., 230, 106.
Duley, W.W., Williams, D.A., 1981, MNRAS, 196, 269.
Dwek, E., Sellgren, K., Soifer, B.T., Werner, M.W., 1980, Astrophys.J., 238, 140.
Forrest, W.J., McCarthy, J.F., Houck, J.R., 1980, Astrophys.J., 240, L37.
Forrest, W.J., Houck, J.R., McCarthy, J.F., 1981, Astrophys.J., 248, 195.
Gillett, F.C., Low, F.J., Stein, W.A., 1967, Astrophys.J., 149, L97.
Gillett, F.C., Forrest, W.J., Merrill, K.M., 1973, Astrophys.J., 183, 87.
Goebel, J.H., 1980, BAAS, 12, 858.
Grasdalen, G.L., 1979, Astrophys.J., 229, 587.
Grasdalen, G.L., Joyce, R.R., 1976, Astrophys.J., 205, L11.
Greenstein, J.L., 1981, Astrophys.J., 245, 124.
Harrington, J.P., Feibelman, W.A., 1982, Astrophys.J., in press.
Harrington, J.P., Seaton, M.J., Adams, S., Lutz, J.H., 1982, MNRAS, 199, 517.
Hazard, C., Terlevich, B., Morton, D.C., Sargent, W.L.W., Ferland, G., 1980, Nature, 285, 463.
Hicks, T.R., Phillips, J.P., Reay, N.K., 1976, MNRAS, 176, 409.
Hildebrand, R.H., Whitcomb, S.E., Winston, R., Stiening, R.F., Harper, D.A., Moseley, S.H., 1977, Astrophys.J., 216, 698.
Hummer, D.G., Kunasz, P.B., 1980, Astrophys.J., 236, 609.
Iben, I., Kaler, J.B., Truran, J.W., Renzini, A., 1982, Astrophys.J., in press.
Jacoby, G.H., 1979, PASP, 91, 754.
Jones, B., Merrill, K.M., Stein, W., Willner, S.P., 1980, Astrophys.J., 242, 141.
Kaler, J.B., Aller, L.H., Czyzak, S.J., Epps, H.W., 1976, Astrophys.J.Suppl., 31, 163.

Koike, C., Hasegawa, H., Manabe, A., 1980, Astrophys.Sp.Sci., 67, 495.
Krishna Swamy, K.S., O'Dell, C.R., 1968, Astrophys.J., 151, L61.
Kwok, S., 1980, Astrophys.J., 236, 592.
McCarthy, J.F., Forrest, W.J., Houck, J.R., 1978, Astrophys.J., 224, 109.
Mathis, J.S., 1978, Proc. IAU Symposium No. 76, ed. Y. Terzian (D.Reidel) p281.
Mathis, J.S., Rumpl, W., Nordsieck, K.H., 1977, Astrophys.J., 217, 425.
Merrill, K.M., Soifer, B.T., Russell, R.W., 1975, Astrophys.J., 200, L37.
Miller, J.S., Mathews, W.G., 1972, Astrophys.J., 172, 593.
Moseley, H., 1980, Astrophys.J., 238, 892.
Mufson, S.L., Lyon, J., Marionni, P.A., 1975, Astrophys.J., 201, L85.
Natta, A., Panagia, N., 1981, Astrophys.J., 248, 189.
O'Dell, C.R., Hubbard, W.B., 1965, Astrophys.J., 142, 591.
Osterbrock, D.E., 1974, PASP, 86, 609.
Page, T.L., 1936, MNRAS, 96, 604.
Panagia, N., 1975, Astr.Astrophys., 42, 139.
Panagia, N., Ranieri, M., 1973, Mem.Soc.Roy.Sci.Liege, Series 6, V, 275.
Perinotto, M., Patriarchi, P., 1980, Astrophys.J., 238, 614.
Perinotto, M., Panagia, N., Benvenuti, P., 1980, Astr.Astrophys., 85, 332.
Persson, S.E., Frogel, J.A., 1973, Astrophys.J., 182, 177 (Erratum: Astrophys.J., 185, 991)
Philipp, H.R., 1977, Phys.Rev., 16, 2896.
Phillips, J.P., Reay, N.K., Worswick, S.P., 1979, Astrophys.Lett., 20, 75.
Pottasch, S.R., Wesselius, P.R., Wu, C.-C., van Duinen, R.J., 1977, Astr.Astrophys., 54, 435.
Renzini, A., Voli, M., 1981, Astr.Astrophys., 94, 175.
Rubin, R.H., 1968, Astrophys.J., 154, 391.
Russell, R.W., Soifer, B.T., Merrill, K.M., 1977, Astrophys.J., 213, 66.
Russell, R.W., Soifer, B.T., Willner, S.P., 1977, Astrophys.J., 217, L149.
Russell, R.W., Soifer, B.T., Willner, S.P., 1978, Astrophys.J., 220, 568.
Russell, R.W., Gull, G., Beckwith, S., Evans, N.J., 1982, PASP, 94, 97.
Scott, P.F., 1973, MNRAS, 161, 35P.
Seaton, M.J., 1955, MNRAS, 115, 279.
Seaton, M.J., 1979, MNRAS, 187, 785.
Shields, G.A., 1978, Astrophys.J., 219, 565.
Sistla, G., Kaftan-Kassim, M.A., 1976, Astrophys.Lett., 17, 49.
Smith, P.P.K., 1981, Nature, 291, 15.
Stephens, J.R., 1980, Astrophys.J., 237, 450.
Telesco, C.M., Harper, D.A., 1977, Astrophys.J., 211, 475.
Terzian, Y., Sanders, D., 1972, Astr.J., 77, 350.
Tokunaga, A.T., Young, E.T., 1980, Astrophys.J., 237, L93.
Torres-Peimbert, S., Pena, M., 1981, Rev.Mex.Astr.Astrof., 6, 301.
Willner, S.P., Jones, B., Puetter, R.C., Russell, R.W., Soifer, B.T., 1979, Astrophys.J., 234, 496.

KWOK: Would you comment on the difference between grains in novae and in PN?

BARLOW: Most novae showing dust emission appear to have featureless infrared continua, although recently Aitken and Roche have discovered a 10 μm "silicate" emission feature in the spectrum of Nova Aquilae 1982.

KALER: Feibelman and I are completing a study of A 78. We can fit the IUE spectrum to a model or a black-body at 90 000 - 100 000 K, and there is little evidence for dust surrounding the star. Given the strong infrared emission, we conclude that the dust must be distributed in such a way that the starlight is not absorbed.

BARLOW: This is very interesting. We at UCL have also acquired IUE spectra of A 30 and A 78 in view of the great importance of Greenstein's discovery of anomalous extinction towards A 30.

HOUCK: The failure of the search for the 25 μm feature of carbonates may be due either to radiative transfer or crystallographic effects. Carbynes seem not to exist in meteorites but I think that the chemists still believe in them.

BARLOW: Although the existence of carbynes is theoretically possible, as I understand it there is no definitive laboratory evidence for their existence.

ZUCKERMAN: To determine the total mass of dust associated with a PN, it is best to observe at submillimeter wavelengths where the intensity of emission is only weakly dependent on the dust temperature. A group of astronomers (Jaffe and Hildebrand, Chicago; Gatley, UKIRT; Roellig and Werner, NASA-Ames; Sopka and myself, Maryland) has used ^{3}He bolometers on UKIRT to measure 380 μm fluxes from infrared giant stars and PN. We have observed about a dozen such objects and found that, along with IRC + 10216 and OH 0739 - 14, the planetary and proto-planetary nebulae NGC 7027, GL 2688 and GL 618 are strong 380 μm sources. There is clearly a large amount of cool dust around these objects since, for example, GL 2688 is as bright as IRC + 10216 at 380 μm, whereas the latter is at least 100 times more intense at much shorter wavelengths (e.g. 5 μm).

ALLER: To what extent can we be sure that the carbon particles are classical graphite structures? Do we have any theoretical or experimental data which enable us to predict detailed structures of microscopic particles?

BARLOW: Laboratory samples of amorphous carbon tend to be composed of random ensembles of graphite microcrystals. Whether 10^{-6} cm radius carbon particles will be graphite structured or truly random collections of carbon atoms is hard to predict, but, until laboratory data are available on truly amorphous carbon particles, I suspect that graphite data will tend to be used.

ZUCKERMAN: About ten years ago, the Caltech infrared group published some beautiful maps of NGC 7027 which showed that the infrared emission at, say, 10 μm came from the same region as the radio emission. Are you now saying that this infrared radiation is not emitted by dust grains located within the ionized gas?

BARLOW: No. The reduced intensity of the C IV resonance lines in the ultraviolet spectrum of NGC 7027 shows that dust is present in the ionized gas. This dust will presumably be the hottest dust and emit much more strongly at 10 μm than the cool (95 K) dust supposed to exist in the surrounding neutral shell.

SOME RECENT RESULTS FROM UV OBSERVATIONS

M.J. Seaton,
Department of Physics and Astronomy,
University College London, Gower Street, London WC1E 6BT

Abstract.

UV observations with IUE have led to recognition of the importance of di-electronic recombination in PN, for the calculation of ionization equilibria and for production of the features C II λ1335 and C III λ2297.

The determination of C abundances is discussed. In high-excitation nebulae one must allow for absorption of C IV λ1549 by dust. It has been shown that $C/O \leqslant 0.5$ for five PN with a silicate IR feature and that $C/O \gtrsim 1$ for four with a SiC feature.

Dust opacities τ_D have been deduced from observed ratios (C IV λ1549)/(C III λ2297) and (C II λ1335)/(C I λ4267). The values obtained, $\tau_D \simeq 0.1$, are consistent with the observed strength of thermal infra-red emission.

Earlier UV observations of PN were made by Pottasch et al. (1978) using ANS. Results obtained using IUE support neither their downward revisions of star temperatures nor their claim that a number of stars with T_Z(He II) >> T_Z(H I) are optically thick for H I.

From IUE observations it is concluded that the planetary K 648 in M 15 has C/O = 2.4 and that its central star has a temperature of 30 000 K and a luminosity of 1700 $L_\odot$.

The present contribution is concerned with some IUE work with which I have been associated but makes no attempt at a comprehensive review of IUE observations of planetary nebulae.

D. R. Flower (ed.), Planetary Nebulae, 129–139.

1. CARBON ABUNDANCES.

The determination of C abundances is of interest because C can be enriched by He burning. It is convenient to consider C/O ratios, for which the solar value is C/O = 0.6 (Lambert, 1978). The determination of C/O ratios is also of interest in connection with the interpretation of IR spectra, since objects with C/O < 1 are expected to have IR features different from those in objects with C/O > 1 (see review by Barlow in the present volume). While we require C/O ratios to an accuracy of much better than a factor of two, different methods in use a few years ago were giving results discrepant by as much as two orders of magnitude!

1.1 C lines observed with IUE.

IUE observations give fluxes in the collisionally-excited lines

C II] λ2326, C III] λ1908 and C IV λ1549. (1.1)

The C^+/C^{2+} ratio for IC 418 obtained from λ2326 and λ1908 was found to be larger than had been predicted from ionization models, by a factor of 5, and Harrington *et al*. (1980) suggested that $C^{2+} \rightarrow C^+$ di-electronic recombination might be important. This was confirmed by Storey (1981 and review in the present volume) who shows that di-electronic recombination increases C^+/C^{2+} and C^{2+}/C^{3+} ratios and also leads to production of a number of spectrum lines, including

C II λ1335 and C III λ2297. (1.2)

The C II λ1335 multiplet is observed in IC 418 both using high dispersion (Clavel *et. al*. 1981) and using low dispersion at a position offset from the central star (Adams and Seaton, 1982a). It is also observed in other objects but is generally weak due to absorption by interstellar C II. IC 418 is a favourable case, because it has a radial velocity which is sufficiently large to displace the wavelength of the C II emission from that of the interstellar C II absorption.

The C III λ2297 line is observed in the spectra of a number of planetaries but is fairly weak and, in low-dispersion spectra, is blended with He II λ2307 and partially blended with [O III] λ2321 and C II] λ2326. For NGC 7009 and NGC 7662 results obtained from low-dispersion observations (Harrington *et al*. 1981) are confirmed by high-dispersion observations (Harrington *et al*., 1981, and Flower *et al*. 1982).

1.2 Resonance lines and dust absorption.

C II λ1335 and C IV λ1549 are resonance multiplets and optical depths for scattering are of order 10^4 for λ1335 in low-excitation nebulae and for λ1549 in high-excitation nebulae. Scattering of photons in resonance lines enhances the probability of their being absorbed by internal dust. The mean probability of photon escape will be denoted by f_e.

For IC 418, Clavel et al. compared the ratio of fluxes

$$F(\text{C II } \lambda1335)/F(\text{C II } \lambda4267), \quad (1.3)$$

corrected for interstellar absorption, with the ratio of emissivities

$$j(\text{C II } \lambda1335)/j(\text{C II } \lambda4267), \quad (1,4)$$

calculated using effective recombination coefficients of Storey (1981) for λ1335 and of Pengelly (1963) for λ4267. They found (1.3) to be about half of (1.4) and concluded that $f_e = 0.5$ for λ1335 in IC 418.

The best evidence for absorption of λ1549 by dust comes from comparing

$$F(\text{C IV } \lambda1549)/F(\text{C III } \lambda2297) \quad (1.5)$$

with

$$\int j(\text{C IV } \lambda1549)dV/\int j(\text{C III } \lambda2297)dV, \quad (1.6)$$

where the integrals in (1.6) are over the observed volume. Now the emissivities for λ1549 and λ2297 are both proportional to $N_eN(C^{3+})$ and it follows that (1.6) depends only on the mean electron temperature, T_e(C IV), of the region in which most of the carbon is C^{3+}. For NGC 7009 and 7662, Harrington et al. (1981, 1982) find (1.5) to be smaller than (1.6), calculated for any reasonable value of T_e(C IV); they estimate that $f_e \simeq 0.3$ for λ1549 in both of these objects.

Evidence for absorption of λ1549 has also been deduced from comparing

$$F(\text{C IV } \lambda1549)/F(\text{C III] } \lambda1908) \quad (1.7)$$

with

$$\int j(\text{C IV } \lambda1549)dV/\int j(\text{C III] } \lambda1908)dV, \quad (1.8)$$

calculated using models (see, for example, the abstracts by Peña and Torres-Peimbert, and by Köppen and Wehrse, in the present volume). While persuasive,the evidence from (1.7) is less compelling than that from (1.5). Since the emissivity for λ1549 is proportional to $N_eN(C^{3+})$ while that for λ1908 is proportional to $N_eN(C^{2+})$, the ratio (1.8) is more sensitive to the structure of a model than is the ratio (1.6). This point is illustrated by the work of Harrington et al. (1982) who give two models for MGC 7662 with different assumptions concerning the spatial distribution of density and the spectral distribution of stellar flux.

1.3 Results for C abundances.

We can now understand why large discrepancies arose in earlier C abundance determinations, and we can obtain results of greatly improved accuracy.

Abundances of C^{2+} were obtained by Torres-Peimbert and Peimbert (1977, TPP) from observations of C II λ4267 interpreted using the effective recombination coefficient of Pengelly (1963), which Storey (1981) shows to be reliable. Although discussion continues as to whether mechanisms other than recombination contribute to excitation of λ4267 (see review by Harrington in the present volume), the main source of error in the C^{2+} abundances of TPP was probably the use of earlier photographic measurements of the λ4267 flux, which tended to be too large and hence to lead to an over-estimate of the C^{2+} abundance.

Abundances of C^{3+} obtained by Pottasch, Wesselius and van Duinen (1978, PWD) from observations of C IV λ1549 with ANS were under-estimated due to neglect of absorption of λ1549 by dust. Both TPP and PWD used ionization correction factors from models which did not take account of di-electronic recombination in calculating total recombination rates. For nebulae of higher escitation, in which most of the carbon is C^{2+} or C^{3+}, TPP over-estimated the ratios C/C^{2+} while PWD under-estimated the ratios C/C^{3+}. In consequence of these various sources of error, the C abundances of TPP were in many cases much larger than those of PWD (this was discussed by Harrington et al., 1981).

For high-excitation nebulae, in which much of the C is C^{3+}, it is difficult to obtain accurate C abundances, unless one has observations of λ2297 or sufficient confidence in predictions of models. For nebulae of medium excitation accurate results can be obtained from λ1908 and λ2326, given good estimates of T_e(C II) and T_e(C III), or from λ4267 if the observations and interpretation can be trusted. The most accurate C abundances are obtained from IUE observations of C II] λ2326 and [O II] λ2470 in low-excitation nebulae, since the C^+/O^+ ratio from these two lines is insensitive to T_e and to reddening.

In a UCL-Durham collaboration, IUE observations have been made of nine planetary nebulae selected from those for which IR observations have been made (see Aitken and Roche, 1982); five show a silicate feature and four show SiC. They are all fairly low-excitation objects, with λ1549 weak or absent. The C/O ratios are given in Table 1 and it is seen that all the objects with a silicate feature have $C/O \lesssim 0.5$ while those with a SiC feature have $C/O \gtrsim 1$. Further discussion is given by Barlow in this volume.

Table 1.

Objects with Silicate feature			Objects with SiC feature		
Nebula	C/O	Ref.	Nebula	C/O	Ref.
SwSt 1	0.5	(a)	NGC 6572	1.1	(f)
Hb 12	0.3	(b)	IC 2501	0.9	(g)
IC 4997	0.4	(c)	NGC 6790	0.8:	(g)
He 2-131	0.3	(d)	IC 418	1.3	(d)
M 1-26	0.5	(e)			

References: (a) Cohen, Flower and Goharji, this volume; (b) Flower and Penn, this volume; (c) Flower (1980); (d) Adams and Seaton (1982b); (e) Adams and Barlow, this volume; (f) Flower and Penn (1981); (g) Adams (1982)

2. DUST OPACITIES.

2.1 Theory.

The problem of radiative transfer with scattering in a spectrum line and with absorption has been solved by Hummer and Kunasz (1980) for plane parallel models. Figure 1 shows, for a mid-plane source, the escape probability f_e against scattering optical depth τ_S for two values of dust optical depth, τ_D = 0.1 and 0.2 (these results are for a ratio of natural to Doppler widths of $a = 4.7 \times 10^{-4}$, typical of values for planetary nebulae). It is of interest that, for $\log(\tau_S)$ between about 4.0 and 5.5, f_e increases as τ_S increases; this is because an increase in the scattering opacity can lead to a shorter path-length through the absorbing dust before escape occurs in the near line-wing.

2.2 Determination of dust opacities.

Harrington *et al*. (1982) construct a model of NGC 7662 which gives agreement with observations for F(λ2297)/F(λ1908) and comparing the observed value of (1.5) with the calculated value of (1.6) they deduce that f_e = 0.3 for λ1549. It follows that the dust optical depth is τ_D = 0.10. From f_e = 0.5 for λ1335 in IC 418, Clavel *et al*. obtain τ_D = 0.08.

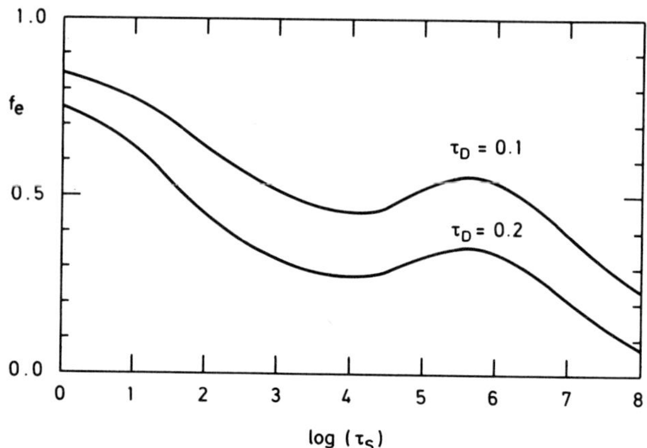

Figure 1. The escape probability f_e, for a mid-plane source, against scattering optical depth τ_S for two values of dust optical depth τ_D (from Hummer and Kunasz, 1980).

2.3 Thermal infra-red emission.

Let F(Lyα) be the Lyα flux which would be observed if there were no absorption; it can be calculated using fluxes observed in other hydrogen lines. Let F(TIR) be the total thermal infra-red emission, obtained from observations. Table 2 gives contributions to F(TIR)/F(Lyα) calculated by Harrington et al.(1982) for NGC 7662, assuming that $\tau_D = 0.1$ and that τ_D is independent of wavelength. The contributions are: absorption of C IV λ1549; absorption of Lyα; absorption of radiation from the stellar continuum (giving a contribution to F(TIR) of F(star){1-exp($-\tau_D$)} where F(star) is the total stellar flux). Adding these contributions gives a value of F(TIR)/F(Lyα) in close agreement with that from the measurements of Moseley (1980), which provides a good check on the value used for τ_D.

Table 2. Thermal infra-red emission in NGC 7662.

Contribution to F(TIR)/F(Lyα)	
C IV	0.4
Lyα	0.6
Star continuum	2.1
Total calculated	3.1
Observed	2.7

2.4 The C IV doublet.

The collision strengths for excitation of the components of the C IV doublet, $\lambda1548.2$ and $\lambda1550.8$, are proportional to the statistical weights and, allowing for small differences in excitation energies, the ratio of emissivities is

$$j(\lambda1548.2)/j(\lambda1550.8) = 2.00 \exp(-160/T_e) \qquad (2.1)$$

with T_e in K. The oscillator strengths for the two lines are also in the 2 to 1 ratio of statistical weights and it is seen from Figure 1 that a change in τ_S by a factor of 2 gives only a small change in f_e. It follows that, even if there is absorption by dust, we would not expect the ratio of fluxes,

$$R = F(\lambda1548.2)/F(\lambda1550.8) \qquad (2.2)$$

to differ significantly from the ratio of emissivities.

There has been some discussion about values of R obtained from IUE observations (see abstracts by Peña and Torres-Peimbert, and by Feibelman, in the present volume). Care is needed in view of the instrument's rather limited dynamic range. Flower et al.(1982) obtained a high-dispersion spectrum of NGC 7662 with an exposure time optimised for the measurement of R (long enough to give good signal-to-noise yet not so long as to give saturation or ITF truncation). Using extended-source data-extraction procedures they obtained $R = 1.92 \pm 0.03$ (error estimate only for noise) compared with 1.98 for the ratio of emissivities.

3. STELLAR PHOTOMETRY.

High accuracy is required if information about temperatures of central stars is to be deduced from optical and UV photometry. The first problem is that the flux distributions, dependence of F_λ(star) on λ, are insensitive to temperature (for the hotter stars the Rayleigh-Jeans limit is approached). Subsidiary problems arise from the need to correct for interstellar absorption and to subtract nebular continuum fluxes from observed fluxes (which in some cases leads to severe cancellation). One can define a colour temperature, T_C, if the dependence of F_λ(star) on λ is the same as that of $B_\lambda(T_C)$. Alternatively, one can fit F_λ(star) to fluxes of stellar atmosphere models.

Zanstra temperatures T_Z(H I) and T_Z(He II) have been used extensively for the placement of stars on the HR diagram. It has usually been assumed that those stars with T_Z(He II) > T_Z(H I) have nebulae optically thin for H I and that T_Z(He II) provides the best estimate of their effective temperatures. It is important to establish whether these assumptions are consistent with optical and UV photometry.

Using broad-band UV measurements from ANS, Pottasch et al.(1978, PWWFD) obtained values of T_C for a number of stars which were smaller than T_Z(He II) and closer to T_Z(H I). They considered that there was a need for a downward revision of many star temperatures, which had "rather far reaching" consequences for the placement of the stars on the diagram.

Subsequent studies have been made using IUE, and more recent results from optical photometry, and several such studies are reported in the present volume. A program of work at UCL has involved the use of IUE spectra from the World Data Center supplemented with new IUE observations to correct for regions of saturation or of under-exposure. All raw data have been scrutinized and fluxes have been re-extracted with correction of ITF and other errors in previous extractions. Results are presented, at this Symposium, by Clegg and Seaton for 20 hotter stars and by Adams and Barlow for 9 cooler stars.

One result which emerges clearly is that, for a number of stars, PWWFD adopted temperatures which are much too low. The case of NGC 7662 is discussed in detail by Harrington et al.(1982) who show that the central star has a temperature much larger than the 48 000 K adopted by PWWFD. Two other examples are: NGC 3242, for which T_Z(H I) = 46 000 K, T_Z(He II) = 93 000 K, PWWFD adopted 50 000 K and Clegg and Seaton obtain T_C = 90 000 K; and NGC 1535 for which T_Z(H I) = 28 000 K, T_Z(He II) = 73 000 K, PWWFD adopted 40 000 K and Clegg and Seaton obtain T_C = 65 000 K. A number of nebulae which PWWFD assumed to be optically thick for H I are almost certainly optically thin.

4. THE PLANETARY K 648 IN THE GLOBULAR CLUSTER M 15.

The planetary nebula K 648 in the globular cluster M 15 is discussed by O'Dell et al. (1964), Peimbert (1973) and Hawley and Miller (1978). From the angular diameter θ = 3".5 measured by Johnson et al.(1979) one obtains an ionized mass of $M(H^+) = 0.10\ M_\odot$ (in the Discussion following my talk it was noted that θ had been over-estimated by Johnson et al. and that a better estimate for $M(H^+)$ is $0.03\ M_\odot$).

Aurière et al. (1982) have made IUE observations of K 648 and have measured U, B and V for the stars in the IUE aperture. Using fluxes in C II] λ2326 and C III] λ1908 they obtain C/O = 2.4. This indicates that C enrichment has occurred but the unusually large value of C/O does not indicate an unusually large amount of enrichment, since the O abundance is low.

The H I Zanstra temperature of the central star is 29 000 K and the colour temperature from optical and UV observations is 33 000 K. Taking the effective temperature to be 30 000 K and the distance of M 15 to be 10 kpc, the luminosity of the central star is 1700 $L_\odot$. Figure 2 shows K 648 on the log(L), log(T) diagram for M 15; results for the cluster stars have been obtained from the V,(B-V) diagrams of Sandage (1970) and of Aurière and Cordini (1981), assuming the stars to radiate like black-bodies.

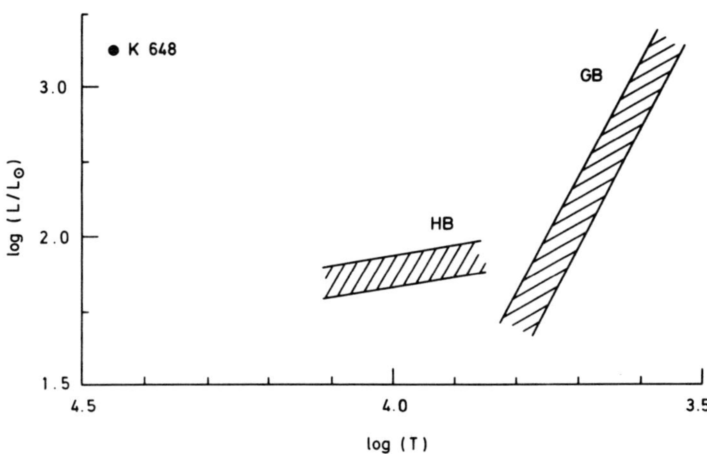

Figure 2. The log L, log T diagram for M 15 showing K 648, the horizontal branch, HB, and the giant branch, GB.

References.

1. Adams S, 1982, Mon. Not. R. astr. Soc., to be submitted.
2. Adams S and Seaton MJ, 1982a, Mon. Not. R. astr. Soc., to be submitted.
3. Adams S and Seaton MJ, 1982b, Mon. Not. R. astr. Soc., 200,7P.
4. Aitken DK and Roche PF, 1982, Mon. Not. R. astr. Soc., 200,217
5. Aurière M, Adams S and Seaton MJ, 1982, Mon. Not. R. astr. Soc., to be submitted.
6. Aurière M and Cordoni J-P, 1981, Astr. Astrophys., 100,307.
7. Clavel J, Flower DR and Seaton MJ, 1981, Mon. Not. R. astr. Soc., 197, 301.
8. Flower DR, 1980, Mon. Not. R. astr. Soc., 193, 511.
9. Flower DR and Pen CJ, 1981, Mon. Not. R. astr. Soc., 194, 13P.
10. Flower DR, Penn CJ and Seaton MJ, 1982, Mon. Not. R. astr. Soc., in press.
11. Harrington JP, Lutz JH, Seaton MJ and Stickland DJ, 1980, Mon. Not. R. astr. Soc., 191, 13.
12. Harrington JP, Lutz JH and Seaton MJ, 1981, Mon. Not. R. astr. Soc., 195, 21P.
13. Harrington JP, Seaton MJ, Adams S and Lutz JH, 1982, Mon. Not. R. astr. Soc., 199, 517.
14. Hawley SA and Miller JS, 1978, Astrophys. J. 220, 609.
15. Hummer DG and Kunasz PB, 1980, Astrophys. J., 236, 609
16. Johnson HM, Balick B and Thompson AR, 1979, Astrophys. J., 233, 919.

17. Lambert DL, 1978, Mon. Not. R. astr. Soc., 182, 249.
18. Moseley H, 1980, Astrophys. J., 238, 892.
19. O'Dell CR, Peimbert M and Kinman TD, 1964, Astrophys. J., 140, 119.
20. Peimbert M, 1973, Mém. Soc. Roy. Sci. Liège, 5, 307.
21. Pengelly RM, 1963, Thesis, London (see also Seaton MJ, 1978, Planetary Nebulae (ed Y. Terzian), p.131, Reidel).
22. Pottasch SR, Wesselius PR and van Duinen RJ, 1978, Astr. Astrophys. 70, 629.
23. Pottasch SR, Wesselius PR, Wu C-C, Fieten H and van Duinen RJ, 1978, Astron. Astrophys., 62, 95.
24. Sandage A, 1970, Astrophys. J., 162, 841.
25. Storey PJ, 1981, Mon. Not. R. astr. Soc., 195, 27P.
26. Torres-Peimbert S and Peimbert M, 1977, Rev. Mex. Astron. y. Astrophys. 2, 181.

NUSSBAUMER: I have two comments: (1) observers find that in symbiotic stars the C IV doublet ratio is often not 2 : 1, but smaller (e.g. Nussbaumer and Schild, 1981, Astron. Astrophys. 101, 118). (2) Kindl and Nussbaumer have followed up the work of Flower on IC 4997 and confirm his findings: C/O ≈ 0.5 and low C/H and O/H abundance ratios. In addition, we find N to be underabundant, relative to the Sun, by a factor 5 and Si to be underabundant by a factor of more than 100.

SEATON: As far as I know, the doublet ratio could be almost anything in symbiotic stars! I think it very unlikely to much differ from 2 for emission from PN envelopes. For these cases, it is necessary to make a careful check on observational errors where values significantly different from 2 are obtained.

ZUCKERMAN: Concerning your discussion of K 648, there is a large discrepancy between the momentum carried by the radiation field of the central star and that of the ionized gas (assuming a typical PN expansion velocity), in the sense that radiation pressure is inadequate to drive the expansion of the nebula. Therefore, either the central star was much more luminous in the recent past, or radiation pressure is not the mechanism that drives the outflow. The momentum discrepancy is sufficiently large to remain even if the ionized mass ($\approx 0.1\ M_\odot$) that you mentioned is several times over estimated.

KAHN: In connection with Zuckerman's remark, the hot stellar wind will shock against the nebular gas and then be bottled up at a temperature $\approx 10^6$K. This confined gas will exert a much larger pressure on the nebula than the radiation directly from the star.

WEIDEMANN: The mass of the nucleus of K 648, from its luminosity and the core mass / luminosity relation, is $0.55\ M_\odot$. It cannot have been higher on the AGB, since K 648 is population II, for which we expect exactly this final mass. Do you know the radius of the nebula, in order to check that the nucleus fits the Schönberner track?

SEATON: The distance of M15 is about 10 kpc and the angular diameter I quoted was 3.5 arc sec.

SCOTT: I have a comment on the radio flux density of K 648. Birkinshaw et al. have measured 4.0 ± 1.7 m Jy with the Cambridge 5 km telescope - a quarter of the early VLA measurement. The source was also unresolved (≲ 2 arc sec).

POTTASCH: The VLA measurements of Johnson et al. were made with the early, incomplete array. More recent measurements show that their flux (16 m Jy) was overestimated and should be about 3.5 m Jy. Likewise, the diameter was overestimated. The main body of the nebula is probably not larger than 1 arc sec, and most of the emission is within a diameter of 0.7 arc sec. The nebular mass is 2×10^{-2} to 3×10^{-2} $M_\odot$.

IBEN: The lower limit to the luminosity of the carbon stars in the Magellanic Clouds is approximately 3×10^3 $L_\odot$. One would expect this to map into a lower limit to the plateau luminosity of PN nuclei progeny. Is there any possibility that your estimate of 1.7×10^3 $L_\odot$ could actually be larger by a factor 2?

SEATON: I think that revision by a factor 2 is unlikely but I could not exclude the possibility.

SECTION II

PHYSICAL PROCESSES IN PLANETARY NEBULAE

RECENT ADVANCES IN ATOMIC CALCULATIONS AND EXPERIMENTS OF INTEREST IN THE STUDY OF PLANETARY NEBULAE

C. MENDOZA
Department of Physics and Astronomy, University College London, Gower Street, London WC1E 6BT.
Present address: Departamento de Fisica y Matematicas, Universidad Simon Bolivar, PO Box 80659, Caracas, Venezuela.

ABSTRACT

Recent advances in the calculation and measurement of transition probabilities, electron excitation rate coefficients and photoionization cross sections relevant to the study of planetary nebulae are discussed. A compilation of these parameters is also presented.

1 INTRODUCTION

Planetary nebulae are, in a sense, rich and challenging atomic physics laboratories. The low electron temperatures and densities of the nebular envelope give rise to a variety of atomic processes which are difficult to reproduce and measure in more accessible laboratories. The atomic data required to interpret these phenomena are therefore mainly obtained by calculation, where a clear understanding is necessary followed by an enormous computational effort. In the last decade several sophisticated computer packages have been developed to calculate atomic data with these points in mind. Advances in computer technology, such as the introduction of the CRAY-1 vector processor, have enabled a high degree of accuracy to be reached for the simpler ions, the possibility to treat consistently ions of the second row of the periodic table, and to make firm introductory attempts in the study of the larger ions such as those of iron.

In the present paper we review briefly recent developments made in the calculation and measurement of transition probabilities, electron excitation rate coefficients and photoionization cross sections. A compilation of these parameters is given in the Appendix.

2 TRANSITION PROBABILITIES

The calculation of accurate transition probabilities for forbidden and semi-forbidden lines, such as those found in nebular spectra, must

D. R. Flower (ed.), Planetary Nebulae, 143–172.

include the relativistic interaction and electron correlation effects. For light ions, where relativistic effects are small, the relativistic interaction can be treated as a perturbation by considering the Hamiltonian

$$H = H_{NR} + H_{BP} \tag{1}$$

where H_{NR} is the non-relativistic Hamiltonian and H_{BP} is the Breit-Pauli relativistic correction[1]. The relativistic wave function $\psi_i(R)$ can then be expanded in terms of the non-relativistic wave functions $\psi_j(NR)$ thus

$$\psi_i(R) = \psi_i(NR) + \sum_{j \neq i} \psi_j(NR) \frac{\langle \psi_j(NR) | H_{BP} | \psi_i(NR) \rangle}{E_i(NR) - E_j(NR)} \ldots \tag{2}$$

Electron correlation effects are usually included by the method of configuration interaction (CI). The NR wave functions are linearly expanded in a configuration basis of the form

$$\psi(NR) = \sum_k c_k \phi_k \tag{3}$$

where the configuration functions ϕ_k are built from one-electron orbitals. The convergence of the CI expansion can be slow and the transition probabilities of forbidden and semi-forbidden transitions are usually sensitive to the number of configurations in the CI expansion. In most cases configurations containing one-electron pseudo-orbitals, i.e. artificial orbitals adjusted variationally to improve term energy separations or fine-structur energy splittings, must be considered. Furthermore, if a high degree of accuracy is desired small semi-empirical adjustments cannot be avoided. For instance, Zeippen et al.[2] introduce semi-empirical term energy corrections such that $\psi(R)$ are calculated with the "exact" NR energies in equation (2), and calculate the A-values with the experimental energy level separations.

In the past these interactions were mainly treated by semi-empirical methods with a fair amount of success. Work such as that by Garstang[3], Czyzak and Krueger[4] and McKim-Malville and Berger[5] on forbidden lines, and Laughlin and Victor[6] on intercombination lines have provided reliable results for many years. The present trend is to carry out large CI calculations where such interactions are included more rigorously.

Perhaps the most significant recent theoretical development made in connection with transition probabilities relevant to planetary nebulae is that by Eissner and Zeippen[7] on the transition probabilities for forbidden lines within the np^3 configuration of the nitrogen and phosphorus isoelectronic sequences. They have shown that higher-order relativistic corrections to the magnetic dipole operator[8] (up to order $\alpha^2 Z^2$), which are usuall negligible, must be taken into account in order to obtain accurate A-values for the low members of the sequence, and thereby ending a long-standing discrepancy between the observed and theoretical values for the density sensitive line intensity ratio

$$r(N_e) = I(^2D^o_{\frac{5}{2}} - {}^4S^o_{\frac{3}{2}})/I(^2D^o_{\frac{3}{2}} - {}^4S^o_{\frac{3}{2}}) \quad (4)$$

in the high density limit. For $N_e \to \infty$

$$r(\infty) = \frac{3}{2} A(^2D^o_{\frac{5}{2}} - {}^4S^o_{\frac{3}{2}})/A(^2D^o_{\frac{3}{2}} - {}^4S^o_{\frac{3}{2}}), \quad (5)$$

where A(i-j) is the transition probability, and thus provides a useful comparison between observation and theory. In table 1 we compare $r(\infty)$ for N^o, O^+ and S^+ calculated in this way (MZ) with previous theoretical results (G) and observations (OBS).

ION	OBS	MZ	G
N^o	$\lesssim 0.51$	0.54	0.65
O^+	0.35	0.35	0.42
S^+	0.45	0.44	0.39

TABLE 1. Observed and theoretical values for $r(\infty)$ (see text). OBS(ref. 9, 10), MZ(ref. 11, 12) and G(ref. 3).

Also in connection with forbidden lines, Nussbaumer and Rusca[13], Zeippen[11] and Mendoza and Zeippen[12] have carried out extensive calculations of A-values for transitions within the np^q configurations (n=2-3; q=2-4; Z=6-28), examining the importance of CI, relativistic corrections and semi-empirical adjustments. They have improved substantially on the earlier semi-empirical results compiled by Garstang[3] and McKim-Malville and Berger[5]. Nussbaumer and Storey[14] are systematically studying the ions of Fe.

There have been a considerable number of detailed calculations on the intercombination lines of astrophysical interest. The work of Glass and Hibbert[15] and Nussbaumer and Storey[16] on the Be-like ions may be chosen to illustrate the main features of these calculations. They demonstrate the high sensitivity of the transition probabilities to the values of small coefficients c_k of the CI expansion, and stress the need to use large configuration bases containing pseudo-orbitals and to make semi-empirical term energy corrections. Although different numerical methods are used the agreement is better than 10%, in marked contrast with recent work by Cowan *et al.*[17] on f-values for astrophysical intercombination lines where limited configuration bases are used and an accuracy of only 50% is claimed.

Even though a perturbative treatment of the relativistic interaction for light ions can be regarded as satisfactory, there are also numerical methods based on the more formal solution of the Dirac equation. Hata and Grant[18] have used the multi-configuration Dirac-Fock method, which allows for correlation effects, to compute the lifetime of the astrophysically important 2^3S state of He^o. They obtain a value of 8.8×10^3 sec in close agreement with the experimental result[19] of $9\pm3 \times 10^3$ sec. This method has also been used recently[20] to compute a large number of E1, E2 and M1 transition probabilities within the n=2 complex with moderate accuracy for low Z.

The experimental techniques employed to measure lifetimes for allowed transitions are in general not suited for forbidden transitions and, consequently, there have been few experimental results. Corney and Williams[21] have measured the radiative decay constant of the $^1S_0$ metastable state of O^o by the pulsed afterglow method. They obtain $\Gamma_{expt} = 1.31\pm0.05$ sec^{-1} in excellent agreement with preliminary results by Mendoza and Zeippen of $\Gamma_{th} = 1.29$ sec^{-1} using the CI method. There is also a new interesting technique being developed by Nightingale[22] to measure transition probabilities for forbidden lines of singly-ionized species. He is trying to measure the A-value for the $^1S_0 - {^1D_0}$ ($\lambda5754$) transition of N^+ by the long-path absorption spectroscopy method. Due to the small ionic absorption coefficients optical paths greater than 1 Km are required which he obtains by using a plasma source ten metres long and a multiple-pass optical laser system.

3 ELECTRON EXCITATION RATE COEFFICIENTS

Some of the problems encountered in the calculation of electron impact excitation cross sections of positive ions of astrophysical interest have been discussed by Mendoza[23]. The electron excitation rate coefficient for a transition between levels $j \to i$ $(i>j)$ at an electron temperature T_e can be expressed in terms of the effective collision strength $\Upsilon_{ij}(T_e)$ (see Appendix), where

$$\Upsilon_{ij}(T_e) = \int_0^\infty \Omega(i,j)\,\exp(-\varepsilon_i/kT_e)\,d(\varepsilon_i/kT_e), \qquad (6)$$

ε_i is the energy of the electron with respect to the ith level, k is the Boltzmann constant and $\Omega(i,j)$ is the collision strength. For the low temperatures found in nebulae the cross section (i.e. $\Omega(i,j)$) must be calculated at low energies ($\varepsilon_i \lesssim 1$ Ryd) where it is dominated by threshold effects such as resonances and the opening of new reaction channels. To be able to reproduce these effects detailed descriptions of the target ion and the collision process are needed. Moreover, for light ions relativistic effects are usually neglected and LS coupling assumed; cross sections between fine-structure levels are then calculated by algebraic transformations of the reactance matrix[24].

The total wave function of the N-electron target ion + incoming electron is expanded in terms of the target eigenfunctions by the close-coupling (CC) expansion

$$\Psi(LS) = \sum_i \mathcal{A} \chi_i \theta_i + \sum_j c_j \Phi_j . \tag{7}$$

Each term in the first summation gives rise to an interaction channel; χ_i are the target eigenfunctions, θ_i the incident-electron functions and $\mathcal{A}$ is an antisymmetrising operator. Φ_j take the form of bound-state configurations for the (N+1)-electron system and account for short-range correlations effects. The variational principle leads to a set of coupled integro-differential equations of the general form

$$\left\{ \frac{d^2}{dr^2} - \frac{\ell_i(\ell_i+1)}{r^2} + \frac{2Z}{r} + \varepsilon_i \right\} F_i(r) + \sum_{i'} W_{ii'} F_{i'}(r) = 0 \tag{8}$$

where $F_i(r)$ is the ith channel electron radial functions. The inter-channel coupling potentials $W_{ii'}$ give rise to the threshold effects mentioned above, and sufficient channels must be included in the CC expansion to account for the effects present in the region of interest. In earlier methods such as the Coulomb-Born and distorted wave the radial functions $F_i(r)$ were calculated neglecting this coupling, and in the exact resonance approximation[26] the important quadrupole p-wave interactions were introduced perturbatively. At present the coupled equations are solved numerically by different methods such as the Linear Algebraic (IMPACT)[27], the R-matrix[28] and the Non-iterative Integral Equation (NIEM)[29] methods. The Linear Algebraic method, in particular, can also be used to calculate bound-state energies for the (N+1)-electron system which leads to an indirect estimate of the accuracy of the cross sections.

Advances in this field have been boosted by the introduction of supercomputers. It is now possible to study ions of the first and second rows of the periodic table with accurate CI target wave functions and sufficient target states in the CC expansion to ensure convergence. There have also been attempts to treat the ions of Fe with some consistency[14]. The carbon and silicon sequences have been studied extensively; in figure 1 we plot $z^2 T_{ij}(10000)$ (which tends to a constant limit as $z \to \infty$) as a function of z, where z is the effective charge of the ion, for the ^3P - ^1D and ^1D - ^1S forbidden transitions of the np^2 configurations. It can be seen that the older results obtained with the exact resonance and distorted wave approximations can show large differences from recent close-coupling results.

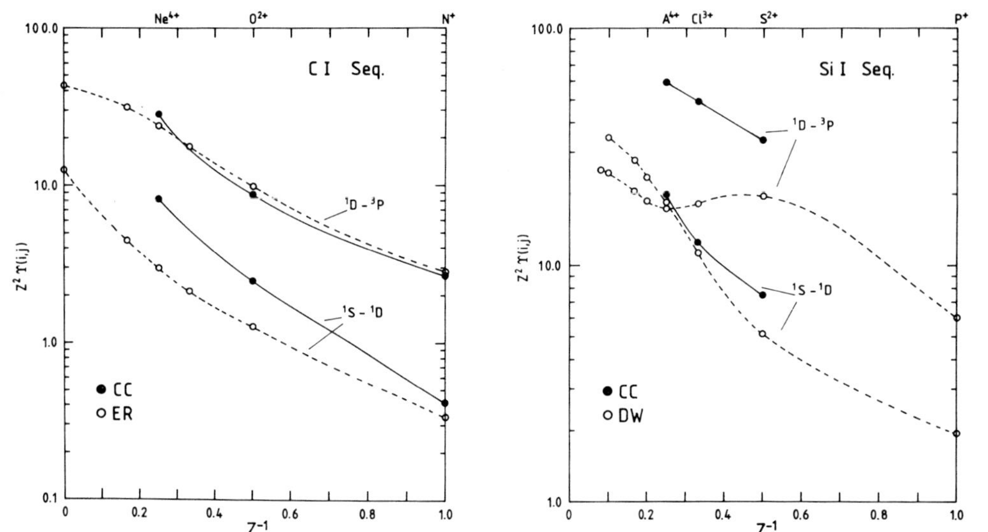

FIGURE 1. $z^2T(10000)$ as a function of z (the effective charge of the ion) for the ^3P-^1D and ^1D-^1S forbidden transitions of the carbon and silicon sequences. CC - recent close-coupling results; ER - exact resonance approximation (ref.26); DW - distorted wave results (ref.35).

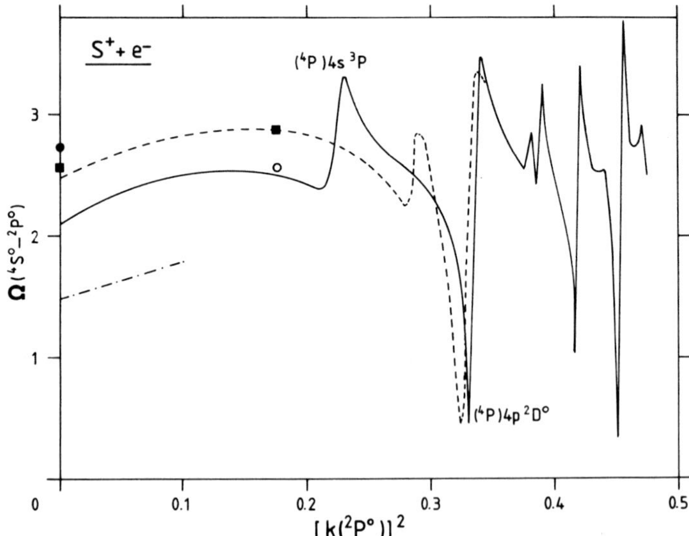

FIGURE 2. Total collision strength for the $^4S^0$-$^2P^0$ transition of S^+. Work by Pradhan[36] in 3CC+CI (filled circles) and DW (filled squares) approximations; DW results by Krueger and Czyzak[35] (dot-dashed curve); recent results by Mendoza[34] in a 6CC+CI approximation using the R-matrix method (solid curve) and Linear Algebraic method (dashed curve).

For some cases, such as the forbidden transitions of O^{2+} (ref. 30, 31), Ne^{4+} (ref. 32, 33) and S^+ (ref. 34), there are comparisons using different close-coupling methods which gives a more realistic indication of the accuracy of the collision rates. In general the agreement is very good, but some transitions show discrepancies greater than the error margin assigned by the authors ($\lesssim 10\%$). This is mainly due to shifts in the resonance positions as illustrated in figure 2 for the $^4S^o - ^2P^o$ transition of S^+. Mendoza[34] has carried out comparative calculations for this transition using the Linear Algebraic and R-matrix methods. Although the same approximation was used in both calculations (6CC+CI) the target orbitals employed by these two packages are calculated by different methods: the former uses orbitals calculated in a Thomas-Fermi statistical model potential whereas the latter uses Hartree-Fock orbitals. The earlier results by Krueger and Czyzak[35] and Pradhan[36] are also shown.

In recent years there has also been progress in the relativistic treatment of the scattering problem. For intermediate-weight atoms the relativistic effects can be included by using the Breit-Pauli Hamiltonian; Scott and Burke[37] have combined this approach with the R-matrix method to calculate excitation cross sections for Fe XXIII finding large discrepancies with the non-relativistic case. For heavier ions it becomes necessary to base the scattering problem on the Dirac equation; Norrington and Grant[38] have included the Dirac Hamiltonian in the R-matrix formalism and tested their code with the electron scattering from Ne^+. They find excellent agreement with previous non-relativistic calculations in support of the approximation mentioned above.

The most outstanding experimental work on electron impact excitation of multiply-ionized ions is that by Dunn and co-workers on the resonance transition of the Li, Na and K isoelectronic sequences using the crossed beams technique[39,40,41]. They obtain absolute cross sections for C^{3+} and N^{4+} which agree with close-coupling calculations to better than 10%. This method has been refined recently to pick up the resonance structure found in the theoretical excitation cross section of Mg^+ (ref. 42), and preliminary results show satisfying agreement.

4 PHOTOIONIZATION CROSS SECTIONS

Experiments show that in most cases the photoionization continuum of light atoms is dominated by special features such as resonances, Cooper minima and delayed maxima. A similar situation is found by theory for the lowly-ionized species.

At present there are two tendencies in the calculation of photoionization cross sections for the large number of atoms and ions needed in astrophysics. On the one hand, there are detailed calculations which include electron correlation effects and aim at bringing out the complicated resonance structure and features of the cross section. Methods such as the close-coupling (described in the previous section) and Many-Body Perturbation Theory[43] are used. On the other hand, there are calcu-

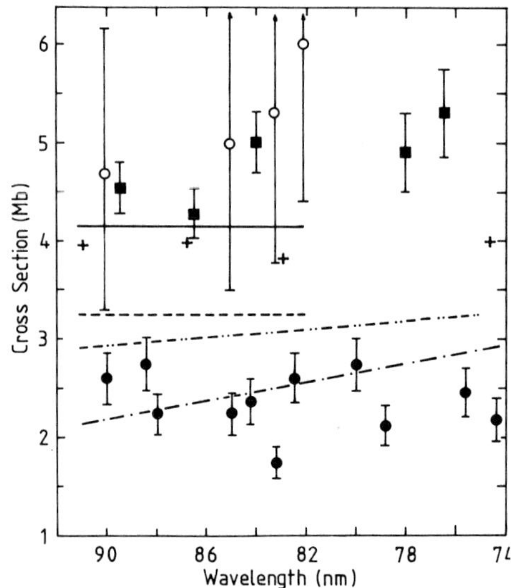

FIGURE 3. Present state of knowledge of the photoionization cross section of the ground state of O^{0} in the threshold region. Experiments by Cairns and Samson[47] (open circles); Comes et al.[48] (filled circles); Kohl et al.[49] (filled squares). Theoretical results by Henry[50] in the length (dot-dash-dotted curve) and velocity (dot-dashed curve) formulations; close-coupling results by Taylor and Burke[51] in the length (solid curve) and velocity (dashed curve) formulations; close-coupling results by Pradhan and Saraph[52] in the length formulation (crosses). Figure originally by Kohl et al.

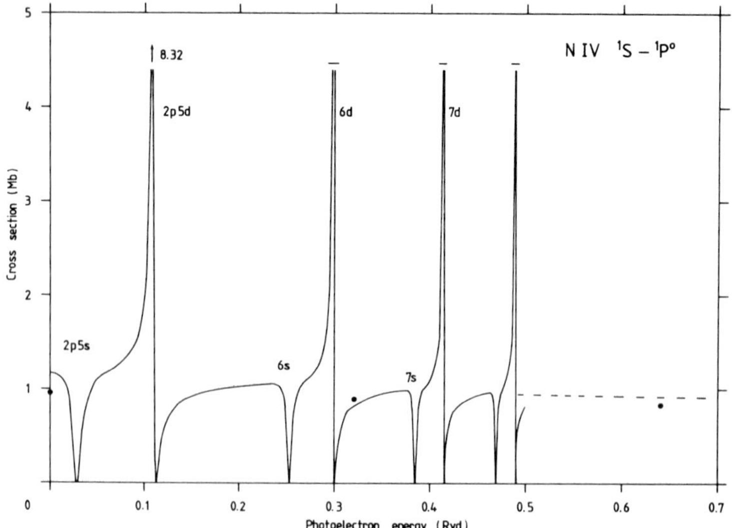

FIGURE 4. Photoionization cross section of the ground state of N^{3+}. Solid curve - close-coupling results by Butler et al. (to be published); dashed curve - average over resonances; filled circles - quantum defect results.

lations based on central potential approximations[44,45] or semi-empirical methods such as the one-channel Quantum Defect Theory[46] which only attempt to obtain the general behaviour of the cross section over a broad energy range, and these results are widely used in astrophysics.

In figure 3 we summarize the present state of knowledge of the photoionization cross section of the ground state of O^{o} in the threshold region. It has taken almost a decade for experiment and theory to converge to a common result. The central potential calculation of Reilman and Manson[44] (not shown) grossly over-estimates the cross section in this region. In figure 4 the photoionization cross section of the ground state of N^{3+} is shown. It can be seen that correlation effects (resonances) are still conspicuous even for a fairly stripped ion, but the background of the cross section is obtained satisfactorily by the Quantum Defect method. Reilman and Manson[44] also obtain the background accurately in this region.

There is good agreement between theory and experiment for the rare gases, C^{o}, N^{o}, O^{o} and $A\ell^{o}$, but there are long-standing discrepancies for the alkalis. For ions there are no absolute experimental results, but it is possible for some numerical methods, such as the Linear Algebraic, to compute f-values at the same time. These can be compared with experiment to obtain an indirect estimate of the absolute value of the cross section.

There is also recent work on the relativistic effects of the photoionization cross section. Taylor and Scott[53] have calculated the resonance structure between the J=3/2 and 1/2 levels of the A^{+} ground state obtaining good agreement with experiment. Chang[54] has included the Dirac Hamiltonian in the R-matrix formalism and calculated the photoionization cross section of the ground state of Ne^{o}. He finds close agreement with previous non-relativistic results.

5 FUTURE WORK

There remains a lot of work to be done in this interesting field. We can summarize the priority areas as follows.

(i) Radiative and collision rates for transitions to the n=3 states of He^{o}.

(ii) Fine-structure collision rates for the intercombination lines of ions of the B isoelectronic sequence.

(iii) Transition probabilities and fine-structure collision rates for ions of the Mg and Aℓ sequences.

(iv) Comparative CI calculations of transition probabilities for forbidden lines of the np^{q} configurations, particularly of neutrals, using different types of orbitals. At the moment there is only a detailed comparison for O^{2+} (ref. 55).

(v) Fine-structure collision rates for the IR forbidden lines of ions of the first and second row.

(vi) Collision rates for ions of the second tow such as Si, Cℓ and A.

(vii) Radiative and collision rates for the larger ions.

(viii) Detailed photoionization cross sections of atoms and ions, particularly from excited states and inner shells. Presentation of the data is also an important factor here.

ACKNOWLEDGEMENTS

I am indebted to Professor MJ Seaton for his constant encouragement throughout the course of this work and to Drs. KM Aggarwal (Belfast), MA Hayes (UCL) and K Giles for allowing me to include some of their results in the compilation before publication. I am grateful to Dr. KA Berrington (Belfast) for assistance with the Daresbury Laboratory Data Bank. Finally, I acknowledge enjoyable and fruitful collaboration with Drs. CJ Zeippen (Meudon) and K Butler (UCL).

REFERENCES

1. Jones M, 1970, J.Phys.B, 3, 1571; 1971, 4, 1422.
2. Zeippen CJ, Seaton MJ and Morton DC, 1977, Mon.Not.R.astr.Soc., 181, 527.
3. Garstang RH, 1968, "Planetary Nebulae", eds. Osterbrock & O'Dell, (Dordrecht; Reidel), p.143.
4. Czyzak SJ and Krueger TK, 1963, Mon.Not.R.astr.Soc., 126, 177; 1965, 129, 103.
5. McKim-Malville J and Berger RA, 1965, Planet.Space Sci., 13, 1131.
6. Laughlin C and Victor GA, 1974, Astrophys.J., 192, 551; 1979, 234, 407
7. Eissner W and Zeippen CJ, 1981, J.Phys.B, 14, 2125.
8. Drake GWF, 1971, Phys.Rev.A, 3, 908.
9. Kaler JB, Aller LH, Czyzak SJ and Epps HW, 1976, Astrophys.J.Suppl.Ser 31, 163.
10. Dopita MA, Mason DJ and Robb WD, 1976, Astrophys.J., 207, 102.
11. Zeippen CJ, 1982, Mon.Not.R.astr.Soc., 198, 111.
12. Mendoza C and Zeippen CJ, 1982, Mon.Not.R.astr.Soc., 198, 127; 199, 1025; in press.
13. Nussbaumer H and Rusca C, 1979, Astron.Astrophys., 72, 129.
14. Nussbaumer H and Storey PJ, 1978, Astron.Astrophys., 70, 37; 1980, 89, 308; 1982, in press.
15. Glass R and Hibbert A, 1978, J.PhysB, 11, 2413.
16. Nussbaumer H and Storey PJ, 1978, Astron.Astrophys., 64, 139; 1979, 74, 244.
17. Cowan RD,Hobbs LM and York DG, 1982, Astrophys.J., 257, 373.
18. Hata J and Grant IP, 1981, J.Phys.B, 14, 2111.
19. Woodworth JR and Moos HW, 1975, Phys.Rev.A, 12, 2455.
20. Cheng KT, Kim YK and Desclaux JP, 1979, At.Dat.Nucl.Dat.Tables, 24, 1
21. Corney A and Williams OM, 1972, J.Phys.B, 5, 686.
22. Nightingale MPS, 1982. Ph.D. thesis, University of London.
23. Mendoza C, 1981, Rev.Mex.Astron.Astrofis., 6, 285.
24. Saraph HE, 1972, Comput.Phys.Commun., 3, 256.
25. Seaton MJ, 1975, Adv.Atom.Molec.Phys., 11, 83.
26. Saraph HE, Seaton MJ and Shemming J, 1969, Phil.Trans.R.Soc.A, 264, 7

27. Seaton MJ, 1974, J.Phys.B, 7, 1817.
28. Burke PG, Hibbert A and Robb WD, 1971, J.Phys.B, 4, 153.
29. Smith ER and Henry RJW, 1973, Phys.Rev.A, 7, 1585.
30. Eissner W and Seaton MJ, 1974, J.Phys.B, 7, 2533.
31. Baluja KL, Burke PG and Kingston AE, 1981, J.Phys.B, 14, 119.
32. Giles K, 1979, Mon.Not.R.astr.Soc., 187, 49P.
33. Baluja KL, Burke PG and Kingston AE, 1980, J.Phys.B, 13, 4675.
34. Mendoza C, 1982, to be published.
35. Krueger TK and Czyzak SJ, 1970, Proc.R.Soc.Lond.A, 318, 531.
36. Pradhan AK, 1976, Mon.Not.R.astr.Soc., 177, 31.
37. Scott NS and Burke PG, 1980, J.Phys.B, 13, 4299.
38. Norrington PH and Grant IP, 1981, J.Phys.B, 14, L261.
39. Taylor PO, Gregory D, Dunn GH, Phaneuf RA and Crandall DH, 1977, Phys. Rev.Lett., 39, 1256.
40. Gregory D, Dunn GH, Phaneuf RA and Crandall DH, 1979, Phys.Rev.A, 20, 410.
41. Taylor PO and Dunn GH, 1973, Phys.Rev.A, 8, 2304.
42. Mendoza C, 1981, J. Phys.B, 14, 2465.
43. Kelly HP, 1964, Phys.Rev, 136, B896.
44. Reilman RF and Manson ST, 1979, Astrophys.J.Suppl.Ser., 40, 815; 1981, 46, 115.
45. Hofsaess D, 1979, At.Dat.Nucl.Dat.Tables, 24, 285.
46. Peach G, 1967, Mem.R.astr.Soc., 71, 13.
47. Cairns RB and Samson JAR, 1965, Phys.Rev, 139, A1403.
48. Comes FJ, Speier F and Elzer A, 1968, Z. Naturforsch, 23a, 125.
49. Kohl JL, Lafyatis GP, Palenius HP and Parkinson WH, 1978, Phys.Rev.A, 18, 571.
50. Henry RJW, 1967, Planet.Space Sci., 1967, 15, 1747.
51. Taylor KT and Burke PG, 1976, J.Phys.B, 9, L353.
52. Pradhan AK and Saraph HE, 1977, J.Phys.B, 10, 3365.
53. Taylor KT and Scott NS, 1981, J.Phys.B, 14, L237.
54. Chang JJ, 1977, J.Phys.B, 10, 3195.
55. Baluja KL and Doyle JG, 1981, J.Phys.B, 14, L11.

APPENDIX

COMPILATION OF TRANSITION PROBABILITIES, ELECTRON EXCITATION RATE COEFFICIENTS AND PHOTOIONIZATION CROSS SECTIONS

We include in this section a selected and critically evaluated compilation of transition probabilities, electron excitation rate coefficients and photoionization cross sections for use in the study of planetary nebulae. We have attempted to present the data in a clear, practical and self-explanatory manner, but find it necessary to discuss briefly the general arrangements of the tables.

1 INDICES

Table 1 is an index of the ions and transitions for which A-values and electron excitation rate coefficients are included in this compilation. It also contains the numbers of the relevant tables (tables 2 - 9), the methods used, estimates of accuracy and the reference sources (referenced at the end of the Appendix). Similarly, table 10 contains a selected bibliography of photoionization cross sections; for each ion, the initial stages for which cross sections have been obtained, the states of the final ion, the method, accuracy and the reference source are given.

2 METHODS

The following abbreviations are used:
CI: Configuration interaction method.
nCC: n-state close-coupling approximation.
MV: Matrix variational method.
ER: Exact resonance approximation.
DW: Distorted wave approximation.
CBII: Unitarised Coulomb-Born approximation.
MP: Polarization model potential calculations.
CP: Central potential calculations.
MBPT: Many-body perturbation theory.
TDHF: Time-dependent Hartree-Fock method.
SE: Semi-empirical calculations.
TABLES: Values obtained from NBS tables.
INT(EXT): Interpolated (Extrapolated).
COMP: Compilation of several methods.
EXPT: Experiment.

3 ACCURACY

The following scheme is used:
A: Uncertainties within 10%
B: " " 20%

C: Uncertainties within 30%
D: " " 40%

A + sign means "much better than". Accuracy ratings given in this work should only be treated as a rough guide as it is very difficult to critically evaluate these data due to the general lack of experimental results, theoretical comparisons and even error estimates provided by the authors themselves. Furthermore, it is not practical within the present context to give estimates of uncertainties for every transition within a multiplet or configuration, and in some cases they can be significantly different. For instance, for the ns^2 - nsnp transitions of the Be and Mg sequences, it is the transition probability for the intercombination line $^1S_0 - {}^3P^o_1$ that shows the greater uncertainties whereas the A-values for the other transitions are probably correct to 5%; for the forbidden transitions within the np^q configurations the accuracy of the M1 transition probabilities is appreciably greater than that of the E2 type; also the accuracy of the photoionization cross section of the ground state of an ion can differ from that of the excited states, and the error in the cross sections can also vary with the energy region particularly where resonances are present.

4 TRANSITION PROBABILITIES AND EXCITATION RATE COEFFICIENTS

Tables 2 - 9 are arranged in isoelectronic sequences. For each transition they list: the observed energy level separation $\Delta E_{ij}(cm^{-1})$, the transition probability $A_{ij}(sec^{-1})$, and the effective collision strength $\Upsilon_{ij}(T_e)$ as a function of electron temperature. The ith level always corresponds to the upper level unless ΔE appears with a negative sign (the A and Υ values are always for the downward transition). The electron de-excitation rate coefficient, q_{ij}, can be obtained from the relation

$$q_{ij}(T_e) = \frac{8.63 \times 10^{-6}\, \Upsilon_{ij}(T_e)}{\omega_i\, T_e^{1/2}} \qquad (cm^3 sec^{-1})$$

where ω_i is the statistical weight of the ith level and T_e is the electron temperature in ^oK. The excitation rate coefficient is given by

$$q_{ji} = (\omega_i/\omega_j)\, q_{ij}\, \exp(-\Delta E_{ij}/kT_e) \qquad (i>j)$$

where k is the Boltzmann constant (1/k = 1.43883 cm K).

When only one value of Υ is given it is assumed to be temperature independent to within the accuracy of the calculation. A value of the effective collision strength bracketed by arrows (← Υ →) corresponds to the value for the whole multiplet (LS coupling); in this case the Υ's for the fine-structure transitions can be obtained from the ratios of the statistical weights of the multiplet levels. For the $^2P^o_J - {}^4P_{J'}$ transition of the B and Aℓ isoelectronic sequences both the total Υ for the multiplet and that for each fine-structure transition are given, as in some cases they have been computed by different methods (eg N^{2+}) or, as explained by the author (eg O^{3+}), the sum of the fine-structure components does not add up to the total LS value given.

Finally, the notation $a \pm b$ signifies $a \times 10^{\pm b}$.

NOTE: The author kindly requests users to quote the original sources whenever data from this compilation are referenced.

ION	TRANSITION(S)		A_{IJ}			Υ_{IJ}		
		TABLE	METHOD	ACC	SOURCE	METHOD	ACC	SOURCE
H^o	n=2	2	TABLES	A+	1	CC	B+	126
He^o	n=2	2	COMP	A	1,2,3,4,5	5CC	C+	6
C^o	$2p^2\ ^1S,\ ^1D,\ ^3P$	6	CI	A	7	CC+CI	C	8
C^o	$2s2p^3\ ^5S^o_2 - 2s^22p^2\ ^3P_J$	6	CI	B	39	MV		9
C^+	$2s2p^2\ ^4P_J - 2s^22p\ ^2P^o_{J'}$	5	CI	B+	10	8CC+CI	B	11
C^+	$2s2p^2\ ^4P - 2s^22p\ ^2P^o$	5				5CC+CI	B+	12
C^{2+}	$2s2p\ ^1P^o_1,\ ^3P^o_J,\ 2s^2\ ^1S_0$	4	CI	A	13	6CC+CI	A	14
C^{3+}	$2p\ ^2P^o_J - 2s\ ^2S_{\frac{1}{2}}$	3	TABLES	A+	1	EXPT;5CC	A+	15;16
N^o	$2p^3\ ^2P^o,\ ^2D^o,\ ^4S^o$	7	CI	A	17	8CC+CI	A	18
N^o	$^2P^o_J - ^2D^o_{J'}$	7	CI	A	17	CC	C	19
N^+	$2p^2\ ^1S,\ ^1D,\ ^3P$	6	CI	A+	7	CC+CI	A	20
N^+	$2p^2\ ^3P_J - ^3P_{J'}$	6	CI	A+	7	ER	C	21
N^+	$2s2p^3\ ^5S^o_2 - 2s^22p^2\ ^3P_J$	6	CI	B+	22	7CC+CI	B+	12
N^{2+}	$2s2p^2\ ^4P_J - 2s^22p\ ^2P^o_{J'}$	5	CI	B+	23	DW	C	23
N^{2+}	$2s2p^2\ ^4P - 2s^22p\ ^2P^o$	5				6CC+CI	A	24
N^{3+}	$2s2p\ ^1P^o_1,\ ^3P^o_J,\ 2s^2\ ^1S_0$	4	CI	A	25	INT	B+	
N^{4+}	$2p\ ^2P^o_J - 2s\ ^2S_{\frac{1}{2}}$	3	TABLES	A+	1	2CC	A+	26,27
O^o	$2p^4\ ^1S,\ ^1D,\ ^3P$	8	CI	A	28	6CC+CI	A	18
O^o	$2p^4\ ^3P_J - ^3P_{J'}$	8	CI	A	28	MV	B+	29
O^+	$2p^3\ ^2P^o_J,\ ^2D^o_{J'},\ ^4S^o_{\frac{3}{2}}$	7	CI	A+	17	5CC+CI	A	30
O^{2+}	$2s2p^3\ ^5S^o,\ 2s^22p^2\ ^1S,\ ^1D,\ ^3P$	6	CI	A+	31	12CC+CI	A	32
O^{2+}	$2p^2\ ^3P_J - ^3P_{J'}$	6	CI	A+	31	12CC+CI	A	33
O^{3+}	$2s2p^2\ ^4P_J - 2s^22p\ ^2P^o_{J'}$	5	CI	B	34	7CC+CI	A	34
O^{4+}	$2s2p\ ^1P^o_1,\ ^3P^o_J,\ 2s^2\ ^1S_0$	4	CI	A	25	6CC+CI	A	14
O^{5+}	$2p\ ^2P^o_J - 2s\ ^2S_{\frac{1}{2}}$	3	TABLES	A+	1	CBII	A	26

TABLE 1. Index of transition probabilities, A_{ij}, and effective collision strengths, Υ_{ij}, for transitions in ions of interest in the study of planetary nebulae. For each ion the number of the relevant table (tables 2 - 9), the method used, an estimate of the accuracy and the reference source (referenced at the end of the appendix) is given.

ION	TRANSITION(S)	TABLE	A_{IJ} METHOD	A_{IJ} ACC	A_{IJ} SOURCE	Υ_{IJ} METHOD	Υ_{IJ} ACC	Υ_{IJ} SOURCE
Ne^{+}	$2p^5\ ^2P^o_{1/2} - ^2P^o_{3/2}$	9	CI	A+	28	ER(adj.)	A	20
Ne^{2+}	$2p^4\ ^1S,\ ^1D,\ ^3P$	8	CI	A+	28	4CC+CI	A	35
Ne^{2+}	$2p^4\ ^3P_J - ^3P_{J'}$	8	CI	A+	28	ER	C	21
Ne^{3+}	$2p^3\ ^2P^o_J,\ ^2D^o_{J'},\ ^4S^o_{3/2}$	7	CI	A+	17	9CC+CI	A	36
Ne^{4+}	$2p^2\ ^1S,\ ^1D,\ ^3P$	6	CI	A+	7	12CC+CI	A	37
Ne^{4+}	$2s2p^3\ ^5S^o_2 - 2s^22p^2\ ^3P_J$	6	CI	A	39	12CC+CI	A	37
Ne^{4+}	$2p^2\ ^3P_J - ^3P_{J'}$	6	CI	A+	7	ER	C	21
Ne^{5+}	$2s^22p\ ^2P^o_{3/2} - ^2P^o_{1/2}$	5	TABLES	A+	1	ER	C	21
Ne^{6+}	$2s2p\ ^1P^o_1,\ ^3P^o_J,\ 2s^2\ ^1S_0$	4	CI	A	25	6CC+CI	A	38
Na^{2+}	$2p^5\ ^2P^o_{1/2} - ^2P^o_{3/2}$	9	CI	A+	28	ER	C	21
Na^{3+}	$2p^4\ ^1S_0,\ ^1D_2,\ ^3P_J$	8	CI	A+	28	ER	C	21
Na^{4+}	$2p^3\ ^2P^o_J,\ ^2D^o_{J'},\ ^4S^o_{3/2}$	7	CI	A+	17	ER	C	21
Mg^{o}	$3s3p\ ^3P^o_1,\ ^3P^o_J,\ 3s^2\ ^1S_0$	4	CI,MP	B	39,40	DW,CC		41,42
Mg^{+}	$3p\ ^2P^o_J - 3s\ ^2S_{1/2}$	3	MP	A+	43	4CC	A	44
Mg^{3+}	$2p^5\ ^2P^o_{1/2} - ^2P^o_{3/2}$	9	CI	A+	28	ER	C	21
Mg^{4+}	$2p^4\ ^1S_0,\ ^1D_2,\ ^3P_J$	8	CI	A+	28	ER	C	21
Mg^{5+}	$2p^3\ ^2P^o_J,\ ^2D^o_{J'},\ ^4S^o_{3/2}$		CI	A+	17	ER	C	21
Si^{o}	$3p^2\ ^1S_0,\ ^1D_2,\ ^3P_J$	6	CI	A	45	DW		46
Si^{+}	$3s3p^2\ ^4P - 3s^23p\ ^2P^o$	5	CI	B	47	7CC+CI	B	48
Si^{2+}	$3s3p\ ^1P^o_1,\ ^3P^o_J,\ 3s^2\ ^1S_0$	4	CI,MP	B	39,40	12CC+CI	B+	49
Si^{3+}	$3p\ ^2P^o_J - 3s\ ^2S_{1/2}$	3	1CC+MP	A+	28	DW	B	26,50
Si^{5+}	$2p^5\ ^2P^o_{1/2} - ^2P^o_{3/2}$	9	CI	A+	28	ER	C	21
S^{o}	$3p^4\ ^1S_0,\ ^1D_2,\ ^3P_J$	8	CI	A	51			
S^{+}	$3p^3\ ^2P^o_J,\ ^2D^o_{J'},\ ^4S_{3/2}$	7	CI	A+	52	6CC+CI	A	53
S^{2+}	$3p^2\ ^1S,\ ^1D,\ ^3P$	6	CI	A+	45	12CC+CI	B+	53
S^{2+}	$3p^2\ ^3P_J - ^3P_{J'}$	6	CI	A+	45	7CC+CI	B+	53
S^{3+}	$3s3p^2\ ^4P_J - 3s^23p\ ^2P^o_{J'}$	5	CI	B+	54	6CC+CI	B+	54
S^{4+}	$3s3p\ ^1P^o,\ ^3P^o,\ 3s^2\ ^1S$	4	CI,MP	B	39,40	5CC+CI	B+	56
S^{4+}	$3s3p\ ^3P^o_J - ^3P^o_{J'}$	4	CI	A+	39	DW(EXT)		55
S^{5+}	$3p\ ^2P^o_J - 3s\ ^2S_{1/2}$	3	1CC+MP	A+	28	DW	B	26,50

TABLE 1. (continued)

ION	TRANSITION(S)		A_{IJ}			Υ_{IJ}		
		TABLE	METHOD	ACC	SOURCE	METHOD	ACC	SOURCE
Cl^o	$3p^5\ {}^2P^o_{\frac{1}{2}} - {}^2P^o_{\frac{3}{2}}$	9	CI	A+	28			
Cl^+	$3p^4\ {}^1S_0,\ {}^1D_2,\ {}^3P_J$	8	CI	A+	51	DW	D	57
Cl^{2+}	$3p^3\ {}^2P^o_J,\ {}^2D^o_{J'},\ {}^4S^o_{\frac{3}{2}}$	7	CI	A+	52	DW	D	57
Cl^{3+}	$3p^2\ {}^1S,\ {}^1D,\ {}^3P$	6	CI	A+	45	12CC+CI	B	53
Cl^{3+}	$3p^2\ {}^3P_J - {}^3P_{J'}$	6	CI	A+	45	DW	D	57
Cl^{4+}	$3s^2 3p\ {}^2P^o_{\frac{3}{2}} - {}^2P^o_{\frac{1}{2}}$	5	CI	A+	28	DW	D	57
A^+	$3p^5\ {}^2P^o_{\frac{1}{2}} - {}^2P^o_{\frac{3}{2}}$	9	CI	A+	28	DW	D	57
A^{2+}	$3p^4\ {}^1S_0,\ {}^1D_2,\ {}^3P_J$	8	CI	A+	51	DW	D	57
A^{3+}	$3p^3\ {}^2P^o_J,\ {}^2D^o_{J'},\ {}^4S^o_{\frac{3}{2}}$	7	CI	A+	52	DW	D	57
A^{4+}	$3p^2\ {}^1S,\ {}^1D,\ {}^3P$	6	CI	A+	45	12CC+CI	B	53
A^{4+}	$3p^2\ {}^3P_J - {}^3P_{J'}$	6	CI	A+	45	DW	D	57
A^{5+}	$3s^2 3p\ {}^2P^o_{\frac{3}{2}} - {}^2P^o_{\frac{1}{2}}$	5	CI	A+	28	DW	D	57
K^{2+}	$3p^5\ {}^2P^o_{\frac{1}{2}} - {}^2P^o_{\frac{3}{2}}$	9	CI	A+	28	DW	D	57
K^{3+}	$3p^4\ {}^1S_0,\ {}^1D_2,\ {}^3P_J$	8	CI	A+	51	DW	D	57
K^{4+}	$3p^3\ {}^2P^o_J,\ {}^2D^o_{J'},\ {}^4S^o_{\frac{3}{2}}$	7	CI	A+	52	DW	D	57
Ca^+	$4p\ {}^2P^o_J - 4s\ {}^2S_{\frac{1}{2}}$	3	MP	A	43	EXPT	A	26,58
Ca^{3+}	$3p^5\ {}^2P^o_{\frac{1}{2}} - {}^2P^o_{\frac{3}{2}}$	9	CI	A+	28	DW	D	57
Ca^{4+}	$3p^4\ {}^1S_0,\ {}^1D_2,\ {}^3P_J$	8	CI	A+	51	DW	D	57
Fe^o	Forbidden transitions		CI		59			
Fe^+	Forbidden transitions		CI	D	60	4CC+CI	D	60
Fe^{2+}	Forbidden transitions		SE		61	5CC		62
Fe^{3+}	Forbidden transitions		SE		63,64			
Fe^{4+}	Forbidden transitions		SE		61			
Fe^{5+}	Forbidden transitions		CI	C	65	DW	C	65
Fe^{6+}	Forbidden transitions		CI	B	66	DW	C	66

TABLE 1. (continued)

PARM	$T_e(10^4K)$	2s-1s	2p-1s
ΔE		82258.9	82259.2
A			4.699+8
T	0.5	0.244	0.403
	1.0	0.280	0.485
	1.5	0.300	0.563
	2.0	0.311	0.631

(i)

PARM	$T_e(10^4K)$	$2\,^3S-1\,^1S$	$2\,^1S-1\,^1S$	$2\,^3P^o-1\,^1S$	$2\,^1P^o-1\,^1S$
ΔE		159850.3	166271.7	169081.3	171129.2
A		1.13-4	5.13+1	1.76+2	1.799+9
T	0.5	0.0686	0.0342	0.0126	0.0081
	0.75	0.0719	0.0393	0.0169	0.0122
	1.0	0.0736	0.0435	0.0213	0.0164
	2.0	0.0781	0.0612	0.0382	0.0347
	3.0	0.0817	0.0792	0.0526	0.0543

(ii)

PARM	$T_e(10^4K)$	$2\,^1S-2\,^3S$	$2\,^3P^o-2\,^3S$	$2\,^1P^o-2\,^3S$	$2\,^3P^o-2\,^1S$	$2\,^1P^o-2\,^1S$	$2\,^1P^o-2\,^3P^o$
ΔE		6421.4	9230.9	11278.8	2809.6	4857.5	2047.9
A		1.51-7	1.022+7	1.29	2.7-2	1.976+6	
T	0.1	1.61	3.08	0.312	0.967	1.07	2.47
	0.2	1.97	5.47	0.464	1.15	2.87	2.83
	0.5	2.39	14.0	0.750	1.41	9.35	3.42
	0.75	2.49	21.9	0.906	1.55	14.5	3.75
	1.0	2.52	29.5	1.03	1.66	19.1	4.05
	2.0	2.45	57.0	1.28	1.89	33.2	4.80
	3.0	2.32	80.0	1.36	1.94	44.1	5.08

(iii)

TABLE 2. Energy level separations ΔE (cm^{-1}), transition probabilities A (sec^{-1}) and effective collision strengths $T(T_e)$ for: (i) 1 → 2 transitions of H^o; (ii) 1 → 2 transitions of He^o; (iii) transitions within n=2 of He^o.

ION	PARM	$T_e(10^4K)$	$^2P^o_{3/2}-^2S_{1/2}$	$^2P^o_{1/2}-^2S_{1/2}$	ION	PARM	$T_e(10^4K)$	$^2P^o_{3/2}-^2S_{1/2}$	$^2P^o_{1/2}-^2S_{1/2}$	ION	PARM	$T_e(10^4K)$	$^2P^o_{3/2}-^2S_{1/2}$	$^2P^o_{1/2}-^2S_{1/2}$
C^{3+}	ΔE		64591.7	64484.0	Mg^+	ΔE		35761.0	35669.4	Ca^+	ΔE		25414.4	25191.5
	A		2.65+8	2.63+8		A		2.55+8	2.54+8		A		1.48+8	1.45+8
	T	0.5				T	0.5	← 15.6 →			T	0.5	← 15.6 →	
		1.0	← 8.88 →				1.0	16.5				1.0	17.5	
		1.5					1.5	17.2				1.5	19.2	
		2.0	8.95				2.0	17.7				2.0	20.8	
N^{4+}	ΔE		80721.9	80463.2	Si^{3+}	ΔE		71748.6	71287.5					
	A		3.38+8	3.36+8		A		9.26+8	9.15+8					
	T	0.5	← 6.61 →			T	0.5	← 16.9 →						
		1.0	6.65				1.0	17.0						
		1.5	6.69				1.5	17.0						
		2.0	6.72				2.0	17.1						
O^{5+}	ΔE		96907.5	96375.0	S^{5+}	ΔE		107137	105874					
	A		4.09+8	4.02+8		A		1.75+9	1.70+9					
	T	0.5	← 4.98 →			T	0.5	← 11.8 →						
		1.0	5.00				1.0	11.9						
		1.5	5.03				1.5	11.9						
		2.0	5.05				2.0	11.9						

TABLE 3. **Energy level separations ΔE (cm^{-1}), transition probabilities A (sec^{-1}) and effective collision strengths** $T(T_e)$ **for the ns - np transition of the Li (n=2), Na (n=3) and K (n=4) isoelectronic sequences.**

ION	PARM	$T_e(10^4K)$	$^3P^o_2-{}^1S_0$	$^3P^o_1-{}^1S_0$	$^3P^o_0-{}^1S_0$	$^1P^o_1-{}^1S_0$	$^3P^o_1-{}^3P^o_0$	$^3P^o_2-{}^3P^o_0$	$^3P^o_2-{}^3P^o_1$
C^{2+}	ΔE		52447.1	52390.8	52367.1	102352.0	23.7	80.1	56.4
	A		5.19-3	9.59+1		1.79+9	2.39-7		2.41-6
	T	0.5	←	1.12	→	3.85	0.848	0.579	2.36
		1.0		1.01		4.34	0.911	0.677	2.66
		1.5		0.990		4.56	0.975	0.776	2.97
		2.0		0.996		4.69	1.03	0.867	3.23
N^{3+}	ΔE		67416.3	67272.3	67209.2	130693.9	63.1	207.1	144.0
	A		1.15-2	5.77+2		2.38+9	4.53-6		4.03-5
	T	0.5	←	0.904	→	3.20			
		1.0		0.852		3.46			
		1.5		0.817		3.58			
		2.0		0.798		3.65			
O^{4+}	ΔE		82385.3	82078.6	81942.5	158797.7	136.1	442.8	306.7
	A		2.16-2	2.25+3		2.92+9	4.54-5		3.89-4
	T	0.5	←	0.733	→	2.66			
		1.0		0.721		2.76			
		1.5		0.674		2.82			
		2.0		0.639		2.85			
Ne^{6+}	ΔE		112711.5	111717	111264.9	214951.6	452	1446.6	995
	A		5.78-2	1.98+4		4.08+9	1.69-3		1.32-2
	T	0.5	←	0.129	→	1.39			
		1.0		0.172		1.56			
		1.5		0.205		1.63			
		2.0		0.228		1.66			
Mg^{o}	ΔE		21911.2	21870.5	21850.4	35051.3	20.1	60.8	40.7
	A		4.13-4	1.80+2		4.93+8	1.45-7	4.08-12	9.10-7
	T								
Si^{2+}	ΔE		53115.0	52853.3	52724.7	82884.4	128.6	390.3	261.7
	A		1.20-2	1.26+4		2.60+9	3.82-5	3.20-9	2.42-4
	T	0.5	←	6.90	→	5.48			
		1.0		5.43		5.82			
		1.5		4.80		6.21			
		2.0		4.41		6.54			
S^{4+}	ΔE		84155.2	83393.5	83024.0	127150.7	369.5	1131.2	761.7
	A		6.59-2	1.26+5		5.13+9	9.07-4	1.63-7	5.96-3
	T	0.5	←	0.911	→	7.30	0.272	0.400	1.24
		1.0		0.910		7.30			
		1.5		0.914		7.29			
		2.0		0.905		7.27			

TABLE 4. Energy level separations ΔE (cm^{-1}), transition probabilities A (sec^{-1}) and effective collision strengths $T(T_e)$ for the ns^2 - nsnp transitions of the Be (n=2) and Mg (n=3) isoelectronic sequences.

ION	PARM	$T_e(10^4K)$	$^4P-{}^2P^o$	$^2P^o_{3/2}-{}^2P^o_{1/2}$	$^4P_{1/2}-{}^2P^o_{1/2}$	$^4P_{1/2}-{}^2P^o_{3/2}$	$^4P_{3/2}-{}^2P^o_{1/2}$	$^4P_{3/2}-{}^2P^o_{3/2}$	$^4P_{3/2}-{}^4P_{1/2}$	$^4P_{5/2}-{}^2P^o_{1/2}$	$^4P_{5/2}-{}^2P^o_{3/2}$	$^4P_{5/2}-{}^4P_{1/2}$	$^4P_{5/2}-{}^4P_{3/2}$
C^+	ΔE			63.42	43003.3	42939.9	43025.3	42961.9	22.0	43053.6	42990.2	50.3	28.3
	A			2.29-6	5.53+1	6.55+1	1.71	5.24	2.39-7		4.32+1	3.49-14	3.67-7
	T	0.4	3.25										
		1.0	3.17	1.25									
		1.5	3.09										
		2.0	2.97										
N^{2+}	ΔE			174.4	57187.1	57012.7	57246.8	57072.4	59.7	57327.9	57153.5	140.8	81.1
	A			4.77-5	3.39+2	3.64+2	8.95	5.90+1			2.51+2		
	T	0.4	2.00	0.701	0.0952	0.0616	0.139	0.175	0.695	0.0800	0.390	0.397	1.26
		1.0	2.03										
		1.4	2.07										
		2.0	2.11										
O^{3+}	ΔE			386.3	71440.0	71053.7	71571.4	71185.1	131.4	71755.9	71369.6	315.9	184.5
	A			5.20-4	1.22+3	1.24+3	3.24+1	2.36+2	5.07-5		9.37+2		1.02-4
	T	1.0	1.27	2.33	0.113	0.087	0.174	0.234	0.989	0.122	0.506	0.592	1.82
		1.5	1.39										
		2.0	1.48	2.40	0.128	0.106	0.197	0.274	1.05	0.147	0.568	0.652	1.97
Ne^{5+}	ΔE			1306.6									
	A			2.02-2									
	T			0.433									
Si^+	ΔE			287.3	42824.4	42537.0	42932.7	42645.4	108.3	43108.0	42820.7	283.6	175.3
	A			2.17-4	4.55+3	3.00+3	1.32+1	1.62+3			2.40+3		
	T	0.5	5.28										
		1.0	5.14										
		1.5	5.05										
		2.0	4.97										
S^{3+}	ΔE			950.2	71180	70230	71524	70574	344	72071	71121	891	547
	A			7.73-3	5.50+4	3.39+4	1.40+2	1.95+4			3.95+4		
	T	1.0		6.42	0.51	0.66	0.87	1.47	3.04	0.95	2.53	2.92	7.01
		1.59		6.41	0.47	0.62	0.82	1.38	2.82	0.90	2.39	2.68	6.49
		2.51		6.38	0.44	0.59	0.77	1.30	2.59	0.85	2.26	2.40	5.85
$C\ell^{4+}$	ΔE			1490.8									
	A			2.98-2									
	T			1.052									
A^{5+}	ΔE			2207.5									
	A			9.66-2									
	T			0.798									

TABLE 5. Energy level separations ΔE (cm^{-1}), transition probabilities A (sec^{-1}), and effective collision strengths T(E) for the ns^2np - $nsnp^2$ transitions of the B (n=2) and Aℓ (n=3) isoelectronic sequences.

ION	PARM	$T_e(10^4K)$	$^1D_2-^3P_0$	$^1D_2-^3P_1$	$^1D_2-^3P_2$	$^1S_0-^3P_1$	$^1S_0-^3P_2$	$^1S_0-^1D_2$	$^3P_1-^3P_0$	$^3P_2-^3P_0$	$^3P_2-^3P_1$	$^5S^o_2-^3P_1$	$^5S^o_2-^3P_2$
C^o	ΔE		10192.6	10176.2	10149.2	21631.6	21604.6	11455.4	16.4	43.4	27.0	33718.8	33691.8
	A		7.77-8	8.21-5	2.44-4	2.71-3	2.00-5	5.28-1	7.93-8	1.71-14	2.65-7	6.94	1.56+1
	T	0.05	←	0.0625	→	← 0.0172 →		0.0620				← 0.150 →	
		0.1		0.125		0.0339		0.0877				0.212	
		0.5		0.603		0.149		0.196				0.475	
		1.0		1.14		0.252		0.277				0.671	
		1.5		1.60		0.320		0.340				0.822	
		2.0		1.96		0.365		0.392				0.950	
N^+	ΔE		15316.2	15267.5	15185.4	32640.1	32558.0	17372.6	48.7	130.8	82.1	46735.9	46653.8
	A		5.35-7	1.01-3	2.99-3	3.38-2	1.51-4	1.12	2.08-6	1.16-12	7.46-6	4.8+1	1.07+2
	T	0.5	←	2.64	→	← 0.352 →		0.405	0.401	0.279	1.13	← 1.27 →	
		1.0		2.68		0.352		0.411				1.28	
		1.5		2.72		0.359		0.418				1.29	
		2.0		2.73		0.365		0.425				1.27	
O^{2+}	ΔE		20273.3	20160.1	19967.1	43072.5	42879.5	22912.4	113.2	306.2	193.0	60211.8	60018.8
	A		2.74-6	6.74-3	1.96-2	2.23-1	7.85-4	1.78	2.62-5	3.02-11	9.76-5	2.12+2	5.22+2
	T	0.5	←	2.02	→	← 0.248 →		0.516	0.517	0.257	1.22	← 1.05 →	
		1.0		2.17		0.276		0.617	0.542	0.271	1.29	1.18	
		1.5		2.30		0.299		0.638	0.553	0.281	1.32	1.22	
		2.0		2.39		0.314		0.634	0.556	0.288	1.34	1.24	
Ne^{4+}	ΔE		30291.5	29879.1	29181.4	63501.2	62803.5	33622.1	412.4	1110.1	697.7	87950.7	87253.0
	A		2.37-5	1.31-1	3.65-1	4.21	6.69-3	2.85	1.28-3	5.08-9	4.59-3	2.37+3	6.06+3
	T	0.5	←	1.70	→	← 0.284 →		0.581	0.244	0.122	0.578	← 1.19 →	
		1.0		1.78		0.248		0.518				1.51	
		1.5		1.85		0.240		0.550				1.53	
		2.0		1.92		0.238		0.602				1.51	
Si^o	ΔE		6298.9	6221.7	6075.7	15317.3	15171.2	9095.5	77.1	223.2	146.0	33248.9	33102.9
	A		4.70-7	7.93-4	2.25-3	3.13-2	9.02-4	1.14	8.25-6		4.21-5		
	T												
S^{2+}	ΔE		11320	11023	10488	26866	26331	15843	297.2	832.5	535.3	59401	58866
	A		5.82-6	2.21-2	5.76-2	7.96-1	1.05-2	2.22	4.72-4	4.61-8	2.07-3		
	T	0.5	←	9.07	→	← 1.16 →		1.42	2.64	1.11	5.79		
		1.0		8.39		1.19		1.88	2.59	1.15	5.81		
		1.5		8.29		1.21		2.02	2.38	1.15	5.56		
		2.0		8.20		1.24		2.08	2.20	1.14	5.32		
Cl^{3+}	ΔE		13767.6	13275.6	12425.7	32055.8	31205.9	18780.2	492.0	1341.9	849.9		
	A		1.54-5	7.23-2	1.79-1	2.47	2.62-2	2.80	2.14-3	2.70-7	8.25-3		
	T	0.5	←	5.10	→	← 2.04 →		0.935	0.475	0.400	1.50		
		1.0		5.42		2.27		1.39					
		1.5		5.88		2.32		1.73					
		2.0		6.19		2.30		1.92					
A^{4+}	ΔE		16299.4	15535.5	14270.2	37148.6	35883.3	21613.1	763.9	2029.2	1265.3		
	A		3.50-5	2.04-1	4.76-1	6.55	5.69-2	3.29	7.99-3	1.24-6	2.72-2		
	T	0.5	←	4.37	→	← 1.17 →		1.26	0.257	0.320	1.04		
		1.0		3.72		1.18		1.25					
		1.5		3.52		1.11		1.24					
		2.0		3.42		1.03		1.23					

TABLE 6. Energy level separations ΔE (cm^{-1}), transition probabilities A (sec^{-1}) and effective collision strengths $T(T_e)$ for the ns^2np^2 and $nsnp^3$ transitions of the C (n=2) and Si (n=3) isoelectronic sequences.

ION	PARM	$T_e(10^4K)$	$^2D^o_{5/2}-^4S^o_{3/2}$	$^2D^o_{3/2}-^4S^o_{3/2}$	$^2P^o_{3/2}-^4S^o_{3/2}$	$^2P^o_{1/2}-^4S^o_{3/2}$	$^2D^o_{5/2}-^2D^o_{3/2}$	$^2P^o_{3/2}-^2P^o_{1/2}$	$^2P^o_{3/2}-^2D^o_{5/2}$	$^2P^o_{3/2}-^2D^o_{3/2}$	$^2P^o_{1/2}-^2D^o_{5/2}$	$^2P^o_{1/2}-^2D^o_{3/2}$
N^o	ΔE		19224.5	19233.2	28839.3	28838.9	-8.71	0.386	9614.8	9606.1	9614.5	9605.7
	A		7.27-6	2.02-5	6.58-3	2.71-3	1.27-8		6.14-2	2.76-2	3.45-2	5.29-2
	T	0.05	9.18-3	6.12-3	3.2-3	1.6-3						
		0.1	2.30-2	1.53-2	8.53-3	4.27-3						
		0.5	0.155	0.103	0.0597	0.0298	0.128	0.0329	0.162	0.0856	0.0626	0.0601
		1.0	0.290	0.194	0.113	0.0567	0.269	0.071	0.266	0.147	0.109	0.097
		2.0	0.476	0.318	0.189	0.0947	0.465	0.153	0.438	0.252	0.190	0.157
O^+	ΔE		26810.7	26830.2	40467.5	40468.6	-19.5	-1.1	13656.8	13637.3	13657.9	13638.4
	A		3.82-5	1.65-4	5.64-2	2.32-2	1.20-7	2.08-11	1.17-1	6.14-2	6.15-2	1.02-1
	T	0.5	0.795	0.530	0.265	0.133	1.22	0.280	0.718	0.401	0.290	0.270
		1.0	0.801	0.534	0.270	0.135	1.17	0.287	0.730	0.408	0.295	0.275
		1.4	0.808	0.538	0.274	0.137	1.14	0.292	0.740	0.413	0.299	0.279
		2.0	0.818	0.545	0.280	0.140	1.11	0.300	0.755	0.422	0.305	0.284
Ne^{3+}	ΔE		41234.6	41279.5	62441.3	62434.6	-44.9	6.7	21206.7	21161.8	21200.0	21155.1
	A		4.84-4	5.54-3	1.27	5.21-1	1.48-6	2.68-9	4.00-1	4.37-1	1.15-1	3.93-1
	T	0.6	0.843	0.562	0.308	0.154	1.37	0.323	0.867	0.482	0.347	0.327
		1.0	0.838	0.559	0.313	0.156	1.36	0.343	0.900	0.509	0.368	0.336
		1.4	0.834	0.556	0.312	0.156	1.36	0.355	0.908	0.515	0.373	0.339
		2.0	0.824	0.550	0.309	0.155	1.33	0.370	0.909	0.516	0.374	0.339
Na^{4+}	ΔE		48313.5	48359.3	73236.4	73201.9	-45.8	34.5	24922.9	24877.1	24888.4	24842.6
	A		1.46-3	2.69-2	4.27	1.76	1.55-6	3.64-7	9.18-1	1.29	1.41-1	9.55-1
	T		0.551	0.368	0.239	0.120	0.696	0.438	0.502	0.279	0.201	0.190
S^+	ΔE		14884.8	14853.0	24571.8	24524.9	31.8	46.9	9687.0	9718.8	9640.1	9671.9
	A		2.60-4	8.82-4	0.225	9.06-2	3.35-7	1.03-6	0.179	0.133	7.79-2	1.63-1
	T	0.5	4.38	2.92	1.44	0.722	8.15	2.39	4.74	3.36	2.55	1.50
		1.0	4.19	2.79	1.52	0.759	7.59	2.38	4.79	3.38	2.56	1.52
		1.5	4.00	2.67	1.54	0.772	7.11	2.33	4.74	3.32	2.51	1.51
		2.0	3.84	2.56	1.54	0.769	6.71	2.27	4.62	3.21	2.43	1.48
$C\ell^{2+}$	ΔE		18118.6	18053	29907	29812	66	95	11788	11854	11693	11759
	A		7.04-4	4.83-3	0.754	0.305	3.22-6	7.65-6	0.316	0.323	0.100	0.303
	T		1.88	1.26	1.26	0.627	3.19	1.34	3.33	1.91	1.38	1.24
A^{3+}	ΔE		21219.3	21090.4	35032.6	34855.5	128.9	177.1	13813.3	13942.2	13636.2	13765.1
	A		1.77-3	2.23-2	2.11	0.862	2.30-5	4.94-5	0.598	0.789	0.119	0.603
	T		0.854	0.570	0.423	0.212	1.35	0.601	2.50	1.24	0.865	1.01
K^{4+}	ΔE		24249.6	24012.5	40080.2	39758.1	237.1	322.1	15830.6	16067.7	15508.5	15745.6
	A		4.59-3	8.84-2	5.19	2.14	1.42-4	2.96-4	1.21	1.86	0.141	1.25
	T		0.455	0.303	0.173	0.086	0.971	0.281	2.12	0.974	0.665	0.884

TABLE 7. Energy level separations ΔE (cm^{-1}), transition probabilities A (sec^{-1}) and effective collision strengths $T(T_e)$ for the np^3 transitions of the N (n=2) and P (n=3) isoelectronic sequences.

ION	PARM	$T_e(10^4K)$	$^1D_2-{}^3P_0$	$^1D_2-{}^3P_1$	$^1D_2-{}^3P_2$	$^1S_0-{}^3P_1$	$^1S_0-{}^3P_2$	$^1S_0-{}^1D_2$	$^3P_0-{}^3P_1$	$^3P_0-{}^3P_2$	$^3P_1-{}^3P_2$
O^0	ΔE		15640.9	15709.6	15867.9	33634.3	33792.6	17924.7	68.7	227.0	158.3
	A		7.23-7	2.11-3	6.34-3	7.32-2	2.88-4	1.22	1.74-5		8.92-5
	T	0.05	←	0.0058	→	← 0.00065 →		0.0221	0.0008	0.0006	0.0027
		0.1		0.0151		0.00184		0.0310	0.0018	0.0022	0.0076
		0.5		0.124		0.0153		0.0732	0.0112	0.0148	0.0474
		1.0		0.266		0.0324		0.105	0.0265	0.0292	0.0987
		2.0		0.501		0.0607		0.148	0.0693	0.0536	0.207
Ne^{2+}	ΔE		24920.3	25197.9	25840.8	55107.7	55750.6	29909.8	277.6	920.5	642.9
	A		8.51-6	5.42-2	1.71-1	2.00	3.94-3	2.71	1.15-3	2.18-8	5.97-3
	T	0.5	←	1.35	→	← 0.152 →		0.220	0.185	0.131	0.527
		1.0		1.34		0.151		0.236			
		1.5		1.33		0.152		0.262			
		2.0		1.32		0.157		0.284			
Na^{3+}	ΔE		29264	29734	30840	65390	66496	35656	469.3	1575.6	1106.3
	A		2.24-5	1.86-1	6.10-1	7.10	1.05-2	3.46	5.57-3	1.67-7	3.04-2
	T		←	1.17	→	← 0.163 →		0.157	0.177	0.111	0.471
Mg^{4+}	ΔE		33410	34149	35932	75503	77286	41354	738.7	2521.8	1783.1
	A		5.20-5	5.41-1	1.85	2.14+1	2.45-2	4.23	2.17-2	1.01-6	1.27-1
	T		←	1.02	→	← 0.146 →		0.129	0.156	0.0908	0.400
S^0	ΔE		8665.0	8842.6	9238.6	21783.9	22180.0	12941.3	177.6	573.7	396.1
	A		3.84-6	8.16-3	2.78-2	3.50-1	8.23-3	1.53	3.02-4	6.71-8	1.39-3
	T										
Cl^+	ΔE		10657.1	10957.6	11653.6	27182.0	27878.0	16224.4	300.5	996.5	696.0
	A		9.82-6	2.92-2	1.04-1	1.31	1.97-2	2.06	1.46-3	4.57-7	7.57-3
	T		←	3.86	→	← 0.456 →		1.15	0.933	0.443	2.17
A^{2+}	ΔE		12439.8	12897.9	14010.0	32153.6	33265.7	19255.7	458.1	1570.2	1112.1
	A		2.21-5	8.23-2	3.14-1	3.91	4.17-2	2.59	5.17-3	2.37-6	3.08-2
	T		←	4.74	→	← 0.680 →		0.823	1.18	0.531	2.24
K^{3+}	ΔE		14062.9	14712.7	16384.1	36874.9	38546.3	22162.2	649.8	2321.2	1671.4
	A		4.54-5	1.98-1	8.14-1	1.00+1	8.17-2	3.18	1.48-2	1.01-5	1.04-1
	T		←	1.90	→	← 0.292 →		0.798	0.421	0.290	1.16
Ca^{4+}	ΔE		15554.7	16425.6	18830.3	41431.8	43836.5	25006.2	870.9	3275.6	2404.7
	A		8.42-5	4.26-1	1.90	2.31+1	1.45-1	3.73	3.54-2	3.67-5	3.10-1
	T		←	0.904	→	← 0.116 →		0.793	0.202	0.224	0.760

TABLE 8. **Energy level separations ΔE (cm^{-1}), transition probabilities A (sec^{-1}) and effective collision strengths $T(T_e)$ for the np^4 transitions of the O (n=2) and S (n=3) isoelectronic sequences.**

ION	PARM	$T_e(10^4K)$	$^2P^o_{\frac{1}{2}}-{}^2P^o_{\frac{3}{2}}$	ION	PARM	$T_e(10^4K)$	$^2P^o_{\frac{1}{2}}-{}^2P^o_{\frac{3}{2}}$
Ne^+	ΔE		780.4	$C\ell^o$	ΔE		882.4
	A		8.55-3		A		1.24-2
	Υ	0.5	0.362		Υ		
		1.0	0.368				
		1.5	0.375	A^+	ΔE		1431.6
		2.0	0.381		A		5.27-2
					Υ		0.635
Na^{2+}	ΔE		1366.7				
	A		4.59-2	K^{2+}	ΔE		2166.1
	Υ		0.300		A		1.83-1
					Υ		1.78
Mg^{3+}	ΔE		2229.5				
	A		1.99-1	Ca^{3+}	ΔE		3117.9
	Υ		0.300		A		5.45-1
					Υ		1.06
Si^{5+}	ΔE		5094.1				
	A		2.38				
	Υ		0.242				

TABLE 9. Energy level separations ΔE (cm^{-1}), transition probabilities A (sec^{-1}) and effective collision strengths $\Upsilon(T_e)$ for the np^5 transitions of the F (n=2) and Cℓ (n=3) isoelectronic sequences.

PARENT(STATES)	FINAL ION(STATES)	METHOD	ACC	SOURCE
$H^{o}(n\ell)$	H^{+}	Hydrogenic	A+	67
$He^{o}(1\,^1S, 2\,^1S, ^3S, ^1P^o, ^3P^o)$	$He^{+}(1s,2s,2p)$	COMP;EXPT	A	68-71;72-74
$He^{+}(n\ell)$	He^{2+}	Hydrogenic	A+	67
$C^{o}(^3P, ^1D, ^1S)$	$C^{+}(^2P^o, ^4P, ^2D, ^2S, ^2P)$	8CC+CI,5CC+CI;EXPT	B+;C	75,105;76
$C^{+}(^2P^o, ^4P)$	C^{2+}	6CC+CI;CC	;D	77;85
$C^{2+}(^1S, ^3P^o)$	C^{3+}	5CC; TDHF	A;	78;79
$C^{3+}(^2S)$	C^{4+}	5CC	A	78,80
$N^{o}(^4S^o, ^2D^o, ^2P^o)$	$N^{+}(^3P, ^1D, ^1S, ^5S^o, ^3D^o, ^3P^o, ^1D^o, ^3S^o, ^1P^o)$	5CC+CI,8CC+CI;EXPT	B+	81,82;83
$N^{+}(^3P, ^1D, ^1S)$	$N^{2+}(^2P^o)$	6CC+CI;CC	;D	84;85
N^{2+}(ground+excited)	N^{3+}	6CC+CI	A	86
$N^{3+}(^1S, ^3P^o)$	N^{4+}	5CC; TDHF	A;	87;79
$N^{4+}(^2S)$	N^{5+}	5CC	A	80
$O^{o}(^3P, ^1D, ^1S)$	$O^{+}(^4S^o, ^2D^o, ^2P^o)$	CC+CI;CC;EXPT	B+;D;A	88,125;85;89
$O^{+}(^4S^o, ^2D^o, ^2P^o)$	$O^{2+}(^3P, ^1D, ^1S)$	CC	D	85
$O^{2+}(^3P, ^1D, ^1S, ^5S^o)$	$O^{3+}(^2P^o, ^4P)$	8CC+CI;CC	A;D	90;85
O^{3+}(ground+excited)	O^{4+}	6CC+CI	A	91
$O^{4+}(^1S)$	O^{5+}	CP;TDHF	B;	92;79
$O^{5+}(^2S)$	O^{6+}	5CC	A	80
$Ne^{o}(^1S)$	Ne^{+}	2CC+CI;MBPT;EXPT	A	93;94;73,95
$Ne^{+}(^2P^o)$	$Ne^{2+}(^3P, ^1D, ^1S, ^3P^o, ^1P^o)$	5CC+CI	B+	96
$Ne^{2+}(^3P, ^1D, ^1S)$	$Ne^{3+}(^4S^o, ^2D^o, ^2P^o, ^4P, ^2D, ^2S, ^2P)$	7CC+CI;CC	B+;D	96;85
$Ne^{3+}(^4S^o, ^2D^o, ^2P^o)$	$Ne^{4+}(^3P, ^1D, ^1S, ^5S^o, ^3D^o, ^3P^o, ^1D^o, ^3S^o, ^1P^0)$	9CC+CI;CC	B+;D	96;85
$Ne^{4+}(^3P, ^1D, ^1S)$	$Ne^{5+}(^2P^o)$	CC	D	85
$Ne^{5+}(^2P^o)$	$Ne^{6+}(^1S)$	CC	D	85
Na^{o}(ground+excited)	Na^{+}	MP;EXPT	D	97-99;100,101
$Na^{+}(^1S)$	Na^{2+}	MBPT	B+	102,103
$Na^{2+} - Na^{4+}$	$Na^{3+} - Na^{5+}$	CP		92
$Mg^{o}(^1S, ^3P^o, ^1P^o, ^3S)$	Mg^{+}	4CC	A	104
$Mg^{+}(^2S)$	Mg^{2+}	MP	A	43
$Mg^{2+}(^1S)$	Mg^{3+}	MBPT	B+	102,103
$Mg^{3+} - Mg^{4+}$	$Mg^{4+} - Mg^{5+}$	CP		92

TABLE 10. Selected bibliography of photoionization cross sections for ions of interest in planetary nebulae studies.

PARENT(STATES)	FINAL ION(STATES)	METHOD	ACC	SOURCE
$Si^{o}(^{3}P, ^{1}D, ^{1}S)$	$Si^{+}(^{2}P^{o})$	7CC+CI;CC	B+;D	104;106
$Si^{+}(^{2}P^{o})$	Si^{2+}	12CC+CI;MBPT;CC		107;108;106
$Si^{2+}(^{1}S)$	Si^{3+}	5CC;CP	;B	109;92
$Si^{3+}(^{2}S)$	Si^{4+}	1CC+MP;MP	;B	109;110
$Si^{4+}(^{1}S)$	Si^{5+}	MBPT	B+	103
$S^{o}(^{3}P, ^{1}D, ^{1}S)$	$S^{+}(^{4}S^{o}, ^{2}D^{o}, ^{2}P^{o})$	6CC+CI;CC	B+; D	104;111
$S^{+}(^{4}S^{o}, ^{2}D^{o}, ^{2}P^{o})$	$S^{2+}(^{3}P, ^{1}D, ^{1}S)$	10CC+CI;CC	B+; D	104;111
$S^{2+}(^{3}P, ^{1}D, ^{1}S)$	$S^{3+}(^{2}P^{o})$	7CC+CI;CC	B+; D	104;111
$S^{3+}(^{2}P^{o})$	$S^{4+}(^{1}S)$	CC	D	111
$S^{4+}(^{1}S)$	S^{5+}	5CC;CP	;B	109;92
$S^{5+}(^{2}S)$	S^{6+}	1CC+MP;CP	;B	109;92
$Cl^{o}(^{2}P^{o})$	Cl^{+}	MBPT	B+	112
$Cl^{+} - Cl^{4+}$	$Cl^{2+} - Cl^{5+}$	CP		92
$A^{o}(^{1}S)$	A^{+}	COMP;EXPT	A	93,94,113;73,114
$A^{+}(^{2}P^{o})$	$A^{2+}(^{3}P, ^{1}D, ^{1}S)$	CC	D	106
$A^{2+}(^{3}P, ^{1}D, ^{1}S)$	$A^{3+}(^{4}S^{o}, ^{2}D^{o}, ^{2}P^{o})$	CC	D	106
$A^{3+}(^{4}S^{o}, ^{2}D^{o}, ^{2}P^{o})$	$A^{4+}(^{3}P, ^{1}D, ^{1}S)$	CC	D	106
$A^{4+}(^{3}P, ^{1}D, ^{1}S)$	$A^{5+}(^{2}P^{o})$	CC	D	106
$A^{5+}(^{2}P^{o})$	$A^{6+}(^{1}S)$	CC	D	106
$K^{o}(^{2}S)$	K^{+}	MP;EXPT	D	115;116,117
$K^{+}(^{1}S)$	K^{2+}	MBPT	B+	102,118
$K^{2+}(^{2}P^{o})$	K^{3+}	CC	B+	119
$K^{3+} - K^{4+}$	$K^{4+} - K^{5+}$	CP		92
$Ca^{o}(^{1}S)$	Ca^{+}	4CC;EXPT	C	104;120,121
$Ca^{+}(^{2}S)$	Ca^{2+}	MP	B	43
$Ca^{2+}(^{1}S)$	Ca^{3+}	MBPT	B+	102,118
$Ca^{3+} - Ca^{4+}$	$Ca^{4+} - Ca^{5+}$	CP		92
Fe^{o}	Fe^{+}	MBPT;CP;EXPT		122;123;124
$Fe^{+} - Fe^{5+}$	$Fe^{2+} - Fe^{6+}$	CP		123

TABLE 10. (continued)

REFERENCES

1. Wiese WL, Smith MW and Glennon BM, 1966, "Atomic Transition Probabilities", Vol.1 (NSRDS-NBS4).
2. Hata J and Grant IP, 1981, J.Phys.B, 14, 2111.
3. Drake GWF, 1979, Phys.Rev.A, 19, 1387.
4. Drake GWF and Dalgarno A, 1969, Astrophys.J., 157, 459.
5. Lin CD, Johnson WR and Dalgarno A, 1977, Phys.Rev.A, 15, 154.
6. Berrington KA, Fon WC and Kingston AE, 1982, Mon.Not.R.Astr.Soc., 200, 347.
7. Nussbaumer H and Rusca C, 1979, Astron. Astrophys., 72, 129.
8. Péquignot D and Aldrovandi SMV, 1976, Astron. Astrophys., 50, 141.
9. Thomas LD and Nesbet RK, 1975, Phys.Rev.A., 12, 2378.
10. Nussbaumer H and Storey PJ, 1981, Astron. Astrophys., 96, 91.
11. Tambe BR, 1977, J.Phys.B, 10, L249.
12. Jackson ARG, 1973, Mon.Not.R.Astr.Soc., 165, 53.
13. Nussbaumer H and Storey PJ, 1978, Astron. Astrophys., 64, 139.
14. Dufton PL, Berrington KA, Burke PG and Kingston AE, 1978, Astron. Astrophys., 62, 111; + Daresbury Laboratory Data Bank.
15. Taylor PO, Gregory D, Dunn GH, Phaneuf RA and Crandall DH, 1977, Phys.Rev.Lett., 39, 1256.
16. Gau JN and Henry RJW, 1977, Phys.Rev.A, 16, 986.
17. Zeippen CJ, 1982, Mon.Not.R.Astr.Soc., 198, 111.
18. Berrington KA and Burke PG, 1981, Planet. Space Sci., 29, 377.
19. Dopita MA, Mason DJ and Robb WD, 1976, Astrophys.J., 207, 102.
20. Seaton MJ, 1975, Mon.Not.R.Astr.Soc., 170, 475.
21. Saraph HE, Seaton MJ and Shemming J, 1969, Phil.Trans.R.Soc.A, 264, 77.
22. Hibbert A and Bates DR, 1981, Planet. Space Sci., 29, 263.
23. Nussbaumer H and Storey PJ, 1979, Astron. Astrophys., 71 L5.
24. Nussbaumer H and Storey PJ, 1982, Astron. Astrophys., 109, 271.
25. Nussbaumer H and Storey PJ, 1979, Astron. Astrophys., 74, 244.
26. Osterbrock DE and Wallace RK, 1977, Astrophys.Lett., 19, 11.
27. van Wyngaarden WL and Henry RJW, 1976, J.Phys.B, 9, 1461.
28. Mendoza C and Zeippen CJ, 1982, preliminary results.
29. Le Dourneuf M and Nesbet RK, 1976, J.Phys.B, 9, L241.
30. Pradhan AK, 1976, Mon.Not.R.Astr.Soc., 177, 31.
31. Nussbaumer H and Storey PJ, 1981, Astron. Astrophys., 99, 177.
32. Baluja KL, Burke PG and Kingston AE, 1980, J.Phys.B, 13, 829; 1981, 14, 119.
33. Aggarwal KM, Baluja KL and Tully JA, 1982, Mon.Not.R.Astr.Soc., in press.
34. Hayes MA, 1982, Mon.Not.R.Astr.Soc., 199, 49P; + private communication
35. Pradhan AK, 1974, J.Phys.B, 7, L503.
36. Giles K, 1981, Mon.Not.R.Astr.Soc., 195, 63P.
37. Baluja KL, Burke PG and Kingston AE, 1980, J.Phys.B, 13, 4675.
38. Dufton PL, Doyle JG and Kingston AE, 1979, Astron. Astrophys., 78, 318; + Daresbury Laboratory Data Bank.
39. Butler K and Mendoza C, 1982, preliminary results.
40. Lin CD, Laughlin C and Victor GA, 1978, Astrophys.J., 220, 734.
41. Clark REH, Magee NH, Mann JB and Merts AL, 1982, Astrophys.J., 254, 41
42. Fabrikant II, 1974, J.Phys.B, 7, 91.

43. Black JH, Weisheit JC and Laviana E, 1972, Astrophys.J., 177, 567.
44. Mendoza C, 1981, J.Phys.B, 14, 2465.
45. Mendoza C and Zeippen CJ, 1982, Mon.Not.R.Astr.Soc., 199, 1025.
46. Pindzola MS, Bhatia AK and Temkin A, 1977, Phys.Rev.A, 15, 35.
47. Nussbaumer H, 1977, Astron. Astrophys., 58, 291.
48. Mendoza C, 1982, preliminary results.
49. Baluja KL, Burke PG and Kingston AE, 1980, J.Phys.B, 13, L543; 1981, 14, 1333.
50. Flower DR and Nussbaumer H, 1975, Astron. Astrophys., 42, 265.
51. Mendoza C and Zeippen CJ, 1982, Mon.Not.R.Astr.Soc., in press.
52. Mendoza C and Zeippen CJ, 1982, Mon.Not.R.Astr.Soc., 198, 127.
53. Mendoza C, 1982, J.Phys.B, to be published.
54. Dufton PL, Hibbert A, Kingston AE and Doschek GA, 1982, Astrophys.J., 257, 338.
55. Feldman U, Doschek GA and Bhatia AK, 1981, Astrophys.J., 250, 799.
56. Giles K, 1980, Ph.D thesis, University of London.
57. Krueger TK and Czyzak SJ, 1970, Proc.R.Soc.Lond. A, 318, 531.
58. Taylor PO and Dunn GH, 1973, Phys.Rev.A, 8, 2304.
59. Grevesse N, Nussbaumer H and Swings JP, 1971, Mon.Not.R.Astr.Soc., 151, 239.
60. Nussbaumer H and Storey PJ, 1980, Astron. Astrophys., 89, 308.
61. Garstang RH, 1957, Mon.Not.R.Astr.Soc., 117, 393.
62. Garstang RH, Robb WD and Rountree SP, 1978, Astrophys.J., 222, 384.
63. Ekberg JO and Edlén B, 1978, Physica Scripta, 18, 107.
64. Garstang RH, 1958, Mon.Not.R.Astr.Soc., 118, 572.
65. Nussbaumer H and Storey PJ, 1978, Astron. Astrophys., 70, 37.
66. Nussbaumer H and Storey PJ, 1982, Astron. Astrophys., in press.
67. Burgess A, 1964, Mem.R.Astr.Soc., 69, 1.
68. Bell KL and Kingston AE, 1970, J.Phys.B, 3, 1433.
69. Jacobs VL, 1971, Phys.Rev.A, 3, 289; 1974, 9, 1938.
70. Jacobs VL and Burke PG, 1972, J.Phys.B, 5, L67; 1972, 5, 2272.
71. Stewart AL, 1978, J.Phys.B, 11, L431; 1978, 11, 2449; 1979, 12, 401.
72. Marr GV, 1978, J.Phys.B, 11, L121.
73. West JB and Marr GV, 1976, Proc.R.Soc.Lond.A, 349, 397.
74. Woodruff PR and Samson JAR, 1980, Phys.Rev.Lett., 45, 110; 1982, Phys.Rev.A, 25, 848.
75. Burke PG and Taylor KT, 1979, J.Phys.B, 12, 2971.
76. Cantú AM, Mazzoni M, Pettini M and Tozzi GP, 1981, Phys.Rev.A, 23, 1223.
77. Drew JE and Storey PJ, 1982, in progress.
78. Drew JE and Storey PJ, 1982, J.Phys.B, 15, 2357.
79. Watson DK, Dalgarno A and Stewart RF, 1978, Phys.Rev.A, 17, 1928.
80. Pradhan AK, 1982, Phys.Rev.A, 25, 592.
81. Le Dourneuf M, Vo Ky Lan and Zeippen CJ, 1979, J.Phys.B, 12, 2449.
82. Zeippen CJ, Le Dourneuf M and Vo Ky Lan, 1980, J.Phys.B, 13, 3763.
83. Samson JAR and Cairns RB, 1965, J.Opt.Soc.Am., 55, 1035.
84. Nussbaumer H and Storey PJ, 1982, in progress.
85. Henry RJW, 1970, Astrophys.J., 161, 1153.
86. Butler K, J.Phys.B, to be published.
87. Butler K, Lugo L and Mendoza C, 1982, J.Phys.B, to be submitted.
88. Taylor KT and Burke PG, 1976, J.Phys.B, 9, L353.

89. Kohl JL, Lafyatis GP, Palenius HP and Parkinson WH, 1978, Phys.Rev.A, 18, 571.
90. Saraph HE, J.Phys.B, 1982, to be published.
91. Saraph HE, 1980, J.Phys.B, 13, 3129.
92. Reilman RF and Manson ST, 1979, Astrophys.J.Suppl.Ser., 40, 815; 1981, 46, 115.
93. Burke PG and Taylor KT, 1975, J.Phys.B, 8, 2620.
94. Chang TN, 1977, Phys.Rev.A, 15, 2392.
95. Wuilleumier F and Krause MO, 1979, J.Electron Spectrosc.Relat.Phenom., 15, 15.
96. Pradhan AK, 1979, J.Phys.B, 12, 3317; 1980,Mon.Not.R.Astr.Soc., 190, 5
97. Laughlin C, 1978, J.Phys.B, 11, 1399.
98. Aymar M, Luc-Koenig E and Combet-Farnoux F, 1976, J.Phys.B, 9, 1279.
99. Aymar M, 1978, J.Phys.B, 11, 1413.
100. Hudson RD and Carter VL, 1967, J.Opt.Soc.Am., 57, 651.
101. Rothe DE, 1969, J.Quant.Spectrosc.Radiat.Transfer, 9, 49.
102. Chang TN, 1977, Phys.Rev.A, 16, 1171.
103. Chang TN and Olsen T, 1981, Phys.Rev.A, 24, 1091.
104. Mendoza C, 1982, to be published.
105. Hofmann H and Trefftz E, 1980, Astron. Astrophys., 82, 256.
106. Chapman RD and Henry RJW, 1972, Astrophys.J., 173, 243.
107. Le Dourneuf M and Zeippen CJ, 1982, in progress.
108. Daum GR and Kelly HP, 1976, Phys.Rev.A, 13, 715.
109. Mendoza C and Zeippen CJ, 1982, in progress.
110. Shevelco VP, 1974, Opt.Spektrcsk.,36, 14.
111. Chapman RD and Henry RJW, 1971, Astrophys.J., 168, 169.
112. Brown ER, Carter SL and Kelly HP, 1980, Phys.Rev.A, 21, 1237.
113. Kelly HP and Simons RL, 1973, Phys.Rev.Lett., 30, 529.
114. Madden RP, Ederer DL and Codling K, 1969, Phys.Rev., 177, 136.
115. Weisheit JC, 1972, Phys.Rev.A, 5, 1621
116. Sandner W, Gallagher TF, Safinya KA and Gounand F, 1981, Phys.Rev.A, 23, 2732.
117. Hudson RD and Carter VL, 1965, Phys.Rev., 139, A1426.
118. Chang TN, 1979, Phys.Rev.A, 20, 291.
119. Combet-Farnoux F, Lamoureux M and Taylor KT, 1978, J.Phys.B, 11, 2855.
120. Carter VL, Hudson RD and Breig EL, 1971, Phys.Rev.A, 4, 821.
121. McIlrath TJ and Sandeman RJ, 1972, J.Phys.B, 5, L217.
122. Kelly HP and Ron A, 1972, Phys.Rev.A, 5, 168.
123. Reilman RF and Manson ST, 1978, Phys.Rev.A, 18, 2124.
124. Lombardi GG, Smith PL and Parkinson WH, 1978, Phys.Rev.A, 18, 2131.
125. Pradhan AK and Saraph HE, 1977, J.Phys.B, 10, 3365.
126. Aggarwal KM, 1982, Mon.Not.R.Astr.Soc., submitted

IONIZATION EQUILIBRIUM IN MODELS OF PLANETARY NEBULAE

Daniel Péquignot
Observatoire de Meudon
Meudon, France

ABSTRACT : A critical picture of the results obtained by means of photoionization models during the 70's is presented. Some reasons for the relative failure of these modellings are advanced and a new perspective is proposed for future investigations.

1. INTRODUCTION

My talk was supposed to deal with the effects of the newly introduced collisional processes involved in the ionization equilibrium of planetary nebulae. However P. Storey and R. Mc Carroll provided us with a comprehensive account of all the recombination and charge exchange processes which are currently believed to control ionization. Not to mention the discussion of UV spectra by M. Seaton who stressed the importance of low-temperature dielectronic recombinations in producing specific lines of diagnostic value.

Ionization depends primarily on photoionization rates, that is (1) on the frequency distribution of the primary radiation, derived from the theory of stellar atmospheres and may be, in the future, from direct satellite EUV observation, (2) on the photoabsorption cross-sections discussed by C. Mendoza and (3) on the radiation transfer some aspects of which being considered by D. Hummer.

On the other hand, J.P. Harrington will certainly present an unified view of the chain of physical processes leading from the extreme ultraviolet radiation of the star to the radiation finally detected at Earth and explain how the machinery of the self-consistent photoionization models can be used to derive the physical conditions and the information of astrophysical interest.

In view of such an exhaustive and authorized coverage of my subject, I feel free to comment on closely related but more general questions :
- Why do we make models ?
- What have we learnt through models so far ?
- On which conditions can we expect to learn something today or tomorrow

D. R. Flower (ed.), Planetary Nebulae, 173–185.

from detailed modelling of nebulae ?

2. EMPIRICAL AND THEORETICAL APPROACHES

The concept of a model is widespread in physics and needs to be clarified in a concrete situation. Two extreme approaches to planetary nebulae can be distinguished :

(1) Plasma diagnostics leading to empirical models of real nebulae : this approach is useful to derive information of practical interest for astrophysics (elemental abundances, see Aller in this volume, shape of nebulae, etc...). It is empirical in that there is a one to one correspondance between the observed data and the derived informations without meaningful check on consistency. In the important case of elemental abundance determination, the standard corrections for unobserved ions and "temperature fluctuations" can be classified as "semi-empirical" in that they are based to some extent on theoretical arguments.

(2) Ab initio self-consistent calculations leading to theoretical models without reference to particular objects : this approach is useful to depict what type of "complexity" should be added to get a correspondance with real nebulae. In practice there are two levels of self-consistency :

(2a) The fully (magneto-) hydrodynamic time-dependent approach in which the gas density distribution appears as resulting from an evolution under specified conditions (Lazareff, 1981).This includes more specialized studies on the instabilities that may develop in the flow (Capriotti, 1973).

(2b) The static (not hydrostatic) approach in which the gas density distribution is given so that the ionization and thermal aspects only are treated self-consistently. Although there is evidently no static nebula, this approach is not rudimentary because in most cases the energetics is largely dominated by radiation and the conversion of mechanical energy into radiation is not important. Thus there is no simple hierarchy between (2a) and (2b) but rather complementarity because (2b) is virtually decoupled from (2a) and a large number of initial conditions and time-dependent parameters should be specified in (2a).

Several variants of (2b) exist :

(2b1) Purely static and stationnary models : thermal and ionization balance applies everywhere and the only input of energy is through photoionization (conceivably photoexcitation) of the gas by radiation from a source independent of time (e.g., Harrington, 1968 ; Flower, 1968).

(2b2) Stationnary models but with a hydrodynamical and out-of-equilibrium treatment of ionization fronts of specified type and velocity : the study of the fronts can then usually be decoupled from that of the main body of the nebula. After some ups and downs, these models were useful in confirming that departures from the static case assuming constant thermal pressure are unimportant as far as the emission of stationnary fronts is concerned (Harrington 1977).

(2b3) Models with non-stationnary primary radiation source with out-of-equilibrium treatment of ionization and thermal balance in the whole nebula but with no reference to a global self-consistent hydrodynamic evolution of the gas (Harrington and Marionni, 1976 ; Tylenda, 1979). This sophistication is not needed in most cases because the local time-scale to reach equilibrium is normally much shorter than the evolution time-scale and much longer than the pulsation time-scale of the central star. However the study of the consequences of, e.g., a burst (see Tylenda, this volume) or a rather sudden shrinking of the source is of theoretical interest at least to learn the limit of validity of the more common stationary approach (2b1) to which we now restrict our attention.

3. FITTING OF THEORETICAL MODELS TO REAL NEBULAE

Purely theoretical approaches by means of, e.g., grids of ab initio models are of limited value in the case of planetary nebulae because :
- the number of input parameters is large,
- the objects accurately observed are not very numerous,
- each nebula is a unique well-characterized object demanding a specific study.

Very schematically, two distinct attitudes can be adopted when performing the fitting of self-consistent models to a given planetary nebula .
(2c) The notion of the model strictly includes atomic physics data currently available and it is implicitly assumed that these data are adequate to describe accurately and exhaustively the situation. The fitting to observed line intensities is then restricted to astrophysical parameters and a satisfactory fit is taken as convincing evidence that reliable information (stellar flux, abundances, gas density distribution, ...) has been obtained.
(2d) The model is viewed as a place of encounter of several types of data, all of them entailing uncertainties and constraints : atomic data are not a priori and systematically assumed to be perfectly adequate (which is evident) so that a certain degree of symmetry is restored between atomic physics and other fields of physics, as they indirectly express themselves through "likelihood" arguments about astrophysical parameters. Stated otherwise the model provides a framework where conceptual experiments in physics (mainly atomic physics) can on occasion be performed with the help of good observations.

No doubt that position (2c) was and still is the most indicated for a first stage and this may explain in part why it seems to be considered up to now as the only conceivable one by many investigators. But it should be recognized that position (2d) may awaken other reservations such as :
- experiments with models are simply meaningless,
- planetary nebulae are too anecdotal or capricious in their structure to constitute reliable laboratories,
- the best we can do with nebulae is to modestly try to wrest snatches of approximate informations for immediate use in Astrophysics without questioning too much about their meaning,

- astronomers would be presomptuous in trespassing on atomic physics' preserves with their simplistic models and skinny observations,
- this new point of view seems to weaken the overwhelming importance of atomic physics,
- astrophysical results would become conditional, which is neither gratifying nor convenient (and may even increase the length of the papers!),
- the additional freedom would introduce an arbitrariness that does not seem to exist in the more traditional approach (2c),
- a problem with models is that they have so many free parameters that they can fit almost everything and you would introduce as many new ones as you want !

Instead of discussing these critics, an appraisal of the last 15 years modelling is presented followed by some general comments.

4. THE MODELS AS A TEST OF IONIZATION EQUILIBRIUM

A. 1968 : The golden age.

In retrospect the event of the first self-consistent photoionization models of planetary nebulae during the late 60's coincides with their golden age : using a priori the simplest assumptions concerning the astrophysical input parameters and the set of freshly obtained atomic data, the models computed simultaneously by several authors were at once able to explain the most important features of the line spectra of two representative planetaries, namely NGC 7662, a high-excitation matter-bounded nebula, and IC 418, a low-excitation ionization-bounded nebula. Because of the apparent coarseness of the assumptions, the weakness of the low-excitation lines, such as [OII]3727 in NGC 7662 or [OI] 6300 in IC 418, as compared to observation, was not yet considered as a serious problem and, in the euphory of this success was born a credo, whose most achieved expression is probably due to Flower (1968) : "It is believed that the atomic data used are reliable and that the computational methods employed by the computer programmes are sufficiently accurate so that discrepancies between theory and observation are essentially a consequence of invalid physical assumptions". In the context, "physical" excludes atomic physics and refers to any mechanism or feature of "astrophysical" significance not considered in calculations. Stated otherwise, the differences between models and observations are supposed to provide information about the structure of the object under study.

B. 1972 : A confusing situation.

Then what were the findings ? In the framework of static models the only way to increase the concentration of low-charge ions at a given distance to the source is to increase the gas density or consider a zone in which the radiation field is attenuated by shielding.This is the origin of the well-know concepts of "clumping", "filling factor", optically thick "globules" and "shadows".

In its crudest version the concept of clumping means that, on a small scale (that is sufficiently small to be undetectable), the ionized gas fills only a tiny fraction of the whole volume. This concept would deserve a serious theoretical elaboration to be acceptable on hydrodynamical grounds since the overall expansion velocity of planetary nebulae is usually not so much greater than the thermal velocity. The observational evidences for the existence of inhomogeneities are scarce and usually qualitative and refer to "intermediate-scale" (that is an appreciable fraction of the whole nebula) rather than small-scale structures. These structures being only detectable in the light of low-excitation ions, this was taken erroneously as confirming the inference of clumping from models. On the contrary, this only indicates that the ionized gas, as depicted by the best tracer, namely Hβ , is quite evenly distributed.

An outstanding example of small-scale structure is the thin radial filaments best observed at the [NII] wavelength in NGC 7293 (Vorontsov-Vel'yaminov, 1968). This was enough to support one of the most fashionable (and perhaps still alive) way to amplify the low-excitation lines : the head of a filament would be the ionization front of an optically thick globule whose shadow would form the filamentary tail containing low-charged ions (Van Blerkom and Arny, 1972 ; Kirkpatrick, 1972). Although it looks quite attractive, this proposal is not substantiated on closer examination (Hummer and Seaton, 1973 ; Mathis, 1976) because the energy supply of these shadows by the diffuse radiation from the unshielded zone is not adequate in quantity (a small amount of mass can be kept ionized) and quality (the radiation is so soft that the temperature remains low and the emissivity per atom of, e.g., [OII]3727 is divided by 25 from T = 12 000 K to T = 6 000 K). Other critics to this interpretation can be made after detailed inspection of the photographs (Mathews, 1978). In the radiation dominated situation prevailing in this nebula, other energy supplies, such as conduction or turbulence dissipation seem even less likely and one could hardly imagine how shock-wave propagation might be sustained and efficiently heat the ionized gas of the filaments. Thus these filaments, which were considered for a long time as the best observational support to the explanation of the low-excitation line intensities by means of high density condensations, may well constitute the best observational evidence that many other suggestions, made for other nebulae, do not work ; in particular these filaments are not shadows, but they are evidently not ionization fronts either. Now, taking as granted that the gas of the filaments is normally ionized and heated by the star radiation, the enhancement of the low-excitation lines implies that the filaments are denser than their surroundings and the initial assumption of (more or less) optically thin condensations must again be considered. However the hydrodynamical objections too remain and the only way round this apparent contradiction is to assume that the filaments constitute trailing flows with small density enhancements, which, according to the models constructed (for other nebulae) in the early 70's, were unable to reconcile the low-excitation lines with observation.

This sketchy discussion (not performed at that time) already tends to suggest that the problem -the so-called "[OII] problem" of the late

70's - may not be solved by any satisfactory choice of astrophysical parameters. The models exhibited in the early 70's were apparently as much successes in explaining the low-excitation lines of NGC 7662 or similar objects but they neglected the physical content of the astrophysical parameters as well as the guidance of some fragil but significant morphological data (Aller, 1956 ; Vorontsov-Vel'yaminov, 1968), too hastily classified as corroborating the theory.

In their review of 1973, Hummer and Seaton already emphasized the acuteness of the difficulty but the situation was very confused and they concluded that "the question of the temperature of the transition region is one of the major unsolved problems".

C. 1973 : A message.

The introduction by Williams (1973) of the resonant charge exchange reaction

$$O^{+} + H^{o} \longleftrightarrow O^{o} + H^{+} \qquad (1)$$

in model nebulae was a very significant stage because it suddenly allowed to explain the [OI] line intensity in, e.g., IC 418 and therefore to demonstrate that a new atomic physics process could prosaically but efficiently resolve an "astrophysical" problem[2]. This (provisionally ?) ends interesting speculations about the existence of shockwaves preceeding the ionization front in compact planetaries.

The significance of this "message" was not fully appreciated and the pursuit of astrophysical facts hidden behind discrepancies continued.

D. 1976 : The stagnation.

The success of Williams concealed negative repercussions in that the gain in O^{o} concentration within the ionized gas was obtained at the expense of O^{+} so that the [OII] line intensity became more evidently and systematically unaccountable, particulary in high-excitation radiation-bounded planetaries such as NGC 7027. Moreover the early guess that a larger temperature of the transition region could solve the problem lost its last trace of credibility since, in such a case, [OI] would be amplified in a larger proportion than [OII] in view of the spatial distribution of the ions.

Taken together, the success of §C and the rationalizations of §§B and D irresistibly suggest that the question was one of ionization equilibrium of the oxygen ions. The successive attempts with either the gas density distribution or the central star spectrum were not specially convincing but a latent period seemed necessary before leaving the impasse. Some of the causes and consequences of this situation may be as follows.

Excessive attention was focused on a few objects such as NGC 7662, which was then attributed two contradictory roles. On one hand the case is taken as representative and it is sufficient to explain its spectrum to conclude that the modelling (not only the model) is satisfactory. On the other hand the case is particular and it is evidently allowed to take into account the specificity of the object : this specificity is indeed what we are supposed to look for and it results from an analysis which amounts to eliminate the discrepancies with observation. However there is a priori no clear limit to the "plasticity" of the free parameters and, with the presupposition of correct atomic physics (and correct observations), the largest difficulties to fit the data lead invariably to the most fascinating results on the condition that a model can finally be exhibited. If the most extreme caution is not exercised, an unconscious shift towards the most "gratifying" cases, namely those which put up a credible resistance and finally give in, is to be feared. It is very unfortunate that NGC 7662, in all likelihood, belonged to this category at least during the 70's and was also taken as an archetype.

The fact that the models were considered as satisfactory in explaining "typical" nebulae had at least three nefast consequences. Firstly the specialists in atomic physics were not on the look-out for discovering new processes since it is never gratifying to quibble the decimals of a closed matter. Secondly the status of the planetaries which resisted modelling was dramatically changed and it is by no means exaggerated to state that they were most often classified as monsters deprived of real interest (at least for modelling) : in such cases the discrepancies were not eliminated but their existence was either "justified" by a ritual (and platonic) recourse to an inextricably complex geometry or "promizingly explained" by an imaginative (and no less platonic) recourse to exotic mechanisms which had a marked tendency to mimic those proposed to solve the very similar problems faced by very similar models in very different contexts. And actually the third consequence was to give support to the conclusions drawn from the successes and reverses of the photoionized models when applied to active nuclei of galaxies and supernova remnants : no doubt that some of the exotic aspects of these objects during the 70's were due to overinterpretation of the weakness of the models.

At least one positive consequence of considering the subject of static models as a closed matter was to direct attention to other studies (see Section 2 ; Harrington, 1978 ; Harrington, this volume) whose methodology will certainly prove useful in the future.

E. 1977 : The charge exchange reactions.

The previous comments justify a change of perspective. The archetype among monsters was NGC 7027. This high-excitation planetary nebula has every reasons to be radiation bounded. Since the density distribution was considered as the stumbling block, two remarks seem in order : (1) the nebula is obscured by dust but it appears fairly symmetrical at radio wavelengths; (2) a full development of hypothetical condensations may not be expected in this moderately expanded nebula and, at any rate, optically thick condensations would just produce fluctuations in the position of the ionization front.

the ionization equilibrium is governed by the equation :

$$n(O^{++}) \,/\, n(O^{+}) = \zeta / \alpha n_e$$

where ζ is the photoionization rate and αn_e the (electron) recombination rate of O^+ ; (2) from the angular size of the nebula and the (de-reddened) flux of Hβ and He II 4686, a quite accurate - indeed very large - ζ can be computed, which depends neither on the distance of the nebula nor on the effective temperature of the star (taking as granted that this temperature is very high) ; (3) the absorption of radiation by the inner zones of the nebula has a moderate and late effect on ζ because the gas is particularly transparent in the energy range 31.1 ev (potential of O^+) - 54.4 ev (potential of He^+) because of the rapid decline of the cross-section of H° and the virtual absence of He° ; (4) when finally ζ is significantly decreased, the photoionization rate of H° is already depleted and the edge of the Stromgren sphere is virtually reached so that the zone where O^+ can coexist with free electrons is exceedingly small ; (5) the O^+ ions are also absent when hydrogen is mainly neutral because of the reaction between O^+ and H° ; (6) assuming uniform distribution, the mean density of the nebula is much above the critical density for collisional de-excitation of the upper level of [O II] 3727 so that the computed intensity is independent of the assumed density distribution (the relative abundance of O^+ is roughly proportional to n_e while the relative efficiency of the line emission is inversely proportional to n_e) ; (7) in the simple case of a radiation-bounded model with constant density, the intensity of [O II] 3727 is one tenth the observed one when [O III] 5007 is correct ; (8) the difficulty is thus fundamental, especially since it exists also in Seyfert 2 galaxies, and the solution probably deals with atomic physics ; (9) but for oxygen the atomic data involving free electron collisions or photons are among the best known ; (10) thus until a very powerful new mechanism is proposed to selectively amplify [O II] , a complementary mechanism must exist to recombine O^{++} into O^+ : since, in addition, the discrepancy is more serious in high-excitation objects, where O^{++} and H° are known to better coexist, it is tempting to attribute this extra recombination to charge exchanges between O^{++} and H°.

Other less blinding but still evident discrepancies existed. Thus Péquignot, Aldrovandi and Stasinska (1978) adopted the following exploratory point of view : (1) the density distribution is spherically symmetric and is the simplest one compatible with density indicators, (2) the central star has a standard spectrum, (3) the classical atomic data are accurate, (4) the discrepancies between observation and model calculations that can be resolved by modifying the ionization equilibria are systematically attributed to charge exchange reactions with atomic hydrogen and (5) the empirical rate coefficients derived for each reaction are open to test by performing atomic physics calculations and by modelling other nebulae.

In 1977, the charge exchange rates were, for the most part, known qualitatively by means of the asymptotic Landau-Zener rule (Dalgarno and Butler, 1978). By a strange coincidence this rule erroneously predicted a negligible rate for the kee reaction

$$O^{++} + H° \longrightarrow O^{+} + H^{+}$$

and this may have cast some discredit on the principle of this approach

(approach (2d) of §3). By now most empirical rates (Péquignot, 1980) are in rather good quantitative agreement with the theoretical rates (see the fundamental papers : Butler, Heil and Dalgarno, 1980 ; Butler and Dalgarno, 1980a). This indicates that many other atomic data are probably quite reliable and that, for the time being, a good working hypothesis is that the planetary nebulae are not more complex than what was believed before inventing models.

However the situation is not completely satisfactory for :
1) The reaction Ne^{++} + H° does not exist contrary to the empirical result so that the line [Ne II] 12.8μ is not explained.
2) If [C I] 9849 and [S I] 7725 are collisionally excited, there is not enough C° and S° in the ionized zone (Butler and Dalgarno, 1980b).
3) The empirical rates involving multiply charged ions were obtained assuming that the next ions (then not observed) were free from extremely fast charge exchange reactions with H° : this is apparently not the case for, e.g., the important reaction O^{+3} + H° and the excellent agreement for O^{++} and N^{++} + H° may be partly fortuitous.
4) Several other moderate discrepancies between observation and recent (unpublished) models are apparent.

Does it mean that approach (2d) already reached its limits and that it is time to go back to the comforting position (2c) ? More accurate observations are becoming available and more systematic modellings are in progress with many improved atomic data so that a new interesting confront may be expected in a near future. We are free to dream that some discrepancies will survive to tell us something about the physical processes at work in planetary nebulae.

F. 1981 : The low-temperature di-electronic recombinations.

From a theoretical standpoint it was not fully realized to which extent the models - during the 70's - were essentially a check of ionization equilibria[3]. Our ignorance about the collisional processes governing ionization was such that we needed ten years to understand that these processes were at the origin of the failure of the models. In the case of charge exchange reactions an attentive inspection of past results in physics might have lead to consider them since the construction of the first models (see Note 2). This is not the case for the "low-temperature" di-electronic recombinations recently introduced by Storey (1981). A complete table of rate coefficients is not yet available (Storey, this volume) but we already know that this is a new crucial ingredient of models since, for selected ions, the classical radiative recombinations can almost be looked on as perturbations !

Up to very recently, this was probably not much a trouble in the comparison of models to observation because this process seems more effective for ions not easily detected with ground base observations. And it may not be pure coincidence if this process was discovered soon after the event of detailed UV observations (Seaton, this volume). Nonetheless the vanity of the abundance corrections based on theoretical models can be appreciated.

CONCLUDING REMARKS

A model in physics is usually a reasoned simplification of reality intended to make easier the description or the understanding of a particular process and it is evaluated by the richness and exactness of its consequences. Now the (astrophysical) outputs of a model - the "results"- can hardly be considered as "predictions" because no experiment can be performed with the true object and, in a sense, the model is the only reality to handle.

The current models certainly appear very coarse when compared to photographs of most planetaries but the aim of these models is not predominantly to give faithful monochromatic images of nebulae with all their anecdotal accidents (even though it may be the case in the future); rather, models intend to account for observed line spectra with exact physics and thus try to represent these "accidents" in a physically realistic but geometrically schematic manner. Thus how and to which extent the structure of planetary nebulae should be considered as complex is a question that typically falls within the scope of modelling and that should not be stated a priori from superficial inspection of beautiful but qualitative and purposedly exposed photographs. In this connexion it should be stressed that the assumption of overall spherical symmetry (used globally or just locally to compute the diffuse radiation field) has often less consequences on the value of the results than one generally believes and, in most cases, attention should rather focus on the postulates concerning the small scale density distribution. Important developments are expected if monochromatic electronographic images of high angular resolution can be obtained (N.K. Reay, this volume).

In perspective (2d), the intention is clearly not to compete with atomic physics but to draw attention to challenging problems in a suggestive and quantitative manner, namely by giving some sort of "equivalence" between a discrepancy with observation and the correction to atomic data that would be needed to resolve it under the assumption that other astrophysical parameters, e.g., the stellar flux distribution or the gas density distribution, are not particularly exotic when there is no good reason to beleive that they should be. Then the model recovers its prerogatives : a view of the object is proposed and is open to the best possible test, namely that of atomic physics. Evidently only those empirical revisions are allowed that seems to "reasonably" fall within current precision of the atomic data in question and that are sufficiently well motivated on astrophysical grounds. If and only if atomic physics and observation maintain their positions are we founded to consider new physical processes or astrophysical particularities.

The pre-eminence of atomic physics is not disputed but strengthened, as it should, in this approach. As we have seen, many of the "astrophysical discoveries" made by means of models in the domains where they are essential were spurious as a consequence of inadequacies of atomic data and it is time to acknowledge this misfortune of astronomers : a planetary nebula is indeed such a good atomic physics laboratory that it is still extremely difficult to do something else with it ! If approach (2c) was a reasonable first guess, approach (2d) is necessary at least as an intermediate step which will eventually make easier a return to (2c) on sounder basis.

The freedom of models is partly a myth when good and numerous observations are available and real tests of consistency are possible in specific cases. The impression that almost everything can be fitted is often due to the arbitrary choice of the astrophysical parameters when they are taken as formal and completely disposable regardless of the physics they are laden with.

On reflexion a self-consistent model of a nebula may not (or should not) be much more than a super plasma-diagnostic, even so have it to be properly calibrated to act this role. The extreme view which rejects any use of models "before the completion of atomic physics" is evidently absurd and position (2d) appears as a possible resort to help in obtaining astrophysical facts rather than atomic physics artefacts.

The massive supply of UV observations and of new or accurate atomic data prompted a renewal of models. Concerning the use of models for practical purpose, there is certainly a shift of interest ; thus the abundance corrections for unseen ions of C, N, O and Ne are now less crucial than an accurate determination of the temperature in the high-excitation zones. However the theoretical situation is not unlike the one of 1968 in several respects. Firstly the atomic data for heavier elements are not yet well circonscribed, not to say that they are often rudimentary in the case of iron in spite of outstanding efforts, and one must be prepared to face, once again, problems of ionization equilibrium with however the help of lighter elements which can calibrate the astrophysical parameters. Secondly the question of the lighter elements themselves cannot be considered as settled if one pretends to use the models to learn something more than abundances about the "astrophysics" of planetary nebulae. Most of the fascinating ideas debated during the 70's have sufficiently good physical basis that some of them will one day revive, probably in the form of "second order perturbations" to the observables. The past 10 years experience with models should be meditated in order not to fall pitifully into the same pitfalls. The risk does exist since, apart from uncompletedness, many coefficients are not guaranteed to better than a factor two by Mc Carroll and Storey and the recent results presented by Mendoza invite to prudence.

NOTES

1. The survival of such filaments almost certainly implies the existence of collapsed neutral "cores", possibly at a temperature not exceeding a few hundred Kelvin, in rough pressure equilibrium with their surroundings. The detection of the H_2 molecule, e.g. in NGC 6720 (Beckwith et al, 1978), provides support to this view and indicates that the case of NGC 7293 is probably not unique. The filaments have often been compared to comets.

This analogy may well be much deeper than would be suggested by some early interpretations : as an example, it may be that the cold neutral core (the "nucleus") intercepts only a very small fraction of the stellar radiation so that the bulk of the emission arises from some type of large matter-bounded "head" surrounding the (presumably evaporating) core and prolonged by a trailing tail. The detailed analysis of these remarkable structures has not yet been undertaken possibly because it was hampered by the pending question of the low-excitation lines.

2. In fact reaction (1) was invoked as early as 1956 by Chamberlain in a slightly different context. After being forgotten for twenty years, we should acknowledge the extraordinary insight of Chamberlain who resolved an astrophysical dilemma, put forward one year before, by means of an atomic physics theory, discovered one year before, and who even speculates about the possible relevance for nebulae of charge exchange reactions between atomic hydrogen and multiply charged ions.

3. In particular the unavoidable freedom on the elemental abundances allowed to account quite easily for the few and relatively inaccurate informations about temperature. In fact a correct account of both the temperature indicators and the strongest line intensity (usually [O III])demonstrated that the main energy source of planetary nebulae was indeed photoionization by the central star radiation <u>and</u> that the relevant collision strengths were reasonably accurate.

REFERENCES

Aller, L. H. : 1956, Gaseous Nebulae, Chapman & Hall ed.
Beckwith, S., Persson, S.E., Gatley, I. : 1978, Astrophys. J. Let. 219, L23
Butler, S.E., Dalgarno, A. : 1980a, Astrophy. J. 241, 838
Butler, S.E., Dalgarno, A. : 1980b, Astron. Astrophys. 85, 144
Butler, S.E., Heil, T.G., Dalgarno, A. : 1980, Astrophys. J. 241, 442
Capriotti, E.R. : 1973, Astrophys. J. 179, 495
Chamberlain, J.W. : 1956, Astrophys. J. 124, 390
Dalgarno, A., Butler, S.E. : 1978, Com. At. Mol. Phys. 7, 129
Flower, D. : 1968, Astrophys. Lett. 2, 205
Harrington, J.P. : 1968, Astrophys. J. 152, 943
Harrington, J.P. : 1977, Mon. Not. R. Astr. Soc. 179, 63
Harrington, J.P. : 1978, IAU Symp. n° 76 (Terzian ed.) p. 151
Harrington, J.P., Marionni, P.A. : 1976, Astrophys. J. 206, 458
Hummer, D.G., Seaton, M.J. : 1973, Mém. Soc. R. Sci. Liège 5, 225
Kirkpatrick, R.C. : 1972, Astrophys. J. 176, 381
Lazareff, B. : 1981, Thèse d'Etat , Université Paris Sud
Mathews, W.G. : 1978, IAU Symp. n° 76 (Terzian ed.), p. 251
Mathis, J.S. : 1976, Astrophys. J. 207, 442
Péquignot, D. : 1980, Astron. Astrophys. 81, 356
Péquignot, D., Aldrovandi, S.M.V., Stasinska, G. : 1978, Astron.Ast.63, 313
Storey, P.J. : 1981, Mon. Not. R. Astr. Soc 195, 27P
Tylenda, R; : 1979, Acta Astronomica 29, 355
van Blerkom, D., Arny, T.T. : 1972, Mon. Not. R. Astr. Soc. 156, 91
Vorontsov-Vel'yaminov, B.A. : 1968, IAU Symp. n° 34 (Osterbrock, O'Dell)p.256
Williams, R.E. : 1973, Mon. Not. R. Astr. Soc. 164, 111

DOPITA: I worry about the $O^{++} + H \rightarrow O^{+} + H^{+}$ charge exchange rate, as published. My models for either "power law" excited regions or H II regions have the [O III] lines much too weak compared with the [O II] lines. How confident are you that the rate coefficient you use is not too large?

PÉQUIGNOT: To date, I did not experience this problem in my own calculations for fairly well-defined nebulae. On the contrary, [O II] tends to be too weak relative to [O III] in the most recent studies. Thus, I have no reason to suspect an overestimate of the rate coefficient in this case (even though the theoretical or empirical estimates may not be more accurate than a factor 2 to 3). I am afraid that dielectronic recombination of O^{++}, discussed by Storey at this meeting, could worsen your problem.

KÖPPEN: My own work and that of Aller on fitting the spectra of a fairly large number of PN show that we now have the problem that the models predict [O II] λ 3727 to be stronger than observed!

CHARGE EXCHANGE REACTIONS IN ASTROPHYSICAL PLASMAS

R. McCarroll
Laboratoire d'Astrophysique, Université de Bordeaux 1, France
P. Valiron
Groupe d'Astrophysique, Université de Grenoble 1, France
L. Opradolce
Instituto de Astronomia y Fisica del Espacio, Buenos Aires

ABSTRACT

A review is presented of charge exchange reactions of multiply charged ions with atomic hydrogen and helium at thermal-eV energies, typical of the physical conditions encountered in planetary nebulae. The basic features of the processes are analyzed in the framework of the molecular model of atomic collisions. A discussion is given of the different theoretical approaches to the calculation of the collision cross sections. A comparison with recent experimental data is included.

INTRODUCTION

An analysis of the emission line spectrum of the planetary NGC 7027 led Péquignot et al. (1978) to the conclusion that recombination of certain doubly and more highly charged ions takes place primarily by charge exchange with neutral hydrogen or helium. Assuming reasonable empirical estimates of the charge exchange reaction rates, Péquignot (1980a) obtained excellent agreement between the theoretical and observed emission fluxes produced by ions of C, N, O, Ne and S in both the visible and UV spectrum. These conclusions were amply confirmred by further investigation (Péquignot 1980b, Ulrich and Péquignot 1980) of other planetary nebulae (NGC 7662, 6720) and indeed of other similar objects such as the nebulae surrounding the Seyfert galaxies (NGC 3516).

Theoretical work on the physics of charge exchange reactions involving multiply charged ions by McCarroll and Valiron (1975, 1976, 1978, 1979), Christensen et al. (1977), Watson and Christensen (1979), Butler et al. (1980), Butler and Dalgarno (1980) and Gargaud et al. (1981, 1982) are in substantial agreement with the empirical rate coefficients proposed by Péquignot, at least for the stronger reactions with rate coefficients of the order of several 10^{-9} cm^3 s^{-1}. For weaker reactions the problem is more confused. For example, the case of charge exchange in O III/H collisions has a controversial history. Péquignot et al. require a rate coefficient of $\sim 8 \times 10^{-10}$ cm^3 s^{-1} to explain the observed O II line

D. R. Flower (ed.), Planetary Nebulae, 187–197.

intensities, in apparent discrepancy with the early prediction (1978) of $\sim 10^{-13}$ $cm^3 s^{-1}$. However, later detailed calculations by Butler et al. (1980), taking account of the possibility of configuration mixing in the O III ionic core, yield a rate coefficient of $\sim 6x10^{-10}$ $cm^3 s^{-1}$, in good agreement with Péquignot. On the other hand, all theoretical considerations of charge exchange in Ne III/H and Ne III/He collisions (Dalgarno et al. 1980) agree that the reaction rate must be slow (or at least several orders of magnitude smaller than the value required by Péquignot to explain the observed data onNe II.

The existence of such exceptions suggests caution in pursuing an analysis with empirical rate coefficients too far. It is therefore of the utmost importance to dispose of reliable cross section data. Only then, can the observations be interpreted in a consistent way.

Some mention should also be made of the possibility of charge exchange as a possible ionization source in nebular gas. Typical charge exchange reactions of an A^{+q} ion with atomic hydrogen

$$A^{+q} + H \rightarrow A^{+q-1}(n) + H^{+} \qquad (1)$$

produces in most cases an ion A^{+q-1} in an excited state n. As a result, the inverse reaction is not usually of importance since the excited state decays rapidly by spontaneous radiative emission. However, if the reaction (1) can proceed via the ground state of A^{+q-1}, then the inverse process of charge transfer ionization can also occur easily. Baliunas and Butler (1980) have shown how the ionization equilibrium of Si is affected by this process. We may also expect a similar effect on the ionization equilibrium of Fe (Gargaud et al. 1982).

The aim of this review is not to give an exhaustive list of rate coefficients. Rather, it is to explain the underlying physical mechanisms of charge exchange in thermal collisions of multiply charged ions with atomic hydrogen or helium and to give some idea where we may expect new advances in the future. The advent of low energy ion sources (Phaneuf 1981, Huber 1982, Ohtani 1982) open up real possibilities of investigating important astrophysical reactions in the energy range from 10 eV upwards. Since the physical mechanism for charge exchange is essentially the same at all energies from thermal to several 100 eV, there is a real hope in the next few years that theory may be used to extrapolate experimental data down to the astrophysically interesting range of 0.01 - 10 eV.

PHYSICAL MODEL OF THE CHARGE EXCHANGE PROCESS

At thermal to eV energies, where the relative collision velocity is small compared with the characteristic orbital velocity associated with the valence electrons, the molecular model may be used to describe the charge exchange process. Let us consider for illustration reaction (1)

The entry channel (A^{+q} + H) is attractive at large internuclear distances, due to the polarization potential $-\alpha q^2/2R^4$, where α is the polarizability of H and R is the internuclear distance. As a consequence there is no activation barrier and, if conditions are favourable, the cross section may become very large at thermal energies.

For q > 2, the exit channels (A^{+q-1}(n) + H^+ are dominated by the repulsive Coulomb potential (q-1)/R. A curve crossing of the adiabatic potential energy curves of the $(AH)^{+q}$ molecular ion may be expected to occur at a value of R given by

$$\{(q-1)/R\} + \{q^2\alpha/2R^4\} = E_i - E_f = \Delta E(R) \qquad (2)$$

where E_i and E_f are respectively the internal energy of the initial and final reaction products. Of course, because of the Wigner non crossing rule, potential energy crossings between states of the same molecular symmetry are not allowed. As a result, many of the crossings predicted by (2) become avoided crossings. A typical illustration is given in figure 1, where are plotted the various molecular states involved in the charge exchange process in Ar VII/He collisions (Opradolce et al. 1982).

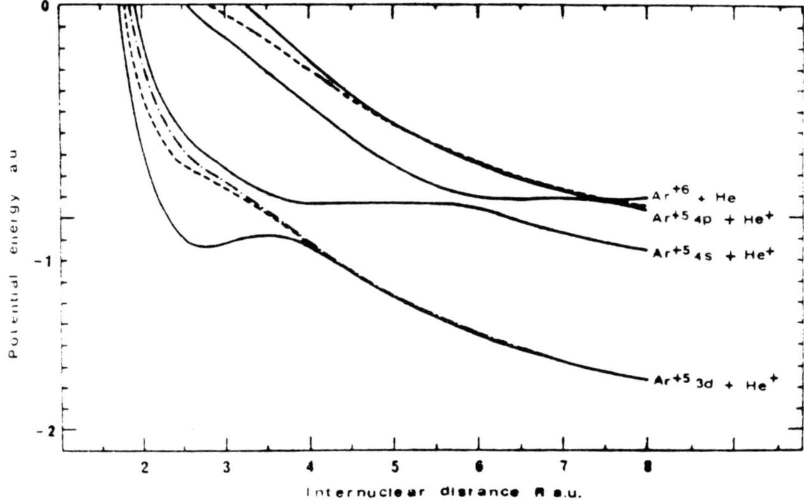

Fig. 1. Singlet potential energy curves of various adiabatic states of the molecular ion $(ArHe)^{+6}$ as a function of the internuclear distance. The solid curves refer to $^1\Sigma$ states, the dashed curves to $^1\Pi$ states and the chain curve to a $^1\Delta$ state. The dissociation products are as labelled.

Since electronic transitions between molecular states occur to the breakdown of the Born-Oppenheimer approximation in the vicinity of the curve crossings, it is clear that charge exchange is a highly selective process. Firstly the reaction must be exoergic, $E_i > E_f$, for otherwise no curve crossing would occur. Secondly, the energy separation $\Delta E(R_x)$ at an avoided crossing located at R_x must be neither too large nor too small. If $\Delta E(R_x)$ is of the order of a few eV or greater, the transition probability is very small and adiabatic conditions prevail. If $\Delta E(R_x)$ is of the order of 10^{-2} eV or less, diabatic conditions prevail. As a general rule electronic transitions take place at low energies only when $0.1 < \Delta E(R_x) < 1$ eV. This condition is only satisfied for crossings which occur at intermediate internuclear distances ($4 < R_x < 10\ a_o$). At higher energies (> 1 keV/amu) these conditions are of course less restrictive since certain molecular states may then be strongly coupled over a wide range of internuclear distances rather than localized in the vicinity of the curve crossings.

It is thus obvious from these simple considerations that charge exchange with highly charged ions takes place mainly via capture into excited Rydberg states. For ionic charge q of the order of 3, 4, capture into states with principal quantum number n = 3 is the most probable and the application of the molecular model is a feasible proposition. For higher q, when capture into Rydberg states with n > 5 becomes possible, the number of interacting molecular states may become exceedingly large; other methods (Janev 1982) are then to be preferred.

It is convenient to distinguish two types of processes (Butler and Dalgarno 1980).

Type I. Monoelectronic processes where electron capture takes place into an orbital (excited or not) without any change of the orbital configuration of the ionic core. Typical examples are

$$N^{+3}(2s^2)\,^1S \;+\; H(1s)\,^2S \;\rightarrow\; N^{+2}(2s^2 3s)\,^2S \;+\; H^+ \qquad (3)$$

$$O^{+3}(2s^2 2p)\,^2P \;+\; H(1s)\,^2S \;\rightarrow\; O^{+2}(2s^2 2p 3p)\,^1P, {}^3P, {}^3D \;+\; H^+ \qquad (4)$$

$$Si^{+2}(3s^2)\,^1S \;+\; H(1s)\,^2S \;\rightarrow\; Si^{+}(3s^2 3p)\,^2P \;+\; H^+ \qquad (5)$$

Type II. Two electron (or configuration mixing) processes, where electron capture into a valence orbital is accompanied by a rearrangement of the ionic core orbitals. Typical examples are

$$N^{+2}(2s^2 2p)\,^2P \;+\; H(1s)\,^2S \;\rightarrow\; N^{+}(2s 2p^3)\,^3D \;+\; H^+ \qquad (6)$$

$$O^{+2}(2s^2 2p^2)\,^3P \;+\; H(1s)\,^2S \;\rightarrow\; O^{+}(2s 2p^4)\,^4P \;+\; H^+ \qquad (7)$$

As a general rule, type II processes are important only when type I processes are weak. This is frequently the case for doubly charged ions when the type I processes are either too exoergic (R_x too small) or too endoergic (no curve crossing). In these cases, such as N III/H and O III/H, type II processes are the dominant ones. However, for a given value of R_x, type II processes are an order of magnitude weaker than type I processes.

POTENTIAL ENERGY CURVES

In the quasi-molecular model of charge exchange, the main problem resides in the determination of the potential energy curves of the molecular states correlated to the entry and all possible exit channels. For type I processes, the dominant processes for ions with $q > 3$, the electronic structure is that of a single electron in an excited state outside of the ground state configuration. Here, frozen Hartree-Fock and model potential methods work well (McCarroll and Valiron 1975, 1976, Christensen et al. 1977), though each method has its limitations. Hartree-Fock methods take no account of electron correlation and model potential potential methods do not allow for adjustments of the core configuration. However, model potential methods offer several advantages over ab-initio methods in that they guarantee the correct dissociation limits and long range interactions; these considerations are of vital importance in the investigation of reactions at eV energies. Neither approach can be expected to treat satisfactorily charge exchange reactions of type II, when more than one active electron is involved. Here, the use of a configuration interaction method is required (Dalgarno et al. 1980).

The critical parameters for charge exchange reactions are the position of the avoided crossings R_x and the energy differences $E(R_x)$ at these positions. (Real crossings between states of different symmetry do not contribute in an important way in the very low energy range.) In general, R_x is well reproduced by all methods which give the correct dissociation energies associated with the entry and exit channels. On the other hand, the determination of $\Delta E(R_x)$ is more delicate. The paucity of published data makes a quantitative comparison of the various methods difficult. However, some idea may be gained by a comparison of the model potential results (McCarroll and Valiron 1975, 1976, 1979, Gargaud et al. 1981, Opradolce et al. 1982) with the empirical formula proposed by Butler and Dalgarno (1980) on the basis of their configuration interaction results. They find that for type I processes, their results are reproduced to within 30% by the formula

$$\Delta E(R_x) = 27.2\, R_x^{\,2} \exp(-\beta R_x) \text{ eV} \qquad (8)$$

where $\beta = 1$ in the case of neutral atomic hydrogen and $\beta = 1.34$ for neutral helium. For the systems listed in Table 1, there is good reason to believe that the model potential results should be accurate. In some cases there is satisfactory agreement between the different methods (N IV/H, B IV/He) but in other cases the discrepancy can sometimes be quite considerable. It is thus clear that an empirical formula, which makes no allowance for the particular symmetry of the final state is unsatisfactory.

COLLISION DYNAMICS

The absence of good experimental data at eV energies has necessitated extensive reference calculations in order to develop simpler models.

Table 1. Energy differences $\Delta E(R_x)$ at avoided Σ-Σ crossings of potential energy curves in various systems.

System	$R_x(a_o)$	$\Delta E(R_x)$ in eV Empirical formula (8)	Model potential
N IV/H	9.0	0.27	0.34
Si III/H	9.7	0.15	0.07
C V/H	7.9	0.60	0.10
	7.4	0.91	1.09
	7.1	1.19	2.01
B IV/He	4.5	1.32	1.99
	7.4	0.07	0.07
Ar VII/He	7.4	0.07	0.37
	5.8	0.39	1.33
	3.6	2.8	4.7

A detailed description of these calculations would be out of place in this review but it is useful to recall the essential steps (Heil et al. 1981, Gargaud et al. 1981).

Basically, a complete description of the collision dynamics requires not only the adiabatic energy curves but also the dynamic coupling matrix elements (both radial and rotational) between the different adiabatic states intervening in the collision. The calculation of these matrix elements is a delicate operation and approximations (often of uncertain validity), such as the Hellmann-Feynman theorem are frequently used to simplify the calculations. This is the case for the calculations of Butler (1981). Even when the matrix elements can be computed without approximation, as in the model potential calculations, great care must be exercised to avoid all possible errors. Once determined, these matrix elements serve to define a diabatic transformation (Smith 1969) which leads to a considerable simplification of the scattering equations.

The smooth variation of the elements of the diabatic potential matrix in the vicinity of the avoided crossings, as compared with the dynamic coupling matrix elements (see figures 2 and 3) would suggest the possibility of their empirical construction directly from the adiabatic potential curves.

This is, of course, the essence of the Landau-Zener (LZ) method, which defines the diabatic transformation using the computed values of $\Delta E(R_x)$ and assuming that the diagonal potential matrix vary linearly. An approximate solution of the dynamical equations in the semi classical

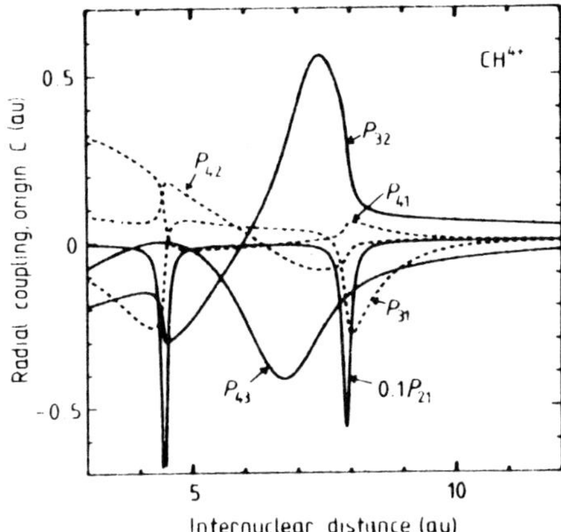

Fig. 2. Radial matrix elements for the Σ states of CH^{+4}. The full curves designate the principal coupling terms P_{21} , P_{32}, P_{43}. The broken curves refer to the less important coupling terms P_{31}, P_{41}, P_{42}.

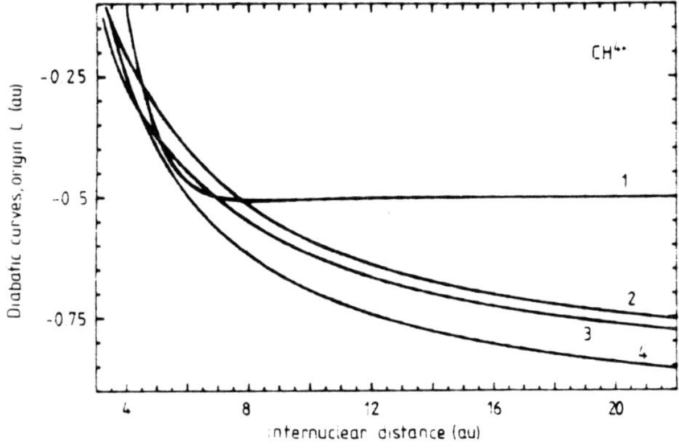

Fig. 3. Diagonal elements of the diabatic potential matrix of CH^{+4} constructed from the radial matrix elements of fig. 2 .

(impact parameter) treatment then yields a total cross section of the form

$$Q = 2\pi \int_0^{b_m} 2w(1-w)\ b\ db \tag{9}$$

where b_m is the maximum value of impact parameter b, for which the crossing point R_x is classically accessible and

$$w = \exp\left[\{-\pi\ \Delta E(R_x)^2\ R_x^2\ \}/\{2(q-1)v\}\right] \tag{10}$$

where v is the radial velocity at R_x. Typical applications of the LZ formula usually assume straight line trajectories, which is valid at energies exceeding a few eV. At thermal energies, it is more reasonable to assume b_m and v to be determined by the polarization potential of the entry channel. This modification enables the LZ approximation to take

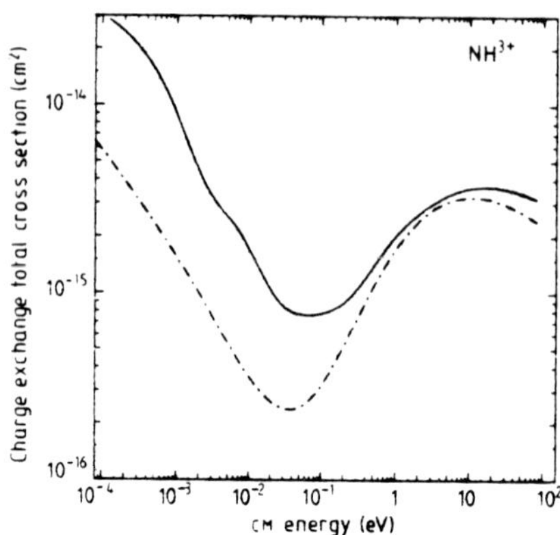

Fig. 4. Charge exchange cross section (in units of cm^2) in N IV/H collisions as a function of energy (in eV). The full curve designates the quantum mechanical calculations. The broken curve refers to the Landau-Zener model with allowance for trajectory effects(Gargaud et al.).

account of orbiting effects, at least in a qualitative manner. The accuracy of the LZ formula can be satisfactory (∿20-30%) when the values of w, which contribute most to the integrated cross section Q are of the order of 0.5. When w becomes very small (adiabatic limit) or approaches 1 (diabatic limit) the LZ formula becomes unreliable. A typical comparison with an exact calculation is given in figure 4 for charge exchange in N IV/H collisions.

It is clear from the preceding paragraph that the LZ formula, although very simple to use, is the result of a series of approximations, whose validity is difficult to ascertain. In actual practice, most of these approximations are unnecessary since there is no fundamental difficulty in solving numerically the quantum mechanical scattering equations once the diabatic potential matrix is known. The only really useful approximation is the empirical construction of the diabatic potential matrix from the adiabatic potentials without having recourse to the calculation of the radial matrix elements. This approach has been followed in the work of McCarroll and Valiron (1976, 1979) for the Si III/H and N IV/H systems, where various parametric forms of the diabatic matrix are proposed. The success of the approach can be judged by a comparison with the results of the exact calculations of Gargaud et al.(1981). More recent calculations, currently in progress, based on a technique developed by Masnou-Seeuws et al. (1982) in another context (see also Faist and Levine 1976), confirm the validity of an empirical construction of the diabatic potential matrix and allow the determination of confidence limits of the method in a given situation. Typical comparisons are shown in Table 2.

The excellent agreement between the exact and empirical diabatic transformation confirms the feasibility of calculating cross sections on the basis of good adiabatic potential curves alone, at least in the energy range of astrophysical interest.

Table 2. Cross sections (in units of a_o^2) at a collision energy (c.m) of 0.8 eV

System	Exact radial coupling	Empirical diabatic matrix
C V/H	55	46
N IV/H	65	58
Si III/H	143	149

RESULTS AND DISCUSSION

The essential problem is to know to what extent the theoretical cross sections can be relied upon. For type I processes, present indications are that the theory can produce results of high accuracy. There is a good consistency between the different theoretical approaches (configuration interaction, frozen core Hartree Fock, model potential) and there is excellent agreement between theory and experiment for those systems where comparison is possible. For example, (see fig. 5) the recent experimental results of Phaneuf (1981) on the system C V/H in the centre of mass energy range from 10 to 500 eV and of Huber (1982) on the system N IV/H in the energy range from 100 to 300 eV agree very satisfactorily with the theoretical results of Gargaud et al.(1981, 1982). Less well tested are the individual cross sections for capture into a specific excited state. These are more sensitive to details of the theoretical model and a real test of the theory must await more detailed experiments. For example, Dalgarno et al. (1981) do not guarantee even the largest capture cross sections for the system O IV/H to more than a factor of three.

As for type II processes, there exists for the moment, only the theoretical data of Butler et. (1980) No comparison with experiment has yet been possible

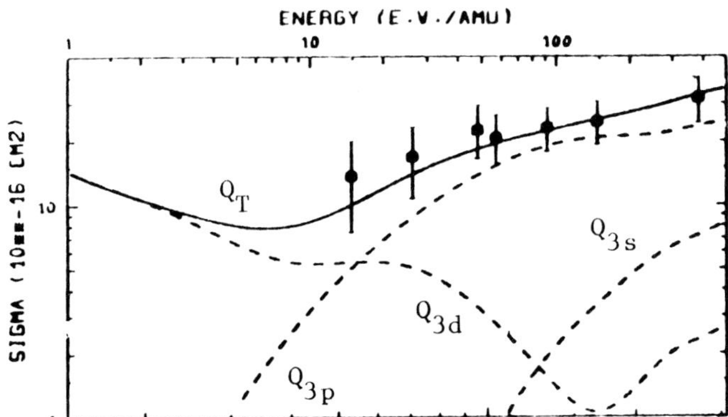

Fig. 5. Charge exchange cross sections in C V/H collisions as a function of the incident ion beam energy.The theoretical results of Gargaud et al. are compared with the experimental results of Phaneuf.

ACKNOWLEDGMENTS

The authors wish to acknowledge helpful contributions to this review from Drs. M. Gargaud and J. Hanssen.

REFERENCES

Baliunas, S.L. and Butler, S.E.: 1980, Astrophys. J. Letters 235, L45
Butler, S.E.: 1981, Phys. Rev. A23, 1
Butler, S.E. and Dalgarno, A.: 1980, Astrophys. J. 241, 838
Butler, S.E., Heil, T.G. and Dalgarno, A.: 1980, Astrophys. J. 241, 442
Christensen, R.B., Watson, W.D. and Blint, R.J.: 1977, Astrophys. J. 213, 712
Dalgarno, A.: 1978, IAU Symposium N° 72 on Planetary Nebulae
Dalgarno, A., Butler, S.E. and Heil, T.G.: 1980, Astron. Astrophys. 89, 379
Dalgarno, A., Heil, T.G. and Butler, S.E.:1981, Astrophys. J. 245, 793
Faist, M.B. and Levine, R.D.: 1976, J. Chem. Phys. 179, 111
Gargaud, M., Hanssen, J., McCarroll, R. and Valiron, P.: 1981, J. Phys. B: Atom. Molec. Phys. 14, 2259
Gargaud, M., Hanssen, J., McCarroll, R. and Valiron, P.: Physics of Electron and Atomic Collisions, Invited Papers XII ICPEAC, (edited S. Datz), p. 707, North-Holland Amsterdam (1982)
Gargaud, M., Hanssen, J., McCarroll, R. and Valiron, P.: Communications au 2e colloque national du conseil français du télescope spatial, p. 77, INAG Paris (1982)
Gargaud, M., McCarroll, R. and Valiron, P.: 1982, Astron. Astrophys. 106, 197
Heil, T.G., Butler, S.E. and Dalgarno, A.: 1981, Phys. Rev. A23, 1100
Huber, B.A.: 1982, Physica Scripta, Topical issue on Physics and Production of Highly Charged Ions, to appear
Janev, R.K.: 1982, Physica Scripta, Topical Issue on Physics and Production of Highly Charged Ions, to appear
McCarroll, R. and Valiron, P.: 1975, Astron. Astrophys. 44, 465
McCarroll, R. and Valiron, P.: 1976, Astron. Astrophys. 53, 83
McCarroll, R. and Valiron, P.: 1979, Astron. Astrophys. 78, 177
McCarroll, R. and Valiron, P.: 1978, J. Physique 39, C1, 52
Masnou-Seeuws, F., Boulmer, J., Maurin, T., Roche, A.L. and Valiron, P.: 1982, J. Phys. B: Atom. Molec. Phys., to appear
Ohtani, S.: 1982, Physica Scripta, Topical issue on Physics and Production of Highly Charged Ions, to appear
Opradolce, L., Valiron, P. and McCarroll, R.: 1982, J. Phys. B: Atom. Molec. Phys., to appear
Péquignot, D.: 1980 a, Astron. Astrophys. 81, 356
Péquignot, D.: 1980 b, Astron. Astrophys. 83, 52
Péquignot, D., Aldrovandi, S.M.V. and Stasinska, G.: 1978, Astron. Astrophys. 63, 313
Phaneuf, R.A.: 1981, Phys. Rev. A24, 1138
Smith, F.T.: 1969, Phys. Rev. 179, 111
Ulrich, M.H. and Péquignot, D.: 1980, Astrophys. J. 238, 45
Watson, W.D. and Christensen, R.B.: 1979, Astrophys. J. 231, 627

TARTER: Does your comment on the lack of validity of the Butler-Dalgarno prescription for ΔE apply to their general methodology or just to the use of the simple formula?

McCARROLL: The limitation of the Butler-Dalgarno prescription for ΔE applies only to the use of their simple formula. The value of ΔE is determined not only by ionization potentials but also by the symmetry properties of the molecular states involved. An empirical formula, which takes no account of such symmetry, cannot predict the correct values of ΔE. Of course, the configuration interaction calculations of Butler are not, in principle, subject to such limitations.

SEATON: Charge transfer can also give rise to observable spectrum lines: O III singlets are discussed by Dalgarno, Heil and Butler (Astrophys. J., 245, 793) and triplets by Dalgarno and Sternberg (Mon. Not. R. Astron. Soc., in press).

PÉQUIGNOT: The importance of charge exchange as an ionizing process should be emphasized. We found that the abundances of several minor species are reduced by a factor of 10 when this process is introduced in the recent model of NGC 7662 by Harrington et al. (see the review of Harrington in this volume). Also, Si^+ is depleted by about a factor of 3.

McCARROLL: I agree entirely. Some remarks on the selective nature of ionization by charge exchange are given in the written version of my talk.

RECOMBINATION PROCESSES

P.J. Storey
Department of Physics and Astronomy,
University College London, Gower Street,
LONDON WC1E 6BT, England.

ABSTRACT

The role of resonances in the photoionization cross-sections of ions is discussed in relation to the radiative recombination coefficient. The calculation of dielectronic contributions to the recombination coefficient is reviewed for nebular conditions, and the importance of autoionizing states close to the first ionization threshold is illustrated. For those ions for which dielectronic recombination is not important at nebular temperatures, an assessment of the accuracy of the best available recombination coefficients is given.

1. RADIATIVE RECOMBINATION AND PHOTOIONIZATION

The most satisfactory approach to the calculation of recombination coefficients makes no distinction between 'radiative' and 'dielectronic' recombination. Consider the radiative capture process

$$X^{+m} \quad + \quad e \quad \rightarrow \quad X_b^{+m-1} \quad + \quad h\nu \qquad (1)$$

resulting in the bound state X_b^{+m-1} and the emission of a photon. Let the radiative capture cross-section be σ_{RC} and the cross-section for the inverse process of photoionization be σ_{PI}. The photoionization cross--section in general has the form of a smooth background interrupted by resonances. The resonances correspond to quasi-bound states of the recombined system X_a^{+m-1}, and Rydberg series of such states converge on the terms of the recombining system. The cross-sections for radiative capture and photoionization are connected by the Milne relation,

$$\sigma_{RC} \;=\; \frac{\omega_b}{2\omega_+}\,\frac{h\nu}{mc^2}\,\frac{h\nu}{\frac{1}{2}mv^2}\;\sigma_{PI} \qquad (2)$$

where ω are statistical weights and v is the electron velocity. The

D. R. Flower (ed.), Planetary Nebulae, 199–209.

rate coefficient for radiative capture is then obtained by integrating σ_{RC} over the free electron velocity distribution. The resonances in the photoionization cross-section correspond to the 'dielectronic' component of the radiative recombination coefficient. The calculation of the total radiative recombination coefficient by this method requires the integration of the photoionization cross-section, including resonances, for all bound states, X_b^{+m-1}.

For the calculation of recombination coefficients at nebular temperatures, only free electron energies up to about 0.2 Rydbergs need be considered. If the photoionization cross-sections $\sigma_{PI}(X_b^{+m-1})$ are strongly enhanced by resonances in this energy range, there will be a significant dielectronic contribution to the radiative recombination coefficient. Such enhancements have been demonstrated for ions of C, N and O (Beigman and Chichkov, 1980, Storey, 1981).

2. H, He AND Li-LIKE IONS

No resonance effects are possible in hydrogenic ions. In He and Li-like ions, the formation of resonance states involves the excitation of an electron from the 1s shell and free electron energies which are not available at nebular temperatures. Dielectronic recombination can therefore be neglected for these ions. Seaton (1980) has reviewed the theory of recombination lines of H and He^+. I shall restrict myself to some comments on the accuracy, in nebular conditions, of currently used recombination coefficients.

2.1 Hydrogen Like

The most complete calculations of recombination coefficients for H are those of Brocklehurst (1970, 1971). The first of these two papers deals with high (principal quantum number, $n \geq 40$) states, making the assumption that the ℓ sub-states are populated according to their statistical weights, $2(2\ell+1)$. The second paper treats the low ($n \leq 40$) states and the populations of ℓ sub-states are calculated explicitly. Intensities of Balmer and Paschen lines are given. Giles (1977) has used unpublished level populations for hydrogen obtained by Brocklehurst to determine the intensities of the infra-red Brackett lines. All these calculations are for Case B of Baker and Menzel (1938). The computer codes of Brocklehurst have recently been revived by Hummer and Storey and generalised to deal with H, He and Li-like ions of arbitrary nuclear charge.

Photoionization cross-sections (and bound-bound transition probabilities) are in principle known exactly for hydrogenic ions and can, in practice, be calculated to any required precision. Brocklehurst (1970) estimates the error in these quantities to be less than 1% in his calculations. Rate coefficients for collisional redistribution of angular momentum and energy are known with less precision. Arguing that a 10% change in collision rates is equivalent to a 10% change in electron density, Brocklehurst (1971) infers a maximum uncertainty of 2% in the intensities of the higher hydrogen Balmer lines. For lines originating from low ($n \leq 10$) states, there is no reason to suppose that the relative intensities given by Brocklehurst (1971) and Giles (1977) for hydrogen lines are in error by more than 1%.

Brocklehurst (1971) also gives intensities for the Pickering and Pfund series of He II. Note that in Table II of his paper, the column headings 10^4, 10^5, 10^6 cm^{-3} should read 10^6, 10^5, 10^4 cm^{-3}. Seaton (1978) has used scaling laws to determine the intensities of the UV Balmer and Paschen lines of He II, and to demonstrate that collisional redistribution of angular momentum is less effective in He^+ than in H. The scaling laws relating the radio recombination lines of He II to those of H I are discussed by Weisheit and Walmley (1977). New calculations for He^+ indicate that a) the results of Seaton (1978) are accurate to about 1% for the He II Balmer line intensities and about 3% for the Paschen lines. b) the uncertainty in the results of Brocklehurst (1971) is about 1%.

In He, the uncertainty due to collisional processes among the excited states is comparable to that for H discussed above. In the work of Brocklehurst (1972), the photoionization cross-sections for He are derived from quantum defect theory using experimental energies for the excited states. The uncertainties in these energies imply an error of less than 1% in the photoionization cross-sections near threshold, and in the resulting recombination coefficients.

3. COMPLEX IONS

3.1 Photoionization

Various approximations exist for the evaluation of photoionization cross-sections for complex ions (see for example the report by C. Mendoza in the present volume). If the wave functions for the X^{+m} + e system are calculated by a method, such as the close-coupling approximation, in which the interaction between open and closed channels is included, resonance effects are automatically incorporated. This approximation is incomplete in certain cicumstances. In a full treatment, the interaction of the ion with the quantised radiation field is included in the evaluation of the photoionization amplitude (Davies and Seaton, 1969). The cases considered here do not generally require this more sophisticated approach.

To a first approximation, the strength of background and resonance contributions to the photoionization cross-section for a bound state are determined by the parentage of that state. For example for states belonging to the $2s^2(^1S)n\ell$ series of C^+, such as $2s^23d\ ^2D$ (Figure 1), there is a strong background contribution, corresponding to photoionization to the $2s^2\ ^1S$ + e continuum, and weak resonance contributions from the $2s2p(^3P^o)n\ell$ series of resonances. The corresponding term diagram of C^+ is shown in Figure 2. For bound states with $2s2p(^3P^o)$ parentage, there is a weak background with prominent resonances. The terms in the ground

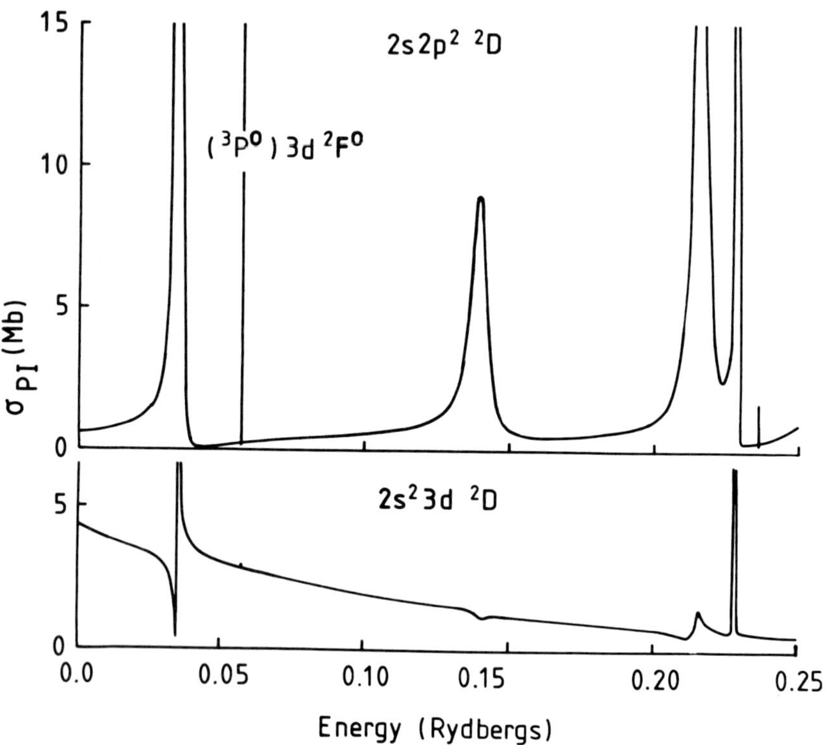

Figure 1. Photoionization cross-sections for C^+ $2s2p^2\ ^2D$ and $2s^23d\ ^2D$ close to the first ionization threshold.

complex ($2s^22p$, $2s2p^2$, $2p^3$), such as $2s2p^2\ ^2D$ have mixed parentage, and there can in general be large resonance contributions from many series. In C^+, the recombination coefficient for the $2s2p^2\ ^2D$ state, and for the ion, is dominated by the $2s2p(^3P^o)3d\ ^2F^o$ resonance.

For most ions of astrophysical interest, photoionization cross-section calculations which incorporate resonance effects have been made only for the ground and some low-lying metastable states. These calculations need to be extended at least to include all terms for which parentage considerations imply a large contribution from resonances. Calculations of this sort are in progress at UCL for C^+, C^{2+} and N^{2+}. An alternative approach consists of evaluating background photoionization cross-sections for parentage allowed processes using, for example the quantum defect method, and adding dielectronic contributions separately. Compilations of coefficients for the non-resonant part of the recombination coefficient have been made by Aldrovandi and Péquignot (1973, 1974, 1976) and Gould (1978). In both cases, a hydrogenic formulation is used from excited states, but the results of Gould are superior in that they make some

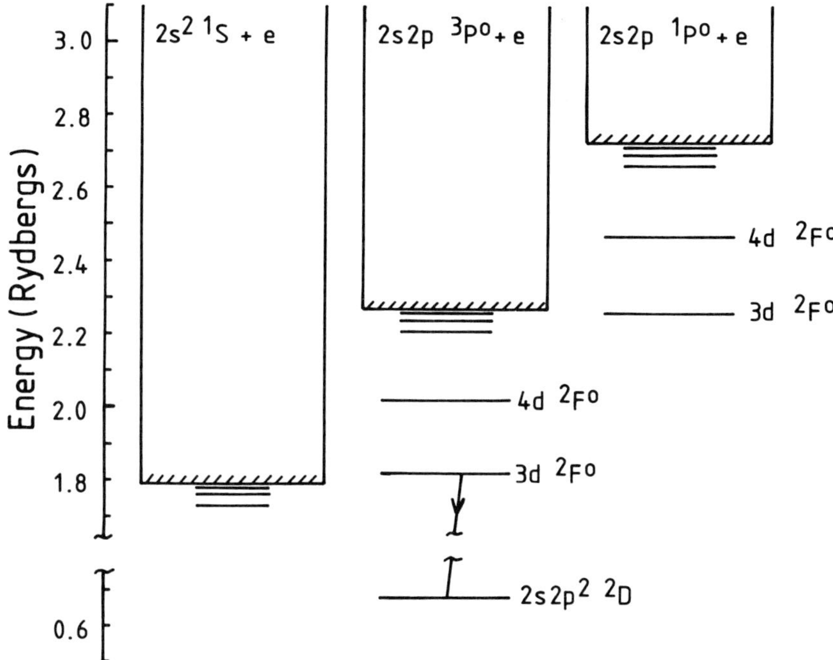

Figure 2. Autoionizing states in C^+.

allowance for their nonhydrogenic nature.

3.2 Dielectronic Recombination

The most widely used approach to the calculation of dielectronic recombination coefficients follows from considering the properties of the quasi-bound states X_a^{+m-1}. Provided angular momentum and parity selection rules are satisfied, these states may undergo a radiationless transition to a continuum state of the same energy,

$$X_a^{+m-1} \rightarrow X^{+m} + e \tag{3}$$

Let the probability of this autoionization be $\Gamma_a^{(A)}$ (s^{-1}). The inverse process is dielectronic capture. In thermodynamic equilibrium, the rates of these two processes must be equal, so

$$N_{TE}(X_a^{+m-1}) \; \Gamma_a^{(A)} = N_e \; N_{TE}(X^{+m}) \; \alpha_c \tag{4}$$

where α_c is the capture coefficient, and the population ratio

$N_{TE}(X_a^{+m-1}) \,/\, N_e\, N_{TE}(X^{+m})$ is given by the Saha equation. In the non TE conditions which prevail in nebulae, the possibility of radiative decay of X_a^{+m-1} with total probability $\Gamma_a^{(R)}$ (s^{-1}) must also be included;

$$N(X_a^{+m-1})\,(\Gamma_a^{(A)} + \Gamma_a^{(R)}) = N_e\, N(X^{+m})\,\alpha_c \tag{5}$$

Combining equations (4) and (5) gives

$$\frac{N(X_a^{+m-1})}{N_e N(X^{+m})} = \frac{N_{TE}(X_a^{+m-1})}{N_e N_{TE}(X^{+m})} \frac{\Gamma_a^{(A)}}{(\Gamma_a^{(A)} + \Gamma_a^{(R)})} = \frac{N_{TE}(X_a^{+m-1})}{N_e N_{TE}(X^{+m})}\, b(X_a^{+m-1}) \tag{6}$$

where $b(X_a^{+m-1}) \equiv b_a$ is a measure of the departure of the population of the autoionizing state from its TE value. The rate of recombination to some bound state X_b^{+m-1}, $N(X_a^{+m-1})\,\Gamma_{ab}^{(R)}$ can then be obtained once b_a and the transition probabilities $\Gamma_{ab}^{(R)}$ are known. The autoionizing state X_a^{+m-1} can be characterised by $(\gamma_p S_p L_p ns\ell;SL)$ where S,L are total spin and orbital angular momentum quantum numbers, $ns\ell$ are principal, spin and orbital angular momentum quantum numbers of the added electron and γ_p represents all other quantum numbers associated with the parent term p. It can be shown (eg Seaton and Storey, 1976) that, in atomic units, $\Gamma_a^{(A)}(n) = C/n^3$ where C is of order unity. In addition, $\Gamma_a^{(A)}$ tends to zero for ℓ large.

3.2.1 Burgess general formula. Burgess (1964) was concerned with electron temperatures characteristic of an ionization balance determined primarily by electron collisions, for which the free electron energies are comparable with the ionization potentials of the abundant ions. Burgess considered resonance series in which the parent term p is connected to the recombining ion ground state (p=1), by an optically allowed transition. He further assumed that the autoionizing states can only decay radiatively via the $p\to 1$ transition in the ion core, the highly excited added electron being considered as a 'spectator'. Following the core transition, the ion is left in a bound state. The total dielectronic recombination coefficient is then obtained by summing over all possible resonance states. Since $\Gamma_a^{(R)}$ is independent of a, the sum converges since $b_a\to 0$ as $\Gamma_a^{(A)}\to 0$ for n,ℓ large. In practice at high temperatures, states with $n\simeq 100$ and $\ell\leq 10$ are important. Burgess (1965) introduced a fit to the dielectronic recombination coefficient, intended for use at high temperatures, in which the energies of the resonance states relative to the first ionization limit are replaced by a constant energy close to the parent term energy difference, ΔE_{1p}, leading to the coefficient falling approximately as $\exp(-\Delta E_{1p}/kT)$ as $T\to 0$. For most ions of interest, $\Delta E_{1p} \gtrsim 0.5$ Rydberg for the first optically allowed transition, and at 10^4K, the general formula gives the dielectronic contribution to the recombination coefficient to be negligible.

3.2.2 Low temperatures. The use of the Burgess general formula is incorrect at nebular temperatures, because a) individual low-lying auto-

-ionizing states are not included, b) for low-lying autoionizing states, the radiative probability $\Gamma_a^{(R)}$ may be enhanced by the decay of the outer electron, c)series of autoionizing states exist for which the p→1 transition is not permitted, which can also decay radiatively due to outer electron transitions. As an example consider recombination to form C^+. In the Burgess model, only the states $2s2p(^1P^o)n\ell$ are included. These are assumed to decay radiatively with the same probability as the C^{2+} $2s2p(^1P^o)\rightarrow 2s^2(^1S)$ transition. In practice, none of these states lie close enough to threshold to contribute significantly to the recombination coefficient. Members of the series $2s2p(^3P^o)n\ell$, however, do lie close to threshold and as mentioned above, the $2s2p(^3P^o)3d\ ^2F^o \rightarrow 2s2p^2\ ^2D$ transition dominates the recombination.

Consider a series of autoionizing states for which the p→1 transition is not permitted, connected to a common bound state , b, by radiative decay of the outer electron. The transition probability , in atomic units, is $\Gamma_{ab}^{(R)}=B\alpha^3 z^4/n^3$, where B is of order unity, z is the recombining ion charge and α is the fine structure constant. For z,ℓ small, $\Gamma_a^{(A)} >> \Gamma_a^{(R)}$ so that $b_a(n)=1$, independent of n. The evaluation of the dielectronic recombination coefficient at low temperatures therefore involves only calculation of the radiative probabilities $\Gamma_{ab}^{(R)}$, since the populations of the autoionizing states are given by the Saha equation. In practice, the most important transitions are usually to the ground complex of the recombined ion, and configuration interaction is therefore important in representing the wave function of the lower state. The rate of recombination from the members of a series

$$N_{TE}(X_a^{+m-1})\ b_a(n)\ \Gamma_{ab}^{(R)} \propto n^{-3} \exp(-E_n/kT) \tag{7}$$

where E_n is the energy of the autoionizing state relative to the first ionization threshold. The sum over a given series therefore converges rapidly and it is normally sufficient to consider only the first few terms explicitly. A correction for higher members can be obtained using the functional form of equation (7).

3.2.3 Recombination lines. At low temperatures, most of the dielectronic recombination involves radiative transitions between a few low--lying resonance states and one or two terms in the ground complex of the recombined ion. Frequently the bulk of the recombination passes through a single such term and a line is strongly enhanced. In the B-like ions, the $2s2p^2\ ^2D$ state is populated, giving rise to emission in λλ1335, 991, 789 in C^+, N^{2+} and O^{3+}. Similarly in the Be-like ions, λλ2297, 1718, 1371 arise from the $2p^2\ ^1D$ state in C^{2+}, N^{3+} and O^{4+}. Many other weaker recombination lines are also generated. In C III, in addition to the strongest lines at λλ1176, 2297, there is for example, a satellite line to C IV λ1550, $2p3d\ ^3F^o \rightarrow 2s3d\ ^3D$ at λ1577 and enhancement of the $2s3p\ ^3P^o \rightarrow 2s3s\ ^3S$ λ4650 transition. For these weaker lines to be utilised, a full treatment of recombination and cascade processes as outlined in section 3.1 is required.

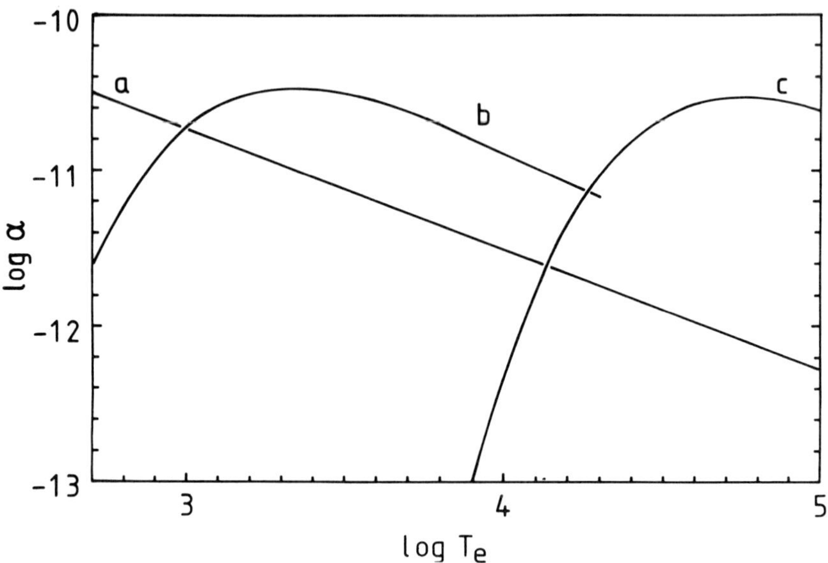

Figure 3. Recombination coefficients for C^{3+} + e as a function of electron temperature. See text for explanation.

3.2.4 Recent work. Since the early work of Burgess, various calculations of total dielectronic recombination coefficients have been made but with the main emphasis on the solar ionization balance. Davis et al. (1977) demonstrate that the general formula may overestimate when autoionizing states lie energetically above a second or higher ionization threshold, due to the additional channels for autoionization. This effect is unlikely to be important at nebular temperatures. A series of papers have been published (Jacobs et al. 1977, 1979, 1980) incorporating this effect and in some cases giving dielectronic recombination coefficients down to 10^4K. The compilation of dielectronic recombination coefficients by Shull and Van Steenberg (1982) consists of fits to these results and to unpublished work of Jacobs for ions of C, N and O. Throughout this work, the assumptions of Burgess (1964), that the recombination only occurs via optically allowed transitions in the recombining ion, have been retained and consequently the coefficients are underestimates at low temperatures , sometimes by orders of magnitude. The same is true of the fits given by Aldrovandi and Péquignot (1973, 1974, 1976), derived from the Burgess general formula. Summers'(1974) extensive tabulations of total recombination coefficients also underestimate at nebular temperatures.

Dielectronic recombination coefficients which take account of the effects discussed in section 3.2.2 have been given by Beigman and Chichkc (1980) for $O^+ \rightarrow O$ recombination, and by Storey (1981) for the recombined ions C^+, C^{2+}, N^{2+}, N^{3+} and O^{4+}. Calculations have also been made for the remaining ions of C, N and O by Nussbaumer and Storey (1983). This work is described in a contributed paper. Storey (1981) also gives effective recombination coefficients for some of the stronger UV recombination lines. Figure 3 shows radiative recombination coefficients for $C^{3+}\rightarrow C^{2+}$ recombination. The non-resonant contribution to the coefficient (a), was calculated from the parameters given by Aldrovandi and Péquignot (1973). Curve (c) is the dielectronic contribution calculated from the Burgess general formula, also taken from Aldrovandi and Péquignot. The values given by Storey (1981), together with some unpublished results for T<7000K are curve(b). The total recombination coefficient is enhanced by about a factor of five at 10^4K by the proper inclusion of resonance effects. Curve (b) falls below (c) at higher temperatures, because the calculations of Storey do not include, in full, the series of autoionizing states in which a permitted transition in the recombining ion is involved. Preliminary calculations show that there is a similar enhancement for $Si^{3+}\rightarrow Si^{2+}$ recombination to that for $C^{3+}\rightarrow C^{2+}$. Work on the ions of Mg, Al and Si is in progress.

Once the electron temperature is low enough that $kT<E_L$, where E_L is the energy of the lowest resonance relative to the first ionization threshold, curve (b) in Figure 3 falls as $\exp(-E_L/kT)$. Below $T_L=E_L/k$, the approach used by Storey is unsuitable. The recombination coefficient can then only be determined correctly from the threshold behaviour of the photoionization cross-sections.

3.2.5 Coupling schemes. The discussion so far has been in terms of LS coupling, in which S, L and parity are conserved in the dielectronic capture process. Usually, for the autoionizing states of interest, $\Gamma_a^{(A)}/\Gamma_a^{(R)} = 10^3 - 10^6$. Relativistic interactions such as spin-orbit coupling, although weak for light ions of low ionization, may lead to $\Gamma_a^{(A)} \gtrsim \Gamma_a^{(R)}$ for states which are excluded from an LS coupling model. Consider the case of $O^+\rightarrow O$ recombination. Above the $O^+\ {}^4S^o$ ground state lie singlet and triplet states of O. In LS coupling, none of the singlets and about two thirds of the triplets are populated by dielectronic capture. The recombination coefficient could be increased by about a factor of two if the spin-orbit interaction is sufficiently strong.

3.3 Summary

The best approach to the calculation of total radiative recombination coefficients at nebular temperatures at present is to add a) the values of Gould (1978) or of Aldrovandi and Péquignot (1973, 1974, 1976) for the non-resonant contributions, b) the values of Aldrovandi and Péquignot for dielectronic contributions from the Burgess general formula and c) the values of Beigman and Chichkov (1980), Storey(1981) and Nussbaumer and Storey (1983) for dielectronic contributions due to low-lying resonances. An ucertainty of about a factor of two has to be reckoned with for the resulting total radiative recombination coefficient.

Photoionization cross-sections which incorporate resonance effects now exist for the ground and some low-lying states of many ions. Work is in progress to obtain recombination coefficients from the cross-sections. As further cross-sections become available, it will be possible to improve on current estimates of the important dielectronic contributions to the radiative recombination coefficient.

REFERENCES

Aldrovandi, S.M.V. and Péquignot, D., 1973. Astron. Astrophys., 25, 137.
Aldrovandi, S.M.V. and Péquignot, D., 1974. Rev. Bras. de Fis., 4, no 3, 491.
Aldrovandi, S.M.V. and Péquignot, D., 1976. Astron. Astrophys., 47, 321.
Baker, J.G. and Menzel, D.H., 1938. Astrophys. J., 88, 52.
Beigman, I.L. and Chichkov, B.N., 1980. J. Phys. B., 13, 565.
Brocklehurst, M., 1970. Mon. Not. R. astr. Soc., 148, 417.
Brocklehurst, M., 1971. Mon. Not. R. astr. Soc., 153, 471.
Brocklehurst, M., 1972. Mon. Not. R. astr. Soc., 157, 211.
Burgess, A., 1964. Astrophys. J., 139, 776.
Burgess, A., 1965. Astrophys. J., 141, 1588.
Davies, P.C.W. and Seaton, M.J., 1969. J. Phys. B., 2, 757.
Davis, J., Jacobs, V.L., Kepple, P.C. and Blaha, M., 1977. J. Quant. Spectrosc. Radiat. Transfer, 17, 139.
Giles, K., 1977. Mon. Not. R. astr. Soc., 180, 57P.
Gould, R.J., 1978. Astrophys. J., 219, 250.
Jacobs, V.L., Davis, J., Kepple, P.C. and Blaha, M., 1977. Astrophys. J., 215, 690.
Jacobs, V.L., Davis, J., Rogerson, J.E. and Blaha, M., 1979. Astrophys. J., 230, 627.
Jacobs, V.L., Davis, J., Rogerson, J.E., Blaha, M., Cain, J. and Davis, M., 1980. Astrophys. J., 239, 1119.
Nussbaumer, H. and Storey, P.J., 1983. Astron. Astrophys., to be submitted
Seaton, M.J., 1978. Mon. Not. R. astr. Soc., 185, 5P.
Seaton, M.J., 1980. "Radio Recombination Lines", ed. P.A. Shaver (D. Reidel), 3.
Seaton, M.J. and Storey, P.J., 1976. "Atomic Processes and Applications", ed. P.G. Burke and B.L. Moiseiwitsch (North-Holland), chapter 6.
Shull, M.J. and Van Steenberg, M., 1982. Astrophys. J. Suppl., 48, 95.
Storey, P.J., 1981. Mon. Not. R. astr. Soc., 195, 27P.
Summers, H.P., 1974. Appleton Laboratory, IM 367.
Weisheit, J. and Walmsley, C.M., 1977. Astron. Astrophys., 61, 141.

PEIMBERT: I am very pleased to see that the computations of recombination coefficients have been repeated and that these are in very good agreement with the previous work by Brocklehurst. These data are needed to determine the pregalactic He/H abundance ratio which is of paramount importance in the study of cosmology and elementary particles. Are the results of the He I numerical study going to be available in the near future?

STOREY: Yes, work is in progress.

NUSSBAUMER: Could you comment on the work recently published in the Astrophys. J. by Shull and Van Steenberg on recombination coefficients.

STOREY: Their non-resonant ("radiative") contributions to the recombination coefficient are taken directly from the paper of Aldrovandi and Péquignot. The dielectronic contributions are from Jacobs and co-workers and are substantial underestimates at nebular temperatures.

PÉQUIGNOT: I am glad to learn that, at least at a qualitative level, the recombination of O^{++} to O^{+} and of N^{++} to N^{+} will be amplified, as indicated by recent models of NGC 7027 (Péquignot, Stasinska and Aldrovandi, 1982, unpublished). Empirical dielectronic recombination coefficients have already been used in a model of the Crab nebula (Péquignot and Dennefeld, 1982, in press).

RADIATIVE TRANSFER PROBLEMS IN PLANETARY NEBULAE

David G. Hummer[†]
Joint Institute for Laboratory Astrophysics, University of Colorado and National Bureau of Standards, Boulder, Colorado, 80309

INTRODUCTION

In view of the enormous importance of the UV observations of planetary nebulae made possible by the IUE, this review will concentrate primarily on the formation of resonance lines in nebulae; an important special case is that of He II Lyα and its role in the Bowen mechanism. Special attention is given to the effects of dust on the line and continuum formation.

RESONANCE LINES

With the IUE a considerable number of resonance lines formed in planetary nebulae have been observed. In interpreting the strengths and profiles of these lines, several factors have to be considered. First, the optical thickness of the nebular shell in these lines will be very large, with values of perhaps 10^3 for lines of metal ions to as much as 10^6 or more for He II Lyα. Under these circumstances the details of the redistribution in frequency are important. Moreover, the effect of absorption can become crucial; the extinction of UV line photons by dust grains is of primary interest. The expansion of the nebular shell with a velocity on the order of the mean thermal velocity is relevant, although probably not crucial in preliminary modeling. Finally, the inhomogeneous distribution of the nebular gas should be taken into account. Because line photons escape preferentially in the least opaque direction, models based on a uniform shell with either planar or spherical geometry could be quite misleading. Although the effects of severely nonuniform gas distributions are expected to be important in many astrophysical situations, no really practical way of attacking this problem has been developed.

[†]Staff Member, Quantum Physics Division, National Bureau of Standards.

D. R. Flower (ed.), Planetary Nebulae, 211–218.

The description of the frequency redistribution mechanism for resonance lines in planetary nebulae in terms of a combination of natural and Doppler broadening appears to be basically correct, as the effect of collisions on redistribution at the low nebular densities is negligible. Thus the use of the so-called R_{II} redistribution function is adequate. For lines of the less abundant species, in which the optical thickness is sufficiently small that transfer occurs within the Doppler core, complete redistribution appears to be a reasonable approximation for our present level of modeling. Techniques for the numerical solutions of these problems are too well known to warrant discussion here.

The effect of dust absorption on the intensities of UV resonance lines is important in interpreting the IUE observations of planetary nebulae. The absorption of line photons by dust or photoionization depends on the mean path length of those photons within the nebula. For a line with an extremely large optical depth this mean photon path length can be many times the geometrical size of the nebula. This problem has been examined on the basis of Monte Carlo calculations by Bonilha, Ferch, Salpeter and Slater (1979), and by Adams (1975) and Hummer and Kunasz (1980), who used an accurate numerical solution of the transfer equation for a wide range of parameters. Subsequently Frisch (1980) obtained asymptotic expressions for the path length and other relevant quantities. In all of these calculations the R_{II} redistribution function was used, as the mean path length is sensitive to the distribution of the free paths between scatterings, and thus to the distribution of frequencies of a photon as it undergoes scatterings. We consider the quantity

$$\rho \equiv \langle \ell \rangle / 2T$$

where $\langle \ell \rangle$ is the mean photon path length and T is the optical thickness of the shell, both measured on the mean optical depth scale. In Figure 1, ρ is plotted versus T for a number of values of the Voigt parameter a. Here, photons are created at the midplane of the shell, although results for a uniform creation model are similar. Although ρ decreases over some range of optical thickness T, for which transfer starts to occur in the wings, the mean path length itself increases monotonically. With complete redistribution, the results for large T are qualitatively different.

From this work it emerged that the fraction of line photons that ultimately escapes depends on the extinction cross section of the dust only through the product of the cross section and the mean path length for no absorption, i.e. on

$$\alpha = n_d \sigma_d \langle \ell \rangle_o / k_{line} \quad .$$

Moreover the numerical results show that for large T, the escaping fraction is a universal function of this parameter α; i.e. results for very different Voigt parameters, optical depths, absorption cross sections, etc., all lie on the same curve as a function of α. The escap-

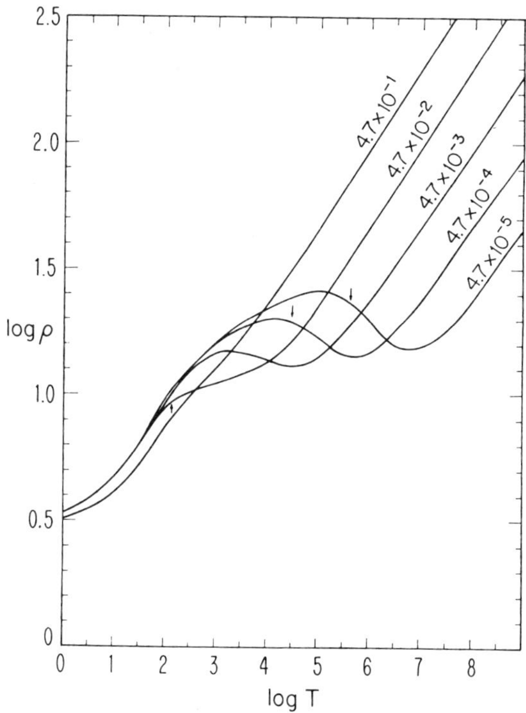

Fig. 1. The ratio ρ of mean path length to the optical half-thickness of the slab, versus the thickness, for indicated values of the Voigt parameter, for a midplane source. The arrows indicate the thickness at which transfer begins to occur in the line wings.

ing fraction decreases much more slowly than a negative exponential, which one might at first expect.

How does a velocity field modify the above results? Bonilha *et al.* (1979) have examined cases with both expansion and dust, and found that for typical expansion rates the escaping fraction was quite insensitive to the expansion rate. These authors note that dust destroys the photons that one might expect the expansion to help most in escaping, i.e. those in or near the line core.

All of the calculations to date have taken the resonance line as a single component, i.e. the multiplet structure was ignored. If the components do not overlap, have a common lower level and distinct upper levels with sufficiently small collisional or radiative coupling, then each component can be treated with the existing theory. "Sufficiently small" coupling means that the product of the largest transition probability and the mean number of scatterings is much less than unity. Expressions and tables for the mean number of scatterings are given by the above authors.

Even if the coupling is negligible, a very curious situation occurs that gives unexpected intensity ratios among the components, in that the stronger component of, say, a doublet may suffer less absorption than the weaker. This may be seen by recognizing that the parameter α can be written as

$$\alpha = \frac{1}{2} \rho_o T_d \quad ,$$

where T_d is the dust optical depth. In the region where ρ decreases with increasing T, ρ and therefore α for the stronger component can be smaller than for the weak component. But smaller α implies a larger escaping fraction, i.e. the stronger component experiences less absorption. This occurs because the distance between scatterings for a photon of given frequency is smaller for the strong component than for the weak component. Since the number of scatterings needed for a photon to escape is only very weakly dependent on the line strength, the path length and hence the probability of absorption of the weaker component is larger than the strong one. Harrington, Seaton, Adams and Lutz (1982) have used this theory to infer from the C IV resonance line in NGC 7662 a dust optical depth of approximately 0.1 for radiation at 1549 Å.

The effects of macroscopic expansion, or more generally, flow velocities may play a role. In nebulae the flow velocities are comparable to the thermal and turbulent velocities, so that the Sobolev approximation is not available, at least in its original form. One possiblity of simplifying transfer problems with complete redistribution in such media has recently appeared through the improvements to the escape-probability approximation. In escape probability methods the source function at a point is related to the probability that a photon will escape from that point. Two kinds of improvements allow such approximations to be used for planetary nebulae. Expressions have been obtained and evaluated by Grachev (1976,1977a,b,1978a,b) and by Hummer and Rybicki (1982a) for the escape probability for an arbitrary ratio of flow to random velocities; previously only the static and the very high velocity or Sobolev limit were used.

The second improvement is the derivation by Hummer and Rybicki (1982b) of a new expression for the source function in terms of the escape probability that is very much more accurate than the previous algebraic relation. Now a simple first-order differential equation and a few integrals must be evaluated for each line, but typical errors of 10% and maximum errors of 25% are found, whereas maximum errors of hundreds of per cent could be obtained previously. This method is limited to complete redistribution, but it should considerably simplify the modeling of most resonance lines, except those of H, He I and He II.

HELIUM II Lyα AND THE BOWEN MECHANISM

Of course, the resonance line par excellence in planetary nebulae is that of ionized helium, which among other things drives the Bowen fluorescent mechanism. Kallman and McCray (1980) have re-examined the Bowen mechanism in a static shell, considering not only the primary cycle, in which He II Lyα at 303.783 Å overlaps O III 303.799 Å and pumps the $2p^2\ ^3P_2$ - $2p3d\ ^3P_2^o$ transition, but also the secondary cycle, in which one of the O III lines at 374.436 Å overlaps N III 374.434, 374.441 Å to pump the $2p\ ^2P_{3/2}$ - $3d\ ^2D_{5/2}$ transition. These authors took as the mean escape probability for He II Lyα the inverse of the asymptotic mean number of scatterings obtained by Harrington (1973) for the R_{II} function, and reduced the resulting depth-independent equation to a simple algebraic problem using the Kneer(1975) approximation to the redistribution function. The O III transfer problem was treated in complete redistribution. The fluorescent efficiencies agreed well with those of Weymann and Williams (1969) and Harrington (1972) for similar values of the parameters. Because of the semi-analytic treatment, results could easily be obtained for a very wide range of parameters. In particular, the O III and N III fluorescent intensities are tabulated as a function of optical thickness. Although the means of the O III and N III fluorescent intensities for a large sample of nebulae agree well with the prototypical results of Kallman and McCray, the considerable variation among the nebulae may be worth analyzing for diagnostics of individual objects.

Although this work includes the ionization of neutral hydrogen and helium by the resonance lines, it does not account for dust absorption specifically. However, as the continuum absorption is parameterized by an equivalent density of hydrogen atoms, it can be interpreted as dust absorption. Kallman and McCray calculated the efficiencies of various processes as functions of the equivalent absorber density. In view of these results, it would be interesting to relate the Bowen efficiencies to the observed intensities of IR dust radiation, bearing in mind, however, that only a small fraction of this radiation comes from He II Lyα.

Another factor of possible interest not treated here is the expansion of the nebula. Because the primary O III line lies in the red wing of the He II line, while the primary N III line lies in the blue wing of the O III 374 Å line, the relative intensities of the O III and N III Bowen lines might give a useful velocity diagnostic.

The recalculation by Saraph and Seaton (1980) of the radiative parameters of O III removes much of the uncertainty in previous discussions of the Bowen mechanism. The primary product of this calculation is the probability $P(\lambda)$ that excitation of O III $2p3d\ ^3P_2^o$ is followed by the emission of a photon of wavelength λ. This provides the coupling strength of the N III cycle to the O III cycle. The calculated relative intensities of the observable O III multiplets agree well with observations and the inferred probability that excitation of

2p3d $^3P_2^o$ by Lyα gives Bowen lines agrees well with the calculations of Harrington (1972).

The fate of He II Lyα photons that escape the region in which they are created has been discussed in some detail by Flower and Perinotto (1980), in connection with the well-known [O II] and [N II] intensity anomaly. Although the authors avoided the transfer problem by parameterizing the fraction of photons escaping the He^{+2} zone, they showed that He II Lyα did not strongly influence the [O II] and [N II] intensities because the Lyα photons were absorbed by H and He before reaching the zone in which these lines are formed. The [O III] and [N III] intensities were increased, however.

CONTINUUM TRANSFER

In modeling planetary nebulae, the transfer problem for the diffuse radiation must be solved in conjunction with the ionization balance and the energy balance. Various procedures have been developed, most of which are refinements of the on-the-spot (OSA) approximation introduced by Zanstra (1951) and developed by Hummer and Seaton (1963). It is not generally recognized that the OSA is a form of the escape-probability approximation with the escape probability set to zero, and that considerably better results can be obtained by using the correct non-zero value. The OSA and the escape-probability method were compared with accurate numerical solutions by Van Blerkom and Hummer (1967), who referred to the escape-probability method as the normalized on-the-spot (NOSA) approximation. It is unfortunate that the utility of the NOSA is not more widely known. Now an even more accurate approximation is available in the second-order escape probability method of Hummer and Rybicki (1982b), which should be nearly exact for continuum problems.

In the early 1970's considerable progress was made in treating spherical-geometry transfer problems. The basic idea was to introduce the so-called Eddington factors, defined as certain ratios of the first three moments of the intensity. In the Eddington approximation these quantities are simply constants, whereas now they are functions of depth that are to be determined by a non-linear iteration procedure. The convergence is very rapid and leads to essentially exact results, although a rather large amount of computational work is required. Following the earliest applications of this method for spherical geometry by Hummer and Rybicki (1971) and by Kunasz and Hummer (1974), for continuum and line problems respectively, Leung (1975) applied this method under the name of quasi-diffusion method to pure dust clouds; a substantial literature now exists on this subject.

One particularly interesting application of the quasi-diffusion or variable-Eddington factor method has been made by Petrosian and Dana (1980) to nebulae containing dust with hydrogen and helium. This paper has two goals, a methodological one of comparing various approximations for the nebular continuum problem, and the physical one of examining in a systematic way the effect of dust, and especially of the dust albedo, on the ionization structure and the temperature of the dust and gas in the nebula. Although this is not the place for an extensive discussion of dust in nebulae, it does appear that Petrosian and Dana have explored the rather extensive parameter space of this problem and that the observable properties of the nebula do depend on the parameters of the dust grains. More to the point of this review, however, is the conclusion that the on-the-spot approximation generalized to include dust absorption is reasonably accurate for purely-absorbing dust, while for dust with a significant non-zero albedo the Eddington approximation seems to be the method of choice, with the modification for extended spherical geometry hardly worth the additional effort.

In conclusion, it appears that transfer theory can account for dust in planetary nebulae, thus providing a bridge between the UV spectrum of the stellar and nebular radiation fields on one hand, and the IR observations on the other.

ACKNOWLEDGMENTS

The preparation of this review was supported in part by National Science Foundation Grant AST80-19874.

REFERENCES

Adams, T.F.: 1975, Astrophys. J. 201, pp. 350-351.

Bonilha, J.R.M., Ferch, R., Salpeter, E.E., and Slater, G.: 1979, Astrophys. J. 233, pp. 649-660.

Flower, D.R. and Perinotto, M.: 1980, Mon. Not. R. astr. Soc. 191, pp. 301-308.

Frisch, H.: 1980, Astron. Astrophys. 87, pp. 357-360

Grachev, S.I.: 1976, Vestn. Leningrad Gos. Univ., No. 1, pp. 128-132.

Grachev, S.I.: 1977a, Astrophysics 13, pp. 95-101.

Grachev, S.I.: 1977b, Vestn. Leningrad Gos. Univ., No. 19, pp. 114-120.

Grachev, S.I.: 1978a, Vestn. Leningrad Gos. Univ., No. 1, pp. 129-135.

Grachev, S.I.: 1978b, Astrophysics 14, pp. 63-69.

Harrington, J.P.: 1972, Astrophys. J. 176, pp. 127-137.

Harrington, J.P.: 1973, Mon. Not. R. astr. Soc. 162, pp. 43-52.

Harrington, J.P., Seaton, M.J., Adams, S., and Lutz, J.H.: 1982, Mon. Not. R. astr. Soc. 199, pp. 517-564.

Hummer, D.G. and Kunasz, P.B.: 1980, Astrophys. J. 236, pp. 609-618.

Hummer, D.G. and Rybicki, G.B.: 1971, Mon. Not. R. astr. Soc. 152, pp. 1-19.

Hummer, D.G. and Rybicki, G.B.: 1982a, Astrophys. J. 254, pp. 767-779.

Hummer, D.G. and Rybicki, G.B.: 1982b, Astrophys. J., in press.

Hummer, D.G. and Seaton, M.J.: 1963, Mon. Not. R. astr. Soc. 125, pp. 437-459.

Kallman, T. and McCray, R.A.: 1980, Astrophys. J. 242, pp. 615-627.
Kunasz, P.B. and Hummer, D.G.: 1974, Mon. Not. R. astr. Soc. 166, pp. 19-55.
Kneer, F.: 1975, Astrophys. J. 200, pp. 367-368.
Leung, C.M.: 1975, Astrophys. J. 199, pp. 340-360.
Petrosian, V. and Dana, R.A.: 1980, Astrophys. J. 241, pp. 1094-1106.
Saraph, H.E. and Seaton, M.J.: 1980, Mon. Not. R. astr. Soc. 193, pp. 617-629.
Van Blerkom, D. and Hummer, D.G.: 1967, Mon. Not. R. astr. Soc. 137, 353-374.
Weymann, R.J. and Williams, R.E.: 1969, Astrophys. J. 157, pp. 1201-1213.
Zanstra, H.: 1951, B.A.N. 11, pp. 341-358.

OSTERBROCK: Do you see any hope of radiative transfer calculations for situations more complicated than plane- or spherical-symmetry, for instance, for a medium containing strong density fluctuations with sizes smaller than the size of the nebula, but not infinitesimal?

HUMMER: A Monte Carlo calculation would probably provide the most immediate solution to this problem. Such a calculation might be used to "calibrate" plane- or spherical-geometry calculations, so that one would have some idea of the magnitude and sign of the error one makes in numerical solutions with a convenient but artificial geometry.

Another approach might be to compute the response of a spherical element of gas to radiation of a given frequency, and then to work out statistically the transfer for a randomly distributed ensemble of such spheres, which could be either stationary or moving, as a kind of "macro-molecule".

PHYSICAL PROCESSES IN NEBULAR SHELLS AND THE INTERPRETATION OF NEBULAR SPECTRA

J. Patrick Harrington
University of Maryland

ABSTRACT

Computed models are now recognized as useful tools for interpretation of the spectra of planetary nebulae. However, even the most detailed models need geometrical parameters such as filling factors which are poorly determined by observations. Some effects may be seen more clearly by modeling the stratification than by just using total fluxes. A simple model for NGC 6720 is presented which reproduces the behavior of [Ne III] λ3869 observed by Hawley and Miller (1977), clearly showing the effects of charge transfer. The behavior of C II λ4267 remains puzzling. Finally, we comment on the interaction of high velocity stellar winds with nebular shells. Non-equilibrium particle distributions at the contact between the shocked stellar wind and the nebula may result in the rapid cooling of the shocked gas.

1. USES AND LIMITATIONS OF NEBULAR MODELS

In the five years since the Ithaca Symposium on planetary nebulae we have seen a major increase in the use of computed models for the interpretation of nebular spectra. One reason for this trend is the realization that such models are the best way to obtain correction factors for the unobserved stages of ionization. Models also provide representative electron temperatures needed to derive the ionic concentrations for the inner zones of high excitation nebulae for which we lack temperature sensitive line ratios. This is of particular importance in the analysis of the ultraviolet data from the IUE satellite, which has had a revolutionary impact on the study of nebular abundances.

The most extensive program has been carried out by Aller and various co-workers. Aller and Czyzak (1982) have summarized results for 41 nebulae. The method of construction of their models is set out in Keyes and Aller (1978). The parameters which are varied to fit the observed nebular spectrum are (i) the chemical composition, (ii) the density, (iii) the stellar spectrum, which can be based on model atmosphere calculations, but which may include *ad hoc* modifications, and (iv) the optical depth in the Lyman continuum, which they express as the ratio of

D. R. Flower (ed.), Planetary Nebulae, 219–227.

the outer radius of the nebula to the radius of a complete Stromgrem sphere.

These models may fit the observed spectrum only approximately, especially when ultraviolet data is included. Thus, in the method described by Shields et. al. (1981), the chemical composition of the best model is not regarded as the final result. Rather, the temperature structure of the model permits the calculation of mean temperatures weighted by the various ions, and these mean temperatures are used to deduce ionic concentrations directly from the observed features. Then the ionization structure of the model provides the corrections for the ions without observed lines and the chemical abundances are found. For many nebulae, this is probably the best approach now available for abundance determinations. The results show the wonderful diversity of these objects. It is no longer possible to speak of "the mean chemical composition of planetary nebulae".

I have been involved in some attempts to model particular nebulae in great detail (Bohlin, Harrington and Stecher, 1978; Harrington _et. al._, 1982; Harrington and Feibelman, 1983). It is not suggested that such models are feasible or even desirable for most planetaries. The primary aim is to attain some assurance that our procedures are sound and that we really are modeling the physical processes in the gas. For it must be emphasized that not only the stellar flux, but also the geometrical structure of the nebula, which is usually treated in a schematic fashon, can have an enormous influence on the predicted spectrum

In our "new, improved" model of NGC 7662 (Harrington _et. al._, 1982) the stellar flux is not arbitrary but is interpolated directly from the model atmosphere sequence of Hummer and Mihalas (1970). The absolute flux level (i.e., stellar radius) is determined by matching the UV continuum as deduced from our IUE observations.

For this model of NGC 7662 (and also for the model of IC 3568 by Harrington and Feibelman, 1983) the density distribution was chosen so that the emissivity of the gas would reproduce observed isophotes. With knowledge of the distance, this fixes the distribution of the gas about the star, and hence the dilution of the radiation at each point. However, in both cases, the density obtained is below that indicated by density sensitive line ratios. It seems that there is small scale structure in these nebulae. We can simulate this by the use of a filling factor so that the density in the filled volume is higher while the average emissivity per unit volume remains the same.

It is also important to determine to what extent structure exists on a larger scale, that is, to what extent there is "clumping" on a scale such that the Lyman continuum optical depth of the nebula varies significantly along different radial directions. Of course we can observe such clumping in many nebulae, but in others it must exist below our limit of resolution.

We used both a filling factor and a clumping factor for the outer zone of our final model of NGC 7662. The ionic concentration of C^{+3} is obtained from the C III λ2297 recombination line, which indicates that internal dust has reduced the flux in the C IV λ1549 resonance line by a factor of 3. For nebulae where λ2297 cannot be observed, it would seem advisable to set the ionization structure by N III] λ1750 and N IV] λ1487 rather than C III] λ1909 and C IV λ1549. This model gives a good fit to almost all the observed spectral features in both optical and UV wavelengths.

But the point I wish to make here is that even when we have based the density distribution of the model on observed isophotes, we have had to use filling and clumping factors which, reasonable though they are, are nevertheless not well determined by observations. And this makes the model less compelling as a unique representation of the physical state of the gas. Pequignot (1980) pointed out, correctly, that an optically thick nebula like NGC 7027 is a better test of the theory than an optically thin one like NGC 7662, for in an optically thick nebula all radial directions terminate in an ionization front and the "clumping" parameter is of diminished importance.

We conclude that by using our current arsenal of physical processes models can be made to match the observed nebular fluxes, but that it is still hard to be positive that we have not used some stellar or geometrical free parameter to compensate for inadaquate physics. So we ask if there is any approach which can sharpen the comparison between observation and theory.

2. STRATIFICATION EFFECTS: NGC 6720

In some cases we can apply more constraints by trying to match not the integrated spectrum, but the spectrum as it varies from point to point in a stratified object. An excellent example is provided by observations of the Ring Nebula, NGC 6720, by Hawley and Miller (1977). They observed regions with greatly differing [N II] and [O II] line strengths and showed that the much used ionization correction formula, $N(N^+)/N(O^+) = N(N)/N(O)$, was valid because it gave the same nitrogen abundance in all regions of the nebula. But they were surprised by the behavior of [Ne III] λ3869. Although the ionization potential of Ne^+ is higher than that of O^+, the [Ne III] line remained strong out to the edge of the nebula while [O III] faded and was replaced by [O II] and [O I]. It has subsequently been suggested by several authors (e.g., Barker, 1980) that this is due to the effects of charge transfer from hydrogen.

I would like to present a simple model of NGC 6720 to demonstrate that this is quantitatively correct. The nebula is represented by a spherical shell with a constant density of $N_H = 800\ cm^{-3}$. The central star is represented by model atmosphere No. 202 of Hummer and Mihalas (1970) with $T_{eff} = 150{,}000K$ and log g = 7. The abundances of C, N, O, and Ne relative to hydrogen are 3.9(-4), 2.0(-4), 7.0(-4) and 1.5(-4), respectively. The stellar luminosity is set at 245 $L_\odot$ ($R_* = .023\ R_\odot$) so that the outer boundry of the model is the edge of the Stromgren sphere. The computer program and atomic parameters used for the calculation are the same as those given by Harrington et. al. (1982). Because of the high stellar temperature and large dilution factor, there is a great deal of neutral hydrogen and charge transfer is extremely effective. The table gives the integrated fluxes from this model for some of the stronger lines. Note the great strength of [O II], [N II], [O I] and [N I].

Ion	λ(A)	I(λ) (Hβ = 100)
He II	4686	30.
C II]	2328	190.
C III]	1909	250.
C IV	1549	50.
[N I]	5200	13.
[N II]	6584	690.
[O I]	6300	60.
[O II]	3727	980.
[O III]	5007	1730.
[Ne III]	3869	230.

Figure 1 shows the emissivity of some of the lines relative to Hβ as a function of radial distance. The effect observed by Hawley and Miller is striking as the [O III] line fades but the [Ne III] line actually increases toward the edge. The effect might be regarded more properly as an [O III] anomaly because in the absence of the $O^{+2} + H^0 \rightarrow O^+ + H^+$ charge transfer process, [O III] would persist to the edge the same as [Ne III] does. The behavior of nitrogen is like that of oxygen because the $N^{+2} + H^0 \rightarrow N^+ + H^+$ rate is also large, while the $Ne^{+2} + H^0 \rightarrow Ne^+ + H^+$ rate is very small. The Ring Nebula seems to be an ideal object to explore these effects and should repay more refined modeling coupled with more extensive observational mapping. But this simple model is enough to demonstrate the reality of charge transfer effects in nebulae more forcefully than modeling the integrated fluxes can.

3. THE C II λ4267 LINE

While we have a good grasp of some stratification effects, observations by Barker (1982) of the same nebula are equally effective in displaying behavior which we do not yet understand. Barker made IUE observations of the Ring at the same positions studied by Hawley and Miller, and also obtained optical observations of the faint C II λ4267 line. From the IUE observations of C II], C III], and C IV he found essentially the same carbon abundance at all positions, C/H = 3.9(-4). Now the C II λ4267 line has been considered a recombination line and used to deduce the abundance of C^{+2} ions. Barker found that in this nebula, the λ4267 line implied abundances up to 10 times higher than those obtained from C III] λ1909,

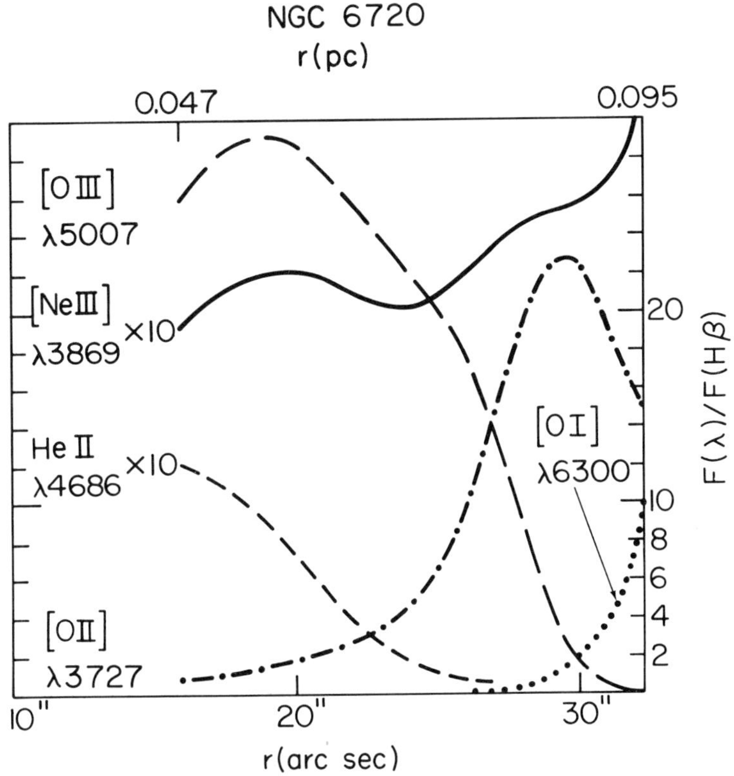

Figure 1. The strength of slected lines relative to Hβ as a function of distance from the central star.

and, what is more significant, this discrepancy was greatest near the central star but decreased to agreement with increasing radial distance.

Such behavior of course suggests fluorescent excitation of C^+ ions by starlight rather than recombination of C^{+2} ions. However, the upper level of the λ4267 transition is 4f $^2F^o$ and cannot be directly excited by absorption from the ground state, which is 2p $^2P^o_2$. Fluorescence would require a process such as the excitation of the nd ^2D levels, where n ≥ 5, followed by cascades to 4f $^2F^o$. Grandi (1976) made approximate calculations and concluded that recombination was an order of magnitude more important than fluorescence for λ4267 in the Orion Nebula. His calculations considered only "a few important levels", however, while one might expect that excitation of the higher n levels would dominate the fluorescent process (Seaton, 1968). This problem needs to be looked at with some care, especially since the λ4267 line is often used as the sole indicator of carbon abundance.

To show just how interesting the problem of C II λ4267 can be, consider the spectra of the hydrogen-poor knots near the nucleus of Abell 30 obtained by Jacoby and Ford (1983). In Knot 3, C II λ4267 is about half as strong as [Ne III] λ3967. If this line is interpreted as a recombination feature, it implies a C/O ratio of over 300! Now the [O III] λ5007/λ4363 ratio gives the temperature of Knot 3 as 16,000K. If, at this temperature, there really were 300 times more C^{+2} than O^{+2}, the C III] λ1909 radiation would be enormous. I don't see how photo-ionization heating could maintain such a temperature in the face of such cooling.

Could the intensity of λ4267 be due to fluorescence? The central star of Abell 30 is in fact much brighter than that of the Ring Nebula. But then why is Knot 4 without the λ4267 line? There are other strange things about these spectra. For example, the [Ne IV] λ4720 lines are far too strong compared with [Ne III] λ3869 for any reasonable Ne^{+3}/Ne^{+2} ratio or neon abundance. But it is important to try to understand such "monsters".

4. STELLAR WINDS AND NEBULAR SHELLS

To round out this collection of comments on physical processes in nebular shells, both understood and mysterious, I want to mention a common situation where interesting physical processes must be occuring, and where the question is whether we can observe the effects. This is the interaction between the nebular shell and the high velocity stellar winds ($v > 1000$ km sec^{-1}) observed in so many cases.

The stellar wind carries mass, momentum and energy. For specific values, consider the case of IC 3568 (Harrington, 1982). The wind velocity is 1840 km/sec and the mass loss rate appears to be at least 4.(-9) $M_\odot$/yr. Thus the kenetic energy carried by the wind is 1.4 $L_\odot$. With an expansion velocity of 7 km/sec, the age of the nebular shell is of the order of 2000 yrs. Clearly the present wind will add very little mass to the nebular shell over such a period of time.

If the wind were able to cool quickly after impact with the nebular shell, it would transfer its momentum to the shell, and this momentum might be large enough to have some influence on the dynamics of the nebular expansion. Compared with the stellar luminosity of 10^3 $L_\odot$, the radiation which would be generated by the cooling wind might seem small. IC 3568, however, is very thin in the Lyman continuum and the nebula captures only a small fraction of the stellar radiation. The Hβ luminosity is just L(Hβ) = 0.8 $L_\odot$. Thus if radiation produced by the wind had a characteristic spectral signature, it could be observed.

Alternatively, the wind, after being heated by encounter with the nebula, may not cool, but instead fill the cavity between the nebular shell and the star with a very hot, tenuous gas. The pressure of this "coronal" gas will be much more effective in accelerating the nebula than the simple transfer of wind momentum. This is the model presented by

Pikel'ner (1968, 1973). A similar picture of the effect of the winds from O-type main sequence stars on the interstellar medium has been worked out in some detail by Weaver et. al. (1977), where the hot, shocked stellar wind blows "interstellar bubbles". An important point is the high temperature of the shocked stellar wind: for an adiabatic shock we have $T = (3/16)(\mu/k)\ v_\infty^2$, which for $v_\infty = 1840$ km/sec yields $T = 5 \cdot 10^7$ K. Since the mean free path of such hot particles is very large, the shock itself would be of the collisionless type. That is, the wind streaming into the ambient material will generate plasma waves which scatter and halt the wind, and the waves in turn heat the electrons to these very high temperatures. For our conditions, this could happen in as little as 10^8 cm. The resulting hot gas is supposed to fill the cavity at a density of $n_H \simeq 1$ cm^{-3}, roughly in pressure equilibrium with the nebular shell. At this temperature and density, the cooling time of the coronal component is much greater than the age of the nebula, so that the only question is whether cooling can be effective at the interface with the nebular gas. In the case studied by Weaver et. al. the coronal gas is seperated from the cool gas by a conduction front. They found that heat loss by conduction was small enough that the solution was quite similar to that obtained neglecting such losses altogether.

But there is a very important difference between the problem studied by Weaver et. al. and the situation in planetary nebulae: the distance from the collisionless shock to the swept up interstellar gas is about 20 pc, while in a planetary nebula, the temperature drop from $5 \cdot 10^7$K to 10^4 K must occur over only about 0.01 pc. This makes a classical conduction front impossible because the mean free path of the electrons will be larger than the front itself. This is the situation discussed by Balbus and McKee (1982) in their study of the evaporation of interstellar clouds in the coronal component of the ISM, except that the geometry is turned inside out. As in their analysis, we might expect the hot electrons to penetrate right into the cool nebular gas. The cooling of the coronal component could be so great that it would cease to exist as such. The Coulomb heating by these electrons could still result in a hot layer at the inner edge of the nebula which might be observable, but the effects on the dynamics would be very different than the scenario proposed by Pikel'ner. This is an area of nebular studies where important theoretical and observational work may yet be done.

REFERENCES

Aller, L.H. and Czyzak, S.J.: 1982, "Chemical Compositions of Planetary Nebulae", Astrophys. J. Suppl. (in press).

Balbus, S.A. and McKee, C.F.: 1982, Astrophys. J. 252, p.529.

Barker, T.: 1980, Astrophys. J. 240, p.99.

Barker, T.: 1982, Astrophys. J. 253, p.167.

Bohlin, R.C., Harrington, J.P., and Stecher, T.P.: 1978, Astrophys. J. 219, p.575.

Grandi, S.A.: 1976, Astrophys. J. 206, p.658.

Harrington, J.P.: 1982, "Mass Loss from the Central Star of the Planetary Nebula IC 3568", in "Advances in Ultraviolet Astronomy: Four Years

of IUE Research", NASA Conference Publication (in press).
Harrington, J.P. and Feibelman, W.A.: 1983, Astrophys. J. (February 1983, in press).
Harrington, J.P., Seaton, M.J., Adams, S., and Lutz, J.H.: 1982, Monthly Notices Roy. Astron. Soc. 199, p.517.
Hawley, S.A. and Miller, J.S.: 1977, Astrophys. J. 212, p.94.
Hummer, D.G. and Mihalas, D.: 1970, JILA Report No. 101, University of Colorado, Boulder, Colorado.
Jacoby, G.H. and Ford, H.C.: 1982, "The Hydrogen Depleted Planetary Nebulae - Abell 30 and Abell 78", Astrophys. J. (in press).
Keyes, C.D. and Aller, L.H.: 1978, Astrophys. Space Sci. 59, p.91.
Péquignot, D.: 1980, Astron. Astrophys. 83, p.52.
Pikel'ner, S.B.: 1968, Astrophys. Letters 2, p.97.
Pikel'ner, S.B.: 1973, Astrophys. Letters 15, p.91.
Seaton, M.J.: 1968, Monthly Notices Roy. Astron. Soc. 139, p.129.
Shields, G.A., Aller, L.H., Keyes, C.D., and Czyzak, S.J.: 1981, Astrophys. J. 248, p.569.
Weaver, R., McCray, R., Castor, J., Shapiro, P., and Moore, R.: 1977, Astrophys. J. 218, p.377.

OSTERBROCK: The agreement of your calculation of the [Ne III] and [O III] distribution in the outer part of NGC 6720 with the measurements of Hawley and Miller is very striking. Am I correct in thinking that you used a physically calculated charge transfer rate in your model work?

HARRINGTON: Yes. The rates used are the same as those in our model of NGC 7662 (Harrington, Seaton, Adams and Lutz, 1982, Mon. Not. R. Astron. Soc., 199, 517) which are mostly due to Butler, Dalgarno and co-workers.

MATHIS: Would not the presence of even a small magnetic field stop the penetration of the "hot" electrons from the shocked wind into the nebular shell? The Larmor radius then determines the distance which the electrons can penetrate.

KAHN: Surely, you cannot have just the fast electrons going forward into the nebular shell because space charge neutrality must be maintained. There would very probably be plasma instabilities at the interface because of the anisotropy in the thermal velocity distribution there.

COHEN: In the light of your knowledge of IC 3568, would you expect dust grains to be present in this nebula? Mike Barlow and I found (at most) a weak 10 μm excess, but I recently detected IC 3568 weakly at 40 μm and strongly at 100 μm. I am wondering whether this radiation might come from cool grains.

HARRINGTON: IC 3568 is completely ionized, as far as we can tell, but I see no reason why these should not be cool grains in the ionized gas. They will be heated mainly by starlight rather than Ly α.

FORD: If you consider a high velocity PN in or near the plane of the Galaxy which sees an external wind of 100 - 150 km s^{-1}, do you expect any observable effects (as in low velocity supernova remnants)?

HARRINGTON: These velocities are, of course, much smaller than those I was considering. It all depends on the density of the wind - if it were high enough, there would be an observable effect.

ALLER: Does the importance of charge exchange depend critically on the details of the model (N(H), $F_\nu(*)$ etc.)? We found that in nebular models of relatively low density and high excitation there could occur large overlap zones of O^{++} and H^o. On the other hand, in compact, low excitation, dense models there was no such overlap. Hence, allowance for this phenomenon has to be on a case by case basis.

HARRINGTON: That is absolutely correct, and worth stressing once more. In NGC 6720, the charge exchange processes are very important. On the other hand, in an optically thin nebula of lower excitation, charge exchange makes only a 25 per cent difference, and in other nebulae even less. This why I see little hope of finding general ionization correction formulae which can be applied to all PN.

ON THE O III/O II PROBLEM IN MEDIUM AND HIGH EXCITATION PLANETARY NEBULAE

Anne Che, J. Köppen
Institut für Theoretische Astrophysik, Universität Heidelberg

Numerical models have been constructed for twelve ionization bounded, medium to high excitation planetary nebulae. In most objects the excitation sensitive line ratio [O III] λ 500.9 nm / [O II] λ 372.7 nm is predicted to be too low as compared to observations. A similar systematic discrepancy is observed for [S III] λ 953.2 nm / [S II] λ 672.0 nm. We investigated the following effects on the ionization structure of the nebulae: $O^{++} + H^{o} \rightarrow O^{+} + H^{+}$ charge exchange reaction, energy distribution of ionizing radiation and density distribution of gas in the nebular shell. The results show that density distribution is the most important factor determining the O III/O II and S III/S II line intensity ratios. While a factor of ten decrease in the charge exchange coefficient is required to explain the systematic discrepancy, a reduction of nebular radius by a few percent - truncated nebula (quasi density bounded model, but nebula still optically thick to Lyman photons) - suffices to produce the correct O III/O II ratio. Also, a density gradient of $n \sim r^{-1}$ to r^{-2} yields much better agreement with observations. Realistic variations in stellar spectrum hardly affects the O III/O II line intensity ratio.

NUSSBAUMER: The newly calculated dielectronic recombination rates of Nussbaumer and Storey (this volume) lead to increased recombination $O^{2+} \rightarrow O^{+}$ as well as $O^{3+} \rightarrow O^{2+}$. It seems likely that the increased rate of $O^{2+} \rightarrow O^{+}$ recombination will worsen your problem, as there is probably insufficient O^{3+} to be turned into O^{2+} by its enhanced rate of recombination.

CHE: That is possible as the recombination and charge exchange rates are of the same order of magnitude.

PLANETARY NEBULAE WITH MASSIVE NUCLEI

R. Tylenda
Laboratory for Astrophysics, Copernicus Astronomical Center,
Chopina 12/18, 87-100 Torun, Poland

Massive central stars ($M > 1 M_o$) of planetary nebulae burn nuclear fuel on a time scale of hundreds or tens of years which is shorter than the recombination time in a typical planetary nebula. Consequently the ionization and thermal structure of a nebula with such a nucleus is expected to be far from equilibrium conditions. The greatest chance of observing such a nebula is when the central star cools down to the white dwarf region. Time-dependent photoionization models suggest the following non-equilibrium effects to be expected at this stage. Firstly, the nebula shows a double shell structure, i.e. a bright, inner ring is surrounded by a faint, extended halo best seen in the HI lines and infrared lines from low-ionization species, such as [Ne II] 12.8 μ. Secondly, the low-excitation emission ([O II], [Ne II], [S II]) is enhanced relative to the high-excitation ([O III], [Ne III], [S III]). Thirdly, different modifications of the Zanstra method result in significantly different temperatures for the central star with a general rule that $T_{HI} > T_{HeII} > T_{HeII/HI}$. The He II Zanstra method gives the most reliable result. Fourthly, the electron temperature derived from the [O III] lines is appreciably higher than that obtained from the [N II] lines. It is suggested that NGC 7027 and NGC 2440 possess massive central stars and that the above time-dependent effects are currently observed in these nebulae.

SECTION III

CHEMICAL ABUNDANCES IN PLANETARY NEBULAE

TYPE I PLANETARY NEBULAE

Manuel Peimbert and Silvia Torres-Peimbert
Instituto de Astronomía, Universidad Nacional Autónoma de México

Abstract. The general properties of PN of Type I are reviewed. A list of 29 PN of Type I is presented, most of them are bipolar. Their bipolar nature might be a direct consequence of the large masses and angular momenta of their progenitor stars. PN of Type I are He and N rich, their observed chemical abundances are compared with theoretical predictions. A group of Type I PN candidates is presented.

I. INTRODUCTION

PN of Type I were defined by Peimbert (1978) based on a handful of objects; it is the purpose of this review to study if PN of Type I form a coherent group with a unique set of physical properties. For example we want to know if they correspond to a given range of precursor masses. If that were the case we would be able to use them to test predictions based on stellar evolution theory (e.g. Renzini and Voli 1981) related to the chemical evolution of the surface layers. Moreover we would be able to use them for the study of the chemical evolution of galaxies (e.g. Serrano and Peimbert 1981).

II. DEFINITION

Primary Criterion

Peimbert (1978; see also Peimbert and Serrano 1980; hereinafter PS) defined PN of Type I as those objects with $N(He)/N(H) \gtrsim 0.14$ or $\log N/O \gtrsim 0.00$. In past photographic work weak lines were systematically overestimated, this effect yields higher He/H ratios and higher electron temperatures which in turn produce spuriously large N^+/O^+ and N/O ratios. The number of photoelectric observations is increasing and with it the quality of the abundance determinations, therefore we decided to relax the definition to include those objects with $N(He)/N(H) \gtrsim 0.125$ and $\log N/O \gtrsim -0.3$ since we consider these objects to belong to a physical group of He and N rich objects.

D. R. Flower (ed.), Planetary Nebulae, 233–242.

TABLE 1
Planetary Nebulae of Type I

Object	He/H	log(N/O)	log(O/H) +12	log(C/O)	Source Abund.	Morph. Kinem.	Source
NGC 650	0.130:	-0.37	8.87	+0.31	1,2	FBV	19,20
NGC 2346	0.130	-0.37	8.63	...	1,2	FB	18
NGC 2371-2	0.122	-0.24	8.83	+0.29	1,2	FB	...
NGC 2440	0.112	+0.21	8.87	+0.17[a]	1	FB	18,20
	0.117	+0.33	8.55	-0.03	3	...	...
NGC 2452	0.111	+0.08	8.59	...	2	FB	...
NGC 2474-5	...	+0.08	...	...	4	FB	21
NGC 2818	0.143	-0.06	8.69	+0.47[a]	1,5	FB	...
NGC 3132	0.127	-0.34	8.78	...	5	S	...
NGC 5189	...	...	...	...	6	FB	22
NGC 5315	0.125	-0.13	8.71	...	5	S	...
NGC 6302	0.186	-0.20	8.70	...	7	FBV	23,24,25
	0.182	+0.22	8.70	-0.70	8	...	...
NGC 6445	0.228	-0.02	...	...	9	FB	...
NGC 6537	>0.140	0.00	...	...	10	FB	26
NGC 6629	0.143:	...	...	...	11	S	27
NGC 6741	0.136	-0.13	...	...	12,13	FB	...
NGC 6751	0.118	-0.23	8.57	...	2,11	F	...
NGC 6778	0.155	+0.03	8.48	+0.53	2,11	FB	...
NGC 6853	0.133	-0.26	8.78	-0.15	13,14	FBV	28
NGC 6894	...	-0.02:	...	...	4	S	27
NGC 7008	0.140	0.00	...	...	10,11	S	27
IC 4406	0.141	-0.32	...	...	4	F	...
Hu 1-2	0.152	+0.21:	8.22:	+0.27:	1,2	FBV	29
M 1-8	0.127	-0.37	8.81	+0.55[a]	1	C	...
M 2-55	0.157	-0.19	...	...	4,15,16	F	...
M 3-3	0.123	+0.17	8.69	+0.58[a]	1	C	...
Me 2-2	0.154	-0.05	8.30:	+0.52[a]	1,17	C	...
Mz 3	0.18:	+0.1:	...	...	18	FB	30
PB 6	0.185	+0.17	8.38:	...	5	F	...
CRL 618	...	-0.10	...	...	18	FBV	31

Notes to Table 1: [a] C^{++}/O^{++} ratio based on λ 4267 of C II; F filamentary; B bipolar, binebulous, biaxial, hourglass; V bipolar velocity field; S relatively smooth; C compact; 1 Peimbert and Torres-Peimbert 1982; 2 Aller and Czyzak 1982; 3 Shields et al. 1981; 4 Kaler 1982a; 5 Torres-Peimbert and Peimbert 1977; 6 Worswick 1980; 7 Aller and Czyzak 1978; 8 Aller et al. 1981; 9 Aller et al. 1973; 10,11,12,13 Kaler 1982b,1980, 1978, 1979; 14 Pottasch 1981; 15 Sabbadin and Hamzaouglu 1981a; 16 this paper; 17 Barker 1978; 18 Calvet and Peimbert 1982; 19 Sabbadin and Hamzaouglu 1981b; 20 Minkowski 1964; 21 Phillips and Reay 1977; 22 Westerlund and Henize 1967; 23 Meaburn and Walsh 1980; 24 Barral et al. 1982; 25 Rodríguez and Morán 1982; 26 Felli and Perinotto 1979; 27 Curtis 1918; 28 Goudis et al. 1978; 29 Torres-Peimbert et al. 1982; 30 Cohen et al. 1978; 31 Carsenty and Solf 1982.

Secondary Criteria

Most He and N rich PN show a very pronounced filamentary structure, and very strong forbidden lines from [O I], [N I], [S II] up to [Ne V]. These objects comprise a subset of almost half of the objects classified by Greig (1971) as B nebulae (binebulous or filamentary). Greig (1972) found from kinematical properties that B nebulae were population I while A (annular) and C (centric) were of population II.

III. CHEMICAL ABUNDANCES

In Table 1 we present a list of He and N rich objects selected mainly from photoelectric observations. The estimated errors are smaller than 0.01 in He/H and smaller than 0.2 in log N/O and in log O/H. When photoelectric data are used from two different sources the average value is given. A previous compilation of PN of Types I and I-II by PS included 19 objects, of those 14 are presented in Table 1 and five have been deleted, the reasons are the following: M2-27 and Cn 2-1 belong to the bulge and could have been formed in a He and N rich medium, for NGC 7293 Peimbert and Torres-Peimbert (1982; hereinafter PTP) obtained He/H= 0.123 and log N/O= -0.43 and for M1-80, 0.095 and -0.48, while Aller and Czyzak (1982) for NGC 6309 obtained 0.120 and -0.91.

Several objects deserve special attention: NGC 650 shows $N(He^{+} + He^{++})/N(H^{+})$= 0.112 and S^{+}/S= 0.14 which is the largest value for the objects observed by PTP, by assuming that $He^{\circ}/He = S^{+}/S$ the value in Table 1 was derived. NGC 650 is the only object in Table 1 corrected for the presence of He° inside the H II zone. For NGC 2440 and NGC 6302 we present two sets of data, the first is based on optical observations alone while the second includes IUE observations. The optical determinations are based on the assumption that $N^{+}/O^{+} = N/O$, the UV observations include N^{++}, N^{+3} and N^{+4}. The UV results for objects of high degree of ionization yield in general somewhat larger N values than those derived optically (e.g. Peimbert 1981).

In Figure 1 we show a plot of He/H versus N/O where the shaded area corresponds to PN of Type II; in this area the objects with lower He and N values are located in the outer regions of our galaxy, while the objects with higher He and N values are located towards the center of our galaxy (PS). The He and N rich objects are represented by filled circles and do not show any tendency related with galactic position, which implies that most of the excess He and N was produced by them (see also PS). In Figure 1 we also show the tracks predicted by Renzini and Voli (1981) for intermediate mass stars, most of the objects fall in the region for $2>\alpha>1$ and $3 \lesssim M/M_{\odot} \lesssim 5$.

In Figure 2 we show a plot of O/H versus N/O where, as in Figure 1, the shaded area corresponds to PN of Type II; in this area the objects with lower O/H and N/O values are located in the outer regions of our galaxy, while the objects with higher O/H and N/O values are located

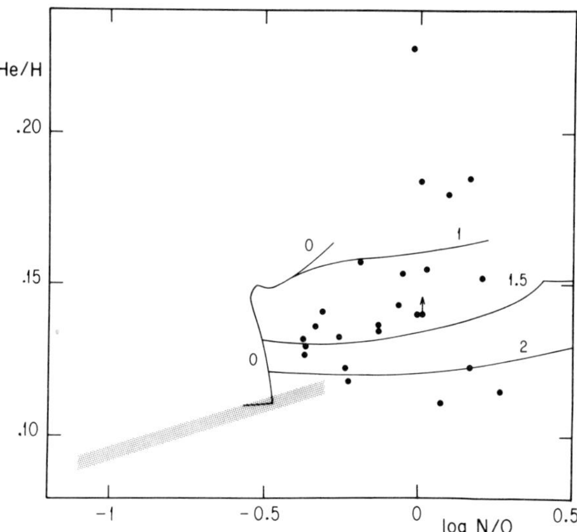

Fig. 1. Observed abundances of Planetary Nebulae. The shaded areas correspond to Type II objects, filled circles to Type I objects from Table 1. Stellar evolution predictions by Renzini and Voli (1981) are shown for different values of α.

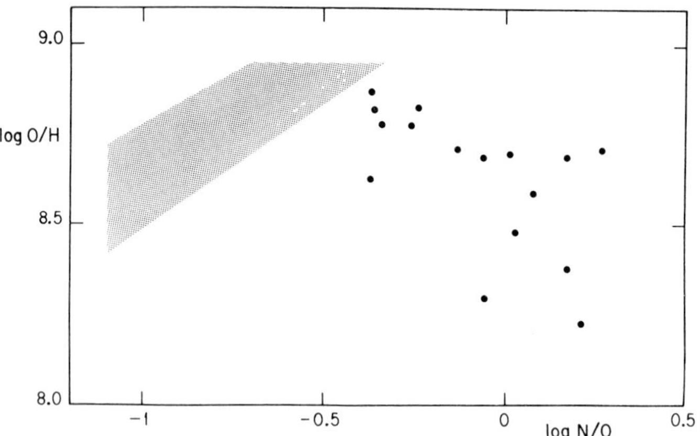

Fig. 2. Relative oxygen to nitrogen abundance. Filled circles correspond to Type I objects from Table 1, shaded area to Type II objects.

towards the center of our galaxy. The He and N rich objects are represented by filled circles, these objects seem to show a correlation in the sense that the higher the N/O ratio the smaller the O/H ratio; this trend was noticed before (Peimbert 1978; Kaler 1980) but no statistical significance was attached to it. From the work by Renzini and Voli (1981) it is found that hot-bottom burning depletes oxygen somewhat in higher-mass stars, however this effect seems to affect only slightly the surface abundances. There are two other possible explanations for the observed correlation in Figure 2: a) that some of the objects are of Population II or that the O/H abundances are not reliable. The two objects with smallest O/H abundance ratios are Hu 1-2 and Me 2-2; from the high radial velocity measurements it seems that Me 2-2 is a Population II object (PS) while recent observations by Torres-Peimbert *et al.* (1982) indicate that Hu 1-2 is not a high velocity object. With respect to the second possibility it should be noted that Hu 1-2 and Me 2-2 are the two objects with highest N_e values observed by PTP and it is possible for a fraction of the H lines to originate in regions with $N_e > 10^6$ cm^{-3} producing an underestimation of the real O/H abundance ratio; this effect is present in CRL 618 and M 2-9 (Schmidt and Cohen 1981; Calvet and Peimbert 1982, hereinafter CP). It is necessary to obtain accurate O/Ar ratios because if O/Ar is underabundant then it is likely that O has been depleted, on the other hand if O/Ar is normal the other two possibilities are still open.

The C/O ratios of NGC 2440, NGC 6302 and NGC 6853 have been derived from UV data and are smaller than 1. There are two effects that can reduce the C/O value as determined from UV data: a) the use of a higher T_e than the real one when the O abundance is derived from optical lines, and b) the high optical thickness in the resonance 1548-1551 Å lines of C IV (e.g. Peña and Torres-Peimbert 1982). Under the assumption that C/O= C^{++}/O^{++} and from the 1909/1663 intensity ratios for NGC 6302 by Aller *et al.* (1981) and Barral *et al.* (1982) we obtain log C/O= -0.35; the H, He, C, N, O abundance ratios for this object are in agreement with models by Renzini and Voli (1981) with $M_i \sim 6$ $M_\odot$ and $\alpha = 2$, NGC 6302 seems to be the most massive well documented object. It is also possible to derive C abundances from the λ4267 line of C II since apparently this line is produced by recombination only (see, Torres-Peimbert *et al.* 1980; Harrington *et al.* 1980). The problem with this method is that λ4267 is very faint and often has been overestimated. Under the crude assumption that C^{++}/O^{++}= C/O from the optical observations by PTP we obtain that for seven Type I PN log <C/O>= +0.45 which together with their He/H values imply progenitors in the $3 < M/M_\odot < 5$ mass range with $2 < \alpha < 1.5$ (Renzini and Voli 1981).

IV. MORPHOLOGY

Most of the objects presented in Table 1 are extremely filamentary and we have marked them with an F. B stands for binebulous, bipolar, biaxial or hourglass, the same object has been called in the literature by one or more of these adjectives (e.g. Westerlund and Henize 1967; Greig 1971; Phillips and Reay 1977; CP). V stands for a velocity field which is bipolar. S stands for relatively smooth and includes 5 objects which are not class B objects according to the classification by Greig; the abundances of these objects are not extremely He and N rich and in some cases are of low accuracy. C stands for compact, we do not have enough resolution to tell if these objects are filamentary or not.

In the last column of Table 1 we present a list of representative references related to the morphological or internal velocity fields of these objects.

The majority of the objects in Table 1 are bipolar, several mechanisms have been proposed to explain bipolarity in PN, we will mention three of them. The three are based on the necessity of the ejection of a shell which is denser near the equatorial plane, due to gravitational braking, differential rotation and radiation pressure (e.g. Phillips and Reay 1977; Sabbadin and Hamzaoglu 1981; CP and references in these papers). CP argue that only stars with $M > 2\ M_{\odot}$, in the main sequence, have enough angular momentum to produce a shell significantly denser in the equatorial plane, they attribute the morphological difference between PN of Type I and others mainly to this fact. Phillips and Reay consider that the principal mechanism determining shell development is gravitational braking; Sabbadin and Hamzaoglu consider that the ansae are produced by Lyman continuum and Lyman α radiation pressure as well as thermal diffusion into vacuum; CP consider that ejection takes place continuously with a velocity increase due to a decrease in the stellar radius and an increase in the escape velocity, two mass loss phases can be distinguished the first one mentioned above which is responsible for the formation of a disc, and a second one which is responsible for the formation of the ansae. This second phase has been studied by Cantó (1980) for several astronomical scenarios; Meaburn and Walsh (1980) were the first to explain the bipolarity of the Type I PN NGC 6302 based on the model by Cantó (1980). The presence of a dense disc, which supports the suggestion by CP for the origin of bipolar PN, has been established for several objects (e.g. CP, Rodríguez and Morán 1982).

The filamentary structure of PN of Type I is probably due to shocks and instabilities produced by denser, and or, faster material catching up with material ejected previously. The filamentary structure is responsible for large variations in N_e. The very hot central stars (see below) together with the density distribution are responsible for the extreme variations in the ionization degree of these objects.

From their bipolar and extremely filamentary structure the following objects are proposed as PN of Type I candidates: NGC 2899, NGC 3699, Hb 5, He 2-29, He 2-76, He 2-111, He 2-114, He 2-207, K 3-46 and M 3-28 (see pictures in Westerlund and Henize 1967; and Perek and Kohoutek 1967). Webster (1978) has studied the velocity field of He 2-111, it resembles that of NGC 6302, moreover the 6583/Hα ratio of He 2-111 is of the order of 3 which probably indicates that it is N rich.

V. MASSES

In addition to the stellar evolution predictions based on chemical abundances that indicate $3 \lesssim M/M_{\odot} \lesssim 5$ for PN of Type I, there is additional evidence in favor of a relatively high mass for these objects.

There are three PN of Type I for which a crude estimate of their masses at the zero age main sequence can be made: NGC 3132, NGC 2346 and NGC 2818. The average mass for these objects turns out to be 2.4 $M_{\odot}$ (PS; Méndez and Niemela 1981; CP). NGC 3132 and NGC 2346 are mild examples of He-N enrichment, therefore the lower mass limit for this process to occur would be around 2.4 $M_{\odot}$.

From the work by Greig (1972) based on the galactic kinematical properties of class B PN it follows that these objects are of Population I.

Kaler (1982a) has placed a few PN of Type I in the HR diagram, they tend to be at higher T^* and lower L^* with respect to other PN which, when compared with theoretical tracks (Paczyński 1971; Iben and Renzini 1982; Schönberner and Weidemann 1981), indicates that their core masses are higher than those of other PN nuclei.

VI. PROTOPLANETARY NEBULAE

There are several bipolar nebulae that have been called protoplanetary nebulae in the literature, in general these objects have $T^* \lesssim 35000$ °K. If these objects are going to develop into Type I PN they should be He-N rich. At their stellar temperatures a substantial fraction of the He atoms is expected to be neutral (e.g. Peimbert *et al.* 1974) and it is very difficult to determine their He/H abundance ratio. On the other hand, it is relatively easy to determine their N/O abundance ratio since most of their N and O is singly ionized inside the H II zone. CP have determined that CRL 618 and Mz-3 are N rich (see Table 1), alternatively it seems that M 2-9 is not. The N/O ratio should be determined for other suspected protoplanetary nebulae with bipolar structure.

VII. CONCLUSIONS

The majority of the PN of Type I are bipolar. An explanation in the sense that their bipolarity is due to the high angular momentum of their precursors has been advanced by CP. There are several independent arguments that indicate that their masses are in the $2 \lesssim M/M_{\odot} \lesssim 6$ range, in particular their H, He, C, N and O relative abundances. The O/Ar ratio should be determined for these objects to study if there has been an oxygen depletion during the evolution of the progenitor star. About 10% to 30% of the PN are of Type I.

It is a pleasure to acknowledge L.H. Aller, I. Iben Jr., J.B. Kaler, S.R. Pottasch, A. Renzini and M.J. Seaton for sending us manuscripts in advance of publication. We are grateful to N. Calvet, J.F. Rayo and A. Serrano for several discussions and comments on this work.

REFERENCES

Aller, L.H. and Czyzak, S.J.: 1978, Proc. Natl. Acad. Sci. USA, 75,pp.1-3.
Aller, L.H. and Czyzak, S.J.: 1982, preprint.
Aller, L.H., Gzyzak, S.J., Craine, E., and Kaler, J.B.: 1973, Astrophys. J. 182, pp. 509-515.
Aller, L.H., Ross, J.E., O'Mara, B.J., and Keyes, C.D.: 1981, Monthly Notices Roy. Astron. Soc. 197, pp. 95-106.
Barker, T.: 1978, Astrophys. J. 220, pp. 193-209.
Barral, J.F., Cantó, J., Meaburn, J., and Walsh, J.R.: 1982, Monthly Notices Roy. Astron. Soc. 199, pp. 817-832.
Calvet, N., and Peimbert, M.: 1982, Rev. Mexicana Astron. Astrof., submitted.
Cantó, J.: 1980, Astron. Astrophys. 86, pp. 327-338.
Carsenty, U., and Solf, J.: 1982, Astron. Astrophys. 106, pp. 307-310.
Cohen, M., FitzGerald, M.P., Kunkel, W., Lasker, B.M., and Osmer, P.S.: 1978, Astron. J. 221, pp. 151-162.
Curtis, H.D.: 1918, Publ. Lick Obs. 13, pp. 57-74.
Felli, M., and Perinotto, M.: 1979, Astron. Astrophys. 76, pp. 69-74.
Goudis, C., Mc Mullan, D., Meaburn, J., Tebbutt, N.J., and Terrett, D.L.: 1978, Monthly Notices Roy. Astron. Soc. 182, pp. 13-25.
Greig, W.E.: 1971, Astron. Astrophys. 10, pp. 161-174.
Greig, W.E.: 1972, Astron. Astrophys. 18, pp. 70-78.
Harrington, J.P., Lutz, J.H., Seaton, M.J., and Stickland, D.J.: 1980, Monthly Notices Roy. Astron. Soc. 191, pp. 13-22.
Iben Jr., I. and Renzini, A.: 1982, preprint.
Kaler, J.B.: 1978, Astrophys. J. 226, pp. 947-962.
Kaler, J.B.: 1979, Astrophys. J. 228, pp. 163-178.
Kaler, J.B.: 1980, Astrophys. J. 239, pp. 78-88.
Kaler, J.B.: 1982a, preprint.
Kaler, J.B.: 1982b, Astrophys. J. in press.
Meaburn, J., and Walsh, J.R.: 1980, Monthly Notices Roy. Astron. Soc. 191, pp. 5-11p.
Méndez, R.H., and Niemela, V.S.: 1981, Astrophys. J. 250, pp. 240-247.
Minkowski, R.: 1964, Publ. Astron. Soc. Pacific 76, pp. 197-209.
Paczyński, B.: 1971, Acta Astr. 21, pp. 417-435.
Peimbert, M.: 1978, in Y. Terzian (ed.), "Planetary Nebulae, IAU Symp. No. 76", Dordrecht: Reidel, p. 215.
Peimbert, M. : 1981, in R.D. Chapman (ed.), "The Universe at Ultraviolet Wavelengths", NASA CP-2171, pp. 557-565.
Peimbert, M., Rodríguez, L.F., and Torres-Peimbert, S.: 1974, Rev. Mexicana Astron. Astrof. 1, 129-141.
Peimbert, M., and Serrano, A.: 1980, Rev. Mexicana Astron. Astrof. 5, pp. 9-18.
Peimbert, M., and Torres-Peimbert, S.: 1982, Rev. Mexicana Astron. Astrof. submitted.
Peña, M., and Torres-Peimbert, S.: 1982, Rev. Mexicana Astron. Astrof. submitted.
Perek, L., and Kohoutek, L.: 1967, "Catalogue of Galactic Planetary Nebulae", Academia Praha.
Phillips, J.P., and Reay, N.K.: 1977, Astron. Astrophys. 59, pp. 91-110.

Pottasch, S.R.: 1981, private communication.
Renzini, A., and Voli, M.: 1981, Astron. Astrophys. 94, pp. 175-193.
Rodríguez, L.F., and Morán, J.M.: 1982, Nature, submitted.
Sabbadin, F., and Hamzaoglu, E.: 1981a, Astron. Astrophys. 94, pp. 25-28.
Sabbadin, F., and Hamzaoglu, E.: 1981b, Monthly Notices Roy. Astron. Soc. 197, pp. 363-368.
Schmidt, G.D., and Cohen, M.: 1981, Astrophys. J. 246, pp. 444-454.
Schönberner, D., and Weidemann, V.: 1981, in I. Iben Jr. and A. Renzini (eds.) "Physical Processes in Red Giants", Dordrecht: Reidel, pp. 463-468.
Serrano, A., and Peimbert, M.: 1981, Rev. Mexicana Astron. Astrof. 5, pp. 109-124.
Shields, G.A., Aller, L.H., Keyes, C.D., and Czyzak, S.J.: 1981, Astrophys. J. 248, pp. 569-583.
Torres-Peimbert, S., and Peimbert, M.: 1977, Rev. Mexicana Astron. Astrof. 2, pp. 181-207.
Torres-Peimbert, S., Peimbert, M., and Daltabuit, E.: 1980, Astrophys. J. 238, pp. 133-139.
Torres-Peimbert, S., Peimbert, M., Rayo, J.F., and Peña, M.: 1982, Rev. Mexicana Astron. Astrof. submitted.
Webster, L.B.: 1978, Monthly Notices Roy. Astron. Soc. 185, pp. 45-50p.
Westerlund, B.E., and Henize, K.G.: 1967, Astrophys. J. Suppl. 14, pp. 154-169.
Worswick, S.P.: 1980, private communication.

DOPITA: Many of your type I candidates have been confirmed by (unpublished) observations of Louise Turtle (née Webster) and myself.

ACKER: I studied relations between chemical, spatial and kinematic properties of 97 PN. Those with high He/H and N/O enrichment (most of which are of morphological type "B"; their nuclei have continuous or type O spectra and their mean excitation class is high) have a low mean distance from the Galactic plane ($\langle|z|\rangle$ = 210 pc) and a very flat velocity ellipsoid. Their spatial and kinematic parameters correspond to an initial mass of about 3 $M_\odot$.

WEIDEMANN: Concerning secondary criteria: what fraction of the central stars are visible?

PEIMBERT: We are just in the process of looking into this problem. For a substantial fraction of these objects, we do not see the central star, but many of them are very far away and we need a good distance to obtain a meaningful upper limit in the H-R diagram.

KALER: Returning to the log(N/O) - He/H correlation, you obtain a better fit if you adopt the Becker and Iben calculation with partial conversion of C to N.

NUSSBAUMER: I recently suggested that acceleration by emerging magnetic flux might be a mechanism producing PN (Astron. Astrophys. 110, L1). Bipolar nebulae might be good candidates for having been produced through magnetic activity in open polar field structures.

TERZIAN: Why did you not include NGC 7027 in PN of type I, since its radio morphology is bipolar and its central mass is estimated to be about 1 $M_\odot$? If I may make a more general comment on PN morphology, although one PN looks different from another, there are some overall similarities and bipolarity is one of these. Almost all high resolution radio maps show some degree of intensity bipolarity.

PEIMBERT: NGC 7027 does not quite meet the He- and N-rich criterion for type I PN. Furthermore, NGC 7027 belongs to those objects which exhibit "very mild" bipolarity. Extreme cases of bipolarity are observed: a well defined disk with ansae present (e.g. NGC 6302, NGC 2440, Mz-3, and He 2-111); weak disk confinement and absence of ansae (e.g. NGC 2474-5).

SULPHUR ABUNDANCES IN THREE HALO PLANETARY NEBULAE

T. Barker
Wheaton College, Norton, Massachusetts, USA

The intensities of the [S II] 6717, 31 Å and [S III] 9532 Å lines have been measured for the first time in the three known extreme halo planetaries. Preliminary estimates of the S/H ratios are: $(4.7 \pm 6.5) \times 10^{-7}$ for 108 - 76°1, $(1.4 \pm 0.6) \times 10^{-7}$ for K 648, and $(2.0 \pm 1.3) \times 10^{-7}$ for 49 + 88°1. The S/H ratio in K 648 (the planetary in M 15) is 0.014 of that measured in the Ring Nebula (Barker, 1980, Ap. J., 240, 99), consistent with the Fe/H abundance of about 1/100 solar found (Cohen, 1978, Ap. J. 223, 487) in the stars in M 15. The average of the S/O ratios in the halo planetaries is an eighth the value in the Ring Nebula, similar to the low Ar/O ratios found previously (Barker, 1980, Ap. J. 237, 482). The implication is either that the abundances of lighter elements such as O have been enhanced by nuclear reactions in the planetary progenitors or that S and Ar were synthesized galactically at a slower rate than lighter elements. In either case, it appears that S and Ar may be more representative of the true heavy metal abundances in planetaries than O.

PEIMBERT: This is an excellent piece of work: the S/H ratio provides us with a powerful constraint when studying the early chemical evolution of the Galaxy.

ALLER: Adjustments of the ionization correction factor in any reasonable manner could not modify the main conclusion: the sulphur abundance is down by two orders of magnitude.

DINERSTEIN: Following on from the previous remark, the ionization correction factors for these particular nebulae are small enough for the associated uncertainty to be very small and not affect your conclusion.

Let me also say that the study of argon and sulphur in PN and H II regions has much to gain from infrared observations, which are currently being pursued by several groups.

PEIMBERT: We find that the C/O ratio in these objects is up by almost an order of magnitude, which indicates that they have ejected mass from regions where complete or nearly complete helium burning has taken place. Therefore, at least part of the O/S and Ne/S excess could be due to the evolution of the progenitors.

BARKER: Yes. On the other hand, Ne/H is extremely low in 49 + 88°1, which you found to have a particularly high carbon abundance.

ELEMENTAL ABUNDANCES IN PLANETARY NEBULAE

James B. Kaler
Astronomy Department
University of Illinois

1. ELEMENTAL CLASSIFICATION AND MEAN ABUNDANCES

To date, we have been able to gather information on the abundances of 16 elements, ranging from helium to iron, relative of course to hydrogen, the seventeenth. Of the lightest 26 elements only the lithium, beryllium, boron trio, aluminum, and the quintet of metals from scandium to magnanese have not been treated. The results of decades of labor on galactic planetaries are presented as succinctly as possible in Table 1, where the elements are shown in order of atomic number. I will take as a general approach that He/H and O/H have readily recognizable gradients and variations, and that the other elements either generally vary in concert with oxygen, or are best studied with respect to that atom. Column (2) classifies the element according to its most prominent behavior. The well-studied ratios that are generally constant, for which a true mean can be derived, are designated "C." Those that are probably constant, but which are not well studied, are noted as "c." The four elements with abundances significantly under solar, which are probably depleted from the nebular gas by grain formation, are called "D." The letters "G" and "E" denote those for which vertical galactic gradients and/or enrichment by the parent star have been clearly established.

The mean ratios that I adopt are given in the third column, with the reference sources in the fourth. In some instances, I take the mean from a specific reference, in others it is derived from the various values given by a group of authors. The papers in each of the groups are listed at the end of the table. They consist of: (1) a group headed by Aller; (2) an ultraviolet group, all of which present results derived from the IUE; (3) a set of 4 papers in which iron abundances were studied; (4) a collection in which the three extreme halo nebulae were examined; and finally (5) a general collection. The fifth column of Table 1 gives alternate values where the term "all ref" refers to all the reference groups following the table. Finally, the solar abundances, taken from the compilations in the Aller group, are given in column 6.

D. R. Flower (ed.), Planetary Nebulae, 245–257.

Table 1. Summary of galactic planetary abundances, and references

Ratio (1)	Type (2)	Abundance (3)	Reference (4)	Other (5)	Solar (6)
He/H	G-E	0.08-0.10-0.22	Kaler (1978a,1979)		0.10
O/H	G	0.6(-4)*-6(-4)	Kaler (1980)		7.4(-4)
C/O	E	0.4-4:	Kaler (1981a)	0.4-3(UV group)	0.6
N/O	G-E	0.13-0.2-2	Kaler (1979)		0.12
F/O	c	8.5(-4):	Aller group		5(-5)
Ne/O	C	0.225±0.01	Kaler (1978b)	0.23 (all ref)	0.15
Na/O	c	3.2(-3)	Aller group		2.6(-3)
Mg/O	D	2(-3)	Harrington & Marionni (1981) Aller group		5.4(-2)
Si/O	D	1.3(-2)	Harrington & Marionni (1981) Aller group		6.0(-2)
P/O	c	4(-4)	Aller group		4.6(-4)
S/O	C	2.7±0.5(-2)	Beck et al.(1981)	2.3(-2)(all ref)	2.3(-2)
Cℓ/O	C	3.3±0.5(-4)	Kaler (1978b)	4.2(-4)(all ref)	4.3(-4)
Ar/O	C	7.0±0.5(-3)	Kaler (1978b)	6.3(-3)(all ref)	5.0(-3)
K/O	c	2.5(-4)	Aller group		1.9(-4)
Ca/O	D	2.6(-4)	Aller group		3.0(-3)
Fe/O	D	2(-3)	iron group		5.1(-2)

* includes extreme halo

C: constant with respect to oxygen -- well studied; c: probably constant with respect to oxygen -- sparsely studied; D: probably depleted; E: enriched by nuclear processes in parent star; G: recognized vertical gradient; G-E: enriched matter compounded by vertical gradient -- 3 values give minimum halo, minimum disk, maximum enriched.

References to Table 1

(1) Aller group: Aller (1978); Aller and Czyzak (1983); Aller and Keyes (1980)*; Aller, Keyes, and Czyzak (1981)*; Aller, Keyes, Ross, and Czyzak (1980)*; Aller, Keyes, Ross, and O'Mara (1981a)*; Aller, Ross, Keyes, and Czyzak (1979); Aller, Ross, O'Mara, and Keyes (1981)*; Shields, Aller, Keyes, and Czyzak (1981)*. (2) UV group: Aller group marked with *; Harrington, Lutz, and Seaton (1981); Harrington, Lutz, Seaton and Stickland (1980); Harrington, Seaton, Adams, and Lutz (1982); Lutz (1981); Marionni and Harrington (1981); Peña and Torres-Peimbert (1981); Perinotto and Benvenuti (1981); Perinotto, Panagia, and Benvenuti (1980); Pottasch, Gathier, Gilra, and Wesselius (1981); Torres-Peimbert, Peimbert, and Daltabuit (1980); Torres-Peimbert and Peña (1981); Torres-Peimbert, Peña, and Daltabuit (1981). (3) Iron group: Garstang, Robb, and Rountree (1978); Nussbaumer and Storey (1978); Shields (1975, 1978); Shields, Aller, Keyes, and Czyzak (1981); (4) Halo group: Barker (1980a); Hawley and Miller (1978a); Peimbert (1973); Torres-Peimbert and Peimbert (1979); Torres-Peimbert, Rayo, and Peimbert (1981); (5) General: Barker (1978a, 1978b, 1980b); Beck, Lacy, Townes, Aller, Geballe, and Baas (1981); Dinerstein (1980); French (1981); Hawley (1978a); Hawley and Miller (1977, 1978b); Marionni and Harrington (1981); Natta, Panagia, and Preite-Martinez (1980); Peimbert and Torres-Peimbert (1971); Price (1981); Torres-Peimbert and Peimbert (1977).

Single averages of course have little meaning for the 4 elements designated G or E, for which I give minima and maxima. In the case of the G-elements, they represent the mean initial halo and initial extreme disk values, where the three extreme halo objects discussed in this symposium by Barker are included only for O/H. For the E-elements they show the mean initial extreme disk abundance and the maximum observed value caused by stellar enrichment processes. Three numbers - minimum halo, minimum disk, and maximum observed - are given for helium and nitrogen, which exhibit both gradients and enrichments. The G and E elements will be discussed in separate sections below; here we look further at the C and D ratios.

There is a strong body of evidence from the references of Table 1 that neon, argon, chlorine, and sulfur are generally constant with respect to oxygen. There are exceptions, notably the extreme halo objects mentioned above, which generally exhibit depressed neon, argon, and sulfur abundances: see Peimbert's discussion in this volume, and a summary by Torres-Peimbert, Rayo, and Peimbert (1981). The problem of gradients for these element ratios merits further study, as there may be variation within the present observational scatter. The means derived are generally better than the solar determinations, and might logically be included in tables of "solar system values."

The study of flourine, sodium, phosphorus, and potassium is not far enough along really to define an accurate mean, and no data on gradients exist except by analogy with other elements. The four elements designated D do seem to show real deficiencies when compared with the solar ratios, probably caused by depletion of the atoms out of the gas phase into grains. The concept of an average value then means little, since depletion factors certainly vary among nebulae.

2. GALACTIC GRADIENTS

A capsule history of galactic composition gradients: Kaler (1970) found that O/H increased with decreasing radial velocity and distance from the galactic plane; Barker (1978a), however, observed no correlation between his set of objects and galactic kinematics; Torres-Peimbert and Peimbert (1977) noted negative radial gradients in He/H, O/H, and N/O; Peimbert (1973) showed that the extreme halo planetary in M15 was deficient in oxygen, by an order of magnitude; this result was confirmed and the work extended to other elements and to the other two extreme halo nebulae by the "halo group" of references following Table 1; and a series of papers by Kaler (1978a, 1979, 1980, 1981b) re-examined the data with regard to both vertical and radial gradients. The principal results of this series are that: (1) initial He/H increases by about 25% from the general halo (the set of high-velocity nebulae exclusive of the 3 extreme objects), to the extreme disk; (2) N/O increases by very roughly 50% in the same manner; (3) O/H increases steadily from the extreme halo (including the three) to the disk in a recognizable series of steps; (4) S/O is constant.

In this set of papers, I contended that radial gradients could not be perceived from the planetaries; that since the majority of nebulae at large radial distances are population II, the radial gradients are only apparent, and are reflections of the vertical gradients. Peimbert (1978) and Peimbert and Serrano (1980), however, claimed to detect radial gradients from a homogeneous set of objects. The vertical gradients in N/O, He/H, and O/H seem to be in little doubt. Almost certainly, the radial gradients are present, as witnessed by the work on diffuse nebulae by, for example, Peimbert, Torres-Peimbert and Rayo (1978), Talent and Dufour (1979), and Hawley (1978b). But it is unclear as to whether they can be detected in planetaries at the present time. For helium and nitrogen, the issue is in addition confused by enrichment processes that are, like population types, related to stellar mass. We need a much larger statistical sample to resolve the problem fully. The very high O/H ratio found by Price (1981) in a bulge planetary provides an interesting direction towards future research.

Another point of contention involves the heavier elements. While Kaler (1981b) claimed constancy for S/O, Hawley and Miller (1978a) and Torres-Peimbert and Piembert (1979) indicated a large deficiency in the extreme halo nebula Ha 4-1, which I saw as a possible ionization effect. Yet the papers of the halo group of references indicate similar deficiencies in neon and argon, for which constancy is assumed in the previous section. These nebulae seem genuinely different, and the "C" indication in Table 1 should for now exclude the 3 planetaries of the extreme halo.

3. ELEMENT ENRICHMENT

It is now very clear that the by-products of nuclear burning find their way into the nebula, having been injected into the hydrogen envelope of the parent AGB star before nebular lift-off. Peimbert and Torres-Peimbert (1971) found nitrogen to be commonly overabundant in planetaries, and Kaler (1974) saw that helium could be heavily enriched. The results are far from complete on carbon, but there is strong evidence that enrichment takes place. This phenomenon gives us a superb opportunity to examine internal processes in stars, and to test the general theories of mass loss and the late stages of stellar evolution. In principle, it should be possible to infer the mass of the parent star from the degree of element enrichment, and thereby possible to relate current properties of nebulae and central stars to initial properties.

The prevailing theoretical view is that element enrichment proceeds in a succession of three convective dredge-up stages, detailed for us by Becker and Iben (1979, 1980, hereafter BI) and Renzini and Voli (1981, hereafter RV), following discoveries and ideas by Iben (1964, 1972, 1975), Iben and Truran (1978), and Perinotto and Renzini (1979). The first stage takes place on the first ascent of the red giant branch, the second for stars over $\sim 3\ M_\odot$ on the AGB, and the third during the thermally pulsing phase of AGB evolution. In the first stage, N is increased, in the second He and N, and in the third, He and C, while in the absence of other processes the net N abundance goes down.

The theoretical predictions of element overabundances can be readily tested with the observations. Peimbert (1978) pointed out that high enrichment rates of nitrogen and helium went together in his type I nebulae, which are derived from the more massive stars of the galactic disk. Kaler, Iben, and Becker (1978) then showed that nitrogen and helium abundances were generally correlated, in numerical agreement with the theory through the first two dredge-up stages, the observed correlation terminating in the type I objects. However, both BI and RV demonstrate that the agreement is not good after the calculations for third dredge up are included: theoretical N/O changes little with increasing He/H, while the observed steadily increases. And Kaler (1981a) noted that C/O did not climb with He/H in accord with BI's theory. Either third dredge-up does not work as predicted, or as pointed out by BI and RV, the excess carbon is reconverted to nitrogen; BI also invoked grain formation to deplete gaseous carbon.

This review affords us with a good opportunity to look at an improved version of the correlations, and a simultaneous examination of N/O, C/O, and He/H, to see whether or not we can find a consistent set of calculations, and make a choice as to the true theoretical scenario. I have revised and improved the abundance determinations from the previously published values (Kaler 1978a, 1979, 1981a) as follows:

Helium. All of the interference filter measurements made of He/H at Illinois were corrected for temperature shifts of the filters, which on the average produced both increases and decreases on the order of 10%. The new values were then re-averaged with He/H derived from the other good data, particularly from Barker (1978c), Torres-Peimbert and Peimbert (1977), and Aller and Czyzak (1979, 1983).

Nitrogen. The major improvement was to add in the N/O ratios derived by observers in the UV group of references of Table 1, and to include the new data by Aller and Czyzak (1979, 1983).

Carbon. This element provides an especially difficult problem. Only one line is extensively observed in the optical, recombination λ4267 CII, and serious errors can be made in extrapolating C/O from an ionic C^{2+}/O abundance, even with the use of models. Better results should be had from the ultraviolet data, where the collisionally excited lines are presumably better understood, and where up to 3 ions can be observed. But the UV analyses have other severe difficulties. The energies of the excited levels are so high that the derived abundances are extremely sensitive to the adopted electron temperature. Kaler (1983a) shows that $T_e(C^{2+})$, derived from λ1909 CIII] and λ4267 CII and the currently accepted atomic parameters, averages 1500 K less than T_e[OIII], and a change of only 1000 K can lead to a factor of two change in abundance. In addition, there is a problem with absorption of the UV line photons by dust (Harrington, Lutz and Seaton

1981). Generally, the optical data give abundances significantly higher than the UV, where differences could be ascribed to a combination of the above problems, in addition to inadequate atomic data, particularly from the recombination cascade analysis.

I approached the problem (Kaler 1981a) by deriving relative C/O ratios from λ4267 CII, an adopted ionization curve, and an empirical correction for electron density, followed by a scaling of the results to those derived from the UV. That way, we could look at a larger statistical sample that should on the average be accurate, albeit with large individual errors. The method assumes that errors in the UV results, caused largely by uncertain electron temperatures, average out in the scaling process. The wealth of new UV carbon abundances now available allows a significant rescaling of the above optical C/O ratios. From comparison with the results of the UV group (Table 1), the Kaler (1981a) C/H and C/O ratios should be raised by a factor of 1.6, or 0.2 in the log. In addition, I have codified the technique into analytic expressions:

$$\log C/H = -3.15 + 0.14 \log t + \log I_c(\lambda 4267) + 0.4465E + 0.6124E^2 + 0.469 \log x \quad (1a)$$

$$\log C/H = 114.73 + 0.14 \log t + \log I_c(\lambda 4267) - 49.66 \log T_* + 5.211 \log T_*^2 + 0.469 \log x, \quad (1b)$$

where equations (1a) and (1b) are to be used for nebulae with and without He II lines respectively. In the above, $t = 10^{-4} T_e$ (electron temperature), I_c is the intensity of λ4267 CII [I(Hβ) = 100] corrected for reddening, $E = He^{2+}/He$, T_* is the central star temperature, and $x = 10^{-4} N_e/\sqrt{t}$, where N_e is the electron density. The final improved abundances were derived from these equations, and include some new data from the references of Table 1. The resultant C/H (or when combined with O/H, the C/O) are still smaller than C^{2+}/H calculated directly from λ4267 and recombination theory, demonstrating the persistent problem between the optical and UV.

The results of the reanalyses are presented in Figures 1 and 2, where log N/O and log C/O are plotted against He/H. In both figures, population I nebulae (see Kaler 1978a) are plotted as filled symbols, population II as open. In Figure 3, the optically derived (scaled) C/O ratios are plotted as circles, the UV-derived ratios as X's. With the improved abundances, the correlation between N/O and He/H is even clearer than before. N/O first increases rapidly with He/H (or may be even independent of He/H), then at He/H ≈ 0.12 it increases at a much slower rate as He/H climbs to a maximum of ≈ 0.20. The C/O ratios, however, show very little trend with He/H: there is too much scatter. Interestingly, the spread for the UV results is the same as that for the optical, implying that if there really is a valid theoretical relationship between C/O and He/H, the two methods are of similar reliability.

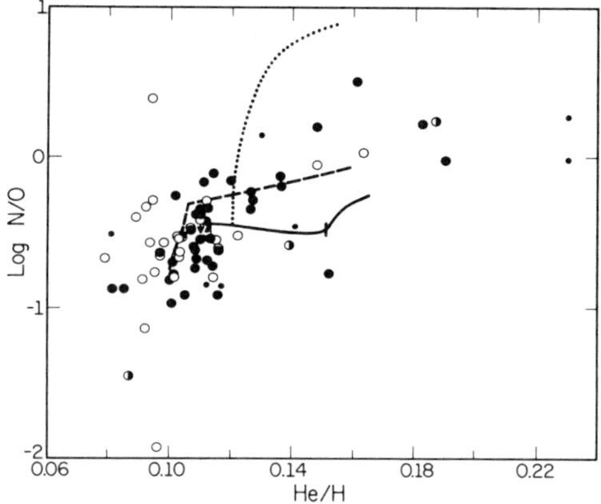

Fig. 1. Log N/O plotted against He/H. Filled symbols: Pop. I; open symbols: Pop. II; solid curve: mean of predicted values from BI and RV through the 3rd dredge-up; dotted curve: RV with envelope burning; dashed curve: BI with conversion of C to N at 1/2 the full rate.

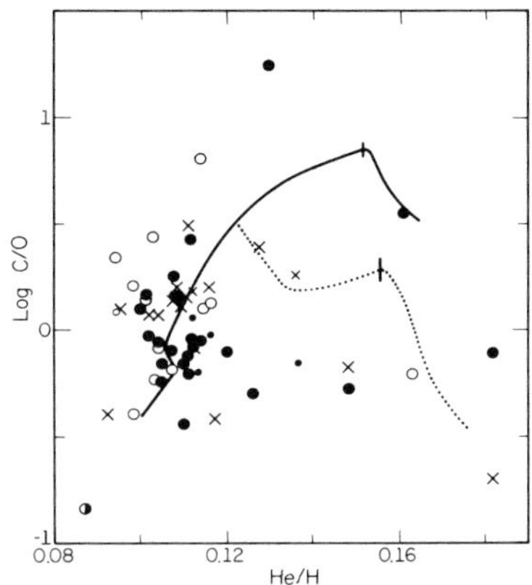

Fig. 2. Log C/O plotted against He/H. The symbols and curves have the same meaning as in Fig. 2. Circles: C/O derived from λ4267 CII; X's: C/O derived from ultraviolet data. The analog to the dashed curve of Fig. 2 lies near the solid curve.

A summary of the theoretical predictions by BI and RV for Population I is shown by the curves in Figures 1 and 2. The solid lines give the average from BI and RV of the abundances expected after all three dredge-up cycles, with no further processes (i.e. excluding RV's burning in the convective envelope). Clearly, the curve in Figure 1 does not fit the well-defined observational relation. If the third dredge-up (which destroys the successful relation demonstrated through the second phase by Kaler, Iben and Becker 1978) is valid then, as BI point out, there must be some mechanism of reconverting the dredged carbon to nitrogen. RV accomplish this by their scheme of envelope burning, which produces the dotted curve, and which fits even more poorly. The best fit is obtained by BI, where they assume that carbon is converted to nitrogen at 1/2 the full rate given by their equation (19), which has little effect on the carbon abundance. This curve, which passes nicely through the points in Figure 1, is similar to, and about 0.1 dex above, the curve through the second dredge-up, and cannot be observationally distinguished from it. Addressing Figure 2, we find that the observations cannot yet distinguish among any of the curves, including that through the second dredge-up (not shown), which is a flat line. Grain formation, as suggested by BI, may certainly be important.

In summary on this topic, the observations will support a theory that incorporates a modest reconversion of carbon back to nitrogen in the convective envelope. Observationally, we must improve the statistics of both the presentations of Figures 1 and 2, the former for comparison with the distribution predicted by an initial mass function, and the latter (for now) to see whether there is any correlation at all. The data, especially in Figure 2, are very strongly affected by observational selection, since we observe only bright nebulae in the UV, or those in the optical with already strong λ4267 CII. Many more observations are needed of λ4267 to reduce the scatter, and to improve the statistics at high He/H. It is particularly important to obtain total CII fluxes free from stratification effects. Theoretically, we need to explain the high He/H ratios that fall beyond the curve, and it would be of great aid if models could be calculated that incorporate the proper physics of C to N conversion during the thermal pulsing phase, in order to avoid the ad hoc assumption used by BI.

4. ABUNDANCES AND THE EVOLUTION OF PLANETARY NUCLEI

Here we shall look more specifically at the relations between abundances and the general evolution of central stars. Perhaps the most exciting development in this area was the discovery in two planetary nebulae of embedded zones or knots that consist of nearly pure helium: Abell 30 (Hazard et al. 1980), and Abell 78 (Jacoby and Ford 1982). Both the central stars exhibit strong mass outflow, as indicated by powerful P-Cygni lines (Greenstein 1981, Heap 1979). The stars have evidently removed their entire hydrogen envelopes, and are now releasing matter from the helium rich cores themselves. Iben et al. (1983) suggest that these stars are remnant cores that suffered a final thermal pulse while they were on the cooling track of their initial passage through the log L - log T plane, which forced them to brighten and repeat their earlier evolutionary tracks. Detailed abundance analyses like that of Jacoby and Ford (1982) will provide further superb means for examining stellar evolutionary processes. We do not know how common objects of this sort may be.

On a more general aspect of the subject, Renzini (1979) and Iben and Renzini (1982) predicted that overabundances in planetaries should correlate with the position of the central star on the log L - log T plane, as a consequence of both stellar evolution and dredge-up theories. Stars of high initial mass will produce nuclei of higher core mass, which because of evolutionary time-scales, will be seen generally on their cooling tracks on the lower left of the log L - log T plane; these are the stars that should have enriched their nebulae the most via convective dredge-up. Even if we ignore the time-scale argument, we would expect that central stars of high core mass should have nebulae with high N/O and He/H. And that is indeed what we find. Figure 3, taken from Kaler (1983b), shows the placement on the log L - log T plane of the central stars of large nebulae ($r > 0.175$ pc) with known N/O. The trend is very clear: half the stars with core masses $> 0.6\ M_\odot$ have nebular N/O ≥ 1, whereas there are none with core mases $< 0.6\ M_\odot$. The true situation may be more complex than predicted, however. Taken at face value, Figure 4 shows a mix of N/O for higher core mass, suggesting that the relations among overabundances, core masses, and initial masses may not be simple and monotonic. This problem is bound up with difficulties in the placement of the stars on the plane, and with the planetary distance scale: see Kaler (1983b) for a detailed discussion.

Finally, we see a strong relationship between overabundances and nebular structure or morphology. Peimbert (1978) noted that the type I nebulae, those with extreme N/O and He/H, "comprise an extreme subset of Greig's (1971) class B," nebulae with a bi-lobed, apparently toroidal structure. This relation is confirmed and broadened by Figure 4, also taken from Kaler (1983b), which shows N/O for the large nebulae plotted against the geometric filling factor, ξ, with only one nebula overlapping with Peimbert's set. This filling factor is a measure of the solid angle with which a nebula appears to surround its star, and is intimately connected with the Greig classification: nebulae with $\xi < 1$ are his class B. Since Kaler (1983b) further shows that the low-ξ nebulae are related to high core mass, the range in N/O for $\xi \approx 0.5$ is analogous to the range seen directly for higher core mass in Figure 3. The nebular abundances show large-scale agreement with theoretical predictions, and general consistency among theories of

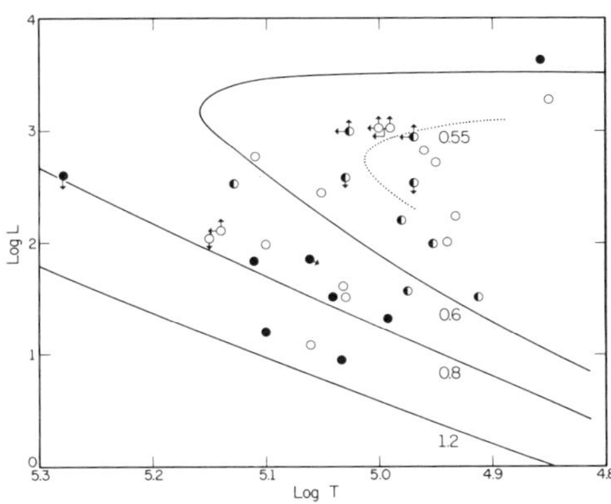

Fig. 3. The central stars of large planetaries ($r > 0.175$ pc) with known N/O on the log L - log T plane. Open symbols: $N/O \leq 0.4$; half-filled sybmols: $0.4 < N/O < 1$; filled symbols: $N/O \geq 1$. Solid curves: evolutionary tracks for 0.6, 0.8, 1.2 $M_\odot$ from Paczyński (1971); dotted curve: extrapolated track for 0.55 $M_\odot$ from Schönberner and Weidemann (1981).

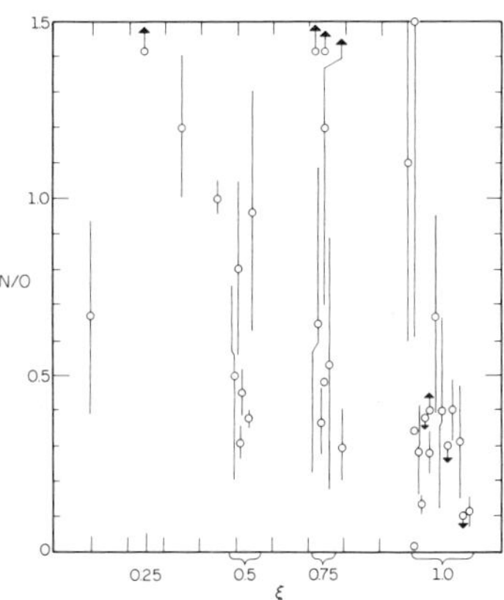

Fig. 4. N/O for large planetaries plotted against ξ, the geometric filling factor, a measure of morphology. Nebulae with $\xi < 1$ are generally Greig class B.

stellar evolution, convective dredge-up, and mass loss. But much more observational and theoretical work is needed before agreement in detail can be attained.

5. EXTRAGALACTIC NEBULAE

No generalized statements comparable to those above can yet be made about abundances in extragalactic planetaries, simply because of the paucity of data and the severe observational selection: only the very brightest nebulae have been observed. The Magellanic Clouds of course are the best studied: see in particular papers by Osmer (1976), Webster (1976), Dufour and Killen (1977), Aller et al. (1981b), and Maran et al. (1982). Results are consistent with those derived from galactic planetaries: O/H is similar to that found in the LMC and SMC diffuse nebulae, and there is a range in He/H and N/O, demonstrating that dredge-up processes are at work. The three nebulae studied by Maran et al. (1982), which show large overabundances in N/O and C/O (the latter particularly in the SMC) relative to the ambient interstellar medium, are shown by Stecher et al. (1982) to have high mass cores ($\approx 1\ M_{\odot}$), consistent with the galactic objects displayed in Figure 3. Looking farther afield, abundances are available for planetaries in M32 (Jenner, Ford, and Jacoby 1979), the Fornax dwarf (Danziger et al. 1978), and NGC 6822 (Dufour and Talent 1980). The latter again displays large N/O (≈ 5) and He/H (≈ 0.19).

As the observations improve, and as we can probe to fainter nebulae, it will be interesting to see how the correlations between abundance ratios might differ among galaxies, and how the dredge-up processes are influenced by different initial abundances. There are interesting possibilities in the comment by Jenner, Ford and Jacoby (1979) that the planetaries in M32 have consistently stronger [NII] lines than do those in other elliptical galaxies. There is certainly no lack of work, both on other galaxies and in our own, to keep us occupied.

REFERENCES

Aller, L. H. 1978, Planetary Nebulae, IAU Symposium 76, Y. Terzian, ed. (Dordrecht: Reidel), p. 225.

Aller, L. H., and Czyzak, S. J. 1979, Ap. Space Sci., **62**, 397.

_____. 1983, Ap. J. Suppl., in press.

Aller, L. H., and Keyes, C. D. 1980, Proc.Nat.Acad.Sci. USA, **77**, 1231.

Aller, L. H., Keyes, C. D., and Czyzak, S. J. 1981, Ap. J., **250**, 596.

Aller, L. H., Keyes, C. D., Ross, J. E., and Czyzak, S. J. 1980, Ap. Sp. Sci., **67**, 349.

Aller, L. H., Keyes, C.D., Ross, J. E., and O'Mara, B. J. 1981a, M. N. R. A. S., **197**, 647.

_____. 1981b, M. N. R. A. S., **194**, 613.

Aller, L. H., Ross, J. E., Keyes, C. D., and Czyzak, S. J. 1979, Ap. Sp. Sci. **64**, 347.

Aller, L. H., Ross, J. E., O'Mara, B. J., and Keyes, D. C. 1981, M. N. R. A. S., **197**, 95.

Barker, T. 1978a, Ap. J., **220**, 193.
_____. 1978b, Ap. J., **221**, 145.
_____. 1978c, Ap. J., **219**, 914.
_____. 1980a, Ap. J., **237**, 482.
_____. 1980b, Ap. J., **240**, 99.
Beck, S. C., Lacy, J. H., Townes, C. H., Aller, L. H., Geballe, T. R., and Baas, F. 1981, Ap. J., **249**,, 592.
Becker, S. A., and Iben, I., Jr. 1979, Ap. J., **232**, 831.
_____. 1980, Ap. J., **237**, 111 (BI).
Danziger, I. H., Dopita, M. A., Hawarden, T. G., and Webster, B. L. 1978, Ap. J., **220**, 458.
Dinerstein, H. L. 1980, Ap. J., **237**, 486.
Dufour, R. J., and Killen, R. M. 1977, Ap. J., **211**, 68.
Dufour, R. J., and Talent, D. L. 1980, Ap. J., **235**, 22.
French, H. B. 1981, Ap. J., **246**, 434.
Garstang, R. H., Robb, W.D., and Rountree, S.P. 1978, Ap. J., **222**, 384.
Greenstein, J. 1981, Ap. J., **245**, 124.
Greig, W. E. 1971, Astr. Ap., **10**,, 161.
Harrington,J.P., Lutz,J.H., and Seaton,M.J. 1981, M.N.R.A.S., **195**, 21P.
Harrington, J. P., Lutz, J. H., Seaton, M. J., and Stickland, D. J. 1980, M. N. R. A. S., **191**, 13.
Harrington, J. P., and Marionii, P. A. 1981, The Universe at Ultraviolet Wavelengths: The First Two Years of IUE, Nasa Conf. Pub. 2171, p. 623.
Harrington, J. P., Seaton, M. J., Adams, S., and Lutz, J. H. 1982, M.N.R.A.S., **199**, 517.
Hawley, S. A. 1978a, Pub.A.S.P. , **90**, 39.
_____. 1978b, Ap. J., **224**, 417.
Hawley, S. A., and Miller, J. S. 1977, Ap. J., **212**, 94.
_____. 1978a, Ap. J.,, **220**, 609.
_____. 1978b, Pub. A.S.P., **90**, 39.
Hazard, C., Terlevich, B., Morton, D. C., Sargent, W. L. W., and Ferland, G. 1980, Nature, **285**, 463.
Heap, S. R. 1979 Mass Loss and Evolution of O-Type Stars (Dordrecht: Reidel), p. 99.
Iben, I., Jr. 1964, Ap. J. (Letters), **140**, 1631.
_____. 1972, Ap. J., **178**, 433.
_____. 1975, Ap. J., **196**, 525.
Iben, I., Jr., Kaler, J. B., Truran, J. W., and Renzini, A. 1983, Ap. J., in press.
Iben, I., Jr., and Renzini, A. 1982, Illinois Ap. Preprint, **IAP 82-2.**
Iben, I., Jr., and Truran, J. W. 1978, Ap. J., **220**, 980.
Jacoby, G. H., and Ford, H. A. 1983, Ap. J., in press.
Jenner, D. C., Ford, H. C., and Jacoby, G. H. 1979, Ap. J., **227**, 391.
Kaler, J. B. 1970, Ap. J., **160**, 887.
_____. 1974, Ap. J. (Letters), **188**, L15.
_____. 1978a, Ap.J., **226**, 947.
_____. 1978b, Ap. J., **225**, 527.
_____. 1979, Ap. J., **228**, 163.
_____. 1980, Ap. J., **239**, 78.
_____. 1981a Ap. J., **249**, 201.

_____. 1981b, Ap. J., **244**, 54.
_____. 1983a, in preparation.
_____. 1983b, Ap. J., in press.
Kaler, J. B., Iben, I., Jr., and Becker, S. A. 1978, Ap. J. (Letters), **224**, L63.
Lutz, J. H. 1981, Ap. J., **247**, 144.
Maran, S. P. Aller, L. H., Gull, T. R., and Stecher, T. P. 1982, Ap. J. (Letters), **253**, L43.
Marionni, P. A., and Harrington, J. P. 1981, The Universe at Ultraviolet Wavelengths: The First Two Years of IUE, NASA Conf. Pub. 2171, p. 633.
Natta, A., Panagia, N., and Preite-Martinez, A. 1980, Ap. J., **242**, 596.
Nussbaumer, H., and Storey, P. J. 1978, Astr. Ap., **70**, 37.
Osmer, P. S. 1976, Ap. J., **203**, 352.
Paczyński, B. 1971, Acta Astr., **21**, 417.
Peimbert, M. 1973, Mem. Soc. Roy. Sci. Liége, Series 6, **5**, 307.
_____. 1978, Planetary Nebulae, IAU Symposium No. 76, Y. Terzian, ed. (Dordrecht: Reidel), p. 215.
Peimbert, M, and Serrano, A. 1980, Rev. Mexicana Astr. Ap.,, **5**, 9.
Peimbert, M., and Torres-Peimbert, S. 1971, Ap. J., **168**, 413.
Peimbert, M., Torres-Peimbert, S., and Rayo, J.F. 1978, Ap. J., **220**, 516.
Peña, M., and Torres-Peimbert, S. 1981, Rev. Mexicana Astr. Ap., **6**, 309.
Perinotto, M., and Benvenuti, P. 1981, Astr. Ap., **101**, 88.
Perinotto, M., Panagia, N., and Benvenuti, P. 1980, Astr. Ap., **85**, 332.
Perinotto, M., and Renzini, A. 1979, ESA/ESO workshop on astronomical uses of the space telescope, Geneva, 12-14 February 1979.
Pottasch, S. R., Gathier, R., Gilra, D. P., and Wesselius, P. R. 1981, Astr. Ap., **102**, 237.
Price, C. M. 1981, Ap. J., **247**, 540.
Renzini, A. 1979, IAU 4th European Regional Meeting in Astronomy, Ap. and Sp. Sci. Lib., **75**, (Dordrecht: Reidel), p. 155.
Renzini, A., and Voli, M. 1981, Astr. Ap., **94**, 175 (RV).
Schönberner, D., and Weidemann, V. 1981, Physical Processes in Red Giants (Dordrecht: Reidel), p. 463, and priv. comm.
Shields, G. A. 1975, Ap. J., **195**, 475.
_____. 1978, Ap. J., **219**, 559.
Shields, G. A., Aller, L. H., Keyes, C. D., and Czyzak, S. J. 1981, Ap. J., **248**, 569.
Stecher, T. P., Maran, S. P., Gull, T. R., Aller, L. H., and Savedoff, M. P. 1982, Ap. J., in press.
Talent, D. L., and Dufour, R. J. 1979, Ap. J., 233, 888.
Torres-Peimbert, S., and Peimbert, M. 1977, Rev. Mexicana Astr. Ap., **2**, 181.
_____. 1979, Rev. Mexicana Astr. Ap., **4**, 341.
Torres-Peimbert, S., Peimbert, M., and Daltabuit, E. 1980, Ap. J., **238**, 133.
Torres-Peimbert, S., and Peña, M. 1981, Rev. Mexicana Astr. Ap., **6**, 301.
Torres-Peimbert, S., Peña, M., and Daltabuit, E. 1981, The Universe at Ultraviolet Wavelengths: The First Two Years of IUE, NASA Conf. Pub. 2171, p. 641.
Torres-Peimbert, S., Rayo, J. F., and Peimbert, M. 1981, Rev. Mexicana Astr. Ap., **6**, 315.
Webster, B. L. 1976, Pub. A. S. P., **88**, 669.

This work was supported by the National Science Foundation.

SEATON: Can we rule out the possibility that C and O are depleted, owing to grains? The interpretation of infrared features (silicates and carbides) rests on the assumption that CO takes all the O for C/O > 1, all the C for C/O < 1. If that is correct, one might expect depletion.

KALER: That is certainly a possibility, which Becker and Iben invoked to explain the lack of agreement with theory.

PAGEL: How much of the excess nitrogen is a primary nucleosynthesis product? Some extragalactic PN have been found to have enormous overabundances of N relative to surrounding H II regions.

PEIMBERT: Most PN show N/O < -0.30 dex and for them it is very likely that most of the excess N is of secondary origin, produced during the first dredge-up episode. For PN of type I, those with N/O > -0.3 dex, a substantial fraction of the excess N is of primary origin, typically about two thirds.

TERZIAN: You indicated that several PN have unacceptably high helium abundances. There certainly are severe theoretical difficulties in explaining such high abundances. Is it possible that the observations are wrong?

KALER: Some may be. One nebula is NGC 7293, which the Peimberts have now found to have lower N/O and He/H than previously believed. Another, NGC 6537, has large errors associated with it; but others, such as NGC 6302, are well observed.

ALLER: We are trying to make detailed measurements of N-rich objects like Hu 1-2, NGC 6537, and NGC 6445 to help clarify the nitrogen problem and He/H ratio, which is indeed important. The carbon abundances derived from either λ 4267 or ultraviolet lines cover a wide range, as your slide showed. We find $\langle \log(C/H) \rangle_{4267} - \langle \log(C/H) \rangle_{UV} \approx 0.1$, so the discrepancy between the λ 4267 and ultraviolet results is lowered but not removed.

EFFECTS OF DUST FORMATION ON CHEMICAL ABUNDANCES

G.A. Shields
Department of Astronomy
University of Texas at Austin

ABSTRACT

Gas-phase abundances of C, Mg, Si, Ca, and Fe have been measured for a number of planetary nebulae on the basis of optical, ultraviolet, and infrared emission-line intensities. The abundances of Si, Ca, and Fe show characteristic depletions of one to two orders-of-magnitude as a result of grain formation. Magnesium shows a similar depletion in the outer parts of several planetary nebulae, but it is undepleted in their inner parts. Carbon is not detectably depleted by grain formation. Efficient condensation of refractory elements can easily occur during the early stages of formation of a planetary nebula; but the observed, residual gas-phase abundances are not understood. Observations of molecules in the envelopes of late-type stars may provide useful clues.

In his pioneering work on galaxies with broad, nuclear emission lines, Seyfert (1943) pointed out similarities of the relative line intensities to those of high excitation planetary nebulae (PN). An exception is that PN have very weak [FeVII] emission, compared with Seyfert galaxies. Since Nussbaumer and Osterbrock (1970) derived roughly solar abundances of iron in Seyfert galaxies, one might guess that the gas-phase abundance of iron in PN is low.

This possibility was confirmed for NGC 7027 by Shields (1974), who used nebular models and the observations later published by Kaler et al. (1976) derive [Fe/H] = -1.4 ± 0.4 relative to a solar value of $12 + \log N(Fe)/N(H) = 7.6$ (Aller 1980). Because NGC 7027 does not show depletions of the abundances of more volatile elements, such as oxygen and argon, the depletion of iron was attributed to incorporation into grains.

These results were extended to five additional, high excitation PN by Shields (1978), who found an average gas-phase depletion [Fe/O] = -1.2; the measured spread of values is ± 0.8, some of which may be measurement error. These objects all have roughly solar abundances of

D. R. Flower (ed.), Planetary Nebulae, 259–264.

oxygen, and there is no tendency for O/H to decrease with decreasing Fe/H. A similar iron depletion is found far the nitrogen-rich planetary NGC 2440 by Shields et al. (1981), based on forbidden lines of [FeV], [FeVI], and [FeVII]. Collision strengths for these ions are given by Nussbaumer and Osterbrock (1970), Nussbaumer and Storey (1978), and Garstang, Robb, and Rountree (1978), respectively.

Calcium shows a similar pattern of depletions. On the basis of [CaV]λ5309 intensities, Aller and Czyzak (1982) give [Ca/H] = -1.3 for each of several groups of PN, including "nitrogen-rich" and "carbon-rich" objects. Shields et al. (1981) find [Ca/H] = -1.5 ± 0.4 for NGC 2440. On the other hand, sodium and potassium are not significantly depleted in these nebulae.

The abundance of silicon can be determined from the collisionally excited, ultraviolet lines SiIII]$\lambda\lambda$1883,1892 and SiIV]$\lambda\lambda$1394,1403. Blending with CIII]λ1909 and OIV]λ1400 makes this measurement somewhat difficult. There is also a strong temperature dependence of the emission coefficients and the familiar problem of allowing for unobserved stages of ionization. Harrington and Marionni (1981) find [Si/H] = -1.0 for three PN including NGC 2440 and indications of a weaker depletion for HU 1-2. This suggests that silicon is generally depleted, but possibly to a somewhat milder degree than for iron and calcium.

The behaviour of magnesium is especially intriguing. Pequignot and Stasinska (1980) found a gradient in the gas-phase abundance of magnesium in NGC 7027. The [MgV]λ2784 line intensity requires a solar abundance of magnesium, whereas the weakness of MgIIλ2800 implies an order-of-magnitude depletion in the outer region where Mg^+ is concentrated. Recent observations of [MgIV] 4.5μ and [MgV] 5.6μ emission by Beckwith et al. (1982) confirm that Mg is not depleted in the high ionization region of NGC 7027. For NGC 2165 and NGC 2440, Harrington and Marrioni (1981) find [Mg/H] $\sim$ 0.0 from [MgV]λ2784 but [Mg/H] $\sim$ -1.5 to -2.0 from MgII. Shields et al. (1981) confirm these results for NGC 2440. Thus, gradients in the Mg depletion appear to be common in high excitation planetaries. However, the low excitation planetary nebula IC 418 has strong MgIIλ2800, consistent with an approximately solar abundance (Harrington et al. 1980)

Abundances of gas-phase carbon are determined from optical permitted lines, some of which are thought to arise primarily from recombination, and from ultraviolet, collisionally excited lines of CIII] and CIV. Results vary from roughly solar in some objects, N(C)/N(O) $\sim$ 0.5, to strongly enhanced, with N(C)/N(O) $\sim$ 2 (e.g., Aller and Czyzak 1982; Kaler 1981; Harrington et al. 1980, 1981). Thus, although carbon in the form of graphite is highly refractory, carbon typically is not substantially locked into grains. This is particularly significant for objects with N(C)/N(O) > 1, since their gas-phase carbon cannot result simply from dissociation of CO molecules when the nebula became ionized.

To summarize the observations, Fe, Ca, Si, and Mg typically show gas-phase abundances one to two orders-of-magnitude less than the solar values, except for magnesium in the inner regions of PN. Carbon does not participate in this depletion.

Are these depletions the result of grain formation? The measurements mostly involve bright, high excitation planetary nebulae of types I or II as defined by Peimbert (1978). These are Population I objects with average heights above the galactic plane of about 150 pc or less and progenitors of mass roughly 1.5 $M_{\odot}$ or more. The abundances of oxygen, neon, and argon are roughly solar. This suggests that the intrinsic abundances of iron, etc., are also solar. Some halo stars do show deficiencies of iron relative to oxygen (e.g., Pilachowski, Wallerstein, and Leep 1980); but these objects are deficient in oxygen, relative to hydrogen. Moreover, Barker (1979) has studied the composition of three extreme halo PN and found argon abundances approximately 1/100 of those in the sun. In the case of K648 in M15, this agrees with the iron abundance in the stars, whereas oxygen is only deficient a factor of 10 with respect to the sun. Thus, the intrinsic abundances of argon and iron appear to vary in roughly constant proportion. Therefore, the fact that argon has roughly a solar abundance in the Type I and II PN indicates that the depletions of the refractory element abundances result from dust formation.

The idea that grains have locked up the refractory elements in PN is consistent with infrared observations of thermal emission from dust heated by the central star and by the nebular Lyman α radiation field (e.g., Becklin, Neugebauer, and Wynn-Williams 1973). Spectral features suggestive of SiC are seen in some objects (Aitken et al. 1979), but the interpretation of other features is unclear (Jones et al. 1980).

The theoretical implications of iron condensation in PN have been considered by Scalo and Shields (1979). If a PN shell of mass about 0.1 solar mass is ejected in a single event, then the temperature falls to the condensation point (1300 K) at a radius of roughly $10^{14.5}$ cm. The hydrogen density at this point is about 10^{12} cm^{-3}. For reasonable grain sizes, the grains should thoroughly accrete any residual gas-phase iron in the time available before the nebula expands substantially. This would leave a gas-phase abundance much lower than observed. Sputtering and grain-grain collisions are ineffective at restoring the observed amount of iron to the gas.

Alternatively, one may assume that the PN is ejected by means of a continuous wind with a mass-loss rate of perhaps 10^{-4} $M_{\odot}$ yr^{-1} for the last thousand years of the evolution of the red giant progenitor. Then densities of about 10^{10} cm^{-3} are likely to prevail at the point where iron condenses. In this case, rather small grains (10^{-6} cm) in large numbers are needed to present sufficient surface area to capture all but a few percent of the iron atoms. Furthermore, the final degree of depletion, achieved when the PN has expanded to low density, depends on the parameters of the nebula in a sensitive fashion.

The observed depletions are fairly uniform, and this seems unlikely to result from a universal ratio of the timescales for depletion and nebular expansion.

Thus, in either the ejection model or the wind model, there is some difficulty in explaining the presence of about 5 percent of the iron in the gas phase. Scalo and Shields (1979) noted that in the wind model, grains of different sizes can acquire, from radiation pressure, relative velocities possibly sufficient for shattering collisions. If roughly one layer of atoms is vaporized per collision, then a gas-phase iron abundance of the observed order can result from an equilibrium between shattering and accretion.

A different explanation was suggested by Shields (1980), who noted that mass lost by a fast wind from the central star might amount to a few percent of the nebular mass (c.f. Kwok et al. 1978). If the gas ejected by the red giant had no gas-phase iron but the white dwarf wind had all its iron in the gas, then the observed abundance might result. Aside from theoretical difficulties with cooling the white dwarf wind and mixing it into the nebula, this model has trouble explaining why gas-phase magnesium has a solar abundance in the cores of PN whereas iron and calcium remain strongly depleted in the same location.

The theoretical picture thus is uncertain, and there is room for other ideas to explain the residual gas-phase abundances of refractory elements in PN. Perhaps the outer layer of atoms on the grain is only weakly bound, so that it is easily sputtered off when the nebula is ionized. Then, for grain sizes of order 10^{-5} cm, a few percent of the total atoms are restored to the gas. This might explain the rough similarity of the depletions for different nebulae and different elements, but it also has trouble explaining the selective lack of depletion of magnesium in PN cores.

Finally, radio observations of molecules in envelopes of late type stars may be relevant (see Morris et al. 1979; and references therein). In IRC+10216, SiO and SiS account for roughly one percent of the total silicon; most of the remainder may be in grains. If this gas later becomes ionized as a PN, dissociation of the molecules could result in a gas-phase abundance of order one percent of the solar amount. This suggests that the gas-phase refractory atoms in PN may be traceable to molecules that escaped incorporation into grains during the formation of the nebula.

This work was supported in part by NSF research grant AST 80-20461, NASA research grant NSG 7232, and an Alfred P. Sloan Research Fellowship. The author acknowledges the kind hospitality of the Lick Observatory and the European Southern Observatory.

REFERENCES

Aitken, D.K., Roche, P.F., Spenser, P.M., and Jones, B.: 1979, Astrophys. J. 233, p. 925.
Aller, L.H.: 1980, unpublished manuscript.
Aller, L.H., and Czyzak, S.J.: 1982, preprint.
Barker, T.: 1980, Astrophys. J. 237, p. 482.
Becklin, E., Neugebauer, G., and Wynn-Williams, C.: 1973, Astrophys. Letters 15, p. 87.
Garstang, R.H., Robb, W.D., and Rountree, S.P.: 1978, Astrophys. J. 222, p. 384.
Harrington, J.P., Lutz, J.H., and Seaton, M.J.: 1981, M.N.R.A.S. 195, p. 21P.
Harrington, J.P., Lutz, J.H., Seaton, M.J., and Stickland, D.J.: 1980, M.N.R.A.S. 191, p. 13.
Harrington, J.P., and Marionni, P.A.: 1981, NASA Conference Publications No. 2171, The Universe at Ultraviolet Wavelenghts, p. 623.
Jones, B., Merill, K.M., and Stein, W.: 1980, Astrophys. J. 242, p. 141.
Kaler, J.B.: 1981, Astrophys. J. 249, p. 201.
Kaler, J.B., Aller, L.H., Czyzak, S.J., and Epps, H.: 1976, Astrophys. J. Suppl. 31, p. 163.
Kwok, S., Purton, C.R., and Fitzgerald, M.P.: 1978, Astrophys. J. (Letters) 219, L37.
Morris, M., Redman, R., Reid, M.J., and Dickinson, D.F.: 1979, Astrophys. J. 229, p. 257.
Nussbaumer, H., and Osterbrock, D.E.: 1970, Astrophys. J. 161, p. 811.
Nussbaumer, H., and Storey, P.J.: 1978, Astron. Astrophys. 70, p. 37.
Peimbert, M.: 1978, IAU Symposium No. 76: Planetary Nebulae (Dordrecht-Holland: D. Reidel Publ. Co.) p. 215.
Pequignot, D., and Stasinska, G.: 1980, Astron. Astrophys. 81, p. 121.
Pilachowski, C.A., Wallerstein, G., and Leep, E.M.: 1980, Astrophys. J. 236, p. 508.
Scalo, J.M., and Shields, G.A.: 1979, Astrophys. J. 228, p. 521.
Seyfert, K.: 1943, Astrophys. J. 97, p. 28.
Shields, G.A.: 1975, Astrophys. J. 195, p. 475.
Shields, G.A.: 1978, Astrophys. J. 219, p. 559.
Shields, G.A.: 1980, P.A.S.P. 92, p. 418.
Shields, G.A., Aller, L.H., Keyes, C.D., and Czyzak, S.J.: 1981, Astrophys. J. 248, p. 569.

TERZIAN: You clearly indicated that Si is depleted and C is not in PN. Can you conclude that the associated grains are primarily silicates and not graphite?

SHIELDS: As much as half the total carbon might be in grains, and that would exceed the amount of silicon. However, it seems unlikely that carbon is depleted to the same degree as iron - the large, gas phase abundance of carbon would then be only one twentieth of the total!

HOUCK: Mg S grains have been proposed to explain the 30 μm feature seen in the spectra of some carbon star shells and PN. Mg S is dissociated at a relatively low temperature, which may explain why the 30 μm feature is not seen in carbon star shells with high grain temperatures. This might also explain why Mg is not depleted in the inner parts of PN - either Mg S grains cannot form or pre-existing grains dissociate.

CLEGG: My impression from the last two talks is that S is not depleted by grains in PN. This is interesting because, in carbon-rich Red Giant shells, observations of CS and Si S suggest that S is depleted significantly on grains.

SHIELDS: The sulphur abundance measurements in PN are somewhat uncertain and may permit a factor of 2 depletion. However, sulphur clearly has not been depleted by about $10^{1.4}$, as in the case of iron.

SECTION IV

ORIGIN OF PLANETARY NEBULAE

RED GIANTS AS PRECURSORS OF PLANETARY NEBULAE

Alvio Renzini
Dipartimento di Astronomia, CP 596, 40100 Bologna, Italy

ABSTRACT

Several physical processes taking place during the red (super)giant phase of intermediate-mass stars have direct observational consequences for the subsequent nebular stage. These processes include: the regular wind and the envelope ejection, the thermal pulses during the AGB phase, the dredge-up processes, and the dust formation in expanding circumstellar envelopes. In this paper it is briefly discussed how such processes affect the mass range of PN nuclei and their evolution, and the PN lifetime, composition and dust content. The last section is devoted to a cursory discussion of PNe which can be generated by binary stars.

1. INTRODUCTION

The nebular phase represents just a chapter in the life of those stars which ultimately become white dwarfs, and therefore the understanding of most properties of PNe and their central stars is necessarily rooted to the study of the previous evolutionary history of these objects. On the other hand, the derivation of physical parameters from the observation of PNe (e.g. chemical abundances) involves a totally different sort of input physics with respect to that currently in use in the case of stars, thus allowing both useful consistency checks and the derivation of quantities which are not accessible to the study of normal stars (e.g. the helium abundance of red giants cannot be directly derived).

Evolutionally speaking, only a few thousand years separate the nebular stage from the red (super)giant phase, when stars are ascending the Asymptotic Giant Branch (AGB), intermittently burning either hydrogen or helium in two separate shells, and in the meanwhile shedding their envelope, thus preparing the condition for the later glowing of the nebula.

D. R. Flower (ed.), Planetary Nebulae, 267–280.

AGB stars are then of particular relevance for PN sudies, to the extent that practically all processes taking place in them have direct *observable* consequences for the PN stage. In this review I will try to elucidate the most important among these causal links between particular processes taking place during the AGB phase, and phenomena which may not become apparent before the nebular stage has been reached. In particular, these processes include: i) the mass loss processes known as red-giant wind and superwind, ii) the superwind quenching at particular phases of the thermal pulse cycles, iii) the dredge-up and envelope burning processes which establish the final nebular composition, and iv) grain formation in the AGB ejecta. Some of these topics have been discussed in other recent reviews (Renzini 1979, 1980a, b, Iben and Renzini 1982a), and they will be only summarized here.

2. THE MASS LOSS PROCESSES AND THEIR CONSEQUENCES

PNe are clearly the manifestation of some mass loss process which operated earlier in the history of their exciting stars. As pointed out by Shklowski (1956), the expansion velocity of PNe indicates that the nebular material has been ejected while the star was a red giant, and subsequent observational and theoretical studies have conclusively confirmed Shkowski's argument. In particular, an impressive body of observational evidences now shows that red giants and supergiants are indeed losing mass, with the appropriate expansion velocities ($\sim$10÷20 km/s) and, in some cases, with the appropriate rate for the production of a PN. However, in a recent series of papers (Wood and Cahn 1977, Renzini and Voli 1981, Renzini 1981a, b, Iben and Renzini 1982a) arguments are given indicating that at least two distinct mass-loss regimes must operate in AGB stars: the regular wind and the so-called superwind.

2.1 Wind and Superwind in AGB Stars

Most red giants (including AGB stars) are losing mass at a rate which is conveniently expressed by the Reimers formula (cf. Reimers 1975) Stars whose mass loss rate (MLR) follows the Reimers expression are considered in the regular *wind* regime, and evolutionary studies show that AGB stars can reach MLR's of at most a few 10^{-6} $M_\odot$/yr in this regime (c.f. Iben and Renzini 1982a). This is largely insufficient to account for the existence of PNe. In fact, combining current estimates of the mass, radius and expansion velocity of PNe, one derives that the MLR during the process leading to the PN ejection should be at least several 10^{-5} $M_\odot$/yr, and possibly much higher. Therefore, the termination of the AGB phase must coincide with a dramatic increase of the MLR, indicating that another, more efficient, mass loss process has to supersede the regular *Reimer* wind. The concise term *superwind* seems appropriate for designating this efficient mass loss process (Renzini 1981a), whose physical nature, though not yet unambiguously identified, is probably connected with the dynamical pulsations of the envelope (e.g. Wood 1974, Tuchman et al. 1979, Tuchman, this volume). Indeed, MLR's ranging from a few 10^{-5} $M_\odot$/yr up to several 10^{-4} $M_\odot$/yr appear to be rather common among infrared and OH maser sources (cf. Knapp et al. 1982), and then such objects are most likely in the superwind regime.

For exploratory purposes, both the wind and the superwind processes are currently parametrized in evolutionary calculations (e.g. Fusi Pecci and Renzini 1976, Wood and Cahn 1977, Iben and Truran 1978, Renzini and Voli 1981), through a numerical coefficient η placed in front of the Reimers formula, and a parameter b entering into the expression for the envelope mass -M_{PN}- at the onset of the superwind regime:

$$M_{PN}(M_H) = b\cdot f(M_H) \tag{1}$$

where M_H is the mass inside the location of the hydrogen-burning shell (the so-called core mass), and $f(M_H)$ is an appropriate function (see for more details: Renzini and Voli 1981, Iben and Renzini 1982a).

M_{PN} is clearly also the mass ejected by the superwind process. Since luminosities and radii of AGB stars are rather sensitive to M_H, M_{PN} is expected to be rather sensitive to M_H, and then, ultimately, to the stellar initial Mass M_i. This expectation is reflected in the adopted parametrization, where (for b = 1) M_{PN} ranges from 0.02 $M_\odot$ for M_H = 0.50 to 1.3 $M_\odot$ for M_H = 1.4. The fact that the amount of superwind material around the central stars of PNe can considerably vary depending on M_i may have important consequences for the PN lifetimes, a point which will be discussed in Section 2.5. Moreover, in this frame the central stars of PNe are then expected to be surrounded by an inner and denser shell (the remnant superwind), and by an outer more tenuous shell (the remnant regular wind) which may extend for several pc. This picture is in qualitative agreement with the observation of nebulae with extended halos (Millikan 1974), as pointed out by Fusi Pecci and Renzini (1976).

2.2 Masses of PN Nuclei and the Mass Range of PN Producers

Coupling evolutionary calculations and parametrized mass loss algoritms, Iben and Renzini (1982a) give the following expression relating the stellar final mass M_f to the initial mass M_i:

$$M_f = \sim 0.53\eta^{-0.082} + 0.15\eta^{-0.35}(M_i - 1). \tag{2}$$

This expression is most accurate for $1/3 < \eta < 2$, and is fairly insensitive to b for $1/2 < b < 1$. Note that the final mass M_f (i.e. the mass of the white dwarf remnant) is also practically identical to the mass of the central star during the PN phase.

Of great interest is also the critical initial mass M_W below which the hydrogen-rich envelope is actually ejected *before* the core mass can reach the Chandrasekhar limit ($\sim 1.4\ M_\odot$), and above which this limit is attained, carbon is ignited in the electron-degenerate core, thus leading to a supernova explosion. Therefore, stars in the mass range $0.85 \lesssim M_i \lesssim M_W$ are those which eventually produce white dwarfs, and then are likely to experience a PN stage. Following the parametrization sketched above, Iben and Renzini (1982a) give:

$$M_W = \sim 1.0 + 9.33\eta^{0.35} - 3.53\eta^{0.27} + 0.8(b - 1.0), \qquad (3)$$

which implies a range for M_W from ~ 4.7 to ~ 8, for η increasing from 1/3 to 2 (with b = 1).

The mass range of potential PN producers is therefore rather extended, and, correspondingly, great quantitative differences are to be expected in the behaviour of PNe and their nuclei, depending on the initial mass of the parent stars.

According to Eq. (2), PN nuclei should have masses ranging from slightly more than 0.50 $M_\odot$ (for M_i = 0.85) up to 1.4 $M_\odot$ (for $M_i = M_W$). The recognition that actual PN nuclei have masses in this range represents a crucial test for the current theory of stellar evolution and the adopted parametrization of the mass loss processes. This point will be discussed in Section 2.4.

2.3 The Superwind Quenching

The onset of the superwind regime marks the beginning of a fast decrease in the envelope mass M_e of AGB stars. Stellar structure calculations indicate that, decreasing the stellar mass, the average location on the HR diagram of an AGB star will at first move to the right, towards larger and larger radii [cf. Eq. (1) in Becker and Iben 1979] ,and, increasing the radius, the superwind instability will most likely be enhanced. This tendency to larger radii (lower effective temperatures) is reversed when M_e falls below a critical value, M_{eD}, marking the departure of the star from the Hayashi line. The value of M_{eD} is very small, and should be in the range from ~ 0.001 to $\sim 0.01\ M_\odot$, depending on the actual value of M_H, as indicated by Paczynski (1971) models. Since the star is now contracting, the superwind instability may become less violent, and suddendly gets quenched leaving a residual envelope mass M_{eR} ($\lesssim M_{eD}$). The theoretical determination of M_{eR} presents severe difficulties. Nonetheless, this quantity plays a crucial role in the subsequent evolution of the star, in particular for what concerns the nebular phase.

It is important to realize that M_{eR} can fluctuate considerably from one star to another, even among stars with virtually identical initial mass. This follows from i) the fact that $M_{eR} \ll M_{PN}$, ii) the hydrodynamical nature of the superwind process, and iii) the possibly episodic character of the superwind, which might consist in a number of discrete ejection events, as indicated by the hydrodynamical models of Wood (1974) and Tuchman et al. (1979). For example, let us take an assembly of stars for which $M_{PN} = 0.2$ (a popular value) and $M_{eD} = 0.001\ M_{\odot}$. If each ejectiontion event removes typically $\sim 10^{-3}\ M_{\odot}$ (as suggested by Tuchman et al.) then some 200 events are required to expel the whole envelope. But clearly nobody can pretend that all such stars will suffer exactly 200 events, instead of, say, 199 or 201! Therefore, M_{eR} will fluctuate from one star to the next by at least a factor of 2, and in some special cases M_{eR} could be considerably smaller than M_{eD}.

Another important aspect of the problem concerns the precise *phase* ϕ during the thermal pulse cycle at which the superwind ceases, where $\phi = 0$ when this happens in coincidence with one helium-shell flash, and $\phi = 1$ when it happens while the star is almost suffering one of such flashes, but not quite. In general, $\phi = \Delta t/\Delta t_{ip}$, where Δt is the time elapsed since the last pulse, and Δt_{ip} is the interpulse period.

It would be rather interesting to establish if there is any preferred value of ϕ, i.e. if envelope ejection and/or superwind quenching take more often place at some particular phase, for instance in coincidence with one flash. However, since speculations on this point can easily go too far, I will assume in the following that the ϕ-spectrum is just flat, but one should keep in mind that this may not be the case.

It is also worth noting that there is no reason why M_{eR} and/or ϕ should not depend (*on average*) on the final mass M_f, and then on M_i. In conclusion, the post-AGB evolution is essentially determined by three parameters, M_f, M_{eR}, and ϕ, the latter two being possibly only marginally correlated with the first one. The effects on the post-AGB evolution of varying each of these quantities is discussed in the coming section.

2.4 The Evolution of PN Nuclei

Naively enough, the transition from the AGB to the region of PN nuclei (which follows the superwind quenching) is often regarded as practically instantaneous. Conversely, as convincingly shown by Härm and Schwarzschild (1975), this transition time can be awfully long if certain conditions are not fulfilled. The transition time t_{tr} is defined as the time interval between the superwind quenching and the instant when the effective temperature of the remnant star reaches 30,000 K, i.e. when the central star becomes hot enough to excite the previously ejected envelope.

Iben and Renzini (1982a) give the following expression for t_{tr}:

$$t_{tr} = \sim 1.6\ 10^{6}\text{yr}\ (M_{eR} - M_{eN})/(M_H - 0.44) \tag{4}$$

where M_{eN} is the envelope mass when the star reaches T_{eff} = 30,000 K, and from the models of Paczynski (1971) one can derive:

$$M_{eN} = \sim 1.8\ 10^{-5} M_H^{-8.23}\ . \tag{5}$$

For illustrative purposes, let us consider the case M_H = 0.6, for which $M_{eN} = \sim 1.2\ 10^{-3}\ M_\odot$. Then Eq. (4) gives t_{tr} = 3000, 8000, and 18,000 yr, respectively for M_{eR} = 1.5, 2.0 and 3.0 $10^{-3}\ M_\odot$. The *age* of a PN (time since the quenching of the superwind) is roughly given by $t_{PN} \simeq R_{PN}/v_{exp}$, where R_{PN} is the observed nebular radius and v_{exp} is the nebular expansion velocity. Observed PNe have ages between ~1000 and ~30,000 yr, the majority clustering around 5000 yr. Clearly, for each observed PN, one must have $t_{tr} < t_{PN}$, and, because of the arguments presented in section 2.3, any fluctuation and/or trend in M_{eR} will necessarily translate into sizable fluctuations/trends of t_{tr}. The PN stage can even be completely bypassed if t_{tr} is too long (for instance, longer than 30,000 yr), since in this case the ejected envelope would disperse before the central star becomes hot enough to excite the nebula (Renzini 1981a). One can then conclude that i) the initial PN radius ($= t_{tr} \cdot v_{exp}$) depends on the residual envelope mass M_{eR}, and ii) quite possibly, there may exist stars which eject their envelope but do not experience an *observable* PN stage.

Another timescale is of capital importance for the understanding of PNe and their nuclei. This is the *fading time* t_f, defined by Iben and Renzini (1982a) as the time taken by the star to fade by a factor of ten in luminosity, after the star has reached the critical temperature of 30,000 K. From the models of Paczynski (1971), Iben and Renzini emphasize that t_f is dramatically sensitive to M_H, the mass of the post-AGB remnants, with $t_f \propto \sim M_H^{-9.6}$! Indeed, a 1.2 $M_\odot$ model takes only 45 yr to fade while a 0.6 $M_\odot$ model takes ~15,000 yr (Iben 1982). From these facts Renzini (1979, 1981a) concluded that initially more massive stars, which according to Eq. (2) should residuate more massive remnants, can only produce PNe with *bolometrically* faint nuclei (say, Log $L/L_\odot < \sim 2$), while low-mass precursors ($\sim M_\odot$) which residuate low-mass remnants ($\sim 0.55\ M_\odot$) can only produce PNe with bright nuclei (Log $L/L_\odot \simeq 3$), since their fading time is longer than 30,000 yr. Obviously, there will be some intermediate initial mass for which t_f is comparable to the nebular lifetime, and for these objects the central stars are expected to experience a sizable decrease in luminosity during the nebular phase. Schönberner and Weidemann (1981) and Schönberner (1981) have questioned these conclusions, and rather argue that irrespective of M_i practically all stars with $M_i < M_w$ leave remnants with $M_f \simeq 0.6$. This result follows from the particular sample of PNe that they have considered, which is selected according to the availability of the V magnitude of the central stars. Since there are quite many PNe with visually undetected central stars (a classical prototype being NGC 7027), their mere existence implies that these stars must have masses considerably above the values preferred by Schönberner and Weidemann. Therefore, their analysis neither invalidates the arguments of Renzini (1979, 1981a), nor implies that all stars with $M_i < M_w$ generate post-AGB stars with essentially the same mass, a claim which, in any case, would be hard to justify in terms of stellar evolution theory and mass loss during the AGB phase. More recently, Kaler (1982) has analyzed a large sample of PNe finding consistency with Renzini's predictions.

We are now left with the discussion of the role played by ϕ. Iben and Renzini (1982a) mention that, particularly in low-mass post-AGB stars, the actual value of ϕ can considerably affect the fading time t_f, an aspect which deserves further numerical experiments. But aside from this, for each set of values (M_H, M_{eR}), the quantity ϕ is expected to control: i) the possibility for the star to suffer a final helium-shell flash after the superwind quenching, and ii) in the case that this happens, the time t_{FF} elapsing from the beginning of the nebular phase (T_{eff} = 30,000 K) to the outbreak of the final flash. As extensively discussed by Iben et al. (1982), when a final flash takes place in the region of PN nuclei the star describes an extended loop in the HR diagram which partly overlaps the previously traced path. Iben et al. argue that some PN nuclei can actually be percurring one of such post-flash loops, rather than being fading for the first time. Indeed, the duration of the loop is of the order of 1/5 Δt_{ip}, and then can be comparable to the nebular lifetime for post-AGB stars in an appropriate mass range. Iben et al. suggest that the PNe A30 and A78 are likely candidates for having central stars in the post-flash phase.

Clearly, PNe can have rejuvenated central stars only if t_{FF} is shorter than the nebular lifetime. The final flash time t_{FF} is a decreasing function of ϕ, and below a critical value t_{FF} is likely to be infinite, i.e. no final flash takes place, while for ϕ approaching unity t_{FF} tends to vanish. Obviously, this behaviour depends on the fact that for each value of M_H, the intershell mass has to reach a threshold value for a flash being initiated, and the smaller ϕ, the smaller the intershell mass, and so the longer the star must wait before suffering the final flash. Then, for ϕ below a critical value the final flash occurs when the nebula has already dispersed. The precise functional relationship $t_{FF}(M_H, M_{eR}, \phi)$ remains to be determined by further laborious stellar model calculations.

Finally, Renzini (1979, 1980b, 1982), Iben and Renzini (1982a), and Iben et al. (1982) argue that the final flash is likely to generate a hydrogen-deficient star, and, in particular, PN nuclei of the Wolf-Rayet variety, like in the case of A30 and A78.

2.5 Nebular Lifetimes and Nebular Radii

Owing to the extended mass range of PN progenitors it would be very surprizing if all PNe were to have the same lifetime. Indeed, Renzini (1981a) argues that PNe cease to be detectable after a time t_{max} since the envelope ejection, with

$$t_{max} \propto \sim M_{PN}^{0.4} \; SB_{min}^{-0.2} \tag{6}$$

where M_{PN} is the mass of the remnant superwind and SB_{min} is the minimum surface brightness for a nebula being included in existing catalogues. Correspondingly, the nebular lifetime t_{PN} is given by $t_{max} - t_{tr}$, and t_{PN} can then be anywhere between zero (when $t_{tr} \geq t_{max}$) and $\sim t_{max}$. Note that t_{max} depends on M_{PN}, and then, in turn, on M_f and M_i. In the frame of the adopted parametrization of the superwind process, t_{max} is expected to increse by roughly a factor of 5, for M_i increasing from 0.85 $M_\odot$ up to M_W. Therefore, it should be quite dangerous to adopt a unique nebular lifetime for stars belonging to different stellar populations, e.g. young disk, old disk, halo.

A correlation is known to exist between the nebular radius and the luminosity of the central star, bigger PNe having, on average, fainter nuclei. This has been often interpreted as evidence that the locus in the HR diagram occupied by PN nuclei is an evolutionary sequence, with fainter PN nuclei being evolved versions of the brighter ones. While this effect is probably present, at least to some extent, it is worth realizing that three other effects can concur, perhaps dominantly, in producing the observed correlation. i) According to the discussion in the preceding paragraph, PNe produced by more massive precursors may grow bigger before disappearing, compared to those generated by less massive precursors, and according to Section 2.4 the former PN nuclei are expected to be fainter than the latter ones. ii) There should exist a trend in t_{tr} with $M_f(M_i)$, and if t_{tr} increases with M_f then the initial nebular radius ($= t_{tr}v_{exp}$) will also increase with M_f, and thus will correlate with the location of the nucleus on the HR diagram. iii) The larger M_{PN}, the larger the expected dust absorption during the early nebular stages, and then massive nebulae may not be detectable in the optical when still too compact. These considerations indicate that the interpretation of the nebular radius/luminosity correlation might not be so straightforward, after all.

3. THE COMPOSITION OF RED GIANT ENVELOPES AND PLANETARY NEBULAE

The nebular composition must reflect that of the stellar envelope at the onset of the superwind regime. In turn, this is the result of various mixing processes having contaminated the stellar envelope, at specific evolutionary stages, with materials having suffered various types of nuclear processing in the stellar interior. These mixing processes include: i) the three *canonical* dredge-up processes described in Iben and Truran (1978), Becker and Iben (1980), Renzini and Voli (1981), and Iben and Renzini (1982a); ii) the so-called Envelope-burning process (cf. Sugimoto 1971, Renzini and Voli 1981, and references therein); and iii) all those mixing processes which may be induced by rotationally-driven instabilities, and which theoretical astrophysicists find hard to model from first principles. Therefore, the study of the composition of PNe provides a very useful tool for checking current evolutionary models, and, in case , for getting new insight into the complicated question of the mixing processes of non-convective origin.

3.1 The Gas Composition

Renzini and Voli (1981) have published theoretical He/H, C/O and N/O values, as predicted by the canonical theory of stellar evolution (i.e. neglecting possible mixings of non-convective origin). These abundance ratios have been computed for various combinations of the parameters η, b and α (the ratio of the mixing length to the pressure scale height), and for quite many values of the initial mass. Not surprizingly, the abundance ratios are found to be very sensitive to M_i, since the various dredge-ups and the envelope burning process have very different efficiency in stars of different mass. Therefore, for the reasons discussed in Section 2.4, a correlation is expected between the nebular composition and the location of the central star on the HR diagram (Renzini 1979, 1981a).

One crucial entry in the theory is the so-called dredge-up law, giving as a function of M_H the amount of intershell material which, following each pulse, is captured by the convective envelope. Renzini and Voli used the Iben and Truran (1978) dredge-up law, which is very tentative for low values of M_H. Moreover, until recently a significant discrepancy was apparent between low-mass AGB models, where the third dredge-up was never active, and the existence of relatively faint carbon stars in the Magellanic Clouds, for which M_H should be as low as $\sim 0.6\ M_\odot$ (cf. Iben 1981a,b, Renzini 1981c). Since most PNe have relatively low-mass precursors (~ 1-$2\ M_\odot$), and then low-mass nuclei [through Eq. (2)] , theory was in trouble in predicting the nebular composition of the most common objects.

However, Iben and Renzini (1982b,c) have eventually succeded in getting the third dredge-up in a low-mass model, thanks to appropriate opacities kindly provided by A.N. Cox and S. Hodson. In fact, following a thermal pulse, the opacity peak around 10^6 K due to incompletely ionized carbon ions triggers the appearance of a semiconvective region in the upper intershell. Moreover, this semiconvective region soon merges with the convective envelope, and intershell material (mostly helium and carbon) is efficiently convected to the surface. These findings are not only relevant for the nebular composition and carbon stars, but also for the nucleosynthesis of s-process elements in low-mass AGB stars (cf. Iben and Renzini 1982c).

Although the third dredge-up process is now known to operate also in low-mass stars, there remains to determine the third dredge-up law for low values of M_H (for, say, $0.50 \lesssim M_H \lesssim 0.80$), which should be obtained by future laborious calculations of evolutionary models. Therefore, until then the Renzini and Voli results for $M_i \lesssim 2.5$ should be used with caution.

Comparisons of Renzini and Voli theoretical abundances with observations have been presented by Peimbert (1981), Aller (1981) and Kaler (1982). By and large, these comparisons look promizing, in particular for the He/H and C/O ratios. Since much new data will certainly be presented at this meeting, I will not discuss these matters any further...

3.2 The Dust Component

The expanding environment of AGB stars, either in the regular wind or in the superwind regime, is certainly one of the ideal sites for the formation and growth of dust particles. The nature of the grains being formed clearly depends on the composition of the circumstellar envelope, and most crucially on the C/O ratio. If a star terminates its AGB phase while still oxygen rich (C/O < 1), then the daughter nebula should only contain oxygen-rich grains, e.g. silicates. When the star begins the superwind phase as a carbon star (C/O >1), the daughter nebula should only contain carbon-rich grains, e.g. silicon carbyde, graphite, soot, etc. In some cases, the envelope may be ejected while the star is a S-type giant (C/O ≃ 1), and correspondingly the PN should be very poor in dust particles at all times. This dichotomy in the dust content of PNe is actually observed (e.g. Aitken and Roche 1982), and the theory of stellar evolution gives the initial mass ranges for the progenitors of carbon-rich and oxygen-rich nebulae (cf. Renzini and Voli 1981).

4. PLANETARY NEBULAE AND BINARY SYSTEMS

Before becoming white dwarfs (WD), stars have necessarirly to experience a high-temperature phase (T_{eff} > ∿30,000 K) during which they are powerful emitters of ionizing photons. This is true irrespectively of the particular process by which WD's are produced. In particular, a WD is formed in binary systems where a primary component less massive than ∿2.2 $M_{\odot}$ fills its Roche lobe while ascending the red giant branch. In this case a helium WD is formed, with a mass less than about 0.5 $M_{\odot}$. Similarly, a carbon/oxygen WD is formed when a primary star initially less massive than about 8 $M_{\odot}$ fills its Roche lobe while ascending the AGB.

According to Tutukov and Yungelson (1980), perhaps 2/3 of all binary stars suffer Roche-lobe contacts of the types mentioned above, and since binaries rival in number single stars, putting two and two together one is forced to conclude that binary-born WD's can be nearly as common as those produced by the wind/superwind processes in single stars. Moreover, since the secondary components can hardly accrete all the material shed by the primaries (cf. Greggio and Renzini 1982, and references therein), binary systems containing a primary evolving towards the WD stage are also expected to be surrounded by an expanding, rather massive and *non-spherically symmetric* shell.

In my opinion, this provides a quite attractive scenario for the production of asymmetric PNe (e.g. the so-called bipolar nebulae). In fact, these binary-born pre-WD's are rather commonly produced, there is gas around to be excited, and there is the source of ionizing photons.

Conversely, the idea that asymmetric PNe are produced by rotating red giants looks quite unattractive when considering that the surface layers of an AGB star rotate several hundred times slower than the initial main sequence rotational velocity. Moreover, significant angular-momentum losses are likely to occur before the envelope ejection, and ultimately the surface rotational velocity of single AGB stars may well be several orders of magnitude lower than ∿20 km/s, the typical expansion velocity of PNe. Clearly, it is not easy to see how such small velocity asymmetry could give rise to highly asymmetric nebulae. A search for binarity among the nuclei of asymmetric nebulae would then be very interesting in this context.

It is worth noting that binarism will also affect much of what said in the previous sections, including the question of the transition time, and the nebular composition. Although binarism introduces more complexities into already complex problems, the possibly common existence of binary-born PNe should always be kept in mind when comparing observations with the evolutionary theory of *single* stars.

Acknowledgments. The author is indebted to Dr.s L.H. Aller and J.B. Kaler for having kindly provided preprints and for very stimulating discussions, to Dr. D. Schönberner for having indicated a mistake in an earlier version of Eq. (4), and, particularly, to Dr. Icko Iben Jr. in cooperation with whom many of these ideas have been conceived and refined.

REFERENCES

Aitken,D.K., Roche,P.F. 1982, M.N.R.A.S. 200, 217

Aller,L.H. 1981, (preprint)

Becker,S.A., Iben,I.Jr. 1979, Ap. J. 232, 831

Becker,S.A., Iben,I.Jr. 1980, Ap. J. 237, 111

Fusi Pecci,F., Renzini,A. 1976, Astron. Astrophys. 46, 447

Greggio,L., Renzini,A. 1982, Astron. Astrophys. (submitted)

Härm,R., Schwarzschild,M. 1975, Ap. J. 200, 324

Iben,I.Jr. 1981a, *Physical Processes in Red Giants*, ed. I. Iben Jr., A. Renzini (Reidel: Dordrecht), p. 3

Iben,I.Jr. 1981b, Ap. J. 246, 278

Iben,I.Jr. 1982, Ap. J. (in press)

Iben,I.Jr., Renzini,A. 1982a, Illinois Astrophysics Preprint IAP 82-2

Iben,I.Jr., Renzini,A. 1982b, Ap. J. Letters (in press)

Iben,I.Jr., Renzini,A. 1982c, Ap. J. Letters (in press)

Iben,I.Jr., Truran,J.W. 1978, Ap. J. 220, 980

Iben,I.Jr., Kaler,J.B., Truran,J.W., Renzini,A. 1982, Ap. J. (in press)

Kaler,J.B. 1982, (preprint)

Knapp,G.R., Phillips,T.G., Leighton,R.B., Lo,K.Y., Wannier,P.G., Wootten, H.A. 1982, Ap. J. 252, 616

Millikan,A.G. 1974, Astron. J. 79, 1259

Paczynski,B. 1971, Acta Astronomica, 21, 417

Peimbert,M. 1981, *Pysical Processes in Red Giants*, ed. I. Iben Jr., A. Renzini (Reidel: Dordrecht), p. 409

Reimers,D. 1975, Mém. Soc. Roy. Sci. Liège, 6e Ser. 8, 369

Renzini,A. 1979, *Stars and Star Systems*, ed. B.E. Westerlund (Reidel: Dordrecht), p. 155

Renzini,A. 1981a, *Physical Processes in Red Giants*, ed. I. Iben Jr., A. Renzini (Reidel:Dordrecht), p. 431

Renzini,A. 1981b, *Effects of Mass Loss on Stellar Evolution*, ed. C. Chiosi, R. Stalio (Reidel: Dordrecht), p. 319

Renzini,A. 1981c, *Phisical Processes in Red Giants*, ed. I. Iben Jr., A. Renzini (Reidel: Dordrecht), p. 165

Renzini,A. 1982, *Wolf-Rayet Stars*, ed. C. de Loore, A.J. Willis (Reidel: Dordrecht), p. 413.

Renzini,A., Voli,M. 1981, Astron. Astrophys. 94, 175

Schönberner,D., Weidemann,V. 1981, *Phisical Processes in Red Giants*, ed. I. Iben Jr., A. Renzini (Reidel: Dordrecht), p. 463
Schönberner,D. 1981, Astron. Astrophys. 103, 119
Shkowski,I.S. 1956, Astron. Zh. 33, 315
Sugimoto,D. 1971, Progr. Theor. Phys. 45, 761
Tuchman,Y., Sack,N., Barkat,Z. 1979, Ap. J. 234, 217
Tutukov,A.V., Yungelson,L.R. 1980, *Close Binary Stars*, ed. M.J. Plavec, D.M. Popper, R.K. Ulrich (Reidel: Dordrecht), p. 15
Wood,P.R. 1974, Ap. J. 190, 609
Wood,P.R., Cahn,J.H. 1977, Ap. J. 211, 499

KWOK: You raised an important point about the transition time, which has to be kept at a reasonably small value. As the transition time is critically controlled by the residual hydrogen envelope mass and this residual mass may vary, depending on the ejection process, would it not be better if the AGB phase is terminated by a wind process alone and not by a sudden ejection? In this way, the residual mass will always be the minimum possible.

RENZINI: The superwind is not necessarily a sudden ejection of the envelope on a dynamical time-scale, although it may consist of a large number of ejection events, as suggested by the models of Wood and Tuchman.

OSTERBROCK: Can you comment on how the transformation from star to PN differs (or otherwise) for binaries and for single stars?

RENZINI: In single stars, the superwind is responsible for the PN ejection, a process which is probably highly spherically symmetric. In binaries, one has Roche lobe overflow from the primary, then matter is stored in an accretion disk, whose size is limited by its Eddington luminosity. So matter must be lost from the system, and I doubt that this will take place in a spherically symmetric way.

REAY: The inner and outer envelopes of A 30 are quite different in form, although both are expanding at a similar velocity. Can you comment as to why this should be so?

RENZINI: From the velocity you have determined for the hydrogen-free knots, one can deduce that some 1500 y ago the nucleus of A 30 was a Red Giant shedding hydrogen-free and carbon-rich material. Most likely, it was a R CrB star, and such stars are known to occasionally eject puffs of matter in a non-spherically symmetric fashion, a process which is responsible for the characteristic deep luminosity minima of these stars.

KALER: How do the high mass stars found in the Magellanic Clouds fit in with the large exponent on M_f? Could you place an error bar on your exponent?

RENZINI: There is indeed an inconsistency between the derived luminosity and the very short times (a few hundred years or less) that these stars (with $M_f \approx 1M_\odot$) should spend at such high luminosities. The expression for the fading time is derived from the old Paczynski tracks, and, although it is of crucial importance to compute further grids of post-AGB sequences, I think that the fading times are not going to change by large factors. It would also be desirable to set an error bar on the derived luminosities and masses.

BEGELMAN: In principle, one could have aspherical mass ejection if, instead of spherical pulsations driving the superwind, one has aspherical modes with zones of upwelling and downdrafts. Do you consider this to be plausible?

RENZINI: Perhaps. I have, however, the feeling that the envelope of AGB stars is spherically symmetric to a high degree.

WANNIER: Why is rotation a less attractive theory for producing asymmetric nebulae?

RENZINI: Because one would expect a lot of angular momentum loss during the whole previous history of the star, in particular, during the wind phase on the AGB, when some 50 per cent of the envelope mass is slowly ejected. Add even a small magnetic field and you derive a formidable torque breaking the whole convective envelope.

ROXBURGH: The solar wind is clearly dominated by the solar magnetic field. Surely, we should expect magnetic fields to be generated by dynamo action even in those slowly rotating but very large Giants. Such magnetic fields could drive and control the mass loss and produce asymmetries.

RENZINI: Maybe, but binaries produce much larger asymmetries.

EVOLUTION OF UNSTABLE RED GIANT ENVELOPES

Y. Tuchman
The Racah Institute of Physics,
Hebrew University of Jerusalem
Jerusalem, Israel

Two main observable phenomena are directly connected with the dynamical behaviour of Red Giant (R.G.) envelopes: The variability of Mira stars and Planetary Nebulae (P.N.) ejection.

According to our dynamical research in R.G. envelopes it turns out that a repetitive shock ejection, which we propose as the mechanism for P.N. formation, is practically the final stage of the Mira phase. Therefore these two phenomena are closely related to each other and should be simultaneously discussed.

Since the characteristical evolutionary time of R.G. stars is much longer than any relevant time typical of their envelopes, the strategy beyond our dynamical calculations was the following: Static models of R.G. envelopes were integrated along their evolutionary track, using the well known luminosity - core mass relationship and assuming a perfect thermodynamical equilibrium.

The dynamical features of these envelopes have been examined by tracing the nonadiabatic thermodynamical variations which are excited in the envelope due to a small radial perturbation. Each dynamical calculation was continued until one of the following situations was reached:

1. The perturbation dies out and the envelope returns to its initial static position.

2. A steady pulsation is clearly established (see fig. 2).

3. A fast mass loss process begins. (see fig. 4).

It turns out that each R.G. passes along its evolution on the Asymptotic branch, all three possibilities in the order listed above. This fact is clearly demonstrated in the following figure: (fig. 1).

D. R. Flower (ed.), Planetary Nebulae, 281–289.

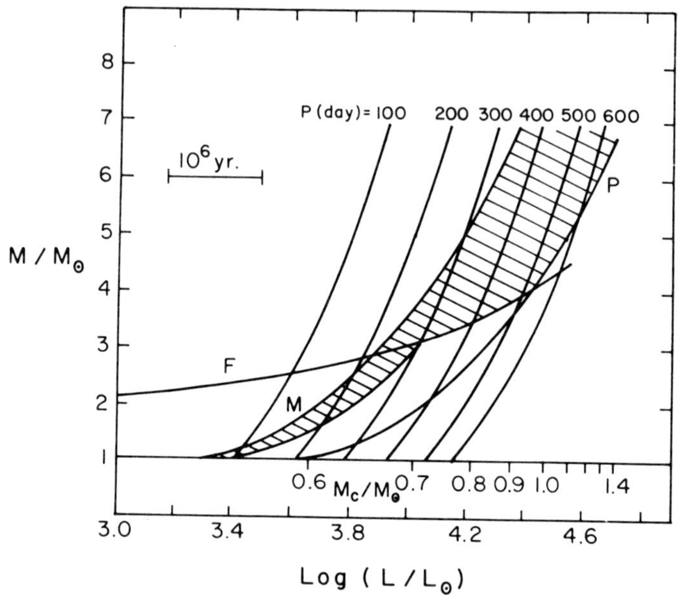

Fig. 1 - Given on the M-M_C (or log L) plane are: equiperiod lines labeled by their respective periods (in days). M,P and F lines (see text). Dashed area is the trucated Mira strip. The segment on the upper left corresponds to an evolutionary time interval of 10^6 years.

Envelopes located to the left of the "M line" are pulsationally stable or are oscillating with an amplitude too small and irregular to be identified as Mira stars. Most of these stars are oscillating in modes higher than the first.

Beyond the "M line" (Fig. 1), at higher luminosities, the envelope is oscillating steadily in the first overtone (Fig. 2). Their periods as well as many other observable features are, as we shall show later, in a good agreement with those observed in Mira variables.

Approaching the "P line" (Fig. 1) the fundamental mode begins to show up, and the envelope oscillates in a mixture of these two modes (Fig. 3). Crossing the "P line" the fundamental mode dominates the pulsation diverges and mass loss process is initiated (Fig. 4).

I will not go into the details of the ejection mechanism since it has been widely described more than two years ago (Tuchman et al 1979.). Briefly, the mass loss process is based upon repetitive shock ejections where the star is losing about 3% out of its prevailing envelope per ejection. The time interval between successive ejections, which is practically the envelope's thermal time scale, is about 30 years. The typical time for losing the entire envelope is therefore about 1000 years.

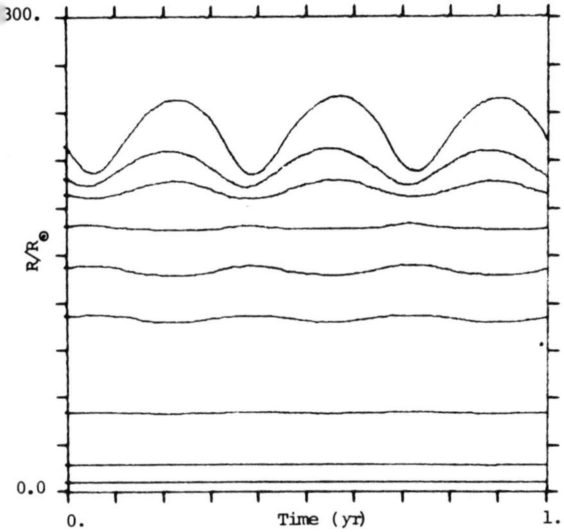

Fig. 3 (see below) Radius variation with time for different mass fractions of a mixed mode variable.

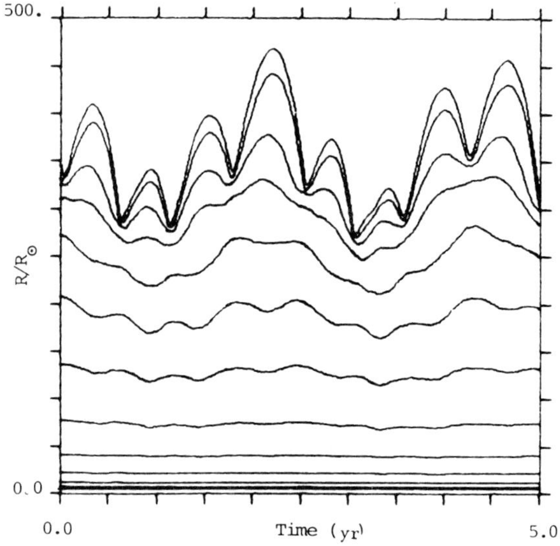

Fig. 2 - Radius variation with time for different mass fractions of a steady first overtone Mira.

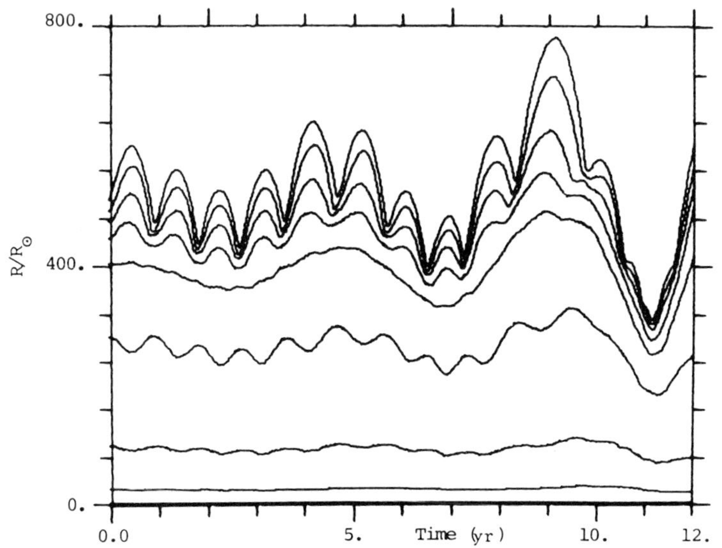

Fig. 4 - Radius variation with time for different mass fractions of a star at a stage of pulsational divergence.

The agreement of this dynamical picture with observation can and should be examined with respect to the following issues:

a) The Mira stage; where we should compare with observation not only the properties of differant variables as singles, but also some important statistical correlations which are observed among them.

b) The ejection process; where we should look for possible influence of the ejection mechanism upon the observed structure of the nebulae.

c) The remnant compact nuclei, mainly their expected and observed mass distributions.

What can be deduced from the specific shape of the Mira strip (Fig. 1)?

First, according to the given equiperiod lines, Mira periods should range from 100 to 600 days, where the surface luminosity and the total mass are, on the average, increasing functions of the period. This is precisely the observed situation (Eggen 1975).

Second, according to Fig. 1, stars oscillating with periods longer than 200 days should have total mass above $1.1M_{\odot}$. This fact might be an explanation for the absence of these stars, with periods higher than 200 days, in globular clusters (Feast 1972).

Another well known correlation, observed among the Mira variables, is concerned with their luminosity profile. In Fig. 5 the luminosity rise time divided by the total period ($\equiv f$) is shown as a function of the period for a sample of Miras presented by Campbell (1955).

According to our dynamical calculations a star with a given mass evolves within the Mira strip from a symmetric light curve ($f \approx 0.5$) to a light curve with a sharp increase towards the maximum and a moderate decline afterwards ($f \lesssim 0.3$). Since Mira stars having short periods naturally happen to exist at the beginning of the strip (Fig. 1) and those with long periods are found mostly close to the end of it (Fig. 1), one should expect precisely the observed correlation as is described in Fig. 5.

Let us pass to the major discrepancies between the presented theoretical picture and the observed data;

Using the time width of the Mira strip together with the theoretical death rate of main sequence stars (Chan and Wyatt 1976) one can predict the expected number distribution of Mira variables as a function of their periods. Such a calculated distribution is shown in Fig. 6.

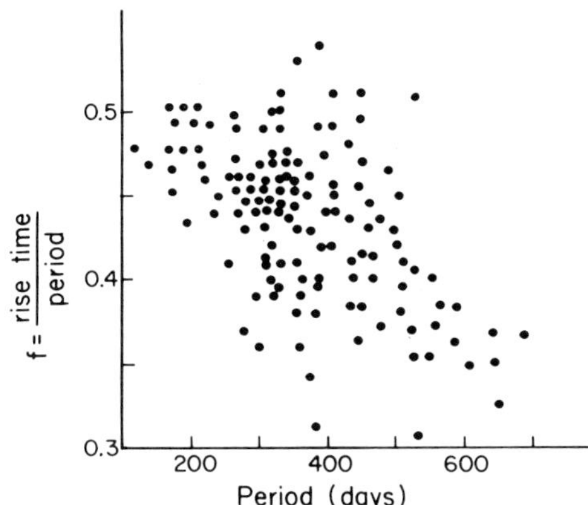

Fig. 5. - Luminosity rise time divided by the total period (f) as a function of period for a sample of Miras (Campbell (1955)).

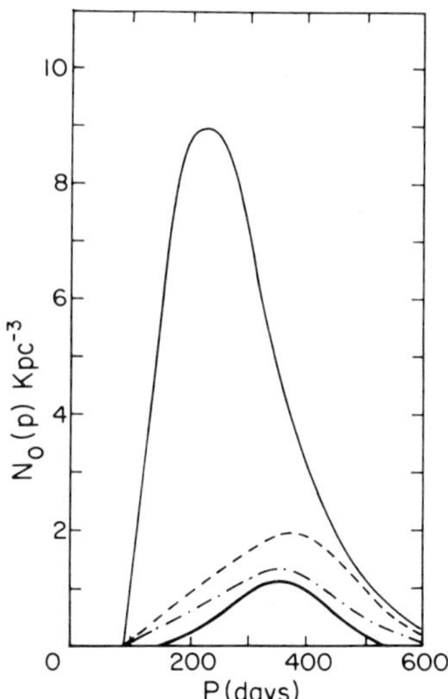

Fig. 6. - Number density N⊝ (Kpc^{-3}) of Miras as a function of period for no steady mass loss and no truncation (solid light line), truncation without steady mass loss (dashed line), truncation with steady mass loss (dot-dashed line) & observed curve (solid heavy line).

The observed distribution as has been prepared by Wood and Cahn (1977) is shown as well (Fig. 6). The discrepancy is obvious and significant First, in the general shape of these curves, where the observed distribution has its maximum near 350 days and has a sharp decrease both towards 100 and 600 days, while our calculated distribution has a peak near 200 days and quite a moderate decline towards the lower periods. Second, and more serious, is the clear disagreement in the total integrated number density, proportional to the area enclosed beneath these curves. According to Wood and Cahn (1977) the observed local number density of Mira variables is 245 per Kpc, while our predicted number ($\sim$1500 per Kpc) is about six times larger.

These two discrepancies can simultaneously be removed if, for some reason, the lifetime of low mass Miras is drastically diminished, which means a truncation in the lower right part of the Mira strip (Fig. 1).

A natural reason for this truncation is provided by the theoretical well known double shell flashes which are believed to occur in the relevant evolutionary stage. If, during such a flash, the surface luminosity exceeds the quiet value, then, there is a possibility for a R.G. to be removed from its position, some where in the Mira strip, to the other side of the "P line" initiating a mass loss process.

In order to carry out a qualitative analysis of this theoretical possibility, one should have the following needed information:

a) The location in the mass luminosity plane, where double shell flashes first occur. (will be denoted "F line")

b) The time dependent luminosity

profile during the flash mainly the height of the luminosity peak and the time interval during which the luminosity exceeds the critical value for mass ejection (The corresponding value at the "P line").

An approximated "F line" that has been compiled from various published results is shown in Fig. 1. Unfortunately, its location is quite uncertain, since it is sensitive to physical and numerical factors, which are far from being clear. In any case it is safe to assume that double shell helium flashes start before the Mira phase at least for stars less massive than $2M_\odot$.

The knowledge concerned with the second point has been for a long time a vague issue as the results found by various authors in literature showed huge discrepancies. Lately, an article devoted precisely to clarifying this point has been published by Wood and Zarro (1981). Using their results, which we believe to be the most reliable at the present time, the Mira strip is truncated in a way shown in Fig. 1. Note that this truncation, which turns out to be independent of the star's total mass, exists only across (to the right) of the "F line".

The number distribution of Mira variables, recalculated for the truncated Mira strip, seems to be much closer to the observed one (Fig. 6), and if the influence of Reimers - type steady mass loss is also added, which means that the star's mass is reduced while evolving through the Mira stage, then the agreement with the observed curve becomes even better.

Having some more confidence with the results concerned with the Mira phase, we may proceed to the next point; The ejection process.

As has been already mentioned earlier, the typical time scale for losing the entire envelope, according to our proposed ejection mechanism, is close to 1000 years, which means a mass loss rate of about $10^{-3} M_\odot/yr$. Using this value and the observed expansion velocities of the nebula (20-50 Km/sec), one can easily obtain the density and the nebular width as a function of the radial distance from the remnant nucleus (for a given mass of the nebula). As an example, for a nebula of $0.1M_\odot$ at a radius of 0.1 Pc the expected density is about 10^4 particles per cubic cm. and the nebular width is about a tenth of its radius. These values are very close to the observed ones.

According to the described dynamical situation there is a high probability for P.N. formation stimulated by helium shell flashes. An obvious question to be asked is whether there is any evidence for this fact in P.N. observational data.

Let us assume that the nebula ejection is indeed initiated by a luminosity peak associated with a helium shell flash. Since the time interval needed for this created nebula to be observed is much longer than the time width of the luminosity peak, there is no chance to observe luminosity variations, corresponding to this helium flash, in the remnant nuclei. On the other hand, since the time width of the luminosity peak is similar to or shorter than the estimated time interval for the nebula creation (∿1000 yr.), it is quite reasonable to assume that in some cases the mass ejection will cease before the entire envelope is lost. The rest of the envelope, in these cases, will probably be ejected only during the next flash. This might be the mechanism for double shell P.N. formation in which the spatial interval between the shells corresponds to the time interval between successive flashes.

The detailed calculations needed to find out whether a certain star will or will not create a multiple shell P.N. is far from being simple. This is mainly due to the fact that the critical value for mass ejection (at the "P line") changes simultaneously with the decrease of the total mass. Thus let me point out the main results:

a) Stars with total mass close to $3M_\odot$ are the expected candidates for double shell P.N. formation. Note that the interflash period in these stars is about 10,000 years (Paczynski (1975)).

b) The calculated statistical occurrence of multiple shell P.N. approach 15%.

These conclusions are compatible with observational evaluations. (Kaler (1974)).

Finally, let us pass to the last stage. The remnant nucleus.

Lately huge progress has been made in analysing the observational data concerned with the evolution of P.N. nuclei. This has been done mainly by Wiedeman and Schonberner (separately (1981), and together (1981)).

According to their investigations the mass distribution of P.N. nuclei is sharp and narrow, concentrated around $0.58M_\odot$.

Our predicted P.N. nuclei mass distribution, based upon the location of the corrected "P line" (Fig. 1) and the mass dependent death rate of Main Sequence stars shows an excellent agreement with observation (Fig. 7).

Just for comparison we show also the distribution one should get by assuming a continuous Reimers-type mass loss as the only mechanism for the nebula ejection (Fig. 7). (assuming θ=1).

This distribution which is identical with the distribution of masses for bright W.D. can be integrated to the past, using a time dependent stellar birthrate function, to get a distribution for all W.D. masses. This distribution turns out to correspond quite well with the observational estimates (Weideman 1980).

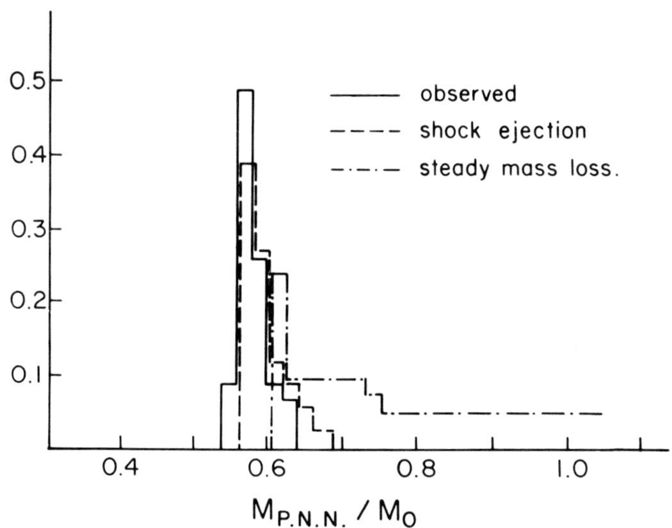

Fig. 7. - Mass distribution of Planetary Nebula nuclei ($M_{P.N.N.}$) calculated from our theoretical model (dashed line). The observed curve (Wiedeman and Schonberner (1981)) (full line). The distribution one should get assuming a continuous Reimers type mass loss is given as well (dot-dashed line).

References

Cahn, J.H., Wyatt, S.P.; 1976, Ap. J., 210, p. 508.

Campbell, L.: 1955, Studies of Long-Period Variables. The American Association of Variable Stars Observers, Cambridge, Mass.

Eggen, O.J.: 1975, Ap. J., 195, p. 661.

Feast, M.W.: 1972, in J.D. Fernie (ed.), Variable stars in Globular Clusters and Related Systems. D. Reidel, Holland.

Kaler, J.B.: 1974, Astron. Jour., 79, p. 59.

Paczynski, B.: 1975, Ap. J., 202, p. 558.

Schonberner, D.: 1981, Astron. Astrophys., 103, p. 119.

Schonberner, D. and Wiedeman, V.: 1981. Physical Processes in Red Giants, Proc. Erice Workshop. I. Iben, A. Renzin, Eds., D. Reidel, Dordrecht, p. 463.

Tuchman, Y., Sack, N., Barkat, Z.: 1979, Ap. J., 234, p. 217

Wiedeman, V.: 1981, IAU coll. No. 59 Effects of Mass loss on Stellar Evolution, C. Chiosi, R. Stalio, Ed. D. Reidel, Dordrecht.

Wood, P.R. and Cahn, J.H.: 1977, Ap. J. 211, p. 499.

Wood, P.R. and Zarro, D.M.: 1980, Ap. J. 247, p.247.

WANNIER: I was particularly intrigued by the approximate 30 y interval between mass ejections, especially in the light of the similar interval between different velocity components in several mass-loss objects. What is the fundamental cause of this extended periodicity?

TUCHMAN: Since each ejection is the result of a huge expansion during which the envelope loses a major part of its thermal energy, the time interval of about 30 y is needed to reconstitute the internal energy of the envelope and is close to the envelope's thermal time-scale.

RENZINI: While the qualitative behaviour of your models is of great interest for the understanding of the superwind, all the quantitative aspects should be viewed with caution. In particular, derived Mira period distributions depend crucially on the relation adopted between effective temperature, luminosity and mass of AGB models and on the treatment of superadiabatic, time dependent convection.

KALER: With the latest data, the mass distribution of planetaries is wider than the one which you presented. This issue is far from being settled.

DEDUCTION OF PLANETARY NEBULAE PROPERTIES FROM LONG PERIOD VARIABLE PRECURSORS

M.S. Bessell, P.R. Wood
Mt. Stromlo and Siding Spring Observatories, Canberra, Australia

Infra-red (JHK) photometry of long period variables (LPV) in the Magellanic Clouds has shown that the LPV's can be divided into core helium burning supergiants and asymptotic giant branch (AGB) stars. Application of the pulsation theory allows masses to be derived for the LPV's while stellar evolution theory allows core masses to be derived for the AGB stars. By considering evolution of the LPV's in the (M_{bol},P) diagram, estimates of planetary nebula mass and planetary nebula nucleus mass are derived as a function of initial mass. Spectra of the LPV's suggest that many low mass planetary nebulae in the Magellanic Clouds should be carbon rich while the more massive nebulae may be nitrogen enhanced.

SERRANO: Did you take into account steady mass loss on the AGB when calculating the evolution lines in your M_{bol} vs. P diagram? If not, the M_{core} vs. M_* relation is changed.

BESSELL: The masses shown are the present masses. If one estimates mass loss from the Reimers formula for $\theta = 1/3$, a present 3.5 $M_\odot$ Mira was a 4 $M_\odot$ Main Sequence star and a 2.0 $M_\odot$ Mira was a 2.2 $M_\odot$ Main Sequence star. For $\theta = 1$, the Main Sequence masses are 4.8 $M_\odot$ and 2.6 $M_\odot$, respectively.

RENZINI: The fact that AGB stars in the Galactic bulge are M-type, in contrast to stars of similar M_{bol} in the Clouds, which are carbon stars, does not necessarily imply that these stars do not dredge up carbon. Being about 30 times more metal-rich than their counterparts in the Clouds, they have to dredge up 30 times more carbon before showing as carbon stars.

BESSELL: That is true, however the dredge-up calculations of Wood indicate that high metallicity inhibits dredge-up at low luminosities, and, therefore, envelope ejection could take place before the star becomes sufficiently luminous for dredge-up to occur.

MASS-LOSS FROM LATE-TYPE STARS: NEW OBSERVATIONAL EVIDENCE

P.G. Wannier, R. Sahai
Owens Valley Radio Observatory, California Institute of Technology, USA

Rapid mass-loss is observed in many late-type stars, yet the mass-loss mechanisms operating are not well understood. A survey of molecular emission from circumstellar shells has been carried out using millimeterwave molecular lines and suggests that radiation pressure alone may be inadequate to explain the observed mass-loss, especially in the case of carbon-rich objects which may display rates in excess of 10^{-5} $M_{\odot}$/yr. Recent near-IR molecular line observations provide evidence for ejected material at several different velocities along the line-of-sight and may indicate the additional mass-loss mechanism at work. Resonantly scattered IR radiation spatially displaced from the central IR continuum source has now been observed for the first time and sheds new light on the IR absorption-line results, providing information about material within 10^{16} cm of the central star. These results are discussed along with recent high-resolution millimeterwave observations.

BEGELMAN: Your point concerning acceleration of the wind by radiation pressure is valid only if the optical depth across the acceleration zone is smaller than unity. If multiple scattering occurred in the acceleration zone, then the amount of momentum which could be extracted from the radiation field would be larger by a factor of about τ, up to a maximum of about c/v_{wind}.

WANNIER: That is correct, though such a high amplification seems unlikely owing to degradation to longer wavelengths upon multiple scattering.

WEHRSE: In our investigations of the CO 2.3 μm lines, we find that the main problem in the interpretation is the mode of line formation (true absorption or scattering) and not the temperature stratification.

WANNIER: In the extended envelopes of high mass-loss objects, the low rotational levels of CO are thermalized. The knowledge of T(r) is necessary in these cases.

BECK: What are the differences in velocity between the observed velocity components?

WANNIER: Typically, as small as 5 km s^{-1} (comparable to the infrared spectral resolution which is used) and up to the expansion velocity (several tens of km s^{-1}). On the other hand, I should point out our results for V Hya, which show significantly larger velocities - several times the expansion velocity.

EFFECTS OF STELLAR MASS LOSS ON THE FORMATION OF PLANETARY NEBULAE

Sun Kwok
Herzberg Institute of Astrophysics
National Research Council of Canada
Ottawa, Ontario, Canada K1A 0R6

I. INTRODUCTION

Over the last decade, stellar mass loss has become recognized as an important factor in the evolution of stars. The magnitude of the mass loss rate is found to be greatest amongst high luminosity stars at both the red and blue sides of the HR diagram. Planetary nebulae (PN), which result from an evolutionary phase during which the parent star traverses from the red side to the blue in a relatively short time scale, are likely to be affected by these processes. In this talk, I shall review the mass loss processes relevant to the PN phase and discuss their affects on the formation and evolution of PN.

II. RED GIANT MASS LOSS

On the basis of the luminosities of the central stars of PN Paczyński (1971) suggested that asymptotic giant branch (AGB) stars undergoing double-shell burning are the immediate progenitors of PN. These progenitors would have core masses between 0.6-1.4 $M_\odot$ and luminosities from 5×10^3 to 5×10^4 $L_\odot$. At about the same time, observers in the infrared and microwave spectral regions discovered that AGB stars are often surrounded by extensive circumstellar (CS) envelopes (Gehrz and Woolf 1971, Wilson and Barrett 1968, Solomon *et al.* 1970). Analyses of the molecular line (OH, CO etc) spectra of such CS envelopes soon reveal that they are produced by continuous stellar winds (Morris 1975; Elitzur, Goldreich and Scoville 1976; Kwok 1976).

The lower half of the luminosity range assigned to PN progenitors is occupied by Mira variables often characterized by 9.7 μm silicate emission (Merrill 1977) and OH/SiO maser emissions. Although mass loss rates can be derived from these observations (and to a lesser extent by optical observations of CS lines), the best estimates are probably those obtained from CO observations. The solution to the CO radiative transfer problem is well developed (Kwan and Hill 1977, Morris 1980, Kwan and Linke 1982) unlike that for the dust continuum case (see e.g. Jones and Merrill 1977) and the molecule distribution does not suffer

D. R. Flower (ed.), Planetary Nebulae, 293–303.

from the ionization structure difficulties of the optical case. The advantage of solving the excitation problem for thermal emission rather than maser emission is obvious. Modern millimeter-wave technology also allows the observation of both the CO $J=1\to0$ and $J=2\to1$ (as well as ^{13}CO) lines, which helps to further constrain the model parameters. The most comprehensive CO observations are by Knapp *et al.* (1982), who observed ∿12 Mira variables, finding mass loss rates ranging from 7×10^{-7} to 6×10^{-6} $M_{\odot}yr^{-1}$. We should note that these are lower limits to the mass loss rate for they are derived assuming all carbon atoms are locked in CO. Nevertheless, these rates are already much greater than the corresponding nuclear burning rates (6×10^{-8} $M_{\odot}yr^{-1}$ for a Mira with a core mass of 0.6 $M_{\odot}$) as well as the rates derived from the Reimers formula (Reimers 1975) which are commonly adopted in stellar evolution calculations (e.g. Renzini 1981).

Although no obvious optical counterparts exist in our Galaxy for PN progenitors in the upper half of the luminosity/mass range, a good case can be made for the many infrared objects discovered in the IRC and AFGL surveys (Neugebauer and Leighton 1969, Price and Walker 1976) as possible candidates. A large number of these IR stars have been found to be OH and CO sources depending on whether the underlying star is oxygen or carbon rich. The optically thick dust envelopes of these stars (evidenced by their low color temperatures and, in the case of oxygen rich objects, by the presence of silicate absorption features) indicate that the mass loss rates ($\dot{M}$) must be very high. Werner *et al.* (1980) estimate $\dot{M}$ to be between 5×10^{-6} - 7×10^{-5} $M_{\odot}yr^{-1}$ for OH/IR sources in the luminosity range of 2×10^{3} - 3×10^{4} $L_{\odot}$. Analysis of CO line emission find $\dot{M}$ up to 10^{-4} $M_{\odot}yr^{-1}$ for AGB stars with luminosities of several times 10^{4} $L_{\odot}$ (Knapp *et al.* 1982). The qualitative correlation between L and $\dot{M}$ suggests stars approaching the tip of the AGB have increasing mass loss rates, although the exact cause (suspected by some to be pulsation related: Wood 1979, Willson 1981) is not understood.

The wind velocities are also found to be higher in supergiants than in Mira variables (Dickinson *et al.* 1975, Cahn and Wyatt 1978). This seems to imply that radiation pressure on grains determines the dynamics of CS envelopes but doubts remain as to whether this mechanism can initiate mass loss (cf. Castor 1981).

Due to the uncertainties in our theoretical understanding of the mass loss mechanism, it is difficult to derive a mass loss formula from first principles at this time. As an alternative, I have attempted to evaluate the effects of mass loss on AGB stars using an empirical formula. The mass loss formula adopted ($\dot{M} = 10^{-13}[L/L_{\odot}]^{2}[M/M_{\odot}]^{-2}$ $M_{\odot}yr^{-1}$) is no more than a empirical representation of the molecular line results, covering the range of mass loss from $\lesssim10^{-6}$ $M_{\odot}yr^{-1}$ for low mass Miras to $\sim10^{-4}$ $M_{\odot}yr^{-1}$ for high-luminosity IR objects. Evolution of stars with masses 1,2,3,5 and 7 $M_{\odot}$ are calculated in Figure 1, starting at the beginning of double-shell burning. The core mass-luminosity relationship given by Iben (1981, eq. [3]) is used. With this mass loss formula, the upper main-sequence-mass limit for a white dwarf is

$\sim 7\ M_{\odot}$, consistent with the observed limit found by Romanishin and Angel (1980).

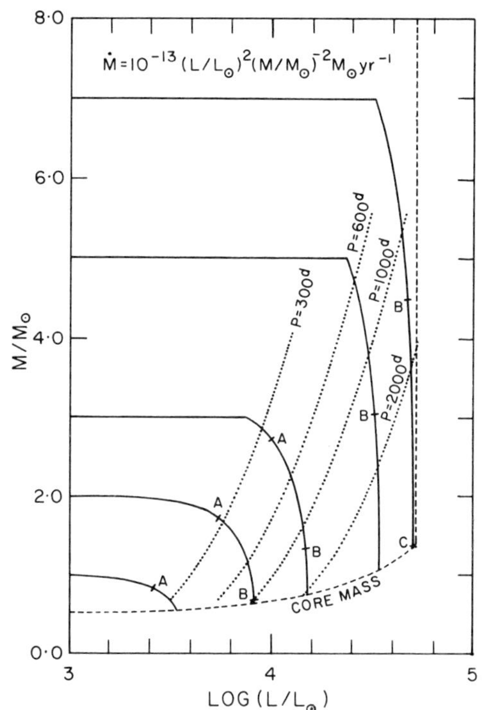

Fig. 1. Evolution curves for AGB stars with initial masses 1, 2, 3, 5 and 7 $M_{\odot}$. The dash line represents the core mass-luminosity relationship. Points A, B and C correspond to locations where $\dot{M}=10^{-6}$, 10^{-5}, 10^{-4} $M_{\odot}yr^{-1}$ respectively.

Constant period curves for fundamental pulsators are also plotted, using the period-mass-radius relation given in Willson (1982) and the luminosity-effective temperature relation given by Wood and Cahn (1977), eq. [6]). Figure 1 shows that the very long period ($P>1000^d$) variables discovered in OH (Herman and Harbing 1981) and in the infra-red (Engles, Schultz and Sherwood 1981) can be produced by high-mass (2-7 $M_{\odot}$) stars undergoing rapid mass loss. Similar plots for first overtone pulsators suggest that short period ($P<300^d$) variables are probably best explained by low-mass ($<3M_{\odot}$) stars pulsating in the first overtone.

Evolutionary curves similar to those presented in Figure 1 can also be obtained by using the Reimers formula with a large coefficient ($\eta=2.5$). However, in this case the maximum mass loss rate by any AGB star under 7 $M_{\odot}$ does not exceed 5×10^{-5} $M_{\odot}yr^{-1}$.

The importance of steady mass loss on the evolution of AGB stars cannot be overemphasized: e.g. according to the mass loss formula adopted in Figure 1, the entire envelope of a 7 $M_{\odot}$ star can be lost in less then 1 million years! *VLA* observations by Bowers, Johnston and Spencer (1981) find *minimum* sizes for the CS envelopes of 20 OH/IR stars to range from 10^3 to 10^4 A.U. OH phase-lag measurement by Jewell, Webber and Synder (1980) find the size of the envelope of IRC +10°011 to be 5×10^3 A.U. Knapp *et al.* (1982) also determine the size of the CO envelope of IRC +10°216 to be $>5\times10^4$ A.U. All these observations suggest that the CS envelopes cannot be the result of a short ($\lesssim 10^3$ yr) episode of sudden activity but rather are the result of continuous steady mass loss. It is entirely possible that steady stellar winds can completely remove the envelopes of AGB stars without invoking an instability strip of the kind proposed by Wood and Cahn (1977)

III. WINDS FROM CENTRAL STARS OF PLANETARY NEBULAE

The importance of winds from central stars of PN in preventing material backfill was recognized as early as 1966 by Mathews (1966). Recent *IUE* observations (reviewed in detail by Heap and by Perinotto in this conference) have shown that winds from central stars are much more common than previously thought. Heap (1982) has suggested that central stars with a luminosity to mass ratio >11,000 will have a wind. Combining this empirical criterion with the Paczyński core mass-luminosity relationship, we have a lower central-star mass limit of $\sim 0.65\ M_\odot$. Although the mass loss rates are still rather uncertain, the wind velocities can be accurately measured and have been found to range from 2000 to 8000 km s^{-1} (Heap 1982). Central stars with higher temperatures are also suspected to generate higher velocity winds. It is easy to show that such high velocity winds carry a significant amount of momentum and energy (compared to those observed in PN shells) and should have an important effect on the dynamical evolution of PN.

IV. INTERACTION OF STELLAR WINDS

The importance of red-giant mass loss in the formation of PN is obviously dependent on the transition time from red giant to PN. Using the minimum envelope masses of red giants on the AGB (10^{-3}-$10^{-2}\ M_\odot$, see Figure 2 in Paczyński 1971) and the nuclear burning rate (6×10^{-8}-$5\times10^{-7}\ M_\odot yr^{-1}$ for stars with core masses between 0.6 and 1.2 $M_\odot$, Paczyński 1971) it can be shown that the transition is relatively rapid, particularly for high mass stars. Renzini (1981) has also convincingly argued that this transition time cannot be longer than the expansion time of PN ($\sim 10^4$ yr) otherwise the nebulae will not be ionized before it disperses into the interstellar medium.

Given the short transition time scale, the extensive circumstellar envelope created by steady mass loss during the AGB should not be neglected in the treatment of the formation process of PN, regardless of the ejection mechanism. A suddenly ejected PN shell can easily sweep up a fraction of a solar mass of wind material over the lifetime of a PN and what we observe as PN can in fact consist mainly of wind material left over from the AGB phase. This possibility has been evaluated by Kwok, Purton and FitzGerald (1978) and by Kwok (1982) who explore the extreme case where there is no sudden ejection and the steady red-giant wind can persist until the exposure of the hot core. UV photons from the core will then exert pressure via resonance lines on the gas and a new fast wind initiated. This new wind will soon interact with the remnant red-giant CS envelope and, like a snow plow, creates a dense shell at the interface of the two winds.

Assuming that PN is made up exclusively of wind material then the mass of the shell (M_S) is no longer a constant over time. In fact,

$$M_s = \left(\frac{\dot{M}}{V} - \frac{\dot{m}}{v}\right) R_s(t) - (\dot{M}-\dot{m})\,t \tag{1}$$

Where $\dot{M}$ and $\dot{m}$ (V and v) are the mass loss rates (velocities) of the red-giant and PN-central-star winds respectively. When the transition time (τ) is taken into account, M_S is given by:

$$M_s(t) \simeq \left(\frac{\dot{M}}{V} - \frac{\dot{m}}{v}\right) R_s(t) - \dot{M}t + \dot{m}\left(t - \frac{V\tau}{v-V}\right) \tag{2}$$

where t=0 is when the red-giant wind stops and t=τ is the time when the central-star wind begins. In the approximation that the collision of the two winds is totally inelastic and all the excess energy is radiated away, R_S quickly approaches an equilibrium velocity V_S:

$$V_s = \frac{(\dot{M}-\dot{m}) + (v-V)(\dot{M}\dot{m}/vV)^{\frac{1}{2}}}{(\dot{M}/V-\dot{m}/v)} \tag{3}$$

Since the central-star wind carries a significant amount of mechanical energy, it is possible that not all of its energy can be radiated away, especially during the later stages of PN expansion when the density is low. A high temperature zone may develop due to shock heating and thermal gas pressure (p) may become an important term in the force equation. Assuming no radiative losses, we have

$$\frac{d}{dt}\left[M_s(t)\frac{dR_s(t)}{dt}\right] = \dot{m}\left(v - \frac{dR_s}{dt}\right) + \dot{M}\left(\frac{dR_s}{dt} - V\right) + 4\pi R_s^2(t)\ p(t) \tag{4}$$

The internal energy of the hot region can be written as:

$$E(t) = 3/2(4/3\pi R_s^3)p(t) \tag{5}$$

and the energy balance of the hot region is

$$\frac{dE(t)}{dt} = \tfrac{1}{2}\dot{m}v^2 - 4\pi R_s^2(t)p(t)\frac{dR_s}{dt} \tag{6}$$

Substituting (1) into (4), we can obtain similarity solutions to (4), (5), (6):

$$\begin{aligned} E &= at \\ R_s &= V_s t \\ p &= ct^{-2} \end{aligned} \tag{7}$$

where V_s is the root of the cubic equation:

$$\left(\frac{\dot{M}}{V} - \frac{\dot{m}}{v}\right)V_s^3 - 2(\dot{M}-\dot{m})V_s^2 + (\dot{M}V-\dot{m}v)V_s - \frac{\dot{m}v^2}{3} = 0$$

and

$$c = \frac{\frac{1}{2}\dot{m}v^2}{6\pi V_s^3}$$

$$a = 2\pi V_s^3 c \qquad (8)$$

Figure 2a shows V_s as a function of $\dot{m}$ for the momentum- and energy-conserving cases. We can see that the derived expansion velocities are comparable to observed values. Robinson, Reay and Atherton (1982) have found evidence for increasing values of V_s with R_s. This could be the result of increasing strengths of the central-star wind or the result of a change from momentum- to energy-conserving approximations.

If we accept the result that $dR_s/dt \doteq$ constant, (2) can be written as:

$$M_s \sim \dot{M}\left(\frac{1}{V} - \frac{1}{V_s}\right)R_s - \dot{M}\tau \qquad (9)$$

This can be compared with the empirical M_s-R_s relationship found by Maciel and Pottasch (1980)

$$M_s(M_\odot) = 1.225\ R_s(\text{pc}) - 0.0123 \quad (10)$$

Comparison of (10) and (9) shows that $\dot{M}\tau \sim 0.0123\ M_\odot$ and $\dot{M}(1/V - 1/V_s) \sim 1.2\times10^{-6}\,(M_\odot yr^{-1})/(km\ s^{-1})$, which can be satisfied by, e.g. $\dot{M} \sim 2\times10^{-5} M_\odot yr^{-1}$, $V \sim 10$ km s^{-1}, $V_s \sim 25$ km s^{-1} and $\tau \sim 600$ yr. Although the consistency of (10) with the interacting winds model may be coincidental, it remains that the existence of

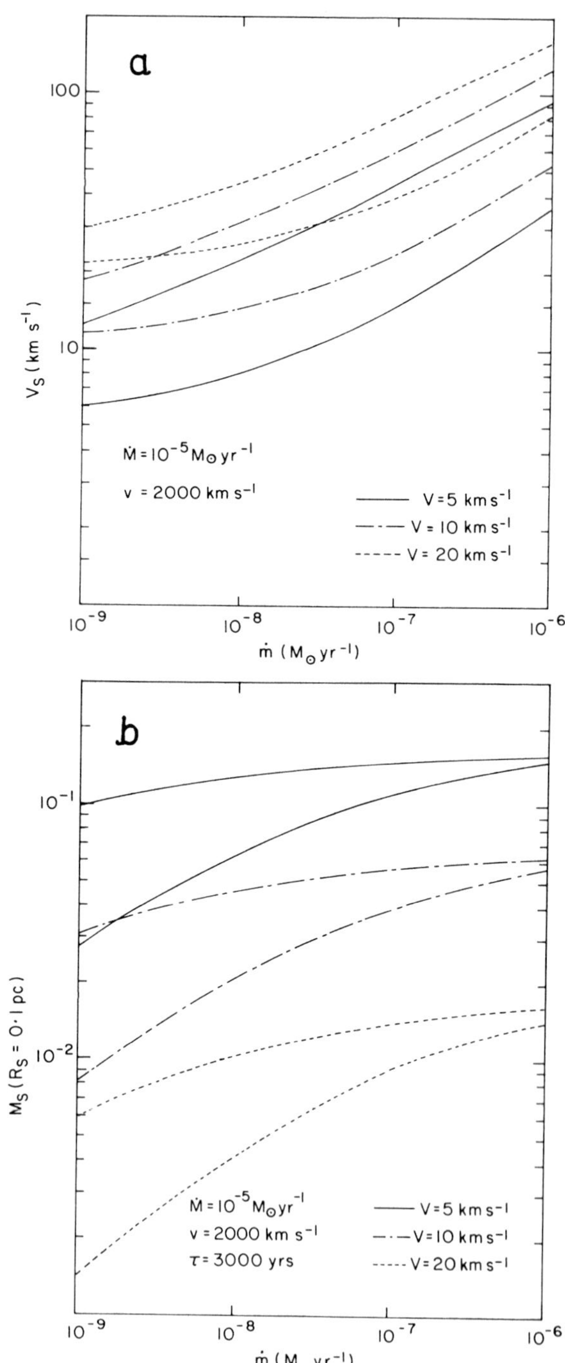

Fig. 2. V_s and M_s as functions of $\dot{m}$. The upper and lower curves of each pair correspond to the energy- and momentum-conserving cases respectively.

such an M_S-R_S relation is difficult to understand in the conventional sudden ejection model.

Figure 2b shows M_S as a function of $\dot{m}$ at the time which $R_S = 0.1$ pc. τ is assumed to be 3000 yr. These curves generally shift upward with increasing $\dot{M}$ and decreasing τ as given by (9).

V. ARE PN FORMED BY SUDDEN EJECTIONS?

Conventional theories of PN formation all rely on sudden ejections due to unstable envelope relaxation oscillations and the problems associated with these models have been discussed in recent reviews by Roxburgh (1978) and Wood (1981). A recent model by Tuchman, Sack and Barket (1979, hereafter TSB) expands on the previous pulsational instability models (Smith and Rose 1972, Wood 1974) and suggests PN is the result of repetitive ejections with exponentially decaying amplitudes over $\sim 10^3$ years. As in Wood and Cahn (1977), TSB also consider PN ejection as an extension of Mira pulsation and a Mira period distribution curve is derived from the instability strip calculated in their model. The model also predicts a higher number of Miras than actually observed (Barket and Tuchman 1980), which can be attributed to the low luminosity limit that they calculate Mira pulsation would occur.

A common prediction of pulsational instability models is that finite amplitude pulsation in the fundamental mode is not possible and the subsequent relaxation oscillations will lead to PN ejection. The truncation of Mira period at $\sim 600^d$ in these models however, fails to account for the existence of long period ($P > 1000^d$) OH/IR sources in our Galaxy and supergiant variables in the SMC, which are almost certainly stable fundamental mode pulsators. IRC 10°216, which has a period of $\sim 650^d$ is also likely to be pulsating in the fundamental mode, yet its CO brightness distribution can be accurately fitted by a wind model to as far as 3' from the star (Knapp *et al.* 1982, Kwan and Linke 1982), suggesting a stable mass loss history of at least 14,000 yr. While an unstable pulsator in the TSB model also has a period of ~ 2 yr (Figure 12, TSB), the runaway ejections of ~ 50 yr separation are simply not observed in IRC 10°216. Even if such shell structures are smoothed by radiation pressure on grains and the resultant grain-gas interactions, the predicted ejection time scale of $\sim 10^3$ yr is too short to explain the extensive CS envelopes of most IR stars.

Furthermore, PN masses calculated by TSB ranges from 0.4 to 6 $M_\odot$, corresponding to progenitor stellar masses of 1-7 $M_\odot$ (see Figure 2 of Barket and Tuchman 1980). It is difficult to reconcile these predicted values to observed masses which lies in the range of 10^{-3}-0.3 $M_\odot$ (Pottasch 1980).

To summarize, although our intuition suggests an impulsive event as the cause of PN, there does not seem to exist adequate theoretical or observational evidence to support it. Envelope instabilities may still occur, but mass lost by a steady wind seems to dominate over any mass ejected over a short time interval. Any instability model however should at least adopt a proper outer boundary condition taking into account the existence of the CS envelope, as it is done in Wood (1979) and Willson and Hill (1979).

VI. PROTO-PLANETARY NEBULAE

The best candidate for a proto-PN (defined as a star in transit between AGB and PN) is probably GL 618 which has a central star of spectral type B surrounded by a CS envelope expanding at $\sim$20 km s^{-1} (Zuckerman 1978). A compact ($\sim$0.2") ionized region is found at the stellar position and the CS envelope is likely to be ionization bounded (Kwok and Feldman 1981). Since its galactic location is in a incompatible with a pre-main-sequence object, its high luminosity (3×10^4[D/2kpc]^2 $L_\odot$) suggest that it is a proto-PN with a central star of $\sim$1 $M_\odot$. With $\gtrsim$2 $M_\odot$ of wind material in the CS envelope, the main-sequence mass of GL 618 is probably >4 $M_\odot$. Shock-excited H_2 emission (as in NGC 7027) has been detected and this is likely to be due to the interaction of the nascent PN shell and the remant red-giant wind.

If GL 618 is indeed a proto-PN then its evolution could be relatively rapid (on a time scale of decades because of the high mass of the central star [Paczyński 1981]). This may offer us an unique opportunity to witness the birth of a PN and test many of our ideas on PN evolution.

VII. CONCLUSIONS

We suggest that AGB evolution is terminated not by a sudden ejection but by a steady wind over a period of >>10^4 yr. Since the precise mechanism responsible for the wind is unresolved, a detailed picture of the transition to PN is still lacking. What is certain, however, is that stellar winds from both red giants and PN-nuclei play significant roles in the formation and evolution of PN. Although a complete understanding of the morphology of PN may involve other complicating factors not described in this paper, it is expected that the interacting winds process will remain a basic component of the PN phenomenon.

REFERENCES

Barket, Z. and Tuchman, Y. 1980, *Astrophys. J.*, 237, 105.
Bowers, P.F., Johnston, K.J., and Spencer, J.H. 1981, *Nature*, 291, 382.
Cahn, J.H. and Wyatt, S.P. 1978, *Astrophys. J. (Letters)*, 234, L79.
Castor, J.I. 1981, in *Physical Processes in Red Giants*, ed. I. Iben and A. Renzini, (Reidel:Dordrecht), p. 285.

Dickinson, D.F., Kollberg, E. and Yngvesson, S. 1975, *Astrophys. J.*, 199, 131.
Elitzur, M., Goldreich, P. and Scoville, N.Z. 1976, *Astrophys. J.*, 205, 384.
Engels, D., Schultz, G.V. and Sherwood, W.A. 1981, in *Physical Processes in Red Giants*, ed. I. Iben and A. Renzini, (Reidel:Dordrecht), p.401.
Gehrz, R.D. and Woolf, N.J. 1971, *Astrophys. J.*, 165, 285.
Heap, S.R. 1982, in *Wolf-Rayet Stars:Observations, Physics and Evolution*, ed. C.W.H. deLoore and A. Willis, (Reidel:Dordrecht), p. 423.
Herman, J. and Harbing, H.J. 1981, in *Physical Processes in Red Giants*, ed. I. Iben and A. Renzini, (Reidel:Dordrecht), p. 383.
Iben, I. 1981, *Astrophys. J.*, 246, 278.
Jewell, P.R., Webber, J.C., and Synder, L.E. 1980, *Astrophys. J. Letters*, 242, L29.
Jones, T. and Merrill, K.M. 1976, *Astrophys. J.*, 209, 509.
Knapp, G.R., Philips, T.G., Leighton, R.B., Lo, K.Y., Wannier, P.G., and Wootten, H.A. 1982, *Astrophys. J.*, 252, 616.
Kwan, J. and Hill, F. 1977, *Astrophys. J.*, 215, 781.
Kwan, J. and Linke, R.A. 1982, *Astrophys. J.*, 254, 587.
Kwok, S. 1976, *J. Roy. Astron. Soc. Can.*, 70, 49.
Kwok, S. 1982, *Astrophys. J.*, 258, 280.
Kwok, S. and Feldman, P.A. 1981, *Astrophys. J.*, 247, L67.
Kwok, S., Purton, C.R., and FitzGerald, P.M. 1978, *Astrophys. J. (Letters)*, 219, L125.
Maciel, W.J. and Pottasch, S.R. 1980, *Astron. Astrophys.*, 88, 1.
Mathews, W.G. 1966, *Astrophys. J.*, 143, 173.
Merrill, K.M. 1977, in *Interactions of Variable Stars with their Environment*, ed. R. Kippenhahn, J. Rahe, W. Strohmeicr, Bamberg, p. 446.
Morris, M. 1975, *Astrophys. J.*, 197, 603.
Morris, M. 1980, *Astrophys. J.*, 236, 823.
Neugebauer, G. and Leighton, R.B. 1969, NASA SP-3047.
Paczyński, B. 1971, *Acta Astron.*, 21, 4.
Pottasch, S.R. 1980, *Astron. Astrophys.*, 89, 336.
Price, S.D. and Walker, R.G. 1976, AFGL-TR-76-0208.
Reimers, D. 1975, *Mem. Soc. Roy. Sci. Leige*, 6e ser. 8, 369.
Renzini, A. 1981, in *Physical Processes in Red Giants*, ed. I. Iben and A. Renzini, (Reidel:Dordrecht), p. 431.
Robinson, G.J., Reay, N.K. and Atherton, P.D. 1982, *Mon. Not. R. Astron. Soc.*, 199, 649.
Romanishin, W. and Angel, J.R.P. 1980, *Astrophys. J.*, 235, 992.
Roxburgh, I.W. in *Planetary Nebulae*, ed. Y. Terzian (Reidel:Dordrecht) p. 295.
Smith, R.L. and Rose, W.K. 1972, *Astrophys. J.*, 176, 395.
Solomon, P.M., Jefferts, K.B., Penzias, A.A., and Wilson, R.W. 1970, *Astrophys. J. (Letters)*, 163, L53.
Tuchman, Y., Sack, N., and Barket, Z. 1979, *Astrophys. J.*, 234, 217 (TSB).
Willson, L.A. 1981, in *Effect of Mass Loss on Stellar Evolution*, ed. C. Chiosi and R. Stalio, (Reidel:Dordrecht), p. 353.
Willson, L.A. 1982, in *Pulsating Stars*, ed. J.P. Cox, in press.

Willson, L.A. and Hill, S.J. 1979, *Astrophys. J.*, 228, 854.
Wilson, W.J. and Barrett, A.H. 1968, *Science*, 161, 778.
Werner, M.W., Beckwith, S., Gatley, I., Sellgrem, K., Berriman, G., and Whiting, D.L. 1980, *Astrophys. J.*, 239, 540.
Wood, P.R. 1974, *Astrophys. J.*, 190, 609.
Wood, P.R. 1979, *Astrophys. J.*, 227, 220.
Wood, P.R. 1981, in *Physical Processes in Red Giants*, ed. I. Iben and A. Renzini, (Reidel:Dordrecht), p. 205.
Wood, P.R. and Cahn, J.H. 1977, *Astrophys. J.*, 211, 499.
Zuckerman, B. 1978, in *Planetary Nebulae*, ed. Y. Terzian, (Reidel: Dordrecht), p. 305.

ROXBURGH: There is evidence from the rotation rates of solar-type stars for a sudden change in angular momentum and mass loss rates. I recently produced an explanation of this phenomenon which can be extended to Red Giants: the mass loss is controlled by dynamo-generated magnetic fields and, as the star slows down, the dynamo switches to a different order mode, allowing a sudden change in mass loss rate.

KWOK: I have so far avoided discussion of mass loss mechanisms. Such mechanisms have been suggested (e.g. Wood, 1979, Astrophys. J. 227, 220; Willson and Hill, 1979, Astrophys. J. 228, 854) that could lead to an increase in the mass loss rate and be responsible for the observed increase in wind strength near the top of the AGB. However, present observations do not support the idea that a sudden (10^3y), large mass loss is responsible for the formation of a PN.

BESSELL: The relative number of radio luminous OH/IR sources to Miras (1 : 60) suggests that the lifetime of a OH/IR source is about 10^4y. To dissipate the large envelope in this time, a very high mass loss rate is required. The long periods of many of these sources ($\simeq$ 1000 d) suggest that they pulsate in the fundamental model, which is unstable, leading to relaxation oscillations and shock ejection. The sharp long period edge to the luminosity-period relation for Miras in the Magellanic Clouds shows that something special happens at these periods. It is reasonable to connect these facts and draw the conclusion that envelope ejection in Miras does not occur over a long time at a high rate of mass loss, but that the switch in pulsation mode from first overtone to fundamental results in rapid mass loss and a short time as a OH/IR source before becoming a PN.

KWOK: Since the very long period variables are likely to originate from high mass ($> 2M_\odot$) stars, it is not surprising to find fewer of them in view of initial mass function and their rapid evolution. In any case, the dynamical time (R/v) of many IR stars greatly exceeds the ejection time scale ($\lesssim 10^3$y) predicted by sudden ejection models. I agree, however, that pulsational mode switching can lead to an increase in the mass loss rate during the later part of the AGB evolution, as suggested by the mass loss formula used in my Fig. 1.

TERZIAN: We have heard suggestions that the PN mass (ionized and neutral gas) can be very large, up to a few $M_{\odot}$. Does this contradict your model?

KWOK: Halos of PN can be explained, in the interacting winds model, as remnants of the Red Giant wind. As can be seen from my Fig. 1, the halo mass may be as high as several $M_{\odot}$.

FAST WINDS IN PLANETARY NEBULAE

F. D. Kahn
Department of Astronomy, The University, Manchester M13 9PL

ABSTRACT

A planetary nebula consists mainly of gas ejected slowly by a red giant. Its dynamics is dominated by the hot central star which is left behind later. In particular a fast wind from this star forms a bubble of hot gas which fills the inner part of the nebula and pushes the envelope into a shell. This shell remains only partly ionized for a considerable time. Its non-ionized part is subject to a Rayleigh-Taylor instability, and is expected to break up into fragments which remain behind in the HII part of the nebula.

I. INTRODUCTION

Good evidence exists that the central stars of planetary nebulae produce fast winds, having speeds of the order of 1000 km s^{-1}; these winds sweep into the gas which had previously been ejected, with much lower speed (10 km s^{-1}), by the red giant which was the progenitor of the nebula and its central star. The phenomena are discussed in some detail by Sun Kwok (1983) in a recent paper. Infra-red observations of circumstellar masers confirm that red giants expel a flow of gas with the right sort of speed (Sun Kwok, 1983). For an early paper on the effect of a fast wind on an HII region see Dyson (1978).

The dynamics of the interaction between the fast wind and the slow envolope is the subject of a thesis by Lazareff (1981). His general model is that the fast wind shocks close to the central star and forms a bubble of hot gas. Material from the slow envelope is swept up into a shell of gas around the bubble. Lazareff considers at some length whether the hot gas can cool appreciably by contact with the HII region which forms on the inner side of the shell. He finds that this effect is present, but not dominant. The energy of the hot gas is therefore largely expended in pushing outward the envelope of slow-moving gas. Lazareff considers various sets of parameters, and finds that by and large the sequence of events is always much the same.

D. R. Flower (ed.), Planetary Nebulae, 305–316.

As it happens the most sensible model is also the simplest to handle. It has two main phases; from time $-t_o$ to time zero the central star ejects a mass M_o of gas, with low terminal speed U. During this period the star is a red giant.

After time t = 0 the central star has evolved to a more compact structure, and has a high surface temperature (T ∿ 50 000 K). It also has a fast wind, which carries off a fraction of a per cent of the energy output. This mechanical luminosity is relatively small, but its importance lies in the fact that the wind soon shocks and turns into a hot gas (T ∿ 3 000 000 K) which is slow to cool.

In first approximation one finds that the shocked shell of gas moves out into the envelope at constant speed. A closer examination shows that the expansion of the HII part of the shell causes the neutral part of the shell to accelerate noticeably. An obvious consequence is that the neutral shell becomes subject to the Rayleigh-Taylor instability. The prediction is therefore that the nebula will lose its simple structure of nested shells, and that more and more pockets of non-ionized gas will be left behind in the HII region as the nebula evolves.

II. THE SIMPLE WIND-DRIVEN MODEL

The primary injection of gas from the central star comprises a mass M_o with terminal speed U, released during time $-t_o < t < 0$. It gives rise to a density distribution

$$\rho = \omega / r^2 \qquad \text{with} \qquad \omega = M_o / 4\pi U t_o \,. \tag{1}$$

From time t = 0 onwards a fast wind blows. The central star is now hot and has a luminosity L, the wind speed is V(>> U) and the rate of input of energy into the wind is ηL. Typical values are as follows:

$$L = 5 \times 10^{36} \text{ erg s}^{-1}, \quad \eta L = 2 \times 10^{34} \text{ erg s}^{-1}, \ M_o = 4 \times 10^{32} \text{ gm},$$
$$U = 10^6 \text{ cm s}^{-1}, \ V = 2 \times 10^8 \text{ cm s}^{-1}, \ t_o = 3 \times 10^{11} \text{ s}.$$

It is generally found, in the case of main-sequence O stars, that the wind energy output is of order V/c times the luminosity (Cassinelli, 1979). This has been taken to apply here also, and explains the choice of the value for η.

The fast wind drives a shock into the primary gas, and sweeps it into a shell. In this simple treatment the shell is taken to be thin; let r be its radius at time t. Another shock facing inwards sits in the fast wind close to the star. The overwhelming bulk of the material in the fast wind is shocked and hot, and expands at a very subsonic speed. The equations governing the motion of the system are:

Energy in the shocked fast wind gas

$$\frac{d}{dt}(2\pi P r^3) = \eta L - 4\pi P r^2 \dot{r} . \qquad (2)$$

Here P is the pressure, and a balance is struck between the energy input from the stellar wind and the work done by the pressure in expanding the shell. There is no allowance made for radiative or conductive losses from the hot gas. This approximation is justified later.

Mass of the swept up shell:

$$\frac{dM}{dt} = 4\pi\omega(\dot{r} - U) = \frac{M_o}{Ut_o}(\dot{r} - U) \qquad (3)$$

Pressure balance at the outer shock:

$$P = \frac{M\ddot{r}}{4\pi r^2} + \frac{\omega}{r^2}(\dot{r} - U)^2 \qquad (4)$$

The two terms on the right hand side come from the acceleration of the gas in the shell and from the transfer of momentum to the newly swept up gas.

There is a simple solution to these equations with

$$r = \lambda U t \quad \text{and} \quad \dot{r} = \lambda U, \qquad (5)$$

$$P = \varpi / t^2 \quad \text{where} \quad \varpi = \frac{(\lambda - 1)^2 M_o}{4\pi\lambda^2 U t_o}, \qquad (6)$$

and

$$M = (\lambda - 1) M_o t / t_o . \qquad (7)$$

The parameter λ satisfies the equation

$$\lambda(\lambda - 1)^2 = \frac{2\eta L t_o}{3 M_o U^2} \equiv N, \text{ say}, \qquad (8)$$

whose solution is tabulated below:

N	0	2	5	10	20	40
λ	1	2	2.44	2.87	3.42	4.12

With the typical values quoted before, N = 10. In this calculation the parameter λ is set equal to 3 , in reasonable approximation. The shocked shell therefore moves outwards at 30 km s^{-1}.

Now to check the various assumptions.

i) The backward facing shock sits in the fast wind close to the star at radius r_c. To produce the correct post shock pressure, which is given by (6), requires that

$$\frac{3\dot{\mathcal{M}}V}{16\pi r_c^2} = \left(\frac{\lambda-1}{\lambda}\right)^2 \frac{M_o}{4\pi U t_o t^2}, \tag{9}$$

where $\dot{\mathcal{M}} \equiv 2\eta L/V^2$ is the rate of injection of mass into the fast wind. There is a small error (of about 6 per cent) in this equation because there is no allowance made for the fact that the newly shocked gas, at r_c, has a finite speed. It follows from (9) that

$$r_c = \left(\frac{3}{2}\right)^{1/2} \frac{\lambda}{\lambda-1} \left(\frac{\eta L U t_o}{M_o V}\right)^{1/2} t, \tag{10}$$

and that the ratio of the radius of the inner shock to that of the shell is

$$\frac{r_c}{r} = \frac{1}{(\lambda-1)}\left(\frac{3\eta L t_o}{2M_o U V}\right)^{1/2} = \frac{3N^{1/2}}{2(\lambda-1)}\left(\frac{U}{V}\right)^{1/2} = \frac{3}{2}\left(\frac{\lambda U}{V}\right)^{1/2} \tag{11}$$

and is always small; with the present assumed physical values it equals 0.18. The hot shocked wind gas therefore occupies all but 0.6 per cent of the volume enclosed by the swept-up shell.

ii) The density of the hot shocked gas is

$$\rho_h = \frac{3\dot{\mathcal{M}}t}{4\pi\lambda^3 U^3 t^3} = \frac{3\eta L}{2\pi\lambda^3 U^3 V^2 t^2} \tag{12}$$

and therefore the temperature is

$$T_h = \frac{\bar{m}P}{k\rho_h} = \frac{\bar{m}\lambda(\lambda-1)^2 M_o U^2 V^2}{6k\eta L t_o} \equiv \frac{\bar{m}V^2}{9kN}. \tag{13}$$

With the present numbers, and with $\bar{m} = 10^{-24}$ gm for a fully ionized gas

$T_h = 3.2 \times 10^6$K.

The cooling of a gas at a temperature in the million degree range can be simply described in terms of the adiabatic parameter κ ($\equiv P/\rho^{5/3}$) by the equation

$$\frac{d}{dt} \kappa^{3/2} = -q \tag{14}$$

(Kahn, 1976), with $q = 4 \times 10^{32}$ ($cm^6\ gm^{-1}\ s^{-4}$) for a gas with the usual cosmic composition. From relations (6) and (12)

$$\kappa^{3/2} \equiv \frac{P^{3/2}}{\rho^{5/2}} = \frac{\pi}{2^{1/2}\ 3^{5/2}} \frac{\lambda^{9/2}(\lambda-1)^3 M_o^{3/2} V^5 U^6 t^2}{(\eta L)^{5/2}\ t_o^{3/2}} . \tag{15}$$

The gas cannot cool effectively when $\kappa^{3/2}$ much exceeds qt, or when

$$t \gg \frac{2^{1/2}\ 3^{5/2} (\eta L)^{5/2}\ t_o^{3/2}\ q}{\lambda^{9/2} (\lambda-1)^3 M_o^{3/2} V^5 U^6} = \frac{81}{2} \frac{\eta L q}{\lambda^3 V^5 U^3} , \tag{16}$$

that is t much exceeds 3.75×10^7s, in the present case, or about one year. In our model it has been assumed that there is a sudden change from a slow wind to a fast wind at time $t = 0$. But it seems unrealistic to think that the switch-over occurs in a time less than a year. Radiative cooling by the hot gas is therefore never important, in practice.

iii) There is heat loss at the "evaporation front" between the hot shocked wind and the HII part of the compressed shell. The effect has been studied by Lazareff (1981) who treated the problem in terms of particle-particle interactions, and did not consider collective plasma effects, other than the need to maintain space-charge neutrality. His conclusion was that there are noticeable energy losses, but that they have no decisive influence on the evolution of the planetary nebula.

It seems probable, though, that mirror and/or firehose instabilities will occur in the hot shocked wind. A large conduction flux in this gas sets up an anisotropic velocity distribution. The local magnetic field is likely to be weak, and so the instabilities occur very easily. Individual charged particles are then scattered by the inhomogeneities that are set up, and the conductive flux is reduced. This effect strengthens Lazareff's conclusion that heat losses at the evaporation front are not significant.

III. THE HII PART OF THE SHELL

So far the compressed shell has been regarded as being thin, but nevertheless it has important structural features. Its inner part will be ionized by the Ly-c flux from the central star. For a stellar temperature in the likely range, the rate of production of Ly-c photons is $j \equiv 10^{10}$ photons per erg emitted. The gas density in the HII shell is

$$\rho_i = \frac{\varpi}{c_i^2 t^2} = \frac{(\lambda-1)^2}{\lambda^2} \frac{M_o}{4\pi U c_i^2 t_o t^2}, \tag{17}$$

where c_i is the isothermal sound speed, say 10 km s^{-1}. If the shell is ionization limited, then the mass of ionized gas M_i is given by

$$jL = \frac{b M_i \rho_i}{m_a^2}; \tag{18}$$

here m_a is the average atom or ion mass. Thus

$$M_i = \frac{4\pi\lambda^2}{(\lambda-1)^2} \frac{jL m_a^2 U c_i^2 t_o t^2}{b M_o}. \tag{19}$$

The fraction of the mass in the shell which is ionized at time t is

$$\xi = \frac{M_i}{M} = \frac{4\pi\lambda^2}{(\lambda-1)^3} \frac{jL m_a^2 U c_i^2 t_o^2 t}{b M_o^2}, \tag{20}$$

The shell has swept up all the primary injection gas at time $t = t_o/(\lambda - 1)$, when equation (20) gives formally

$$\xi = \xi_* = \frac{4\pi\lambda^2}{(\lambda-1)^4} \frac{jL m_a^2 U c_i^2 t_o^3}{b M_o^2}. \tag{21}$$

With the values being used here, $\xi_* = 1.19$. Of course ξ cannot exceed unity: the interpretation has to be that the shell is completely ionized when

$$t = t_i = \frac{t_o}{(\lambda-1)\xi_*} = \frac{0.84 t_o}{\lambda - 1}. \tag{22}$$

Up to that time the primary injection gas beyond the shell would be neutral, but soon after it will be overtaken by an R-type ionization front. For other parameters of the system it is of course possible that the shell sweeps up all the gas before it becomes fully ionized itself.

The neutral gas in the shocked shell will be cool: Lazareff thinks that a temperature of 100K is rather on the high side. In any case this gas will be highly compressed and confined to a thin layer. But the finite thickness of the ionized part of the shell can have important dynamical effects; at time t it equals

$$\Delta = \frac{M_i}{4\pi\lambda^2 U^2 t^2 \rho_i} = \frac{4\pi\lambda^2}{(\lambda-1)^4} \frac{jLm_a^2 c_i^2 t_0^2 t^2}{bM_0^2} . \tag{23}$$

The ratio of the shell thickness to the shell radius is

$$\frac{\Delta}{\lambda Ut} = \frac{4\pi\lambda}{(\lambda-1)^4} \frac{jLm_a^2 c_i^4 t_0^2 t}{bM_0^2 U} = \frac{\xi c_i^2}{\lambda(\lambda-1)U^2} . \tag{24}$$

Here c_i^2/U^2 is about unity, $\lambda(\lambda - 1) = 6$, and so Δ is always small compared with λUt during the phase in which the shell is partially ionized ($\xi < 1$). Nevertheless the gradual thickening of the shell has interesting dynamical consequences.

IV. DYNAMICAL EFFECTS OF SHELL THICKENING

When the HII part of the shell expands it restricts the volume available for the hot shocked stellar wind, and therefore raises the pressure. As a consequence the non-ionized part of the shell accelerates outwards. This effect can be treated as a perturbation of the flow pattern described in Section II. The linearized treatment is strictly valid for early stages of the motion, when t is small. At that time only a small fraction of the mass of the shell is ionized. This has the advantage that the pressure difference across the HII part of the shell can be ignored. Further, ionized gas enters the HII shell via the ionization front, with speed v_i relative to the neutral shell. The speed v_i is important in connection with the stability at this front, but it can be ignored in calculating the momentum flux carried by the newly ionized gas.

Then let the neutral shell be at radial distance $r = \lambda Ut + R_a$, and the hot shocked wind/HII interface at $r = \lambda Ut + R_b$, and let the pressure be $P = \varpi/t^2 + \Pi$. From relation (23)

$$\Delta = R_a - R_b = \frac{4\pi\lambda^2}{(\lambda-1)^4} \frac{jLm_a^2 c_i^4 t_0^2 t^2}{bM_0^2} , \tag{25}$$

The energy equation for the hot shocked wind gas is now

$$\frac{d}{dt}\left[2\pi(\lambda Ut+R_b)^3\left(\frac{\varpi}{t^2}+\Pi\right)\right]=\eta L-4\pi(\lambda Ut+R_b)^2(\lambda U+\dot{R}_b)\left(\frac{\varpi}{t^2}+\Pi\right) \qquad (26)$$

Pulling the first order part out of equation (26) gives, after some simplification, that

$$\lambda Ut^3\dot{\Pi}+5\lambda Ut^2\Pi+5\varpi\dot{R}_b+4\varpi R_b/t=0. \qquad (27)$$

It is clear from relation (25) that R_a and R_b should vary like t^2; relation (27) then shows that Π varies like t^{-1}, and therefore

$$R_b=X_bt^2,\quad \Pi=-\frac{7\varpi X_b}{2\lambda Ut}=-\frac{7(\lambda-1)^2M_oX_b}{8\pi\lambda^3U^2t_ot}. \qquad (28)$$

The momentum equation at the neutral shell is

$$\frac{\varpi}{t^2}+\Pi=\frac{M_o(\lambda-1)t\ddot{R}_a}{4\pi\lambda^2U^2t^2t_o}+\frac{M_o}{4\pi Ut_o}\frac{[(\lambda-1)U+\dot{R}_a]^2}{(\lambda Ut+R_a)^2} \qquad (29)$$

provided that the bulk of shocked gas is still non-ionized. The first order part of equation (29) gives that

$$\Pi=\frac{M_oX_a}{2\pi U^2t_ot}\frac{(2\lambda+1)(\lambda-1)}{\lambda^3} \qquad (30)$$

The comparison between relations (28) and (30) shows that

$$X_b=-\frac{4(2\lambda+1)}{7(\lambda-1)}X_a \qquad (31)$$

and so, from equation (25)

$$X_a=\frac{28\pi\lambda^2jLm_a^2c_i^4t_o^2}{3(\lambda-1)^3(5\lambda-1)bM_o^2}, \qquad (32)$$

The expansion of the HII shell produces an acceleration $2X_a$ of the neutral shell; it is independent of the time t, and can be expressed in terms of the time t_i, when the shell is fully ionized, by

$$f_a = 2X_a = \frac{14\, c_i^2}{3(5\lambda - 1)\, U t_i} \tag{33}$$

With the values adopted in this paper, $t_i = 1.3 \times 10^{11}$s and the acceleration equals 2.6×10^{-6}cm s^{-2}.

V. STABILITY OF THE NEUTRAL SHELL

The gas in the neutral part of the shell is much denser than the HII gas which is accelerating it outwards. Taking c_o (= 1 km s^{-1}) to be the sound speed in the neutral gas, one gets that the thickness of the shell is

$$\delta = \frac{M c_o^2}{4\pi \lambda^2 U^2 \varpi} = \frac{c_o^2 t}{(\lambda - 1) U} \tag{34}$$

The maximum growth rate for the Rayleigh-Taylor instability is

$$\sigma_g \sim \sqrt{f_a / \delta} = \left[\frac{14(\lambda - 1)}{3(5\lambda - 1)}\right]^{1/2} \frac{c_i}{c_o} (t_i t)^{-1/2} \tag{35}$$

The growth is opposed, to a certain extent, by a damping effect associated with the ionization front. Imagine a corrugated front in which a particular portion projects a distance z beyond the mean surface. If n_i is the ion density in the HII region then this portion is illuminated by an additional number $b n_i^2 z$ of Ly-c photons, per unit area and unit time. This extra flux tends to remove the corrugation because the HI gas, with atom density n_o, must supply an extra number of ions; if this effect were present alone, i.e., no Rayleigh-Taylor instability, then we should have

$$n_o\, dz/dt = -b n_i^2 z . \tag{36}$$

The particle densities n_o and n_i are related by the condition of pressure equilibrium which demands that

$$n_i / n_o = c_o^2 / c_i^2$$

and so the damping rate due to this effect is

$$\sigma_d = b n_i c_o^2 / c_i^2 . \tag{37}$$

Since $n_i \equiv \rho_i/m_a$ it now follows from relation (17) that

$$\sigma_d = \frac{(\lambda-1)^2 \; b M_o c_o^2}{4\pi \lambda^2 \; m_a U c_i^4 t_o t^2} \tag{38}$$

or in terms of t_i

$$\sigma_d = \frac{j L m_a c_o^2 t_i t_o}{(\lambda-1) M_o c_i^2 t^2} . \tag{39}$$

Clearly the Rayleigh-Taylor instability and the damping process work in opposite directions. The net effect is that the instability will grow when σ_g exceeds σ_d, that is when

$$\xi \equiv \frac{t}{t_i} > \left[\frac{3(5\lambda-1)}{14(\lambda-1)^3}\right]^{1/3} \left(\frac{jLt_o}{\mathcal{N}_o}\right)^{2/3} \frac{c_o^2}{c_i^2} . \tag{40}$$

Here $\mathcal{N}_o \equiv M_o/m_a$ is the number of atoms + ions in the nebula, typically 2×10^{58}. The first factor on the right hand side in (40) is of order unity, unless $\lambda - 1$ is very small which would imply that the stellar wind is very weak. With our chosen values the condition is $\xi > 0.20$.

There is therefore a substantial period in the evolution of a planetary nebula during which the shocked shell is only partially ionized, and the neutral part of the shell is subject to the Rayleigh-Taylor instability. The fastest growing disturbances have a length scale comparable with the thickness of the shell; in the present case this is of the order of

$$\delta = 10^{15} \text{ cm}.$$

At a typical stage the mass per unit area in the neutral shell is some 3×10^{-4} gm cm^{-2}: the individual neutral fragments therefore have a rather low mass, typically 3×10^{26} gm.

VI. CONCLUSION

A planetary nebula originates in the slow ejection of gas from a red giant. It becomes ionized when the central star has shrunk to a small hot object. If, as seems likely, this star also produces a fast wind, then the structure of the nebula is dominated by the bubble of hot gas that forms around the star. The HII part of the nebula will at first be confined to the inner part of the compressed shell that surrounds the hot bubble. At later stages there are two possibilities: if the shell becomes fully ionized before it has swept up all the gas in the envelope, in this case the remaining gas in the nebula becomes fully ionized soon afterwards. Alternatively the whole envelope is swept up before the shell is fully ionized. This simple picture ignores a Rayleigh-Taylor instability which can break up the non-ionized part of the swept-up shell. The instability sets in at a relatively early stage, perhaps after one-fifth of the time needed to sweep up the whole envelope. It results in the formation of blobs of non-ionized gas which tend to be left behind in the HII region.

There is also the possibility for more violent instabilities to occur at a later stage. Consider a case where the shocked shell reaches the boundary of the envelope while it is still only partly ionized. It now suffers a considerable acceleration because there is no longer any drag from the newly swept up gas. There has not been time to analyse this effect in the present talk, but clearly it must be significant in nebulae which contain a large mass of gas.

ACKNOWLEDGEMENTS

The author thanks Dr. J. E. Dyson for his many helpful comments on this work.

REFERENCES

Cassinelli, J. P.: 1979, Ann. Rev. Astron. Astrophys., *17*, 275.
Dyson, J. E.: 1978, Astron. Astrophys., *62*, 269.
Habing, H. J.: 1983, contribution to this Symposium.
Kahn, F. D.: 1976, Astron. Astrophys., *50*, 145.
Kwok, Sun: 1983, Paper given at this Symposium.
Lazareff, B.: 1981, "Role de la phase coronale dans la dynamique du milieu interstellaire", Thesis, Universite Paris-Sud.

BEGELMAN: What are the prospects for observing soft X-rays from the shocked wind?

KAHN: The hot shocked wind cools quite slowly, mainly by excitation of spectral lines of various impurities in the far ultraviolet. Bremsstrahlung contributes only a small ($\approx$ 1%) fraction to this low cooling rate. So the prospects are not too good.

SEATON: Could one expect to observe a coronal-type spectrum from the hot bubble?

KAHN: For a PN of age 10^3 y, the hot bubble cools by radiation on a time scale of 10^6 y, mainly by electron excitation of various metal ions. For a wind luminosity of, say, 10 $L_\odot$ this gives a luminosity of 10^{-2} $L_\odot$ in the cooling lines. For different times, t, the cooling rate varies like $1/t^2$.

SEATON: What might one expect to be the observable differences in properties of nebulae with and without winds?

KAHN: When there is a fast wind, the nebular gas is partly swept into a shell. For young PN, only the inner part of the shell is ionized. Later the shell should fragment, leaving behind neutral globules, each surrounded by an ionized layer. An individual globule would evaporate in about 10^3 y.

NUSSBAUMER: We made a preliminary investigation of the possibility of observing hot ($T > 10^6$ K) gas in the young PN V1016 Cyg. The observed X-ray flux can be explained by the hot central star ($T_* = 160\,000$ K) alone. A search for forbidden Fe X and Fe XIV in the visual spectrum has been negative. Thus, at present, there are no observational indications of a very hot gas. However, the uncertainties involved in interpreting the "Einstein" X-ray data and the quality of the visual data are such that the existence of such hot gas cannot be ruled out.

KWOK: In the calculations that I did, conduction of the Lorentz gas is assumed to be the major cooling mechanism and the temperature of the hot zone is found to vary as $t^{-2/7}$. Do you obtain similar results?

KAHN: Following Lazareff, I did not include the dynamical effect of this cooling process.

HARRINGTON: Once globules have been produced inside the hot shocked region, they will be important in cooling the gas. This problem merits more careful attention.

KAHN: Yes, the combined areas of the interfaces between the ionized jackets of the globules and the hot gas could be quite large. One should repeat Lazareff's estimate of the heat loss to be expected at such contact surfaces.

NUMERICAL MODELS OF DYNAMICAL AND SPECTRAL EVOLUTION OF PLANETARY NEBULAE

V.A. Okorokov, B.M. Shustov, A.V. Tutukov
Astronomical Council of the USSR Academy of Sciences,
Moscow, USSR

H.W. Yorke
Universitäts-Sternwarte, Göttingen, W. Germany

Numerical models of planetary nebulae (PN) dynamical evolution are calculated under the assumption of spherical symmetry and discussed in the light of infrared, radio and optical observations. The set of hydrodynamical equations is solved simultaneously with equations for nongrey radiative transfer.

Continuous spectra of outgoing radiation are constructed in the range 10^7 - 10^{16} Hz. Some model parameters - mass of central star M_{nucl}, mass loss rate $\dot{M}_w$ and dust to gas density ratio ρ_d/ρ_g - are systematically varied. The mass of the envelope M_{env} is assumed to be 0.2 $M_\odot$. The formation of the envelope is interpreted as the phase of mass outflow with constant $\dot{M}_w$ followed by phase of high velocity wind (up to 2000 km/s). The evolution of the central star is considered as an inner boundary condition.

Density distribution and spectra for one of the model sets are shown in the Figure. Times, beginning from the onset and from the end (in parentheses) of outflow are indicated. Fluxes are normalised at a distance of 3 kpc.

Results of comparison of some models with observations of NGC 7027 and CRL 2688 show the usefulness of the models for the analysis of observable data.

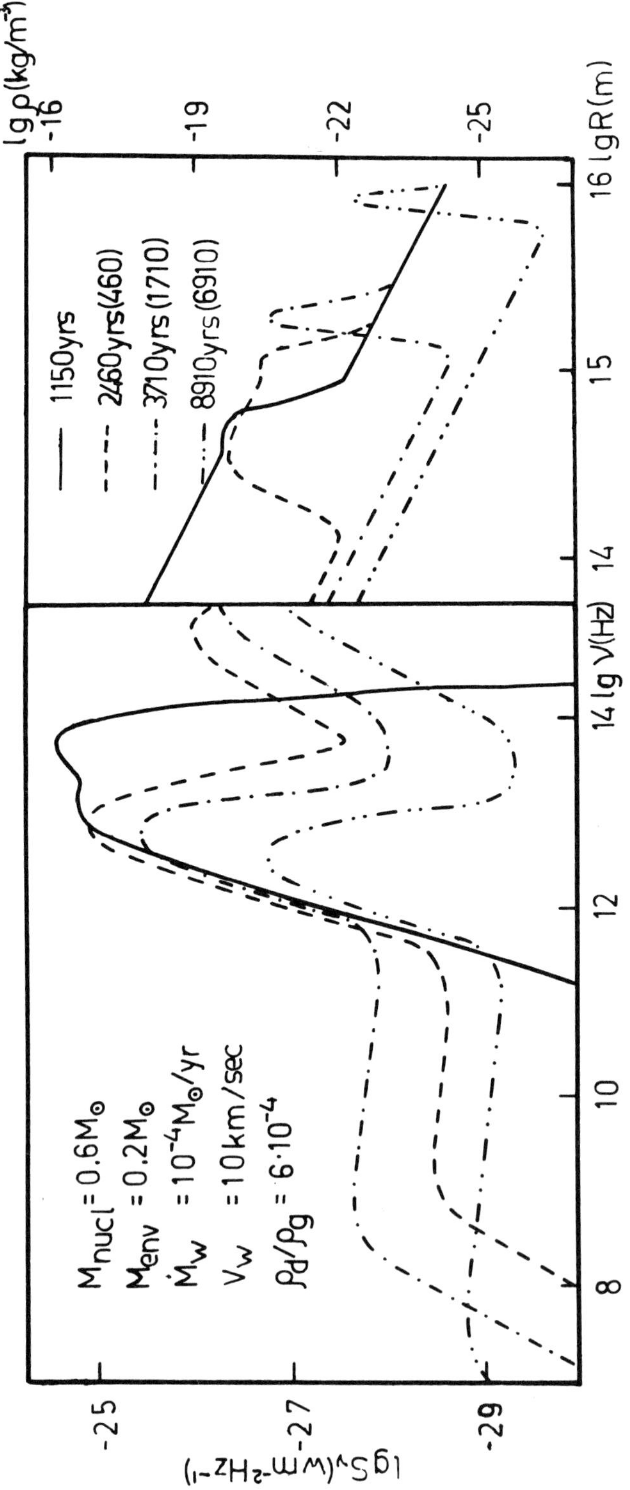

lg ρ(kg/m⁻³)
-16
-19
-22
-25
16 lgR(m)
15
14
1150yrs
2460yrs(460)
3710yrs(1710)
8910yrs(6910)
14 lg ν(Hz)
12
10
8
M_nucl = 0.6M⊙
M_env = 0.2M⊙
Ṁ_W = 10⁻⁴M⊙/yr
V_W = 10km/sec
ρd/ρg = 6·10⁻⁴
-25
-27
-29
lg S_ν(Wm⁻²Hz⁻¹)

NUMERICAL GAS-DYNAMIC INVESTIGATION OF THE WHIMPER MODEL FOR THE FORMATION OF PLANETARY NEBULAE

C.R. Purton
Dominion Radio Astrophysical Observatory, Penticton, B.C., Canada

The suggestion that the shells of planetary nebulae may be formed at the interface between two stellar winds of different velocity ("Not with a bang, but a whimper") (Kwok, Purton and FitzGerald, 1978, Astrophys. J. 219, L 125) is investigated using the 'beam scheme' (Sanders and Prendergast, 1974, Astrophys. J. 188, 489) adapted to a system with spherical symmetry. Initial conditions include the remnant wind from the red giant phase ($\dot{M} = 10^{-5}\ M_{\odot}\ y^{-1}$, $V = 10$ km s^{-1}) and a high-speed wind from the hot nucleus ($\dot{m} = 10^{-6}/10^{-7}\ M_{\odot}\ y^{-1}$ (two tests), $v = 10^3$km s^{-1}). The collision began at a radial distance of 200 AU. A third test at 50 AU indicated that the end result was insensitive to the details of the transition from one mass-loss mechanism to the other.

The shell formed at the interface increased in radius and mass: after several thousand years it had the characteristics of a conventional idealised planetary nebula. Calculated expansion velocities were in the range observed, and increased slowly with time. Temperatures of 2.5 x 10^7 K were found for the gas inside the shell, producing weak but observable (at 1 kpc) X-ray emission. λ 4686 emission produced by collisional ionization was found to be as strong as H_β emission at the later epochs. The radio spectrum at T = 100 y resembled Hb 12, at T = 500 y a conventional 'optically thick' planetary, and after a few thousand years became optically thin.

Stellar mass loss, both components, clearly have a profound effect on the dynamics of a planetary nebula. It may not be necessary to invoke any sudden ejection for the formation of a planetary nebula; it may be the natural consequence of ordinary mass-loss processes in the two stages of the star's evolution.

ISAACMAN: The model predicts the presence of 10^4 K gas with a density of about 10^3 cm^{-3} outside the shell. This seems inconsistent with observations of H_2 in several PN, as H_2 is presumably shock-excited by the expanding shell but dissociates at a few thousand K.

PURTON: The model assumes the gas to be ionized. A more realistic approach would be to follow the ionization front through the gas.

TARTER: I believe that X-ray line cooling will dominate bremsstrahlung at temperatures of 10^6 K. The diffuse emission in X-rays could have an observable effect on high ionization trace species such as Fe XIV.
DOPITA: The enhancement of X-ray cooling of the hot gas in non-equilibrium ionization conditions cannot be neglected. The large ionization time-scale of hydrogen- and helium-like ions ensures that a large amount of collisional excitation of their lines occurs. This raises the emissivity to about 30 times the equilibrium value.

HBV 475 AS A CANDIDATE PROTO-PN

S. Tamura
Astronomical Institute, Tohoku University, Sendai, Japan

The symbiotic nature of this object has been investigated by optical spectroscopy and photometry in the near infrared region. There are three components, which are (i) the ionized expanding envelope consisting of two layers of low and high excitation species, (ii) the late type component corresponding to a blackbody of 2500 K and indicated by TiO absorption band, (iii) the hot remnant star whose temperature is estimated indirectly as 150 000 K or more.

Recently a high dispersion spectrum with self scanning diode array detector was obtained. If we assume circular motions in the Galactic plane, we can estimate the distance as 9 kpc with the aid of the radial velocity. In this spectrum, we can also clearly see Fe I absorption lines (λ 4325.8, 4383.6, 4404.8) shifted to the violet by about 10 Å (= -750 km s^{-1}) as well as broad Hγ and [O III] λ 4363 emission lines. In order to interpret these absorption lines, a single star hypothesis may be preferable.

On the basis of IR photometry and radio data, the dimensions of both the ionized expanding envelope and the late type component can be estimated. It should be considered that the late type component is extremely small and perhaps is embedded in the ionized expanding envelope.

To explain the observed data, a schematic model is proposed. (Papers will appear in Publ. Astron. Soc. Japan).

COMPARISON OF PLANETARY NEBULAE AND SYMBIOTIC STAR EMITTING REGIONS

C.D. Keyes
Department of Astronomy, University of California, Los Angeles

The spectra of symbiotic stars generally display many emission lines seen in moderate to high-excitation planetary nebulae, but are superposed upon a strong continuum characteristic of a cool star, typically of type M. Furthermore, the spatial distribution of symbiotics has been noted to resemble that of planetaries. These similarities suggest that the symbiotic stars and planetary nebulae might have some relationship (causal or otherwise) or that they might arise from similar progenitors.

We have secured and analyzed contemporaneous spectrophotometric IUE images and ground-based scans which cover the entire range, λλ 1200-8000 Å, of several "S-type" (stellar) symbiotic stars, notably AG Peg, RW Hya, AG Dra, YY Her, and V443 Her.

All of the stars studied clearly appear to be binaries. The hot components are all well below the main sequence, but, with the possible exception of AG Dra, are not degenerate stars. Hot component He II Zanstra temperatures are in the range 85 000 - 115 000 K. The cool components are M stars, probably of luminosity class II-III.

Emitting line region electron temperatures are typically 9000 - 15 000 K, and perhaps as high as 20 000 K for YY Her. The electron densities in our sample are in the range 1-50 x 10^9 cm^{-3}. Though the electron temperatures are similar to those of planetary nebulae, the electron densities definitely are not. These results are similar to those found for a few other symbiotics (Altamore et al., 1981, Astrophys. J., 245, 630; Michalitsianos et al., 1980, Nature, 284, 148). As in planetary nebulae, photoionization appears to be the dominant energy-input mechanism, except possibly in the case of AG Dra, which is a soft X-ray source. There is often considerable and variable bound-free and free-free H emission producing a strong contribution to the total energy distribution.

One object, AG Peg, is clearly an evolved system near the second stage of mass transfer, but, significantly, probably has not undergone a planetary shell-ejection episode. The cool component currently is not filling its critical Roche surface. The hot component of YY Her appears to be embedded in a large disk-like structure, and we probably view this system from near the orbital plane.

The observational morphology, eruptive behavior, and our diagnostic results suggest that some S-type symbiotic stars consist of a hot subdwarf embedded in a dense, but small, nebula or "disk-like" structure which presumably must be supported by either cool component mass loss or hot component stellar wind. The symbiotics may be from a stellar population similar to that of planetary nebula progenitors, but no direct evidence exists that S-type symbiotics are the result of planetary nebular eruptions in binary systems.

WADE: In the stars that you have observed, is the Balmer decrement consistent with case B?

KEYES: The Balmer decrement in these stars is similar to that in case B, but it should be realized that there is a strong, narrow "nebular" component and a very broad component to the Balmer lines. Undoubtedly, different physical conditions pertain to the various emitting regions. We are not able to separate the components in our current spectrophotometric observations.

NUSSBAUMER: I should like to emphasize that symbiotic stars are not a homogeneous class. Some may be binaries, others are probably single stars. Some may be related to the nova phenomenon, whereas others may well be PN in a very early stage of evolution; V1016 Cyg is probably the best studied representative of the latter group (e.g. Nussbaumer and Schild, 1981, Astron. Astrophys. 101, 118). These symbiotics may be an important part of the missing link between Red Giants and more evolved PN.

KEYES: I agree that the "D-type" symbiotics, such as V1016 Cyg, seem to be slightly more akin to PN than the "S-type" symbiotics studied here.

MASS LOSS FROM CENTRAL STARS OF PLANETARY NEBULAE

M. Perinotto
Osservatorio Astrofisico di Arcetri, Firenze, Italy

Abstract. Stellar winds have been revealed in a large fraction of central stars of planetary nebulae from P Cygni profiles observed with the IUE satellite. The relevant lines are essentially the resonance lines NV λ1240, Si IV λ 1397, CIV λ 1549 and the subordinate lines OIV* λ 1342, OV* λ1371, NIV* λ1579. Edge velocities are of the order of 1000-3000 km s^{-1} , similar to the case of population I O stars. Detailed determinations of the mass loss rate have been performed for NGC 6543, NGC 2371, IC 2149 and IC 3568 with values between 4.10^{-9} to 7. 10^{-7} M_{o} yr^{-1}. The accuracy of these determinations is not well known. It is however clear from the variety of observed profiles in these and in several other objects that properties of the winds (ionization structure, etc.) varies considerably from object to object and that very likely the mass loss rate will span over a large interval. Some possible consequences of these winds are discussed.

1. INTRODUCTION

Our concern is with relatively fast winds originating in stars known to be already in the phase of planetary nebula.

Such a wind can make itself evident through radiation in the continuum or in the line spectrum. In practice we may have:

a) Free-free emission of the ionized expanding envelope in the radio domain.

b) Similar emission in the infrared (IR) spectral range.

c) P Cygni profiles in the subordinate lines of hydrogen and helium, in particular at Hα and λ4686 He II in the visible region of the spectrum.

d) P Cygni profiles in resonance lines or in subordinate lines of

D. R. Flower (ed.), Planetary Nebulae, 323–335.

abundant heavy ions in the space UV (λ < 3000 A).

2. EVIDENCE OF WINDS IN CENTRAL STARS OF PN

Methods involving the radio or IR continuum did not provide so far, to my knowledge, evidence of winds originating in central stars of planetary nebulae. The reason is soon recognized in the fact that, at present state of technology, methods 1a and 1b are not sensitive enough to the mass loss rate $\dot{M}$, while on the other hand planetary nebulae are relatively distant objects.

Actually an ionized envelope with $T_e = 10^4$ K expanding at a velocity v_∞ is predicted to produce at Earth a radio flux (cf. Panagia and Felli, 1975; Wright and Barlow, 1975),

$$S_\nu = 2.4\ 10^4 \left(\frac{\dot{M}}{M_o yr^{-1}}\right)^{4/3} \left(\frac{v_\infty}{1000\ km\ s^{-1}}\right)^{-4/3} \left(\frac{\nu}{10\ GHz}\right)^{0.6} \left(\frac{D}{Kpc}\right)^{-2}\ Jy. \quad (1)$$

Favorable values of $\dot{M} = 10^{-7}\ M_o\ yr^{-1}$ (we will see this corresponds to $\sim$ the maximum $\dot{M}$ so far quoted for central stars of PN), $v_\infty = 2000$ km s^{-1}, $\nu = 5$ GHz (6 cm), D = 500 pc implies $S_\nu = 0.01$ mJy which is rather below the sensitivity limit of present radiotelescopes. A much higher sensitivity, coupled with high spatial resolution (~0".1) to exclude radiation contributed by the nebula, is then needed to detect weaker winds in more distant objects.

Similar arguments apply to the IR range where in addition one has to detect a faint free-free source over a possibly strong stellar source and must avoid a generally important contribution from heated dust.

Evidence of an expanding atmosphere from the optical spectrum (method 1c) is clear from the blue-shifted absorption components observed in nuclei of WR-type (see Aller, 1976). Evidence in the optical range was also noted in other nuclei of planetaries, as in He 2-131 from P Cygni profiles of various lines (Koelbloed, 1962; Heap, 1977) and in NGC 6891 and 6826 from broad emission lines (Heap, 1977).

The existence of "brisk stellar winds" was also suggested to explain high excitation lines appearing in the optical spectrum (λ 3811, 3834 O VI) (Aller, 1976).

However a definitive evidence of mass loss by stellar winds in central stars of planetary nebulae was obtained with method 1d during the commissioning period of the IUE satellite with the detection of P Cygni profiles of NV λ 1240, CIV λ 1549 and NIV* λ 1719 in the nucleus of NGC 6826 (Heap et al. 1978).

Some detections of P Cygni profiles from IUE spectra and first estimates of properties of associated winds have followed (cf. Heap,

1979; Seaton, 1980; Benvenuti and Perinotto, 1980; Koppen and Werhse, 1980; Perinotto, Benvenuti and Cacciari, 1981). A few detailed analysis of stellar winds from IUE spectra have appeared, on which we will report later on. Clearly much more studies are expected in the near future.

3. NEW INFORMATION FROM A SAMPLE OF LOW RESOLUTION IUE SPECTRA

We have searched a number of released low resolution (~ 7 A) IUE spectra of planetary nebulae and their central stars for the presence of P Cygni - like profiles in order to assess the general properties of the phenomenon. The detection of a stellar P Cygni like profile requires: 1) To reveal the stellar continuum, 2) to identify the profile and 3) to correct it for nebular contamination. Item 1) is achieved for a relatively large number of planetaries so far observed with IUE. Concerning item 2), since P Cygni like profiles vary considerably in the relative importance of the emission and absorption components, limiting cases may require the high IUE resolution (~ 0.15 A) to a proper wavelength setting. The high IUE resolution is also decisive for item 3) since nebular lines are much narrower than stellar or circumstellar features. To exploit 3), with low resolution IUE spectra only, one takes advantage from considering the properties of spectrum of the central star in the optical, the general level of excitation of the nebular spectrum and particularly the comparison of large aperture (~ 220 arcsec square) spectra with small aperture (~ 7 arcsec square) spectra.

We report in Table 1 preliminary results of the inspection of a number of IUE low resolution spectra. The objects in Table 1 have all a stellar continuum visible in the spectra. Ions with lines displaying a relatively clear P Cygni profile are indicated approximately in order of decreasing importance of the phenomenon.

We see from Table 1 that the phenomenon of stellar winds

1) is quite common in nuclei of planetary nebulae;

2) appears fairly ubiquitous among nuclei of PN with different spectral type with the exception of stars having a "continuum" spectrum in the optical;

3) is presumably present in objects of quite different luminosity and gravity, so strongly increasing the range of these parameters in hot stars known to display mass loss.

These facts are evidently quite important not only for the study of these objects, but to investigate the causes of hot star winds in general. Point 1) and 2) are better illustrated in Table 2 showing stars with detected P Cygni profiles versus spectral type.

In WR and Of nuclei the phenomenon appears always present in the

Table 1. P Cygni phenomenon in central stars of planetary nebulae from low resolution IUE spectra.

Object	mag[1)]	Sp Type[2)]	P Cygni[3)]	
NGC 40	11.6 V	WC8	CIV,SiIV	
246	11.9 V	OVI	-	
1360	11.2 V	-	-	
1514	9.4 V	AO+O	NV,OV	
1535	11.6 V	O7,(Sd)O3:	NV,OV	
2022	14.9 V	cont.	-	
2371	14.8 V	OVI	CIV	
2392	10.5 V	O7f, O6f	-	
2867	14.9 P	OVI	-	(d)
3132	8.8 P	A+sdO	-	
3211	-	-	-	(d)
3242	>11.3 -	cont.	-	
4361	12.9 V	O6	-	
5189	14.1 V	OVI	CIV,NV	
6210	11.3 pV	O7f,(sd)O3	NV,OV	
6572	>11.0 V	Of+WR	NV,CIV,SiIV	
6720	14.7 V	cont.	-	
6826	10.2 V	O6fp,O3f	CIV,NV,SiIV,OIV,OV,NIV	
6891	11.1 P	O7f,O3:f	CIV,NV,SiIV,OV,NIV	
6905	13.9 P	OVI	-	
7009	11.5 P	cont.	NV,OV	
7293	∿13 p	cont.	-	
7662	(12.5)P	cont.	-	
IC 351	15 P	cont.	-	(d)
418	9.6 V	O7fp	CIV,SiIV,NIV	
1297	-	-	-	(d)
2149	10.5 pV	O7.5fp,O4:(f)	CIV,NV,SiIV	
2448	-	-	-	
3568	11.4 V	O5f	CIV,NV,OV	
4593	10.8	O7fp	CIV,NV,SiIV,OIV,NIV	
A 30	14.3	O5fep	CIV,NV,OV	
A 36	11.5 V	SdO7	-	
A 78	13.3 V	O5fek	CIV,NV,OV	
J 320	13.5 P	em uncl	NV,OV	
BD+30 3639	10.1 V	WC9	CIV,SiIV	
CD-23 12238 = Me 2-1	-	-	-	(d)
HD 167362	11 -	Of+WR	CIV,SiIV,NIV	
HD 138403 = He 2-131	10.3 V	-,O7(f)eq	CIV,SiIV,NV	
Hu 2-1	-	-	CIV,NV,SiIV,OIV,NIV	

1) From Aller (1976) except NGC 7293 from PK Catalogue and He 2-131 from Heap (1977).
2) From Aller (1976) and Heap (1977).

3) From present work: ions with lines showing P Cyg profile, ordered approximately with decreasing strength of the phenomenon. "d" means that the presence of a stellar continuum is doubtful.

Table 2. Detected P Cygni profiles versus spectral type.

Sp. Type		No. of positive detection / No. of considered nuclei	
-WR	....	2/2	
-Of+WR	....	2/2	
-Of	....	3/4	
-Ofp,e	....	7/7	
-OVI	....	2/5	
-cont.	....	1/7	
			Total 17/27 = 63%
Others:			
O6,O7	....	1/2	
A+O	....	1/1	
A+sdO	....	0/1	
sdO	....	0/1	
unclassified	....	2/6	Total 4/11 = 36%
			Gross Total 21/38 = 55%

UV, about in half cases in OVI stars and never (as reasonably expected) in the "continuum" stars, with the only exception of NGC 7009 that likely is not a bona fide "pure continuum" star. The percentage of positive detection is of 63% for the mentioned nuclei and of 36% for the other 11 objects with miscellanea spectra shown on Table 2. The gross total of detections amounts to 55%. These numbers are subject to revision when more accurate analysis of these and of other objects will be made; likely they may result underestimated.

As for T_{eff} the phenomenon is present in objects as cold as IC 2149 ($T_{eff} \sim 30\,000$ K) and as hot as NGC 2371 ($T_{eff} \sim 100\,000$ K). The two nuclei have $L/L_o \sim 3.5$ and 3.2 respectively. As for the luminosity the phenomenon is present in NGC 6210 (log $L/L_o \sim 2.2$, log $T_{eff} \sim 4.70$) and NGC 6891 (log $L/L_o \sim 3.8$, log $T_{eff} \sim 4.72$). These numbers come from the Zanstra method following Harman and Seaton (1966) after allowing for distances by Acker (1978), except for IC 2149, taken from Perinotto et al. (1981).

The behaviour of the phenomenon varies greatly from object to object in Table 1, as shown by the variety of profiles of the various lines. The edge velocities are of the order of 1000-3000 km s^{-1}, similar to the case of population I O stars. It is clear that a lot of information on the properties of the winds in central stars of planetary nebulae, including insight on the causes of the phenomenon of stellar wind in hot stars and on the reasons why the phenomenon is not present in various cases is shortly expected from the accurate study of the large quantity of material obtained and to be obtained with the IUE satellite.

4. DETAILED ANALYSIS

To my knowledge the following nuclei of planetary nebulae have received relatively accurate studies of wind's properties and associated mass loss: NGC 6543 by Castor, Lutz and Seaton (1981) (CLS), NGC 2371 by Pottasch, Gathier, Gilra and Wesselius (1981) (PGGW), IC 2149 by Perinotto, Benvenuti and Cerruti-Sola (1982) (PBC) and IC 3568 by Harrington (1982) (H). To these works, a value of $\dot{M} \simeq 7\ 10^{-7}\ M_o\ yr^{-1}$ for the nucleus of NGC 6543 by high resolution IUE spectra (Heap, 1981) (given without further details) is to be added.

The observed P Cygni profiles have been interpreted in all these works basically in terms of the Sobolev approximation (wind velocity large compared with the local thermal velocity) of the theory of line formation for a two-level atom in an expanding atmosphere (Lucy and Solomon, 1970; Castor, 1970; Lucy, 1971; Castor, Abbott and Klein, 1975). Based on it, Castor and Lamers (1979) (CL) have produced an atlas of theoretical P Cygni-type line profiles valid for resonance lines and a similar work has been made by Olson (1981) for excited lines.

Recently, calculations have been made in which the Sobolev approximation has been released (Weber, 1981; Leroy and Lafon, 1982a,b). The role of multiple scattering of photons has also been investigated (Panagia and Macchetto, 1982). These progresses in the theory do not seem, at their present stage, very important for a better determination of the mass loss rate in the above objects, relative to the use of the mentioned theory, also because of uncertainties in the relevant stellar parameters (see later on).

In the recalled approximate theory, the optical depth for scattering can be expressed as

$$\tau(v) = \frac{\pi e^2}{mc} f \lambda_o n_i(r) \left(\frac{dv}{dr} \right)^{-1} \qquad (2)$$

where m is the mass of the electron, λ_o is the laboratory wavelength of the line, f its oscillator strength, dv/dr is the velocity gradient in the envelope and $n_i(cm^{-3})$ is the number density of absorbers. The velocity law and the opacity law can be parametrized (CL)

$$w(x) = w_o + (1-w_o)(1- \frac{1}{x})^{\beta} \qquad (3)$$

$$\tau(w) = T(\gamma+1)(1-w_o)^{-1-\gamma}(1-w)^{\gamma} \tag{4}$$

$$x = r/R_* \quad , \quad w = v/v_\infty \tag{5}$$

where R_* is the photospheric radius coincident with the base of the wind where its velocity has a low value w_o (generally assumed $\simeq 0.01$), β is positive to ensure an outward increasing velocity (in agreement with various observational tests) up to a terminal velocity v_∞ , $\gamma \geq 0$ and

$$T = \int_{w_o}^{1} \tau(w)\, dw = \frac{\pi e^2}{mc}\; f\, \lambda_o\, v_\infty^{-1}\, N_i \tag{6}$$

is the total optical depth in the wind due to ions with column density $N_i(cm^{-2})$.

The quantities β, γ , T can be obtained by matching observed profiles with computed ones using e.g. the atlas of CL.

The mass loss rate, under hypothesis of spherical symmetry and homogeneity, can be written

$$\dot{M} = 4\pi r^2\, \rho(r)\, v(r) \quad . \tag{7}$$

Thus one obtains

$$q_i\, \dot{M} = (4\pi\mu m_H)\, \frac{mc}{\pi e^2}\, \frac{R_*\, v_\infty}{\lambda_o f A_i}\; T(\gamma+1)(1-w_o)^{-1-\gamma}(1-w)^{\gamma}\left[x^2 w\, \frac{dw}{dx}\right] , \tag{8}$$

where A_i is the abundance of the element relative to hydrogen and q_i is the fractional ionic abundance. The right side of (8) can be evaluated at any point x. The value of x corresponding to $w = 0.5$ is generally chosen since with this choice the quantity in square bracket depends little on a accurate determination of β.

The above two-level atom theory has been shown to be valid even for the subordinate lines of OIV* λ 1342, OV* λ 1371, NIV* λ 1719, by CLS. In this case one must evaluate the population of the lower level of the transition, that now do not coincide with the total abundance of the ion, via an appropriate radiation temperature.

Although with some formal differences, the authors of the mentioned detailed studies have used this theory to obtain the values of $q\,\dot{M}$ reported in Table 3. A proper determination of the parameters T, β, γ requires high resolution IUE spectra. Such spectra have been used by PBC and H, while CLS and PGGW use low resolution IUE spectra. Actually CLS develop a method which permits the best use of low resolution IUE spectra. The method however assumes $\beta = 1$, $\gamma = 1$ in equations (3) and (4). H, on the other hand, uses expressions (calculated for $\beta = 1$)

that allow to obtain $q\dot{M}$ at the point of the wind where q is maximum, while q in the previous formulation is essentially an average value across the wind.

Despite these differences the values of $q\dot{M}$ in Table 3 should be comparable, as illustrated from the similarity of the $q\dot{M}$ by CLS in NGC 6543 to the $q_m\dot{M}$ deduced for the same object using the data of CLS and the expressions of H. (See Columns 3 and 6 of Table 3). It is to be noted that CLS did prefer to relay on the subordinate lines to deduce $\dot{M}$. This is due in part to problems with profiles of CIV and NV lines in NGC 6543, and mostly to advantages offered from the OIV*, OV* lines in the determination of q_i, since most of the oxygen is believed to be in this object in these two ionization stages. PBC have preferred instead to relay on the resonance lines because: 1) profiles of subordinate lines in IC 2149 show winds developed to a much lower velocity than resonance lines and theoretical profiles appropriate to match these lines were not available, nor the CLS method ($\beta = \gamma = 1$) was here applicable; 2) T_{eff} of IC 2149 is relatively lower than that of NGC 6543 or IC 3568 so that $\dot{M}$ becomes quite sensitive to the exact value of the radiation temperatures. As for the uncertainties in $q\dot{M}$, a factor of 2-3 comes from the photospheric radius which depends on T_{eff} and the adopted distance or luminosity and another factor of 2 from chemical abundances.

The next step to deduce $\dot{M}$ is to determine q. CLS argue that $q(OV) + q(OIV) \simeq 1$ in the wind of NGC 6543 and thus deduce $\dot{M}$ and then the q_i's of the other ions. However the comparison of these "observed" ionizations with the ones calculated using present theories for ionization in the winds raises problems discussed by CLS. Another procedure (used by PBC) is to accept the mean ionization structure adopted by Lamers, Gathier and Snow (1980) who have determined empirical mean ionizations from a group of O4 to B1 population I stars. We underline that whatever will result the most correct way to determine the ionization structure, the last appears quite different in IC 3568 and NGC 6543, the wind being more ionized in the first object than in the second one (much larger values of $q(NV)/q(NIV)$ and $q(OV)/q(OIV)$). Also it must be realized that particularly in nuclei with small radius ($< 0.1\ R_o$) the characteristic time of expansion in the wind becomes smaller than recombination time for the relevant C,N,O ions. Therefore classical computations of the ionization structure in the wind will be inadequate. If we accept errors in q in NGC 6543, IC 2149 and IC 3568 again of a factor of 2, we arrive to a total uncertainty in $\dot{M}$ of an order of magnitude. This would mean that the discussed determinations of $\dot{M}$ might be not really different to each other. However the numbers of Table 3 taken at their face value, the determination by Heap (1981) of $7 \cdot 10^{-7}$ in NGC 6543 and the variety of profiles and strengths of the P Cygni phenomenon observed in many central stars of planetary nebulae, show that it is very likely $\dot{M}$ will vary over a large range, possibly $10^{-6} \div 10^{-9}\ M_o\ yr^{-1}$. We mention finally that $\dot{M}$ in NGC 6543 is not likely to be much larger than $10^{-6}\ M_o\ yr^{-1}$, because of an upper limit of $5\ 10^{-6}$ obtained for it with VLA at 6 cm wavelength (Thompson and Sinha, 1980).

Table 3. Mass loss rates in central stars of PN ($\dot{M}$ in M_o yr^{-1}).

		NGC 6543 (CLS)				NGC 2371 (PGG W)			IC 2149 (PBC adapted)			IC 3568 (H)		
λ	Ion	$q\dot{M}$	q	$\dot{M}_8$	$q_m\dot{M}$	$q\dot{M}$	q	$\dot{M}_8$	$q\dot{M}$	q	$\dot{M}_8$	$q_m\dot{M}$	q	$\dot{M}_8$
1240	NV	-	-	-	3.3-10	-	-	-	1.4-10	1.2-2	1.2	1.4-9	∿0.35	-
1342	OIV*	7.9-8	0.92	9 (OIV*+OV*)	8.4-8	-	-	-	-	-	-	-	-	-
1371	OV*	0.7-8	0.09		0.7-8	-	-	-	-	-	-	3.9-9	∿1.	>0.4
1397	SiIV	6.8-11	8.-4	-	7.2-11	-	-	-	∿2.4-11	3.7-3	∿0.6	-	-	-
1549	CIV	-	-	-	3.6-11	2.8-11	>3.-3	<10	>2.4-10	1.5-2	>1.6	3.5-10	∿0.09	-
1719	NIV*	1.2-8	0.14	-	1.2-8	-	-	-	-	-	-	5.4-10	∿0.13	-
	Adopted $\dot{M}$	$\sim 10^{-7}$				$< 10^{-7}$			10^{-8}			$\sim 4\ 10^{-9}$		

5. DISCUSSION

One wishes to know: a) how the derived mass loss rates for nuclei of planetary nebulae compare with observations in population I stars and with predictions of theories of stellar winds, b) which are the consequences for the evolution of the star and c) for the behaviour of the nebula. To this aims we use the numbers of Table 3 at their face values.

a) Observed mass loss rates in population I stars have been found to fit a single dependence on L of the type $\dot{M} \propto L^{a}$ with $a = 1.73$ (Garmany et al., 1981 from UV study of 30 O stars). Other works, based on larger variety of population I objects found a dependence even on radius and mass (Chiosi, 1981; Lamers, 1981). The last study obtains $\dot{M} \propto L^{1.42}\ R^{0.61}\ M^{-0.99}$. We adopt 0.6 M_o for the PN's nuclei of Table 3, from their position in the HR diagram. The Garmany et al. law predicts values of $\dot{M}$ unacceptably smaller (2 order of magnitude) than the observed ones. The last are instead consistent with the Lamers law, which on the other hand is considered compatible with the radiation driven wind theory. This theory also predicts (Castor, Abbott and Klein, 1975) v/v_{esc} equal to $2 \div 3$. The present values are somewhat larger than that in NGC 6543 (4.0) and IC 2149 (4.2) and smaller in IC 2371 (1.4). Moreover the momentum of the wind exceeds the one of the radiation field by a factor of 5 (using $\dot{M} = 10^{-7}\ M_o\ yr^{-1}$) in NGC 6543, against the prediction of the single scattering radiation driven wind theory. Therefore improvements in the theory seem required by the presently available data on stellar winds in central stars of PN. The fluctuations theory by Andriesse (1979), based on perturbations caused by non thermal processes in the photosphere, predicts values of $\dot{M}$ an order of magnitude smaller than observed in NGC 6543 (accepting $10^{-7} M_o\ yr^{-1}$) and in IC 3568 and therefore is little supported by present data in nuclei of PN.

b) The consequences of the fast stellar winds are expected to be important for the evolution of the central star since the nuclear burning rate along the upper part of the evolutionary track for a 0.6 M_o star by Paczynsky (1971) is $\sim 3\ 10^{-8}\ M_o\ yr^{-1}$ and less in the lower part.

c) The momentum available in the wind during the lifetime of the nebulae is easily seen to be comparable with the one of the nebulae. Therefore the winds are important for the nebular dynamics and they may also produce observable effects in the radiation from the nebula in optically thin cases, i.e. if the kinetic energy transferred from the wind into nebular radiation is comparable with the absorbed UV radiation from the central star (Harrington, 1982). Finally the properties of the observed winds might be used to test the interacting stellar winds theory (Kwok et al., 1978; Kwok, 1982), according to which the presently seen nebula would be build up at the interface of the slow wind of the red giant precursor phase and of the fast wind produced by the exposed hot core. It is unfortunate that properties of the slow wind have at present been detected only in NGC 7027 and IC 418 through observations of the molecular CO cloud (cf. Knapp et al., 1982), because the theory could be simply tested from the measured expansion velocity of the optical nebula and $\dot{M}$, v_∞ of the two winds.

References

Acker, A.: 1978, Astron. Astrophys. Suppl. 33, 367.
Aller, L.H.: 1976, Mém. Soc. R. Sci. Liège, 6 serie, Tome IX, 271.
Andriesse, C.D.: 1979, Astrophys. Space Sci. 61, 205.
Benvenuti, P. and Perinotto, M.: 1980, Proceed. of Second European IUE Conference, Tübingen, ESA SP-157, p. 187.
Castor, J.I.: 1970, Monthly Not. Roy. Astr. Soc. 149, 111.
Castor, J.I., Abbott, D.C. and Klein, R.I.: 1975, Astrophys. J. 195, 157.
Castor, J.I., Lutz, J.H. and Seaton, M.J.: 1981, Monthly Not. Roy. Astr. Soc. 194, 547 (CLS).
Castor, J.I. and Lamers, H.J.G.L.M.: 1979, Astrophys. J. Suppl. 39, 481 (CL).
Chiosi, C.: 1981, preprint.
Garmany, C.D., Olson, G.L., Conti, P.S. and van Steenberg, M.E.: 1981, Astrophys. J. 250, 660.
Harman, R.J. and Seaton, M.J.: 1966, Monthly Not. Roy. Astron. Soc. 132, 15.
Harrington, J.P.: 1982, "Advances in Ultraviolet Astronomy, Four Years of IUE Research", Goddard Space Flight Center, in press (H).
Heap, S.R.: 1977, Astrophys. J. 215, 864.
Heap, S.R., Boggess, A., Holm, A., Klinglesmith, D.A., Sparks, W., West, D., Wu, C.C., Boksenberg, A., Willis, A., Wilson, R., Macchetto, F., Selvelli, P.L., Stickland, D., Greenstein, J.L., Hutchings, J.B., Underhill, A.B., Viotti, R., Whelan, J.A.J.: 1978, Nature 275, 385.
Heap, S.: 1979, in "Mass Loss and Evolution of O-Type Stars", p. 99, Conti, P.S. and de Loore, C.W.H. eds. (Dordrecht:Reidel).
Heap, S.: 1981, "The Universe at Ultraviolet Wavelengths", Goddard Space Flight Center, NASA Conf. Publ. 2171, p. 415.
Knapp, G.R., Phillips, T.G., Leighton, R.B., Lo, K.Y. Wannier, P.G. and Wootten, H.A.: 1982, Astrophys. J. 252, 616.
Koppen, J. and Werhse, R.: 1980, Proceed. of Second European IUE Conference, Tübingen, ESA SP-157, p. 191.
Kwok, S., Purton, C.R. and Fitzgerald, P.M.: 1978, Astrophys. J. (Letters) 219, L125.
Kwok, S.: 1982, Astrophys. J. 258, 280.
Lamers, H.J.G.L.M., Gathier, R., Snow, T.P.: 1980, Astrophys. J. Letters 242, L33.
Lamers, H.J.G.L.M.: 1981, Astrophys. J. 245, 593.
Leroy, M. and Lafon, J.P.J.: 1982a, Astron. Astrophys. 106, 345.
Leroy, M. and Lafon, J.P.J.: 1982b, Astron. Astrophys. 106, 358.
Lucy, L.B. and Solomon, P.M.: 1970, Astrophys. J. 159, 879.
Lucy, L.B.: 1971, Astrophys. J. 163, 95.

Olson, G.L.: 1981, Astrophys. J. 245, 1054.
Panagia, N. and Felli, M.: 1975, Astron. Astrophys. 39, 1.
Panagia, N. and Macchetto, F.: 1982, Astron. Astrophys. 106, 266.
Paczynsky, B.: 1971, Acta Astron. 31, 417.
Perinotto, M., Benvenuti, P. and Cacciari, C.: 1981, Proceed. of IAU Coll. No. 59, "Effects of Mass Loss on Stellar Evolution" eds. Chiosi, C. and Stalio, R., p. 45.
Perinotto, M., Benvenuti, P. and Cerruti-Sola, M.: 1982, Astron. Astrophys. 108, 314 (PBC).
Pottasch, S.R., Gathier, R., Gilra, D.P. and Wesselius, P.R.: 1981, Astron. Astrophys. 102, 237 (PGGW).
Seaton, M.J.: 1980, "Ultraviolet Spectra of Planetary Nebulae and their Central Stars", Proceed. XVII IAU General Assembly Highlights of Astron. 5, 247.
Thompson, A.R. and Sinha, R.D.: 1980, Astron. J. 85, 1240.
Weber, S.V.: 1981, Astrophys. J. 243, 954.
Wright, A.E. and Barlow, M.J.: 1975, Monthly Not. Roy. Astron. Soc. 170, 4.
Koelbloed, D.: 1962, Bull. Astron. Netherl. 16, 163.

MENDEZ: I would like to comment on the spectral types you used. There are two kinds of O VI objects, one with WC characteristics and the other with predominantly absorption lines (e.g. NGC 246). We have recently reclassified the WC "O VI" objects (Méndez and Niemela, IAU Symposium no. 99). I would suggest that WC "O VI" objects ought to be grouped with the other WC's.

COHEN: You showed a viewgraph of the locations in the HR diagram of nuclei with and without winds. Does this plot imply that stars with winds have the lower masses? If so, it would seem to contradict Heap's finding that low mass nuclei do not show winds in IUE low dispersion spectra.

PERINOTTO: No - the diagram shows the wide range in both luminosity and temperature of nuclei showing the P Cyg phenomenon.

HEAP: The reason that I derived a higher rate of mass loss from the nucleus of NGC 6543 than Castor, Lutz and Seaton (CLS) is that I used a much softer velocity law for the wind. CLS assumed a velocity law typical of young O-stars ($\beta = 1$). High resolution spectra of the central star of NGC 6543 show wind profiles that can be reconciled with the theoretical profiles of Castor and Lamers only for a more slowly accelerating wind ($\beta = 4$).

PERINOTTO: The analysis of CLS is also based on the assumption of $\gamma = 1$ in the opacity law.

KALER: Feibelman and I have observed some stars with high gravities and find a P Cyg profile with $v_\infty \approx 10^4$ km s^{-1}. We are still uncertain of the identification, and confirmation is important. If correct, we have $v_\infty/v_{escape} \approx 3$ or 4, up to log g $\approx$ 8.

PERINOTTO: It would be interesting to confirm your finding; the maximum value of v_∞ in the group of objects I discussed is approximately 4000 km s^{-1}.

CLEGG: Do there exist means of determining the kinetic temperature in the central star winds?

PERINOTTO: The electron temperature in the wind is not yet well determined. Calculations of the ionization equilibrium in the wind suggest rather high temperatures, to match the "observed" fractional abundances. However, such high temperatures would imply much stronger emission in the C IV lines than is observed. The C IV emission is consistent with a temperature close to the effective temperature of the central star.

A POWERFUL METHOD FOR DERIVING MASS-LOSS RATES FROM PLANETARY NEBULAE AND OTHER OBJECTS: THE FIRST ORDER MOMENT W_1 OF UNSATURATED P CYGNI LINE PROFILES

Jean Surdej
Institut d'Astrophysique, Universite de Liège, Belgium

For the case of optically thin lines, we show that the relation existing between the first order moment

$$W_1 \propto (E(\lambda)/E_c-1)\ (\lambda-\lambda_{12})d\lambda$$

of a P Cygni line profile and the quantity $\dot{M}$ n (level) (cf. Castor, Lutz and Seaton, 1981), where $\dot{M}$ is the mass-loss rate of the central star and n (level) the fractional abundance of the relevant ion, is in fact independent of any Sobolev-type approximations used for the transfer of line radiation. Consequently, all results established in the context of "very rapidly" expanding atmospheres (see Surdej, 1982) and mainly referring to the non-dependence of W_1 on various physical (underlying photospheric absorption line, limb darkening, collisions, multiplet line transitions, etc.) and geometrical (radial and rotational velocity fields, size of the atmosphere, etc.) effects remain unchanged for arbitrary (e.g. non-Sobolev type) outward-accelerating velocity laws.

Whenever applied with caution, the following relation

$$\dot{M}(-M_\odot/\text{year}-)\ n\ (\text{level}) = -1.19\ 10^{-21}\ v^2_{max}(-\text{km/sec}-)R^*(-R_\odot-)W_1/$$
$$(f_{12}\lambda_{12}(-10^3\text{Å}-)\ A(\text{element})),$$

where the different symbols have their usual meaning, thus provides a very powerful means of deriving mass-loss rates - with a total uncertainty less than 60 per cent - from the measurement W_1 of unsaturated P Cygni profiles observed in the spectrum of planetary nebulae, early as well as late-type stars, quasars, etc.

Castor, J.I., Lutz, J.H. and Seaton, M.J.: 1981, Monthly Notices Roy. Astron. Soc. 194, 547.
Surdej, J.: 1982, Astrophys. Space Sci. 88, 31.

PERINOTTO: Did you try to apply your method to any observed objects?
SURDEJ: Only to NGC 6543, for which we derive the same mass loss rate (to within 50%) as Castor, Lutz and Seaton.

INFLUENCE OF THE STELLAR WIND ON THE NEBULAR IONIZATION IN NGC 1535 AND 4361

J. Adam, J. Köppen
Institut für theoretische Astrophysik, Heidelberg, W. Germany

The high excitation planetaries NGC 1535 and 4361 were observed with IUE satellite in the short wavelength region at high dispersion. In NGC 1535 we found P Cygni profiles of NV 1238, 1242 and OV 1371 lines with a terminal wind velocity of about 2000 km/sec. In NGC 4361 these lines are narrow absorption lines (width 0.5 Å), probably of photospheric origin.

The observed nebular emission line spectrum, from our IUE data and published optical data, is compared with theoretical ionization models. The physical assumptions are: The ionizing flux is taken from NLTE model atmospheres (R.P. Kudritzki, private communication) which had been fitted to the photospheric hydrogen and helium line profiles of the central stars (R.H. Mendez et al., 1981, Astron. Astrophys. 101, 323). The nebulae are homogeneous and density bounded.

The observed UV continuum spectra agree well with the model atmosphere flux distributions.

With an effective temperature of 100 000 K which lies near the hot end of the error bar given by Mendez et al., the observed nebular spectrum of NGC 4361 can be reasonably well reproduced.

However, for NGC 1535 the central star with T_{eff} around 50 000 K does not provide enough He^+ ionizing photons. Thus, C IV, Ar V, Ne IV and Ne V lines also are predicted too weak. The electron temperature of the model is lower than observed, suggesting that the photon distribution between 13.6 eV and 54 eV, responsible for heating, is too cool.

The presence of NV and OV in the stellar wind of NGC 1535 indicates a wind temperature of about 10^5 K. A mass loss rate of 10^{-7} $M_\odot$ y^{-1} yields a sufficient number of He^+ ionizing photons to explain the nebular He II lines. We conclude that in NGC 1535 the hot stellar wind is an important source of photons ionizing the nebula.

HARRINGTON: I find your results quite interesting - we find the same need for additional flux below 54 eV in our study of IC 3568. It seems that those stars in the range $5 \times 10^4 \lesssim T_* \lesssim 8 \times 10^4$ K have this problem. The model atmosphere fluxes appear satisfactory for $T_* > 10^5$ K.

I would like to stress that the best objects in which to study stellar winds are PN. This is because the nebular spectrum gives information on the flux in the far ultraviolet emerging from the stellar atmosphere, particularly for $\lambda < 228$ Å. This information is not available for O-type main sequence stars.

PURTON: In addition to considering the effects of radiation from the hot wind, it may be useful to consider the direct effect of the wind on the nebular shell. The impact may have observable consequences for the ionization of the gas.

SECTION V

CENTRAL STARS OF PLANETARY NEBULAE

NON-LTE MODEL ATMOSPHERE ANALYSIS OF CENTRAL STARS*

R.H. Méndez[1 2 3], R.P. Kudritzki[4] and K.P. Simon[4]

1. INTRODUCTION

This review will be concentrated on the determination of the main atmospheric parameters (T_{eff}, log g, helium abundance) of PN nuclei, and of other subluminous objects, by fitting the observed absorption line profiles with theoretical profiles obtained from non-LTE model atmosphere calculations.

The main motivations for this approach are two. The first is to be able to discuss the evolution of central stars without all the uncertainties related to the use of nebular distances; this requires to place our objects on the log g - log T_{eff} diagram instead of the Hertzsprung-Russell diagram. The second reason is to explore the surface helium abundance of the central stars, which is likely to play an important role in a detailed discussion of their evolutionary status.

The 'model atmosphere approach' has recently become possible after extensive computations of non-LTE model atmospheres by people working in Kiel. The following selected list of references includes: (a) descriptions of the models and complementary non-LTE line formation calculations (Kudritzki, 1976; Kudritzki and Simon, 1978); (b) study of sphericity effects (Kudritzki and Simon, 1978, Gruschinske and Kudritzki, 1979); (c) application of the models to the study of massive O stars (Kudritzki, 1980), subdwarf O stars not associated with planetary nebulae (Hunger et al., 1981; Kudritzki et al., 1982a, Simon, 1982), and central stars of planetary nebulae (Méndez et al., 1981).

2. THE MODELS

The main characteristics of the models can be summarized as follows: they are plane-parallel, in hydrostatic and radiative equilibrium, computed in non-LTE for a variety of effective temperatures, surface gravities and He/H abundance ratios. No metals are included; the atmosphere is assumed to consist of H and He only. Particularly due to the assump-

D. R. Flower (ed.), Planetary Nebulae, 343–357.

tion of hydrostatic equilibrium, in a first stage we have restricted our analysis to those central stars which show predominantly absorption-line spectra. Gruschinske and Kudritzki (1979) have shown that, even for subluminous objects, it is not necessary to consider extended (spherical) hydrostatic atmospheres; in general, for the analysis of the stellar spectrum, we expect to need a significantly extended atmosphere only as a consequence of departures from hydrostatic equilibrium.

This restriction to absorption-line central stars is not so severe as it might seem at first glance, because they are quite frequent. In connection with this, we would like to remark that we have found two of the prototypes of the so-called 'continuous' objects to be absorption-line stars (NGC 3242 and NGC 7009; for the first one see Kudritzki et al., 1981). Therefore, we expect most - if not all - of these objects to be analyzable with present-day hydrostatic models.

3. COMPARISON WITH OTHER MODELS

In view of the complexity of the computer programs, it is always useful to compare the results with other work. Unfortunately, in this case there is not much overlap. In Figure 1 we have a comparison of the H_γ absorption profile computed from the hottest NLTE model of Mihalas (1972) versus the one used in the present work, for T_{eff} = 55,000 K, log g = 4 and y = [N(He)/(N(He) + N(H))] = 0.09. The blue wings differ because Mihalas's profile does not include the HeII absorption at 4338 A; but the red wings are in good agreement. This was expected because the physical processes included in both models are essentially the same.

Figure 2 shows a comparison with the LTE and non-LTE H_γ profiles published by Wesemael et al. (1980) for T_{eff} = 100,000 K and log g = 6. Here even the non-LTE models are not strictly comparable, because Wesemael et al. did not include bound-bound transitions in the statistical equilibrium equations, and considered almost pure H atmospheres. Most of the difference in Figure 2 can be attributed to these two facts. However, this statement should not be overinterpreted to mean that our models do not need improvements; both Mihalas's and our treatment of bound-bound transitions are still somewhat schematic, because in the calculation of occupation numbers, only Doppler profiles have been used, instead of fully Stark broadened profiles (of course, in the final calculation of synthetic profiles the complete broadening functions are used). Work is under way to improve the determination of occupation numbers by including Doppler and Stark broadened profiles.

4. APPLICATIONS

Although the fitting procedure is described elsewhere (see e. g. Méndez et al., 1981), we consider it worthwhile to make a few comments on the internal accuracy we can obtain. Figure 3 shows the H_γ profiles

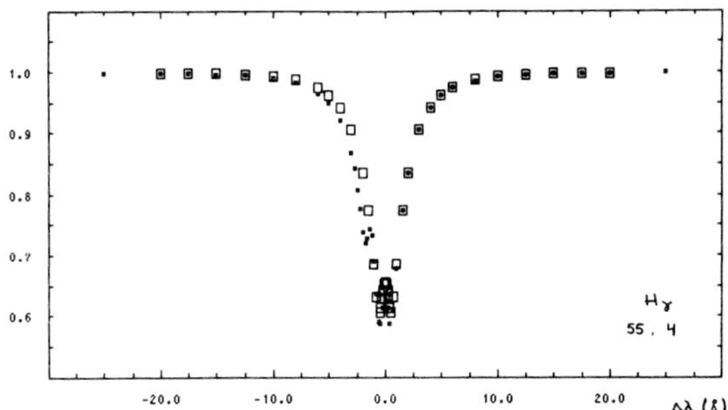

Figure 1: Comparison of H_γ theoretical absorption profiles, for T_{eff} = 55,000 K, log g = 4 and normal He abundances. Large open squares are from Mihalas; small filled squares from Kudritzki.

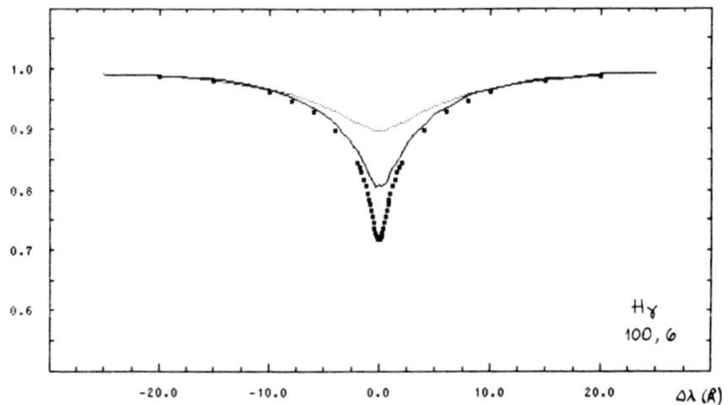

Figure 2: Comparison of H_γ theoretical absorption profiles, for T_{eff} = 100,000 K and log g = 6. The upper (dotted) profile is from an LTE model of Wesemael et al. (1980) for y = 10^{-6}. The lower profile (filled squares) is from their NLTE model for the same y. The full line is from Kudritzki's model for y = 0.01.

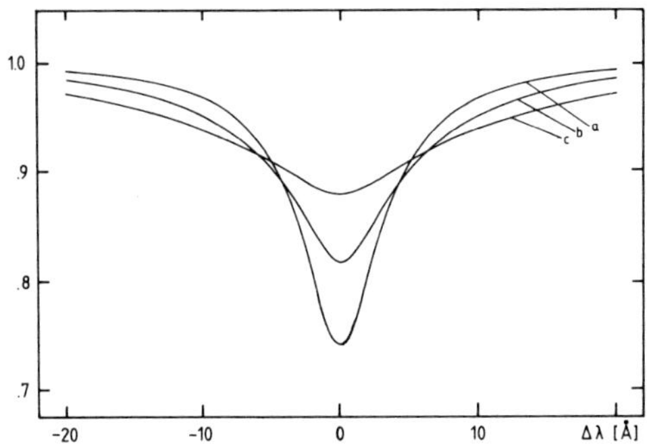

Figure 3: The variation of the theoretical H_γ profile along a curve of constant equivalent width on the log g - log T_{eff} plane,
a: T_{eff} = 55,000 K, log g = 5
b: T_{eff} = 65,000 K, log g = 6
c: T_{eff} = 75,000 K, log g = 7
These theoretical profiles (and all those in the following figures) have been convolved with a Gaussian instrumental profile having a FWHM = 3 A, except when stated otherwise.

for three different positions on the log g - log T_{eff} diagram. A good signal-to-noise ratio enables us to discriminate very easily between low T_{eff}, low g objects and high T_{eff}, high g ones. The situation is not so good at very low gravities; Figure 4 shows that in such cases we loose information on the temperature. One example of this problem is NGC 3242 (Kudritzki et al., 1981). The determination of T_{eff} would be much better if it were possible to use the ionization equilibrium of HeI and HeII; but in most cases the HeI lines (e.g. 4471) are too faint and/or are severely contaminated by nebular emission, and therefore provide only a lower limit for T_{eff}.

On the other hand, the determination of log g and helium abundance is quite accurate. Figure 5 shows the behaviour of HeII 4685 and HeII 4541, which yields a sensitive discrimination between low and high gravity. Figure 6 shows the effects of helium abundance. It is important to remark that the determination of log g and helium abundance is relatively insensitive to uncertainties in T_{eff}.

Our first application of the 'model atmosphere approach' to central stars of planetary nebulae (Méndez et al., 1981) was based on image-tube spectrograms obtained at the Cerro Tololo Inter-American Observatory (CTIO). More recently, we have used the SIT-Vidicon system with the R-C spectrograph of the CTIO 4-m telescope, and also the IDS with the Cassegrain spectrograph of the ESO 3.6 m telescope. A few objects were reobserved (NGC 1360, NGC 1535, Abell 36), and it was encouraging to find that three different observational techniques produce essentially the same results.

Figures 7 to 13 illustrate some of the fits we have obtained, and Table 1 gives the resulting atmospheric parameters for all the objects. Most of the earlier results (Méndez et al., 1981) remain unaffected. The exceptions are:
(a) The Vidicon data for NGC 7293 suggest a higher temperature. This was not surprising, because our single image-tube spectrogram of this central star was not very good. It is interesting to note that this modification of the temperature did not affect our previous determination of log g and the helium abundance.
(b) A better spectrum of NGC 4361 was obtained by adding 10 image-tube spectrograms obtained in 1981 with the R-C spectrograph of the CTIO 4-m telescope. The reductions and additions were made with the PDS microphotometer and associated software of the Kitt Peak National Observatory. The new analysis yields a somewhat smaller helium abundance than before.
(c) The Vidicon data for NGC 1360 yield a somewhat larger helium abundance than before.

It is necessary to remark that in some cases (Abell 36, Longmore 8, Abell 15) we do not obtain a good fit. The Balmer absorption profiles suggest a temperature equal to or lower than the lower limit imposed by the absence of the HeI 4471 absorption. In such cases we have given more weight to the T_{eff} suggested by the He lines; one thing to remember is that, in cases of higher temperature, the He lower limit is not so

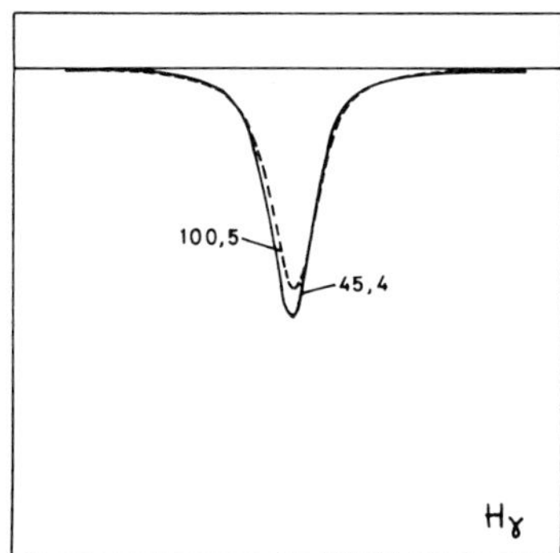

Figure 4: A comparison of H_γ profiles for T_{eff} = 100,000 K, log g = 5, and T_{eff} = 45,000 K, log g = 4. In both cases y = 0.09.

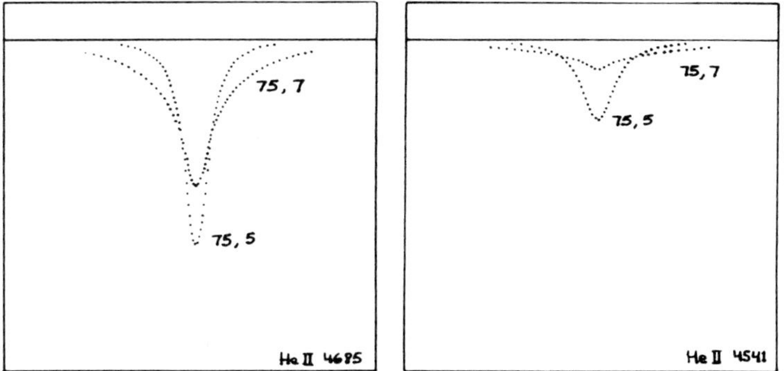

Figure 5: The effect of surface gravity on HeII 4685 and 4541, at T_{eff} = 75,000 K and y = 0.09.

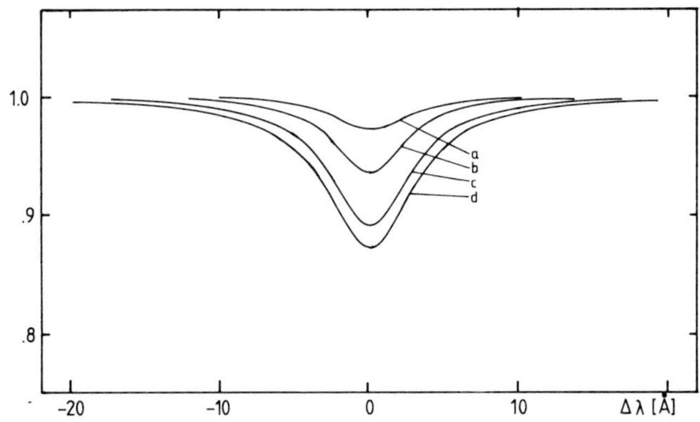

Figure 6: The effect of helium abundance on HeII 4541, at T_{eff} = 65,000 K and log g = 5. a: y = 0.01, b: y = 0.03, c: y = 0.09, d: y = 0.17.

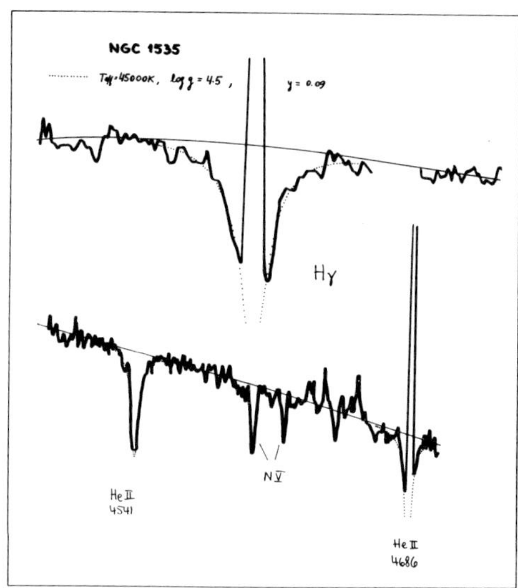

Figure 7: SIT-Vidicon line profiles of NGC 1535, fitted with non-LTE profiles convolved with FWHM = 2 A. The results confirm an earlier analysis by Méndez et al. (1981).

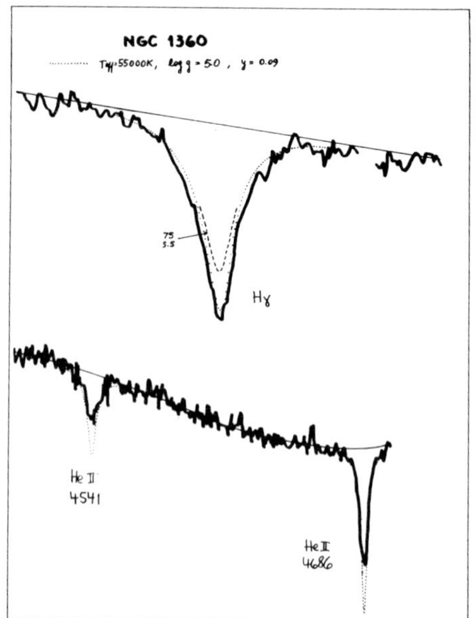

Figure 8: SIT-Vidicon line profiles of NGC 1360. The theoretical profiles are convolved with FWHM = 2 A. The results confirm an earlier analysis by Méndez et al. (1981), except for a slightly larger helium abundance. Notice how the theoretical H_γ profile for T_{eff} = 75,000 K fails to fit the observed profile.

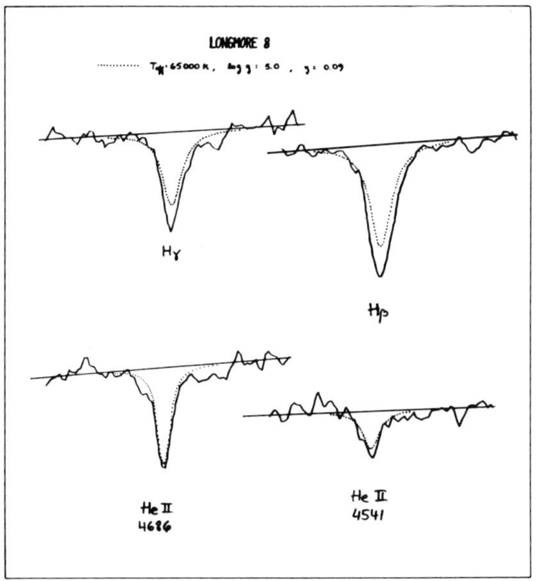

Figure 9: ESO-IDS profiles of Longmore 8. The theoretical profiles are convolved with FWHM = 3.5 A. The Balmer absorption profiles would indicate a lower T_{eff}, which is excluded by the high helium abundance and the lack of a detectable HeI absorption at 4471.

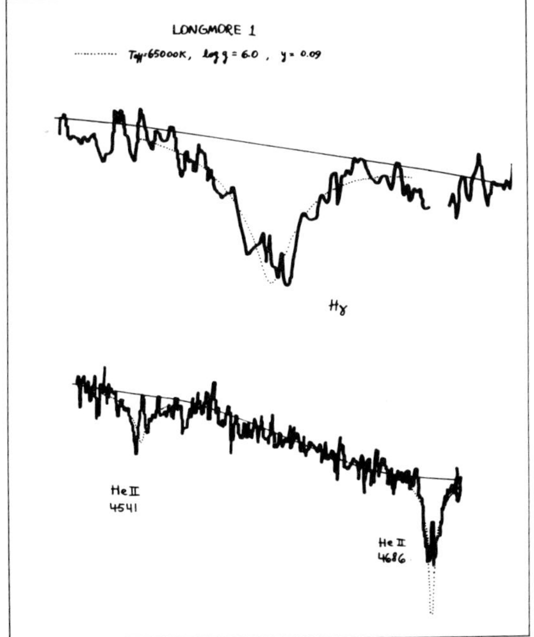

Figure 10: SIT-Vidicon line profiles of Longmore 1. The theoretical profiles are convolved with FWHM = 2 A.

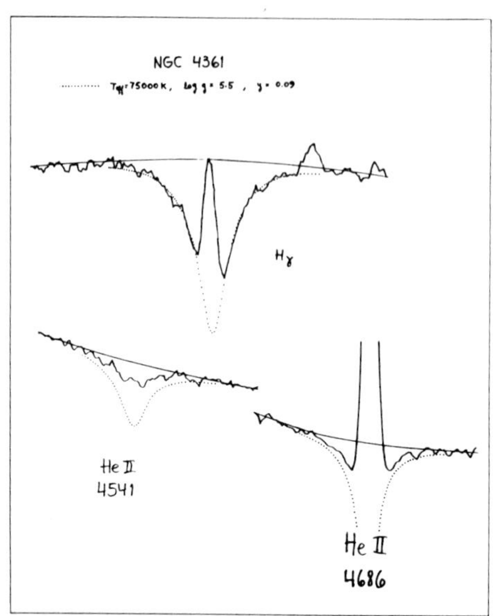

Figure 11: Absorption line profiles of NGC 4361, obtained by digitally adding 10 image-tube spectrograms taken with the CTIO 4-m R-C spectrograph. The theoretical profiles are convolved with FWHM = 3 A. The results confirm an earlier analysis by Méndez et al. (1981), except for a slightly lower helium abundance.

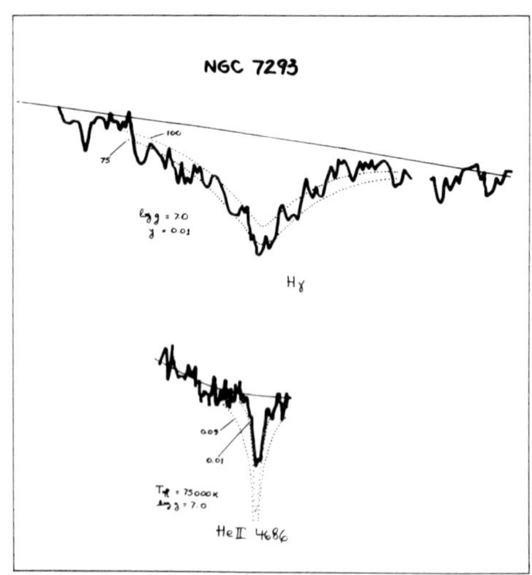

Figure 12: SIT-Vidicon line profiles of NGC 7293. The theoretical profiles are convolved with FWHM = 2 A. The H_γ profile indicates a larger T_{eff} than in the earlier analysis by Méndez et al. (1981), but log g and y remain unaffected.

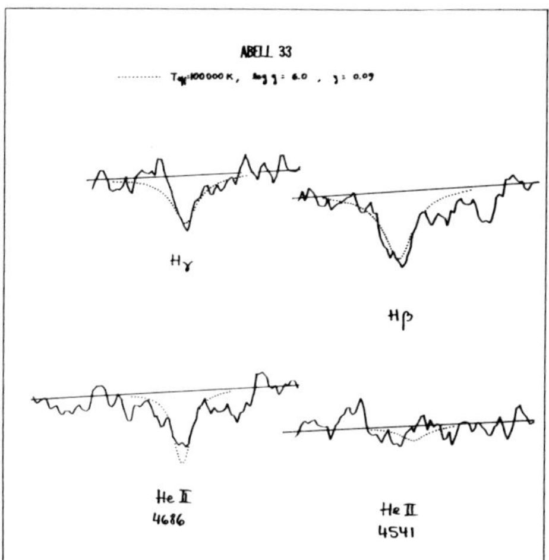

Figure 13: ESO-IDS profiles of Abell 33. The theoretical profiles are convolved with FWHM = 3.5 A.

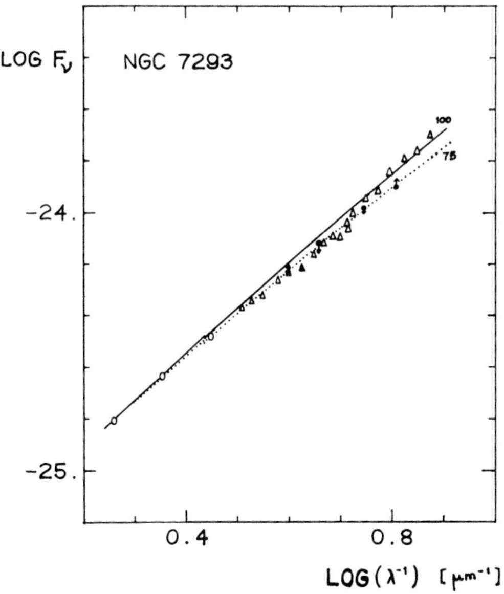

Figure 14: The 'undereddened' continuous energy distribution of the central star of NGC 7293. Open circles: Shao and Liller's (unpublished) UBV photometry. Filled circles: ANS data (Pottasch et al., 1978); the arrows indicate the corrections to the ANS fluxes applied by Bohlin et al. (1982). Triangles: IUE data (Bohlin et al., 1982). Two NLTE continuous energy distributions are plotted, for T_{eff} = 100,000 K and 75,000 K. In both cases log g = 7 and y = 0.01.

critical, and a good fit of a Balmer line (e.g. as for NGC 4361) might well be masking a similar systematic difference, which might be as large as 20,000 K. This problem will be mentioned below in connection with the analysis of the continuous energy distributions. One final remark is that, anyway, we expect the model atmospheres to yield reliable temperature differences; in other words, we can be reasonably assured that objects like NGC 1360, Abell 36 and Longmore 8 have lower T_{eff}s than objects like Abell 33, NGC 7293 and NGC 4361.

5. THE CONTINUOUS ENERGY DISTRIBUTIONS

Once the atmospheric parameters have been determined from line profile fits, we expect the continuous energy distribution derived from the corresponding non-LTE model atmosphere to agree with the observed continuous energy distribution. This critical test has been performed in several cases of lower temperature objects, always with satisfactory results (see e.g. Kudritzki and Simon, 1978; Thé et al., 1980; Kudritzki et al., 1982a), even in the case of Of objects like Zeta Puppis (Kudritzki et al., 1982b).

What is the present situation concerning central stars of planetary nebulae? Perhaps we should start by pointing out that the T_{eff}s usually found in the literature are not directly comparable with ours, because most of them are based on comparisons with black-body energy distributions, which are well known to produce higher temperatures than model atmospheres (see e.g. Figure 13 in Méndez et al., 1981). The magnitude of this difference depends on the fitting procedure used. If the fit extends from 1500 to 6000 A, then the difference is not larger than 15,000 K at T_{eff} = 80,000 K. However, if only the far UV is fitted, say from 1500 to 3000 A, then the difference can exceed 30,000 K; i.e. a non-LTE model with T_{eff} = 65,000 K gives the same slope in the Balmer continuum as a black-body at 100,000 K.

It is important to emphasize that for these high-temperature objects the continuous energy distribution becomes almost insensitive to temperature, which means that a great observational accuracy is required, both for UV and for visual fluxes, and that a very good determination of the interstellar extinction is essential.

Obviously, this kind of determination was impossible before the advent of ANS and IUE. Even with these ultraviolet satellites, we are barely able to deal with this difficult observational problem; an uncertainty of $\pm$10% in the absolute flux calibration can produce, even assuming a perfect visual magnitude, uncertainties of $\pm$20% in T_{eff} at 100,000 K.

Let us now discuss the IUE + ANS energy distribution of the central star of NGC 7293, as shown in Figure 14. In this figure we are assuming no interstellar reddening; compare with Figure 1 of Bohlin et al. (1982), where the same data are dereddened assuming E(B-V) = 0.012. As stated by them, the difference is not very large, and the precision of the slope is limited primarily by the uncertainty in the absolute calibration.

In our opinion, this continuous energy distribution does not permit an accurate determination of T_{eff}; it cannot be fitted with any model at all. The wavelength region λ > 2000 A can be fitted with a T_{eff} slightly below 100,000 K, which is in good agreement with our T_{eff}; but for λ < 2000 A there is an abrupt change in slope which no existing model atmosphere can reproduce. At this high temperature, the effects of line blocking are not expected to be so large. Assuming now that the quoted 10% uncertainty in the IUE fluxes can be used to rectify the energy distribution, we find a reasonable agreement with our line profile fits.

Other central stars in our sample have recently been measured with IUE, and an analysis of their continuous energy distribution has been presented in this Symposium by R.E.S. Clegg and M.J. Seaton. In a few cases the discrepancy is really large, particularly for NGC 1360 and Abell 36. According to their interpretation these two objects would be hotter than NGC 7293, in sharp contradiction with what the line profiles suggest.

Therefore, the situation appears to be rather confusing. At the present time it seems preferable to keep our minds open to all possibilities: the models may need improvements, the flux determinations also, and perhaps both are essentially correct and some objects have an ultraviolet excess. Concerning the model atmospheres, from the discussion on the line profile fits in §4 it would not be surprising to find that a slight shift is necessary in our 'temperature scale'. However, the reasonable agreement found for the energy distribution of NGC 7293 indicates that very probably some other factor is playing an important role.

6. DISCUSSION

The present (hopefully transitory) uncertainty concerning the effective temperatures of central stars makes it advisable to avoid a discussion of Zanstra temperatures; however, as stated above, surface gravities and helium abundances are well determined and deserve some comment. Figure 15 shows the positions of the 11 central stars on the log g - log T_{eff} diagram. It is interesting to notice that the two objects with higher gravities (NGC 7293 and Abell 7), which are presumably more advanced in their evolution towards the white dwarf stage, have surface helium abundances significantly (a factor of ten) smaller than solar. This implies the action of gravitational settling and provides a very strong observational connection between these central stars and DA white dwarfs.

A complete picture of helium abundances in central stars must include those objects which appear to have little or no hydrogen in their

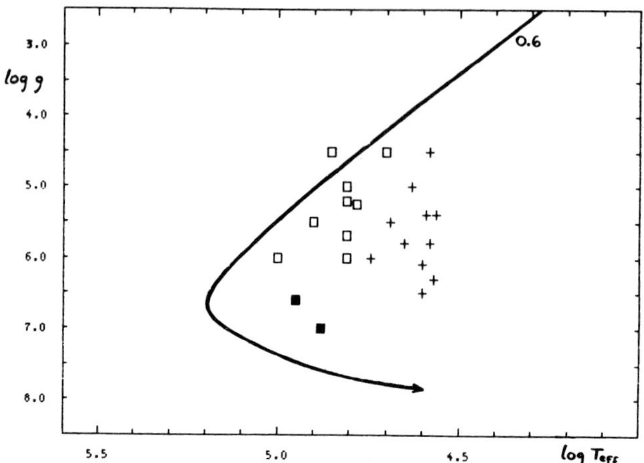

Figure 15: The log g - log T_{eff} diagram for central stars of planetary nebulae (squares) and other hot subluminous stars (plus signs; Hunger et al., 1981). The two filled squares correspond to Abell 7 and NGC 7293, both with y = 0.01. The solid line is a theoretical evolutionary track for a star of 0.6 solar masses descending from the asymptotic giant branch (Schönberner, 1981).

Table 1

ATMOSPHERIC PARAMETERS OF CENTRAL STARS

OBJECT	T_{eff}	log g	$y = \frac{N(He)}{N(He) + N(H)}$
NGC 1360	60^{+15}_{-5}	5.2 ± 0.2	0.06 ± 0.03
NGC 1535	50^{+10}_{-5}	4.5 ± 0.3	0.09 ± 0.03
NGC 3242	70^{+30}_{-20}	4.5 ± 0.5	0.10 ± 0.03
NGC 4361	80 ± 10	5.5 ± 0.3	0.05 ± 0.02
NGC 7293	90 ± 10	6.6 ± 0.3	$0.01^{+0.01}_{-0.005}$
Abell 7	75 ± 10	7.0 ± 0.5	0.01 ± 0.005
Abell 15	65 ± 10	6.0 ± 0.5	0.09 ± 0.04
Abell 33	100 ± 30	6.0 ± 0.5	0.09 ± 0.05
Abell 36	65 ± 10	5.2 ± 0.3	0.13 ± 0.04
Longmore 1	65 ± 10	5.7 ± 0.3	0.10 ± 0.03
Longmore 8	65 ± 10	5.0 ± 0.5	0.11 ± 0.03

spectra: the WC central stars (for a recent reclassification see Méndez and Niemela, 1982) and several objects with predominantly absorption line spectra, e.g. NGC 246, K 1-27, Longmore 3 and Longmore 4 (Méndez and Kudritzki, in preparation). Several authors have pointed out that these objects may be likely progenitors of non-DA white dwarfs. Unfortunately, non-LTE model atmosphere analyses of these objects are not yet possible.

From the non-LTE analyses already performed, it appears reasonable to suggest that the time scale for helium depletion in hydrogen-dominated atmospheres of PN central stars is comparable to or shorter than the time scale for nebular dissipation. A better determination of this time scale from a more numerous sample may help to put constraints on theoretical models of post-AGB (asymptotic giant branch) evolution.

It is also of interest to compare the positions of PN central stars in Figure 15 with the positions of 11 hot subluminous stars not associated with planetary nebulae (Hunger et al., 1981). The clean separation obtained cannot be affected by uncertainties in T_{eff}, or (obviously) in the distances, and is a strong argument favoring the assignment of systematically lower masses to the hot subluminous stars not associated with PN. This hint may lead to some clarification of the late stages in the evolution of low-mass stars. A reasonable working hypothesis (Hunger and Kudritzki, 1981) is to interpret the non-PN hot subdwarfs as stars which are not able to reach the asymptotic giant branch, and evolve directly from the horizontal branch towards the white dwarf stage, providing the lower-mass end of the white-dwarf mass distribution.

REFERENCES

Bohlin, R.C., Harrington, J.P., Stecher, T.P. 1982, Astrophys. J. 252, 635.

Gruschinske, J., Kudritzki, R.P. 1979, Astron. Astrophys. 77, 341.

Hunger, K., Gruschinske, J., Kudritzki, R.P., Simon, K.P. 1981, Astron. Astrophys. 95, 244.

Hunger, K., Kudritzki, R.P. 1981, The ESO Messenger, No. 24, p. 7.

Kudritzki, R.P. 1976, Astron. Astrophys. 52, 11.

Kudritzki, R.P. 1980, Astron. Astrophys. 85, 174.

Kudritzki, R.P., Méndez, R.H., Simon, K.P. 1981, Astron. Astrophys. 99, L15.

Kudritzki, R.P., Simon, K.P. 1978, Astron. Astrophys. 70, 653.

Kudritzki, R.P., Simon, K.P., Hamann, W.R. 1982b, submitted to Astron. Astrophys.

Kudritzki, R.P., Simon, K.P., Lynas-Gray, A.E., Kilkenny, D., Hill, P.W. 1982a, Astron. Astrophys. 106, 254.

Méndez, R.H., Kudritzki, R.P., Gruschinske, J., Simon, K.P. 1981, Astron. Astrophys. 101, 323.

Méndez, R.H., Niemela, V.S. 1982, in IAU Symp. 99.

Mihalas, D. 1972, NCAR Tech. Note TN/STR-76.

Pottasch, S.R., Wesselius, P.R., Wu, C.-C., Fieten, H., van Duinen, R.J. 1978, Astron. Astrophys. 62, 95.
Schönberner, D. 1981, Astron. Astrophys. 103, 119.
Simon, K.P. 1982, Astron. Astrophys. 107, 313.
Thé, P.S., Tjin A Djie, H.R.E., Kudritzki, R.P. and Wesselius, P.R. 1980, Astron. Astrophys. 91, 360.
Wesemael, F., Auer, L.H., Van Horn, H.M., Savedoff, M.P. 1980, Astrophys. J. Suppl. Ser. 43, 159.

* Based partly on observations made at the ESO, La Silla, Chile

[1] Instituto de Astronomia y Fisica del Espacio, C.C. 67, Suc. 28, 1428 Buenos Aires, Argentina

[2] Visiting Astronomer, Cerro Tololo Inter-American Observatory, operated by the Association of Universities for Research in Astronomy, Inc., under contract with the U.S. National Science Foundation

[3] Member of the "Carrera del Investigador Cientifico", Conicet, Argentina

[4] Institut für Astronomie und Astrophysik der Universität München, Scheinerstr. 1, D-8000 München 80, Federal Republic of Germany

MATHIS: What do you think the effects of line blanketing would be in the IUE spectral region, and what would be the effects of a stellar wind on the models?

MÉNDEZ: Concerning line blanketing, I hardly dare to make a prediction, although perhaps we should not expect a large effect. I would certainly like to see models incorporating line blanketing!

As to the second part of your question, some of these stars have winds, but we have restricted our analysis to those objects showing predominantly absorption lines and have used lines which are formed rather deep in the photosphere. Therefore, we do not expect the wind, when present, to affect our results significantly. The situation changes when there is a velocity field deep in the photosphere. Probably for the Of and Ofp central stars, and certainly for WC central stars, we need hydrodynamic models of extended atmospheres which, of course, are not yet available.

HEAP: Have you determined the N or C abundances in any of these stars?

MÉNDEZ: Not yet. Some analysis of resonance ultraviolet lines in the spectra of sdO stars has been done at Kiel. We have IUE high dispersion spectra of a few central stars, but their study has just started.

HARRINGTON: Since you showed our plot of the ultraviolet data for NGC 7293, I think I should comment on the Zanstra temperatures we found (Bohlin, Harrington and Stecher, 1982, Ap. J. 252, 635): T_Z(H I) = 100 000 K and T_Z(He II) = 123 000 K. In deriving T_Z(He II), we assumed a normal helium abundance. With the low helium abundance which you find, T_Z(He II) will surely be below 123 000 K, perhaps close to the value of 100 000 K which you derive from the line profiles.

CLEGG: I wish to draw attention to a result derived from IUE data and reported by several people at the second poster session. Hot subdwarf O stars often show a flux, for $\lambda < 1500$ Å, in excess of black-body or NLTE model atmosphere predictions. NGC 1360 is the best example. Although the IUE flux calibration is a little (perhaps 15%) uncertain in this spectral region, the effect seems to be quite real.

MÉNDEZ: Yes, this discrepancy appears to be serious. A comparison of observed line profiles indicates that objects like NGC 1360 and A 36 have lower temperatures than, e.g. NGC 7293, in contradiction with what is suggested by the observed (IUE) continuous energy distributions.

CLEGG: The problem is that the line blanketing should be treated in NLTE - but this is impossible for 92 elements, each with many ions and energy levels! The problem could be tackled in LTE (although much atomic data is missing), but to consider all lines as being formed in pure absorption would lead to erroneous model temperature structures.

LYNAS-GRAY: What is the interpretation of low helium abundances in two high gravity stars? For sdO stars, the distinction between high and low helium abundance is at $T_{eff} \approx 40\,000$ K.

MÉNDEZ: Both for sdO's and for PN nuclei, the most probable explanation would appear to be gravitational settling. However, the detailed mechanisms at work in each case may differ and the internal structures almost certainly differ. These problems are essentially unsolved at the present time.

RENZINI: How can you distinguish a sdO star from a PN nucleus whose nebula has already dispersed?

MÉNDEZ: The sdO's and PN nuclei do not overlap on the log g / log T_{eff} diagram, and it seems that sdO's cannot be explained as post-AGB objects. We still have no quantitative information on the gravities of the lower temperature central stars because most of them have Ofp or WC spectra, but we would expect all low T_{eff} central stars to have much lower gravities than most of the sdO stars not associated with PN.

WEIDEMANN: In this connection, the separation of the locations of the sdO's and the nuclei of PN in the log g / T_{eff} diagram (Hunger et al., 1981, Astron. Astrophys.; Méndez et al., 1981, Astron. Astrophys.) is evidently due to the fact that sdO's have masses smaller than 0.55 $M_\odot$, as would be expected from stellar evolutionary calculations which show that, e.g. horizontal branch stars less massive than 0.55 $M_\odot$ do not go up to the AGB but move over directly to the White Dwarf region.

EVOLUTION AND MASS DISTRIBUTION OF CENTRAL STARS OF PLANETARY NEBULAE

D. Schoenberner
Louisiana State University Observatory
Louisiana State University
Baton Rouge, Louisiana 70803-4001

and

V. Weidemann
Institut fuer Theoretische Physik und Sternwarte
Universitaet Kiel
Federal Republic of Germany

1. INTRODUCTION

Considerable progress has been made in our understanding of the evolution of the central stars of planetary nebulae (NPN) compared to the situation five years ago at the Ithaca Symposium where Shaviv (1978) and Paczynski (1978) reviewed the subject. Shaviv stressed the necessity to start theoretical calculations with realistic initial models but doubted - in view of the loops in the HR diagram made by flashing stars - if the Harman-Seaton sequence could be taken as a single evolutionary sequence. Paczynski pointed out how strongly the theoretical rate of evolution depends on the stellar mass - a result which had appeared in his earlier calculations (1971) - and expected the existence of more flashing NPN's of the FG Sagittae type among the luminous ($L > 10^4\ L_\odot$) central stars, for which the core mass luminosity relation ($M_c > 0.7\ M_\odot$) combined with the core mass interpulse time relation predicts fairly short (2.10 yrs) intervals between flashing events. Weidemann, however, at the Symposium and shortly thereafter (1977a) concluded in view of the lower effective temperature derived by Pottasch et al. (1978) and the observed narrow mass distribution of white dwarfs around a 0.6 $M_\odot$ combined with the theoretical predicted <u>horizontal</u> tracks from the red giant branch towards the NPN region at a luminosity given by the core mass luminosity relation that the high luminosity part (and also the "upturn") of the Harman-Seaton sequence does not exist. He also proposed an increase in the distances by an average factor of 1.3 compared to the Seaton/Webster (Seaton, 1968) or Cahn/Kaler (1971) scale in order to bring the observed NPN on the 0.6 $M_\odot$ track in the HR diagram and to lower the NPN birth rates to a value compatible with white dwarf birth rates. Renzini (1979) took up Paczynski's remarks and results and studied the relation between

D. R. Flower (ed.), Planetary Nebulae, 359–371.

evolutionary times of NPN and PN expansion carefully thereby predicting visibility ranges for PN as a function of the NPN mass. Since it appeared that - within a typical nebular life-time - only massive NPN could illuminate their nebulae at lower observed NPN luminosities ($M_{NPN} > 0.6\ M_\odot$ for $\log L/L_\odot < 2.6$) he concluded that the NPN do not form a single evolutionary sequence but devised a new scheme according to which the high luminosity NPN have small masses and belong to low mass progenitors whereas the low luminosity NPN have larger masses and belong to higher masses progenitors. He thus predicted differences in population characteristics for both groups of NPN, higher nebular masses and different chemical compositions for the fainter NPN, and the existence of an upturn caused essentially by NPN with masses of 0.7 $M_\odot$ whose nebulae are visible at the comparably highest luminosities of the NPN. A similar result emerged from Haerm and Schwarzschild's (1975) approach. They took off the envelope from AGB-models and got remnants of about 0.65 $M_\odot$ which evolved rapidly into the NPN region. Their lifetimes as luminous stars ($\sim 10^3\ L_\odot$), however, exceeded considerably that of any associated nebula, and again only more massive remnants would be able to explain low luminous NPN.

In the meantime Schoenberner (1979) had calculated evolutionary tracks for low mass stars all the way from central helium burning, through the asymptotic giant branch, without flash suppression, and steady mass loss included, towards the white dwarf region, thereby fulfilling Shaviv's call for more realistic initial models for NPN evolution. The existence of these calculations together with a general correlation between NPN location in the HR diagram and nebular radii - which had been established in the past - prompted Schoenberner and Weidemann (1981) to study the empirical material in order to check Renzini's predictions. They confirmed the general correlation between luminosity and nebular radii, but found by a comparison of nebula expansion time scales with theoretical time scales of NPN evolution the surprising result that the mass distribution of NPN appears to be even narrower than that of white dwarfs, essentially confined to $0.55 \leq M/M_\odot \leq 0.64$, and that within the observed sample there were only very few NPN with masses above 0.64 $M_\odot$. This implies that the HR positions of the NPN present essentially an evolutionary sequence and thus explains the general correlation between luminosities and nebular radii. However, even within the narrow mass range in which most NPN occur, differential effects of the Paczynski-Renzini scheme are considerable, in the sense that a 0.64 $M_\odot$ NPN evolves much faster than a 0.55 $M_\odot$ NPN. The upturn is absent since stars in that mass range reach already the observed low luminosities. For these low mass NPN thermal pulses are expected to be a rare event during a PN lifetime (interpulse time $\sim 10^5$ yrs) which makes Shaviv's (1978) and Paczynski's (1978) predictions irrelevant. In a more fully elaborated version of his investigation Schoenberner (1981) reached further conclusions, concerning the role of thermal pulses in the ejection of PN, the remnant hydrogen envelopes on NPN, the modus of burning and the time scales involved. Of highest importance with respect to the evolution on the asymptotic giant branch and the general picture of late stages of stellar evolution is the fact that enrichment of PN - as observed and studied in numerous papers in the past - with helium and other burning products occurs already at the low masses derived, in contrast to the canonical theory of stellar evolution, which predicted enrichment only for higher core masses above 0.8 $M_\odot$. This fact is in line with similar conclusions about lower luminosities of Mira's (considered to be progenitors of PN) and carbon stars. The questions and problems in this context are numerous: they are dealt with extensively in a forthcoming review by Iben and Renzini (1983), to which we refer.

In the following we shall present the essential steps and results of our investigation and the implications. Then we will discuss objections and outline future research.

2. EVOLUTION AND MASSES OF THE CENTRAL STARS

We first consider the empirical relation between absolute optical luminosities, M_V, and nebular radii, R_N, based on Cahn and Kaler (1971) distances, photometry by several sources (for details and quotations: see Schoenberner, 1981), and corrected for interstellar absorption. Figure 1 shows that relation for a local ensemble (Cahn and Wyatt, 1976) in order to avoid selection effects in favor of NPN with high luminosities. A gap appears at $M_V \sim 5$, which is present also in the total ensemble, and in older material (O'Dell, 1974) and more objects are below that gap. Uncertainties in the distances can not change the essential structure of the diagram. A radius histogram shows the

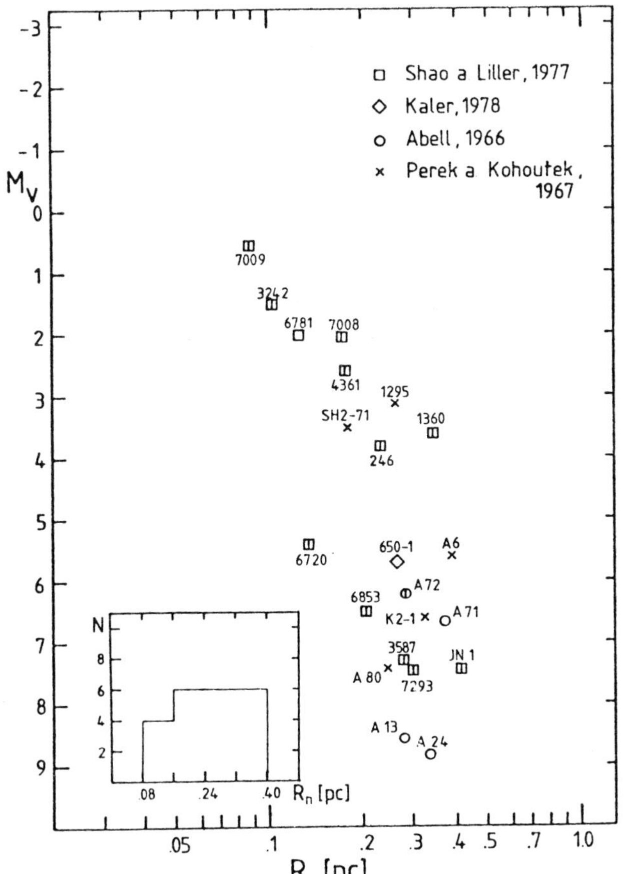

Figure 1. Absolute visual magnitudes for 22 central stars with optically thin nebulae from a local ensemble vs. nebular radii, together with a radii histogram.

number of NPN per radius interval to be nearly constant and thus suggests constant expansion velocity.

In Fig. 2 a larger ensemble is superposed with theoretical evolutionary tracks computed by Schoenberner (1979). In doing this, we converted the bolometric luminosities of the models into absolute magnitudes M_v and plotted these vs. ages. It is assumed that the nebular shell is formed in a time very short compared to the nebular lifetime, and that the remnant is able to maintain its thermal equilibrium. The NPN ages are counted from $T_{eff} = 10^{3.7}$. The evolution of these post-AGB models is accelerated by mass loss (stellar wind) only for $T_{eff} < 10^4$ K (for more details, see Schoenberner, 1981,1982). We want to emphasize the great sensitive of the fading time on the NPN mass (= core mass of the AGB-progenitor), as displayed in the Figure. The reason is the large increase of luminosity and the large decrease of available fuel as one goes from low to higher NPN masses (Schoenberner, 1982). The fading of these models is much faster than in Paczynskis calculation, and they reach lower luminosities during the PN lifetime. A 0.6 $M_\odot$ NPN may thus be found (within a PN event) at log $L/L_\odot$ = 2.0. The reason for that difference has been explained in detail by Schoenberner (1981). The tie-in with the observed NPN is made by the assumption of constant nebular expansion, with v = 20 km/s, changes for different expansion velocities are minor, as indicated by the arrows in the lower left corner. The thick broken line excludes NPN for which the nebulae are probably optically thick, so that the Shklovsky distances (Cahn and Kaler, 1971) do not apply. The distances are here increased by a factor of 1.2 compared to the CK distances. The thin broken lines give effective temperatures taken from the model calculations.

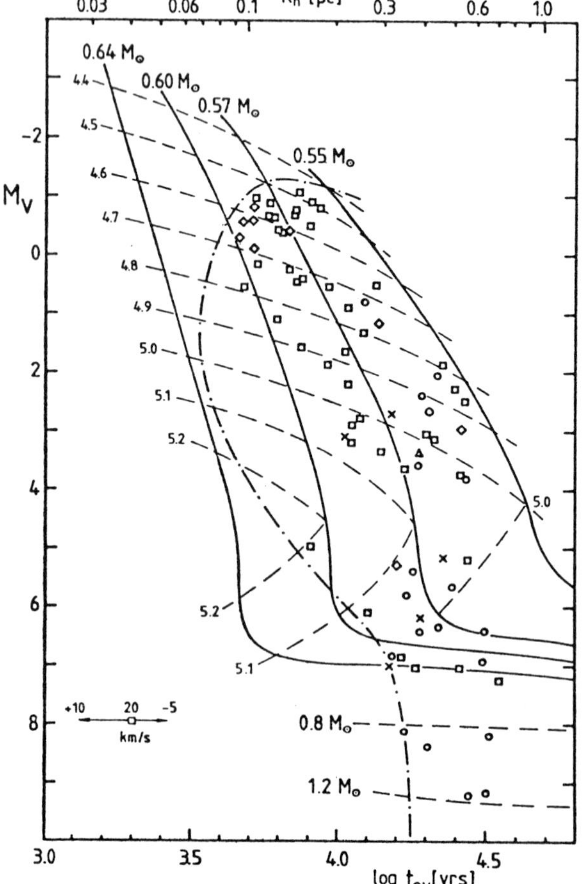

Figure 2. Absolute visual magnitudes vs. ages of observed central stars and post-AGB models. For explanations, see text.

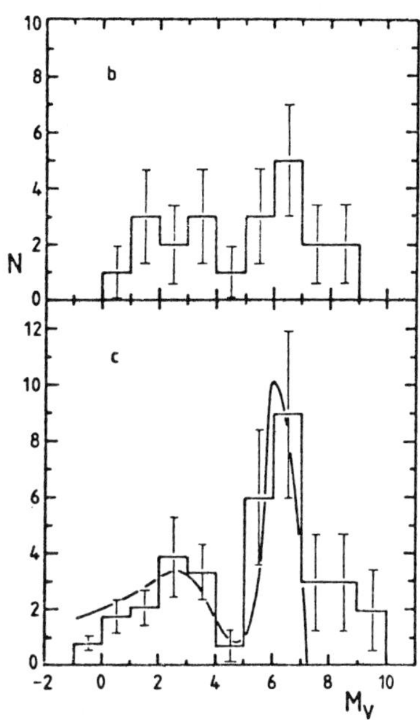

Figure 3. Above: Local ensemble of Fig. 1; Bottom: Ensemble of Fig. 2, corrected empirically for different sample volumina. The curved line gives the theoretical prediction of a 0.57 $M_\odot$ model.

The Fig. 2 shows that all displayed NPN are covered by the evolutionary tracks from 0.55 $M_\odot$ to 1.2 $M_\odot$. The gap is now explained by accelerated fainting during the extinction of the hydrogen shell source after a phase of quiet hydrogen burning. The important conclusion is that the PN-ejection and the subsequent central star evolution take place between two successive thermal pulses, with the outcome that no further pulse will (in most cases) occur. The observed positions indicate evolution of the NPN with a luminosity independent narrow mass distribution. There are only very few NPN with probable masses above 0.64 $M_\odot$, concluded from their (uncertain) positions at the lowest luminosities. Evolutionary tracks with thermal pulses, at the moment of PN ejection or during PN lifetime, show a completely different behaviour (Schoenberner, 1981). Not only would the empirical material not display a luminosity independent mass distribution, but also the gap could not be explained. A comparison of theoretical and empirical luminosity functions (Fig. 3) which is independent of the nebular expansion velocity, and from the choice of the zero point of NPN evolution, also agreees only for NPN evolution with quiet hydrogen burning. The luminosity function (local ensemble) shows more objects at faint luminosities, contrary to what would be expected if these NPN were the products of high mass (and therefore rare) progenitors.

The mass distribution, as evident from Fig. 2, corrected for selection effects, is extremely narrow (Fig. 4). We can identify only about 25% of the faint NPN ($M_V > 5$) with a possible mass in excess of 0.64 $M_\odot$ (this fraction varies sensitively with the assumed distance scale!). The maximum is at exactly the same value (0.58 $M_\odot$) which was found for the DA white dwarfs (Koester, et al., 1979), however, the white dwarf distribution appears to be broader (broken lines). Although the true WD mass distribution may be even narrower (due to observational uncertainties) there are definitely white dwarfs with $M < 0.55$ $M_\odot$. We thus presume that the progenitor of white dwarfs with $M < 0.55$ $M_\odot$ do not go through the PN stage, but evolve directly from the horizontal branch to the WD region, thereby crossing the sdO region (Hunger et al., 1981).

The steep decline towards higher NPN masses should be real even if there may be more higher mass NPN which were not in our ensemble, hidden

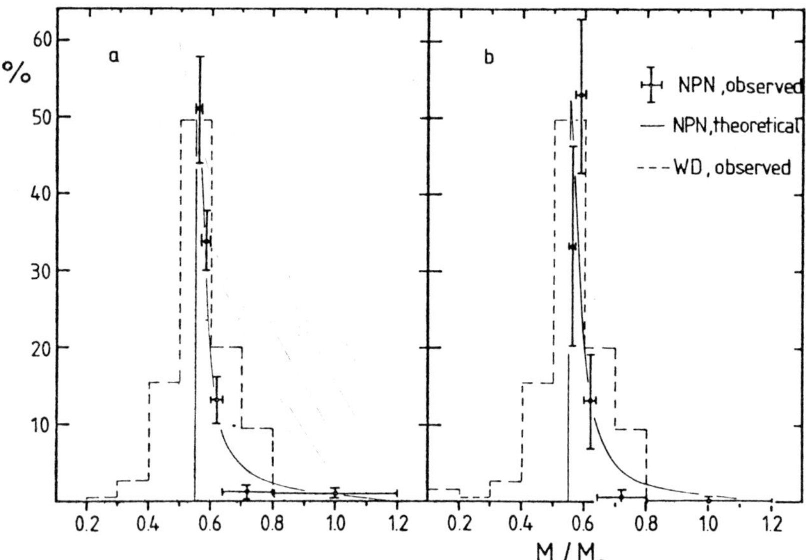

Figure 4: Left: Mass distribution, for the (corrected) ensemble of Fig. 2; Right: The same for the local ensemble. For comparisons, the observed (DA) white dwarf distribution (Koester et al., 1979) as well as the predicted according to the Barkat and Tuchman (1980) ejection scheme.

in massive PN's or undetected or with unmeasured central stars. The reason is: already the 0.60 $M_\odot$ model predicts 3.5 times more NPN with $M_v > 5$, than with $M_v < 5$. More massive NPN stay for almost the total PN-lifetime at luminosities $M_v > 7$. They must be really rare in order to account for the paucity of observed objects in that luminosity range, even if their number has been considerably underestimated. Note that this follows only from the appearance of the NPN luminosity function (Fig. 3) and the differentially evolutionary behaviour of post-AGB models with different masses. Theoretically, a very narrow mass distribution can be predicted, if PN's are ejected according to the shock mechanism proposed by Barkat and Tuchman (1980) combined with steady mass loss on the AGB with a Reimers scale factor $\eta \geq 1$.

Consequences for the initial/final mass relation in stellar evolution have been discussed by Weidemann (1981). That this function runs very flat for progenitor masses from the galactic turnoff up to at least 2.5 $M_\odot$ has been concluded already from white dwarfs which occur in clusters with known turn-off masses (see Weidemann, 1977b, 1979). Recent investigations by Koester and Reimers (1981), Reimers and Koester (1982) have demonstrated that white dwarfs do occur up to progenitor masses of 7 $M_\odot$ with WD masses up to 1.2 $M_\odot$. Thus there remains little doubt that the majority of all stars evolve through the PN stage and leave remnants with small masses around 0.6 $M_\odot$. More massive NPN's or white dwarfs are as rare as progenitor stars with, say $M > 4\ M_\odot$, i.e., about 20% or less, depending on the initial mass function and the past rate of star formation (see Koester and Weidemann, 1980). In the white dwarf case efforts to confirm higher-than-average masses (Weidemann and Koester, 1980, Schulz and Wegner, 1981, Weidemann, 1982) brought as yet meager results, but further studies can be made.

In the NPN case one has to search for and investigate especially low luminosity objects - as predicted by the visibility criteria of Renzini. Spectroscopic analysis has been possible up to now only for a few fairly bright central stars. The result by Mendez et al. (1981) confirm the surface gravities and temperatures expected by our method, which is independent of NPN temperature.

In the cases in which reliable temperature determinations are available a consistency test can be made. If the theoretically derived temperature differs, a change in the distance can always enforce agreement. Schoenberner (1981) showed that the smaller distances proposed by Acker (1978) are inconsistent in that the ages of the central stars and the ages derived from the nebular diameter are more discrepant.

Enrichment - and therefore dredge-up - evidently occurs at the small core masses corresponding to the range of our observed ensemble. And the low excitation Pop II objects - in Kaler's notation - do not form a separate group by mass. Schoenberner (1981) also finds no correlation between excitation parameter and temperature of the exciting star (see his Fig. 15) and concludes that mechanical heating of the surface layers must be responsible for the high He ionization.

3. DISCUSSIONS

Before we start with a discussion of several objections which have been raised or could be made against the presented evolutionary scheme, let us make a few remarks. The precise knowledge of effective temperatures of NPN - besides the distances - is essential for the determinations of meaningful bolometric luminosities. Up to now, these temperatures are known - if at all - only for a very limited number of objects. Our scheme is independent of temperature and uses only visual luminosities in order to study NPN evolution. We have to adopt, however, a mean value for PN expansion velocities. But this assumption is not as bad as it appears at the first glance: Only very few of measured PN expansion velocities exceed the mean of 20 km/s by more than a factor of two on either side (Robinson et al., 1982), whereas the fading time of NPN varies already by a factor of 10 between $0.55\ M_\odot$ and $0.64\ M_\odot$. Moreover, when using the luminosity function of NPN, we are even free from expansion velocities and absolute timescales. We thus believe that our approach is at present more reasonable than any representation which uses temperatures and luminosities.

Let us now first discuss the reliability of the evolutionary tracks. Additional computations have been made for low mass stars in the meantime by Kovetz and Harpaz (1981) and Iben (1982) which reproduce the Schoenberner (1979) results, especially the fact that a 0.6 $M_\odot$ NPN reaches low luminosities during a nebular lifetime. The reasons for the different results of Paczynski (1971) have been explained by Schoenberner (1981), see also Iben and Renzini (1982) . The luminosity at the horizontal part of the tracks correspond to that given by the core mass luminosity relation for quiet hydrogen burning at the asymptotic giant branch. Since in a post flash phase - before the hydrogen shell has been fully reactivated - the luminosity is lower for a considerable fraction of the interpulse time, one could object to the use of these tracks if PN ejection would occur with equal probability sometimes during the interpulse interval. But, as shown by Schoenberner (1981), only post-AGB models with a fully activated H-shell and a degraded He-shell are able to account for the observed central star luminosity function. Additionally, it is reasonable to assume that the star has to exceed a threshold luminosity in order to enter the region of enhanced mass loss rates, i.e., the PN-ejection region. Then it has to gain at least the preflash luminosity when recovering from that flash. This occurs when about half of the interpulse time has been elapsed. But then the hydrogen burning shell is already fully activated, and the helium already degraded. The remaining luminosity increase is then $\log L/L_\odot \sim 0.1$ at $\log L/L_\odot \sim 3.3$ and $\log L/L_\odot \sim 0.04$ at $\log L/L_\odot \sim 3.7$, respectively. The use of our tracks may then lead to an underestimate of NPN masses (on the average) by only 0.001 $M_\odot$ and 0.003 $M_\odot$, respectively. The core-mass luminosity relation itself is well established and agrees numerically between different authors as shown by the following values for $M_c = 0.6\ M_\odot$, $\log L/L_o$ = 3.71, 3.78, 3.79, 3.80, 3.72 for Gingold (1974), Schoenberner (1979), Kovetz and Harpaz (1981), Wood and Zarro (1981), and Iben (1982), or 3.68 for the more crude models of Paczynski (1971).

Important for the appearance of the tracks in Fig. 2 is the fading time, i.e, the time a post-AGB model needs to reach a limiting luminosity during a PN lifetime. This time depends on the available fuel, which, in turn, depends on the envelope composition, for the metal poor post-AGB models of Gingold (1974) and Iben (1982) have much higher envelope masses at appropriate positions in the HR-diagram. For instances, the 0.6 $M_\odot$ model of Iben (Z = 0.001) is burning twice the amount of hydrogen (with the same luminosity) than the corresponding Schoenberner (1979) Pop I model (Z = 0.021), thus needing also twice the time to evolve through corresponding positions in the HR-diagram. Fortunately, the limiting luminosity does not depend on the envelope mass. Thus the 0.6 $M_\odot$ track of Iben (1982) has to be placed in Fig. 2 as follows: displaced by 0.3 dex to the right from the original 0.6 $M_\odot$ track. This then would shift the NPN mass distribution peak from 0.58 $M_\odot$ to 0.61 - 0.62 M_o. We think, however, that a Pop I composition is appropriate for most PN and that the application and labelling of the tracks in Fig. 2 is therefore justified. As long as any composition differences are not as drastic as in the example above, we can neglect them.

It is exactly the high sensitivity of the fading time on M_c which secures the narrow mass distribution derived, even if individual distances, magnitudes or nebular expansion velocities are uncertain. This leads us to the second objection: the distances used are based on the Shklovsky method and therefore uncertain. True, but as long as these uncertainties are within a factor of, say 2, the pattern of Figs. 1-3 will not be changed. Vice versa, the narrow confinement of the NPN with optically thin nebulae points to the applicability of the Shklovsky method (see Schoenberner, 1981). This is supported by a recent study of Daub (1982). The consistence test has shown that the smaller distances and PN radii, derived by Acker (1978), give incompatible expansion and NPN evolution ages. Expansion velocities would have to be lowered by factors between 5 and 10, or the NPN evolution accelerated by the same factors in order to reach agreement. Both possibilities must be ruled out: expansion velocities, if not 20 km/s, do not differ very much (see Robinson et al., 1982, for a recent study), again the corresponding changes in the derived mass distribution of Fig. 2 are minor. Continued mass loss of NPN, partly derived from IUE observations, is not sufficiently strong as to accelerate the evolution, however may be important for the thickness of remnant hydrogen layers on the final white dwarfs. If it were strong enough to accelerate the NPN evolution significantly, this would result in even lower NPN masses (the tracks in Fig. 2 would be shifted to the left).

The final objection concerns selection effects. Our ensemble has been selected by existing photometry of the central stars. Indeed, there may be many cases in which NPN are not visible, or too faint to be measured. Especially Pottasch (1981) has selected 12 objects with very faint NPN, $m_v \sim 18-19$, and has presented an HR diagram which shows several hot NPN at effective temperatures far beyond the upper limit of our ensemble, 150 000 K. Aside from the fact that his temperature determinations are very uncertain, the existence of hot NPN with high luminosities is in contradiction to evolutionary tracks which predict such high temperatures only for massive NPN with short evolutionary time scales which pass this region before they are able to ionize the nebulae. Furthermore, the visual magnitudes are extremely faint and uncertain in these cases. In 5 of his 17 objects photometry by Kaler (1978) and Kohoutek and Martin (1981) yields luminosities which are several magnitudes higher. However, it is almost certain that there are more NPN at faint luminosities which have not yet been detected. A striking example is 158 + 17°1 (Purgathofer and Weinberger, 1980). Observational efforts should be made to increase the local ensemble and to measure the luminosities and other parameters of these faint NPN. Similarly, efforts to determine nebular abundances should be concentrated on a more local ensemble, in order to confirm or refute our result that enrichment occurs independent of the position in the HR diagram.

Finally, we want to discuss the role of thermal pulses during the PN lifetime in more detail. When an AGB-star ejects its envelope between successive thermal pulses, then the remnant may experience a thermal pulse as long as its hydrogen shell remains active. Now the ratio between the lifetime of the H-shell during the NPN-evolution and the thermal pulse period is about 0.1 for masses above 0.6 $M_\odot$, independent of mass. Stars with cores below 0.58 M_o do not experience full amplitude flashes, and the flash period increases with core mass (Schoenberner, 1979), contrary to the interpulse period-core mass relation (valid only for well developed thermal pulses). Thus, for NPN near the lower mass limit of 0.55 $M_\odot$ we expect a 30-40% chance for them to experience a final thermal pulse (Schoenberner, 1982). Altogether, we estimate that 15% of all NPN will undergo a (final) thermal pulse, and that pulse - not necessarily with full amplitude - will result in an expansion of the PN-nucleus to giant dimensions for some 10^3 years (see for details Schoenberner, 1979). To see such an event during a PN lifetime (3.10^4 yr) is rather unlikely: about only 1 out of 100 NPN is to be expected to display a rapid evolution through the HR-diagram. In fact, up to now only one object is known which can be identified to be in this evolutionary phase: the late type central star FG Sge. Its distance is known (Herbig and Boyarchuk, 1968), and its luminosity of $10^{3.6}$ $L_\odot$ corresponds well to a post-AGB model with, say, 0.6 $M_\odot$, which displays rapid evoution to low effective temperatures which is driven by a final thermal pulse (Schoenberner, 1979). The expansion age of the PN is about 5000 yrs, and its ionization indicates that the central star was at least as hot as 55000 K some 10^2 years ago (Harrington and Marionni, 1976). We conclude that in this case the formation of the PN - and the creation of a contracting NPN - took place shortly <u>before</u> the onset of a thermal pulse (the interpulse period is $\sim 10^5$ yrs). Thus the very existence of the FG Sge appears to be an observational proof that i) thermal pulses really exist, ii) the PN formation occurs during the interpulse phase of quiet hydrogen burning, i.e., during the thermal equilibrium phase of an AGB-star.

During the post-flash recovering phase, a FG Sge-like nucleus will evolve through the same region of the HR-diagram and with a similar fading time than normal nuclei. Thus 15 or 20% NPN out of the ensembles in Figs. 1 and 2 might be in this stage, but we cannot detect them because that fraction is too small. If the final flash occurs at very high effective temperatures of the NPN, say at $T_{eff} \gtrsim 10^5$ K, the convective helium shell cuts into the hydrogen envelope, and a more or less complete mixing of the hydrogen occurs (Schoenberner, 1979, Iben et al., 1982). It might be possible to explain the existence of two peculiar PN, A 78 and A 30, which show helium enriched matter, by such an event. But their position in Fig. 2 can also be explained by assigning them a nucleus of ~ 0.55 $M_\odot$. Such a model evolves slowly enough to allow for high central star luminosity and large PN radius. The only method to distinguish between both modes of evolution is provided by a precise temperature determination of the nucleus because the post-flash cooling track of the Iben et al. (1982) 0.6 $M_\odot$ model displays - during the PN-lifetime - higher effective temperatures (~ 0.1 dex) than a 0.55 $M_\odot$ track with quiet hydrogen burning. In any way such a mixing flash should be rather rare: From evolutionary timescales one may estimate that only 30% of all final thermal pulses occur at $T_{eff} \gtrsim 10^5$ K, or with other words, only a few percent of all NPN are expected to experience a flash driven mixing.

4. FINAL REMARKS

Although we have shown that a general correlation between M_v and nebular radius does exist, and that the HR positions present in essence an evolutionary sequence we want to point out that even within the narrow NPN mass range derived there are important differential effects which explain, for example, Kaler and Hartkopf's (1981) finding of two contrasting Abell PN. Within our scheme, their results for Abell 43 and Abell 50 can be easily interpreted by minor mass differences. From the data given we obtain M (Abell 43) = 0.56 $M_\odot$ and M (Abell 50) = 0.60 $M_\odot$. Indeed the more massive NPN has reached lower luminosities within a shorter timescale than the less massive comparison objects. Our scheme predicts also effective temperatures as well, and we find $T_{eff} \sim 110000$ K for A 50, while Kaler and Hartkopf determined the Zanstra temperatures to be 104000 K and 125000 K for H and He II, respectively. We have a similar agreement for A 43. We conclude that our scheme is consistent - even without knowing individual expansion velocities - and we encounter "with a magnifying glass" within our ensemble exactly the differential effects which were fundamental for the set up of Renzini's evolutionary scheme. Furthermore, A 50 is considered to be a Pop II object according to Kaler and Hartkopf (1982) and consequently we could not accept any substantially higher mass for its nucleus than that derived above. The mere fact that a low mass Pop II NPN appears as faint as $M_v \sim 7$ supports our statement that most of the faint nuclei have indeed such low masses.

Clearly, it will be rewarding to continue investigations of this kind and to improve the observational material on which further tests of the present interpretation of NPN evolution can be made.

REFERENCES

Acker, A.: 1987, Astron. Astrophys. Suppl. 33, 367.
Barkat, Z., Tuchman, Y.: 1980, Astrophys. J. 237, 105.
Cahn, J. H., Kaler, J. B.: 1971, Astrophys. J. Suppl. 22, 319.
Cahn, J., Wyatt, S. P.: 1976, Astrophys. J. 210, 508.
Daub, C. T.; 1982, preprint.
Gingold, R. A.: 1974, Astrophys. J. 193, 177.
Haerm, R., Schwarzschild, M.: 1975, Astrophys. J. 200, 324.
Harrington, J. P., Marionni, P. A.: 1976, Astrophys. J. 206, 458.
Herbig, G., Boyarchuk, A. A.: 1968, Astrophys. J. 153, 397.
Hunger, K., Gruschinske, J., Kudritzki, R. P., Simon, K. P.: 1981, Astron. Astrophys. 95, 244.
Iben, I., Jr.: 1982, Astrophys. J., submitted.

Iben, I., Jr., Kaler, J. B., Truran, J. W., Renzini, A.: 1982, Astrophys. J. Lett., submitted.
Iben, I., Jr., Renzini, A.: 1983, Ann. Rev. Astron. Astrophys. preprint.
Kaler, J. B.: 1978, Astrophys. J., 226, 947.
Kaler, J. B., Hartkopf, W. I.: 1981, Astrophys. J. 249, 602.
Koester, D., Reimers, D.: 1981, Astron. Astrophys. 99, L8.
Koester, D., Schulz, H., Weidemann, V.: 1979, Astron. Astrophys. 76, 262.
Koester, D., Weidemann, V.: 1980, Astron. Astrophys. 81, 145.
Kohoutek, L., Martin, W.: 1981, Astron. Astrophys. 94, 365.
Kovetz, A., Harpaz, A.: 1981, Astron. Astrophys. 95, 66.
Mendez, R. H., Kudritzki, R. P., Gruschinske, J., Simon, K. P.: 1981, Astron. Astrophys. 101, 323.
O'Dell, C. R.: 1974, IAU Sympos. No. 66, "Late Stages of Stellar Evolution." R. J. Tayler, Ed., Reidel, Dordrecht, p. 213.
Paczynski, B.: 1971, Acta Astronomica 21, 417.
Paczynski, B.: 1978, IAU Sympos. No. 76, "Planetary Nebulae," Y. Terzian Ed., Reidel, Dordrecht, p. 201.
Pottasch, S. R.: 1981, Astron. Astrophys. 94, L13.
Pottasch, S. R., Wesselius, P. R., Wu, C. C., Fieten, H., Van Duinen, R. J. 1978, Astron. Astrophys. 62, 95.
Purgathofer, A., Weinberger, R.: 1980, Astron. Astrophys. 87, L5.
Reimers, D., Koester, D.: 1982, Astron. Astrophys. (submitted).
Renzini, A.: 1979, Stars and Star Systems. B. E. Westerlund Ed., Reidel, Dordrecht, p. 155.
Robinson, G. J., Reay, N. K., Atherton, P. D.: 1982, Monthly Notices Roy. Astr. Soc. 199, 649.
Schoenberner, D.: 1979, Astron. Astrophys. 79, 108.
Schoenberner, D.: 1981, Astron. Astrophys. 103, 119.
Schoenberner, D.: 1982, in preparation.
Schoenberner, D., Weidemann, V.: 1981, Physical Processes in Red Giants. I. Iben, Jr., and A. Renzini, Eds., Reidel, Dordrecht, p. 463.
Schulz, H., Wegner, G.: 1981, Astron. Astrophys. 94, 272.
Seaton, M. J.: 1968, Astrophys. Letters 2, 55.
Shaviv, G.: IAU Sympos. No. 76, Planetary Nebulae, Y. Terzian Ed., Reidel, Dordrecht, p. 195.
Weidemann, V.: 1977a, Astron. Astrophys. 61, L27.
Weidemann, V.: 1977b, Astron. Astrophys. 59, 411.
Weidemann, V.: 1979, White Dwarfs and Variable Degenerate Stars. H. van Horn, V. Weidemann Eds., Univ. Rochester, p. 206.
Weidemann, V.: 1981, Effects of Mass Loss on Stellar Evolution. C. Choisi, R. Stalio, Eds., Reidel, Dordrecht, p. 339.
Weidemann, V.: 1982, in preparation.
Weidemann, V., Koester, D.: 1980, Astron. Astrophys. 85, 208.
Wood, P. R., Zarro, D. M.: 1981, Astrophys. J. 247, 247.

SEATON: I do not think you should be so shy of temperatures! We can make good progress in determining temperatures of individual PN nuclei. Their distances represent a much harder problem.

RENZINI: I expect that the real mass distribution is somewhat wider towards higher masses for the simple reason that your sample is compiled according to the availability of M_V. Since there are many PN with visually undetected nuclei, and since the more massive a PN nucleus the fainter it will appear, your sample suffers from this obvious selection effect.

SCHÖNBERNER: The mass distribution which I presented is valid for *observed* nuclei of a *local* ensemble. Only further observations will show if this sample is representative.

PHYSICAL PROPERTIES OF THE CENTRAL STARS OF PLANETARY NEBULAE IN THE MAGELLANIC CLOUDS

T.P. Stecher, S.P. Maran, T.R. Gull
Laboratory for Astronomy and Solar Physics, NASA-Goddard Space Flight Center, Greenbelt, Maryland 20771, USA

L.H. Aller
Department of Astronomy, University of California, Los Angeles, USA

M.P. Savedoff
Department of Physics and Astronomy, University of Rochester, New York, USA

Absolute flux distributions in the ultraviolet continua of the central stars of three planetary nebulae of known distances have been derived from observations made with the International Ultraviolet Explorer. The observations confirm the existence of planetary nebulae nuclei with masses of $\approx 1 M_{\odot}$ and indicate that the progenitors of the nebulae were carbon stars near the theoretical upper luminosity threshold of $M_{bol} = -6.5$. The derived masses, luminosities, and temperatures of the three stars indicate that they are currently on horizontal tracks in the HR diagram and probably have not yet attained their maximum luminosities. The present luminosities ($\approx 4 \times 10^4 L_{\odot}$) each are well above the Eddington luminosity for an 0.6 $M_{\odot}$ star. The derived properties of the stars and associated nebulae (LMC P40, SMC N2, SMC N5) are consistent with a nebular ejection mechanism that involves radiation pressure on carbon grains.

EDITOR'S NOTE: Dr. Stecher preceded his scheduled paper by a brief presentation of the detection of an emission line object, possibly a PN, in the globular cluster M5. The N IV] λ 1487 transition had been observed with the IUE satellite. The discussion below refers to the scheduled paper.

PEIMBERT: Are SMC N2 and SMC N5 carbon-rich?

STECHER: Their carbon abundance is similar to that of Galactic PN, but the oxygen abundance is low so they are "carbon-rich". The enrichment relative to the interstellar abundance is a factor of 40, showing that carbon has been made in the stars and convected to the surface.

MATHIS: Interstellar extinction is important in the IUE wavelength range. Did you use a "Galactic" or "LMC" extinction law?

STECHER: We have used the results of work by four different groups to obtain the extinction curve. The shortest wavelength point of the continuum would change by 30% if a "Galactic" curve were used.

IUE OBSERVATIONS OF CENTRAL STARS

Sara R. Heap
Goddard Space Flight Center, NASA
Greenbelt, Md. U.S.A.

1. INTRODUCTION

Despite similar evolutionary histories and a common ultimate fate as white dwarfs, central stars of planetary nebulae have surprisingly diverse spectral properties. Their visual spectral types encompass all varieties known for hot stars, including Wolf-Rayet, O and Of, subdwarf O, white-dwarf, and continuous (Aller 1968, 1976), and O VI-emission types (Smith and Aller 1969, Heap 1982). Their spectroscopic temperatures range from less than 30,000 °K (e.g. He 2-138, Mendez and Niemela 1979; the WC 11 stars, Houziaux and Heck 1982) to upwards to 150,000 °K or more (e.g. NGC 246, Heap 1975; Abell 30, Greenstein 1981). Their atmospheres range from demonstrably helium- and carbon-rich (e.g. the WR stars, Barlow and Hummer 1982, Benvenuti _et al_. 1982) to apparently normal (e.g. the Of stars, Heap 1977a,b), to helium-poor (e.g. the nascent white dwarfs in Abell 7 and NGC 7293, where gravitational settling appears to have already taken effect, Mendez _et al_. 1981).

Perhaps this diversity is not surprising when you consider what a wide range in initial mass (approximately 1 to 5 M⊙) gets funneled through the central-star sequence. Still, saying that it's not surprising is not the same thing as understanding or verifying it, so I would like to use as the focal point of this talk the question: is (initial) mass a factor in determining the spectral properties of central stars?

To answer this question, it is not necessary (or possible!) to make exact determinations of the mass of each central star. Instead, it is adequate to make mass distinctions among central stars; that is, to show that one group of central stars is more massive than another. We can therefore take advantage of the mass-discrimination inherent

D. R. Flower (ed.), Planetary Nebulae, 375–390.

in evolutionary plots, as first developed by Schönberner (1981). These plots consist of a superposition of observed and theoretical data. The observed data on the plot are the stellar absolute magnitude as a function of nebular radius. Since planetaries have a near-common expansion velocity (assumed here to be 20 km/s), the nebular radius serves as a measure of age since ejection. The theoretical data consist of a stellar mass - age grid, which is derived from atmospheric models and evolutionary models for central stars (Paczynski 1971, Schönberner 1979).

For our observational sample, we shall make use of about sixty galactic planetaries and three planetaries in the Magellanic Clouds whose ultraviolet spectra have been obtained at low-dispersion with the IUE. These observations were obtained by astronomers world-wide (most of whom are here today) over the last four years and are stored in the World Data Centers. Central stars observed by the IUE tend to be optically bright, so most of their spectra have also been studied in the visual and are known to exhibit the full range in spectroscopic properties outlined earlier. The IUE sample incorporates a mixture of galactic populations, as shown in Figure 1 by their spatial and kinematic distribution. In contrast to Pottasch's (1981) sample of faint, presumably more massive central stars, there is a slight leaning of the IUE sample toward the halo population. Only one star in Pottasch's sample, the nucleus of IC 2165, is represented in this IUE sample.

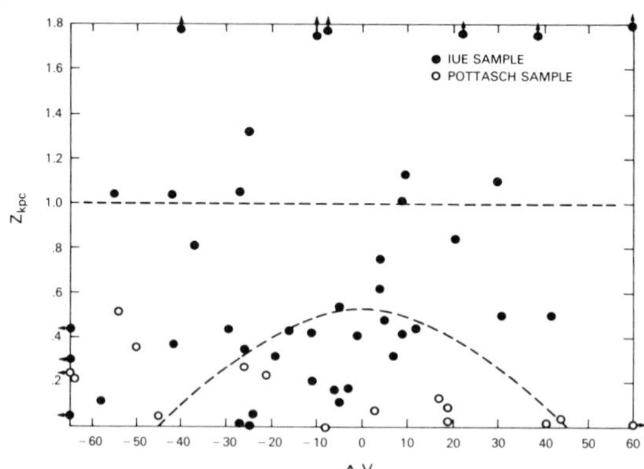

Figure 1. Height Above the Galactic Plane vs. the Deviation of Radial Velocity From That for a Circular Galactic Orbit. The dashed lines show (arbitrary) boundaries of the disk and halo populations.

The IUE sample does overlap greatly with Schönberner's (1981) sample. It also covers a somewhat broader cross-section of stars by including central stars in large, high-excitation nebulae studied by Kaler (1981) and by retaining central stars in small, optically thick nebulae, some of which have been studied by Kohoutek and Martin (1981). The main reason for retaining stars in optically thick nebulae was to try to get around selection effects that tend against massive stars ($M > 0.65$ $M_\odot$) in their luminous phase of evolution. The only problem with these objects is that, being ionization-bounded, their observed radii (and hence, age) are smaller than their actual radii (age).

2. THE MASS DISTRIBUTION OF CENTRAL STARS SAMPLED BY THE IUE

The primary observational data for an evolutionary plot are the stellar apparent magnitude corrected for interstellar reddening, distance, and nebular angular diameter. Since the main purpose in constructing an evolutionary plot is to make inter-comparisons among central stars, it is essential to eliminate or reduce systematic errors that would influence the location on the plot of one group of stars with respect to another group , so it is attractive to take advantage of improvements in stellar apparent magnitude offered by the IUE. The attractions of IUE fluxes are consistancy of data obtained by what has turned out to be a very stable spectrophotometer, and an enhanced contrast of the star with respect to the nebula. This contrast enhancement is brought out by a small entrance aperture that rejects extensive nebular contributions and by access to short wavelengths where the stellar flux is at its strongest with respect to nebular continuum emission. As an example, Figure 2 compares the contributions made by the star and surrounding nebula to the observed flux distribution of NGC 2867, a small, bright planetary with a WC/O VI-type nucleus (Smith and Aller 1969). While the visual flux of the star is swamped by nebular emission, the ultraviolet flux of the star is approximately equal in strength to that of the nebula at 1300 Å. The chief disadvantage of IUE fluxes for this purpose is their susceptibility to errors in the amount of interstellar extinction. Errors in the extinction coefficient, c, are magnified six-fold in their effect on stellar magnitude at 1300 Å, so care is required in deriving the appropriate correction, and caution is required in interpreting an ultraviolet evolutionary plot, particularly at the low-temperature region where mass and evolutionary effects are not well discriminated.

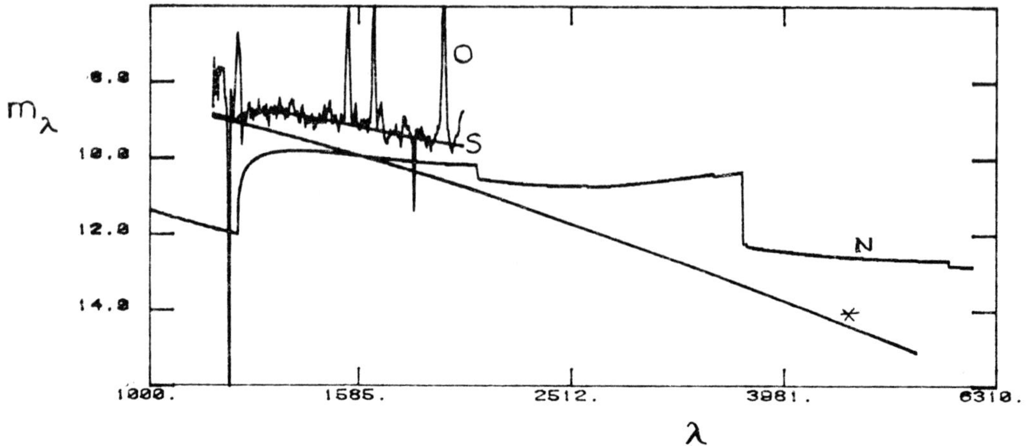

Figure 2. Contributions to the Spectrum of NGC 2867. The ordinate is apparent magnitude [m=-2.5log(flux)-21.1] corrected for reddening. Annotations of the spectra are: O = observed spectrum from the IUE, N = calculated nebular continuous spectrum, * = stellar spectrum as deduced from a blackbody fit to (O-N), and S = sum of the nebular and stellar fluxes.

Also in an effort to reduce systematic errors, while still retaining stars embedded in either optically thick or optically thin nebulae, I have adopted Cudworth's (1974) method of determining distance. This method assumes that the nebular absolute magnitude is a constant for optically thick nebulae, and that the nebular ionized mass is a constant for optically thin nebulae.

An evolutionary plot using ultraviolet stellar magnitudes is shown in Figure 3. Superposed on the observed data is a mass-temperature grid extended from the visual (Schönberner 1981) to the ultraviolet with the use of black-body colors. The immediate inference to be made from this plot is that central stars in optically thin nebulae (identified by Cudworth as having a radius larger the 0.07 pc) are constrained to a sequence implying a narrow distribution about 0.58 M☉ in stellar mass. This finding is in agreement with Schönberner's results, which were based on visual stellar magnitudes and Cahn-Kaler distances. However, central stars in young, optically thick nebulae do not form a <u>bright</u> extension to the sequence, as they do in Schönberner's plot, but instead give the appearance of a <u>high-mass</u> extension to the sequence. This new result is a consequence of two factors that work together to reduce the stellar luminosity of young central stars with respect to older central stars on the evolutionary plot: the suppression of nebular contamination from the observed flux, and the use of Cudworth's distance scale. Among this group of young central stars in optically thick nebulae are three central stars in the Magellanic Clouds (shown as squares) for which Maran et al. (1982) derived large stellar masses (>1.0 M☉) by independent determinations. Although it is difficult to see how such massive stars could be found in their bright stage, their location with galactic planetaries adds force to the interpretation that stars in optically thick nebulae are among the most massive of the sample.

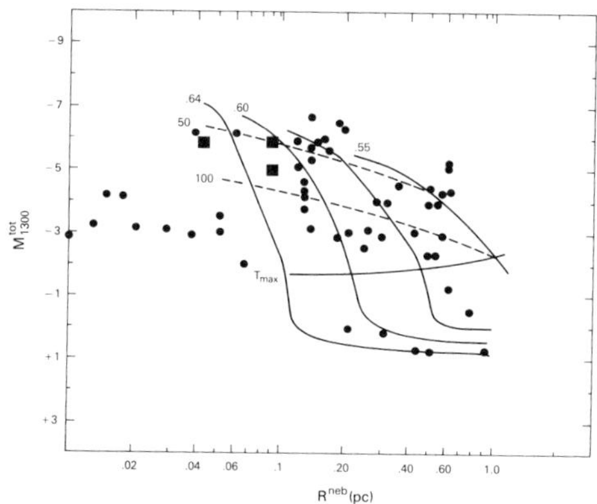

Figure 3. Evolutionary Plot for the IUE Sample of Central Stars in the galaxy (filled circles) and in the Magellanic Clouds (squares). The solid lines show the course of evolution for a star of .65, .60, .57, and .50 M☉. The two dashed curves show lines of constant temperature = 50,000 °K and 100,000 °K.

Confirmation of this distribution of stellar masses is highly desirable and should be posssible in the near future. The process of confirmation involves a test for self-consistency: a star's temperature, as read off the mass-temperature grid, should match independent determinations such as its Zanstra temperature, optical color temperature, or spectroscopic temperature. This test not only checks the placement of the M-T grid on the plot, and in doing so, checks the inferred mass of the sample, but the test also -- and most important for our purposes -- checks the shape of the M-T grid, which is essential for distinquishing mass effects from evolutionary (temperature) effects. While desirable, it is too early to carry out a test of this kind, because atmospheric models, upon which both the mass-temperature grid and all independent temperature determinations depend, are deficient in that they do not yet take into account the effects of photospheric line-blanketing or alteration of the flux distribution by a wind. As many bolometrically luminous central stars have winds while faint ones do not , we can expect that models that take these effects into account (Hummer 1982; Abbott and Hummer, this volume) will yield an M-T grid with a considerably different shape than the one shown in Figure 3. We can also expect that these model flux distributions, along with observed ultraviolet fluxes from the IUE, will improve the quality of independent temperature determinations.

3. THE MAPPING OF CENTRAL-STAR MASS INTO INITIAL MASS

With such a narrow mass-distribution in the IUE sample of central stars, you might not expect to distinguish mass-effects within the sample. Nevertheless, such effects can be seen (Figures 4 and 1) by the fact that, with few exceptions, stars of high mass, as inferred from the evolutionary plot, are confined to the galactic plane and, by implication, have progenitors of high initial mass, while the low-mass central stars are found in the halo and therefore have low-mass progenitors.

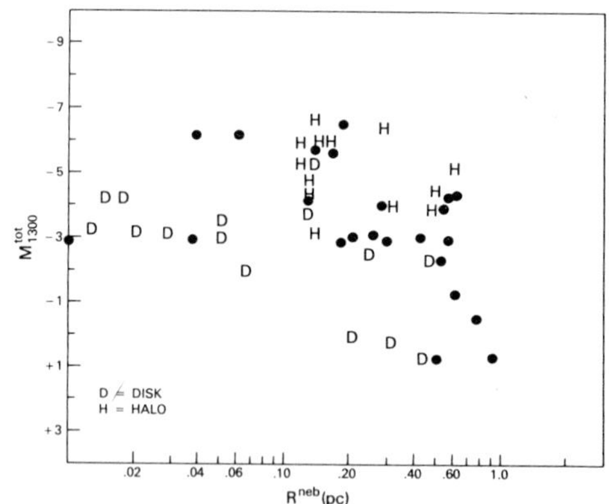

Figure 4. Identification of Halo and Disk Planetaries.

Weidemann (1981) has presented evidence that the amount of mass lost by a red-giant is a strong function of initial mass, with the high-mass giants shedding the largest amount. The segregation of central stars by population type supports this conclusion and, furthermore, implies that the mechanism(s) of mass-loss in central-star progenitors is so highly tuned that, regardless of the initial mass of the progenitor, the final remnant falls neatly into a narrow range in mass, with the high-mass (disk) and low-mass (halo) progenitors producing central stars falling at the high- and low-mass ends of the narrow mass range.

The fact that the high-mass central stars are the optically fainter of the sample suggests that still higher-mass objects (M > 0.7 Mo) would be difficult to detect and would be under-represented in any "random" sample of central stars, such as this one.

4. DIFFERENTIATION OF CENTRAL STARS BY SPECTRAL TYPE

The segregation of central stars by population type in the evolutionary plot implies that very small differences in stellar mass among central stars map into large differences in initial mass. There is therefore some justification in examining an evolutionary plot to pursue the question whether initial mass and consequent evolutionary history determines the spectral type of a central star .

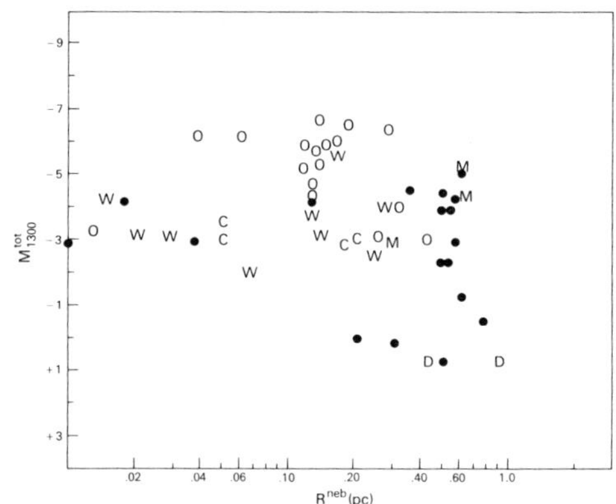

Figure 5. Identification of Stellar Spectral Type. The notation for spectral type is: W=Wolf-Rayet, O=O, Of, and sdO, M=mixed absorption/O-VI emission (cf. Heap 1982), D=white dwarf, and filled circles = unknown.

The evolutionary plot does, in fact, show (Figure 5) some segregation by spectral type in a way that can probably, but not necessarilly, be ascribed to mass. The principal distinction is between the O and WR stars. The O stars are optically bright with M(1300)= -6, while the WR stars are usually about 3 magnitudes fainter. Whether this separation is strictly a mass effect is not clear for two reasons. One reason is that both classes of stars have members with winds for which the M-T grid in Figure 3 does not apply. The other is that both groups have members associated with optically thick nebulae, whose actual sizes are highly uncertain. But if the observed radii are within, say, a factor of two of the true radii, then the data suggest that, by-and-large, WR stars are more massive than O stars and that O stars evolve into hot subdwarf-O stars.

One surprise is the lack of distinction between WR and continuous-type central stars. Some continuous-type stars, such as the nuclei of NGC 7662, IC 2165 (in optically thick nebulae) or NGC 6778 and NGC 2022 (in optically thin nebulae) are found in the same region of the evolutionary plot as WR stars. This brings up the question whether such stars really have continuous spectra or whether their spectral features have not been noticed due to masking by the nebula, or due to low-resolution and/or noisy spectrograms. This is an essentially unanswered question at the present time. However, recent spectroscopic studies (Mendez et al. 1981) of stars previously classified as continuous have detected spectral features, and a high-dispersion spectrum of the nucleus of NGC 7662 with the IUE reveals weak, broad emission at C IV 1550, indicative of a weak, high-velocity wind. (The existence of this feature should be confirmed by re-observation.)

In summary, IUE data support Schönberner's assertion that most central stars have masses around 0.58 M⊙, although some young, faint stars may be considerably more massive. Despite the narrow distribution in mass, distinctions can be made among stars in this sample, with the following conclusions: (1) low and intermediate-mass stars retain their identity in the sense that stars of higher initial mass produce higher-mass central stars, and (2) Wolf-Rayet stars may be among the most massive stars of the sample.

The reduction of the IUE data was carried out with Dr. Harry Augenson. The nebular contribution of the observed flux of NGC 2867 (Figure 2) was calculated with the use of Dr. J.P. Harrington's computer program, CONTIN.

REFERENCES

Abbott, D.C. and Hummer, D.G.: 1982, this volume.

Aller, L.H.: 1968, "Planetary Nebulae": IAU Symposium No. 34, p. 339.

Aller, L. H. : 1976, Mem. de la Soc. Roy. des Sci. de Liege, 6eme Ser., p 271.

Barlow, M. J. and Hummer, D. : 1982, "Wolf-Rayet Stars": IAU Symp. No. 99, C. de Loore, ed. (D. Reidel: Dordrecht).

Benvenuti, P, Perinotto, M. and Willis, A.J.: 1982, "Wolf=Rayet Stars", C. de Loore, ed., (D. Reidel: Dordrecht.)

Cudworth, K.: 1974, Astron. J. 79, 1384.

Greenstein, J. L.: 1981, Astrophs. J. 245, 124.

Heap, S. R.: 1975, Astrophys. J. 196, 195.

Heap, S.R.: 1977, Astrophys. J. 215, 609,864.

Heap, S.R.: 1982, "Wolf-Rayet Stars", C. de Loore, Ed., (D. Reidel: Dordrecht).

Hummer, D.: 1982, Astrophys. J. 257, 724.

Houziaux, L. and Heck, A.: 1982, "Wolf-Rayet Stars", C. de Loore, Ed., (D. Reidel: Dordrecht).

Kaler, J.: 1981, Astrophys. J. 250, L31.

Kohoutek, M. and Martin, W.: 1981, Astron. Astrophys. 94, 365.

Maran, S.P., Stecher, T.P., Gull, T.R., Aller, L.H., Savedoff, M.: 1982, "Advances in UV Astronomy: Four Years of IUE Research", Y. Kondo, Ed., NASA

Mendez, R. and Niemela, V.: 1979, Astrophs. J. 232, 496.

Mendez, R., Kudritzki, R.B., Gruschinske, J., Simon, K.P.: 1981, Astron. Astrophys. 101, 323.

Paczynski, B.: 1970, Acta Astron. 21,417.

Pottasch, S.: 1981, Astron. Astrophys. 94, L13.

Schönberner, D.: 1979, Astron. Astrophys. 79,108.

Schönberner, D.: 1981, Astron. Astrophys. 103, 119.

Smith, L.F. and Aller, L.H.: 1969, Astrophys. J. 157, 1245.

Weidemann, V.: 1981, "Effects of Mass-Loss on Stellar Evolution", C. Chiosi, R. Stalio, Eds., (D. Reidel: Dordrecht).

APPENDIX: UV Spectra of Selected Central stars Observed By the IUE. The spectra are corrected for reddening and normalized at 1830 Å.

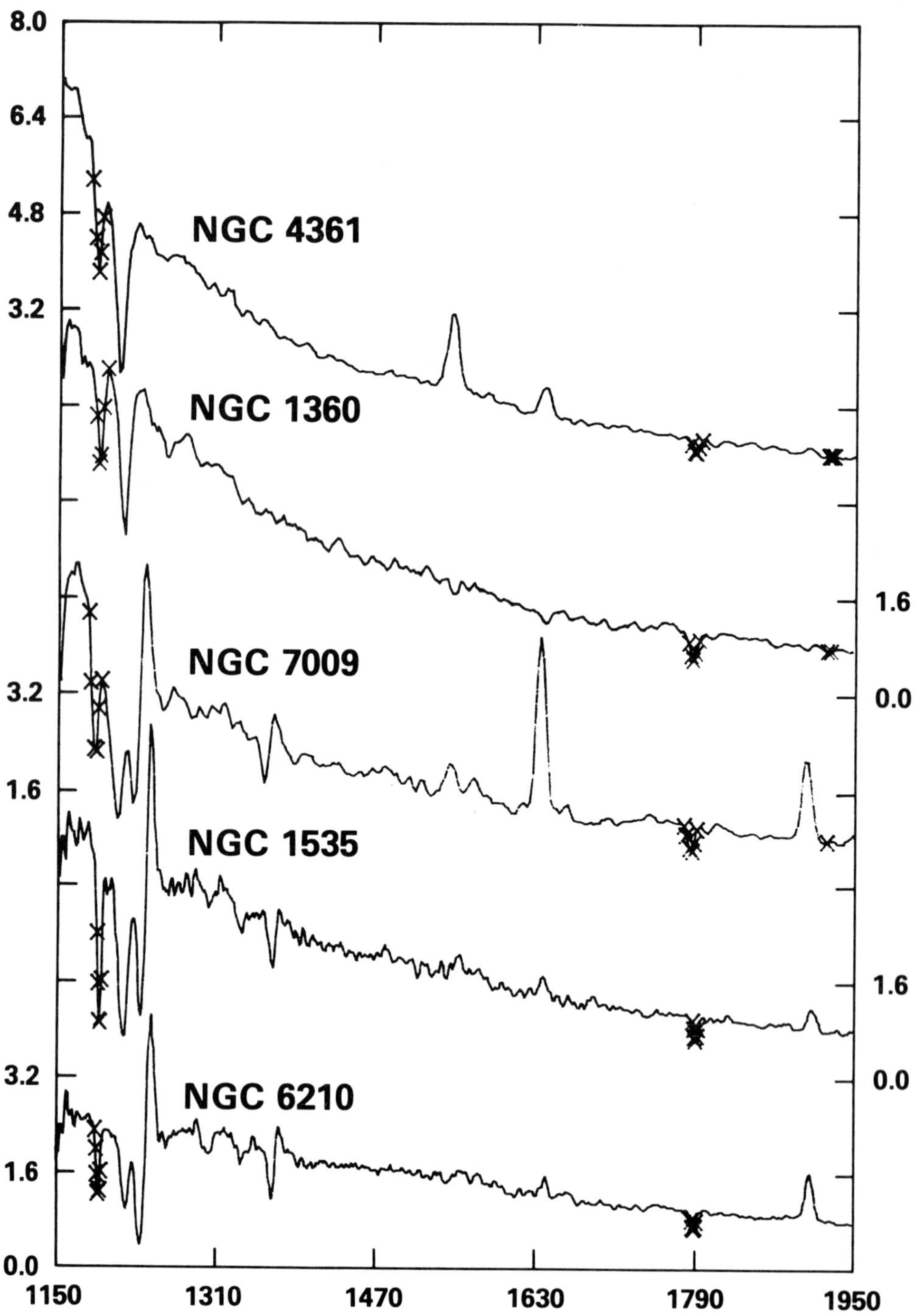
8.0
6.4
4.8
3.2
NGC 4361
NGC 1360
1.6
0.0
NGC 7009
3.2
1.6
NGC 1535
1.6
0.0
3.2
NGC 6210
1.6
0.0
1150
1310
1470
1630
1790
1950

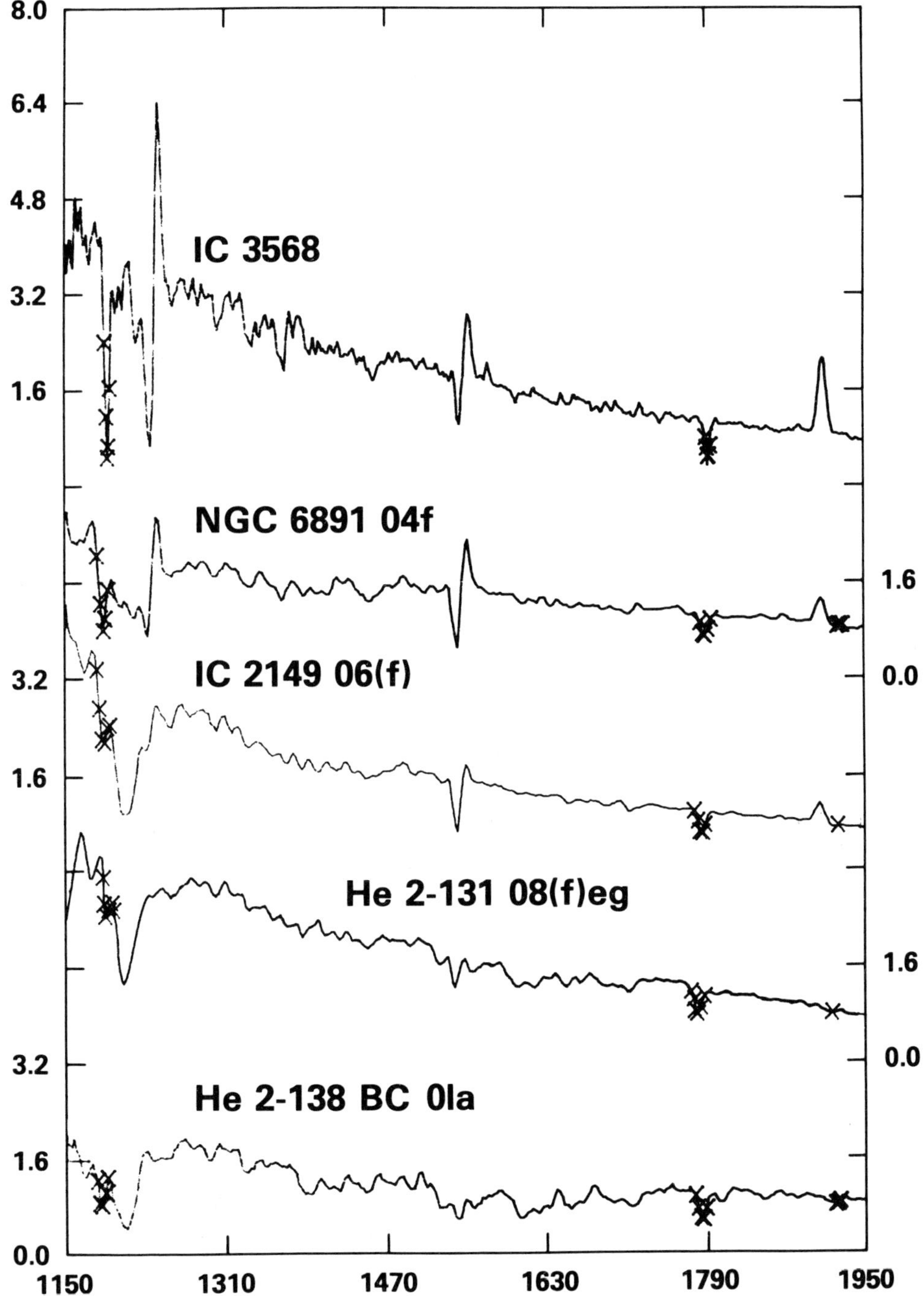

8.0
6.4
4.8
3.2
1.6
IC 3568
NGC 6891 04f
1.6
IC 2149 06(f)
3.2
0.0
1.6
He 2-131 08(f)eg
1.6
3.2
0.0
He 2-138 BC 0la
1.6
0.0
1150
1310
1470
1630
1790
1950

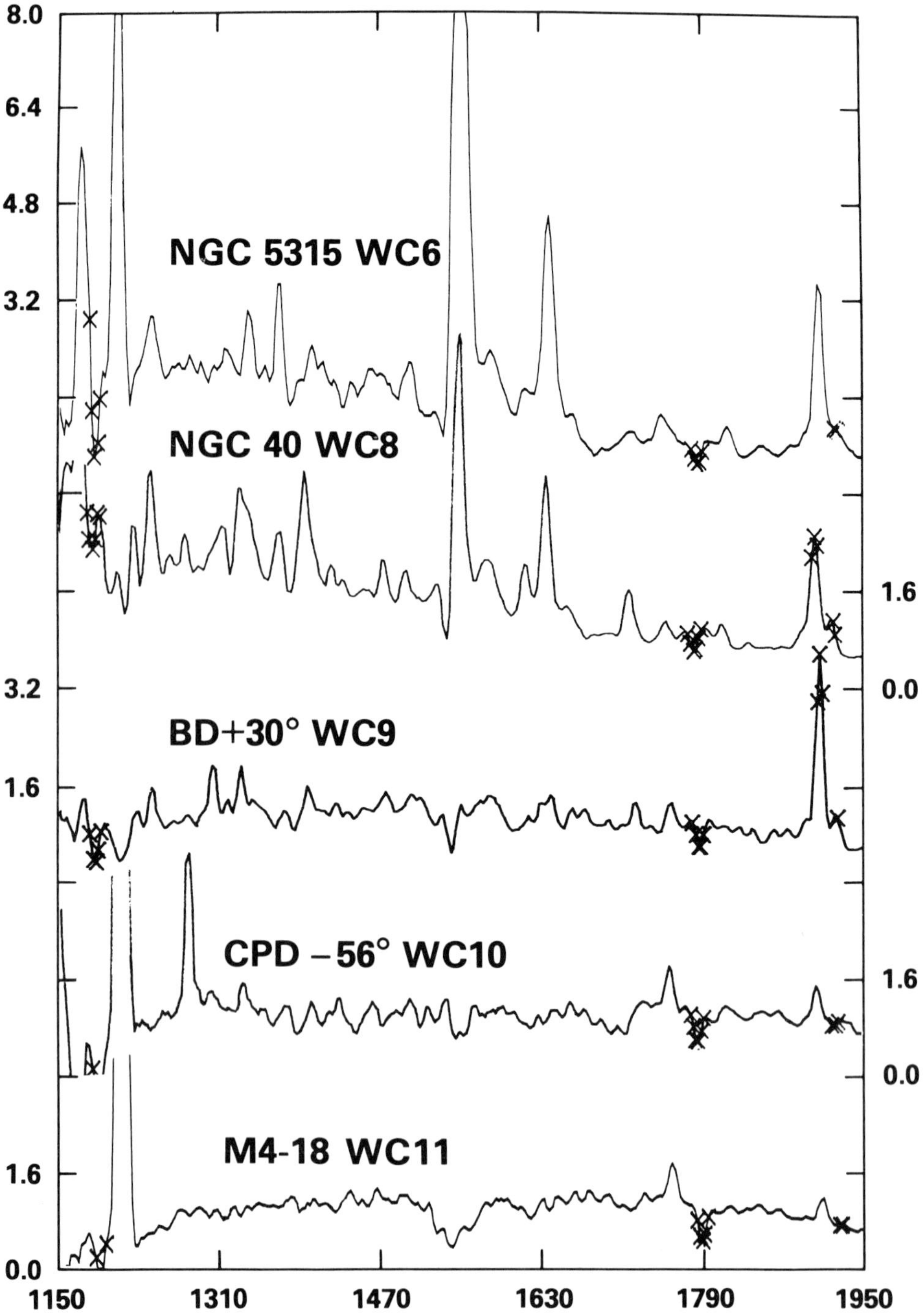
8.0
6.4
4.8
3.2
NGC 5315 WC6
NGC 40 WC8
1.6
3.2
0.0
BD+30° WC9
1.6
CPD −56° WC10
1.6
0.0
M4-18 WC11
1.6
0.0
1150
1310
1470
1630
1790
1950

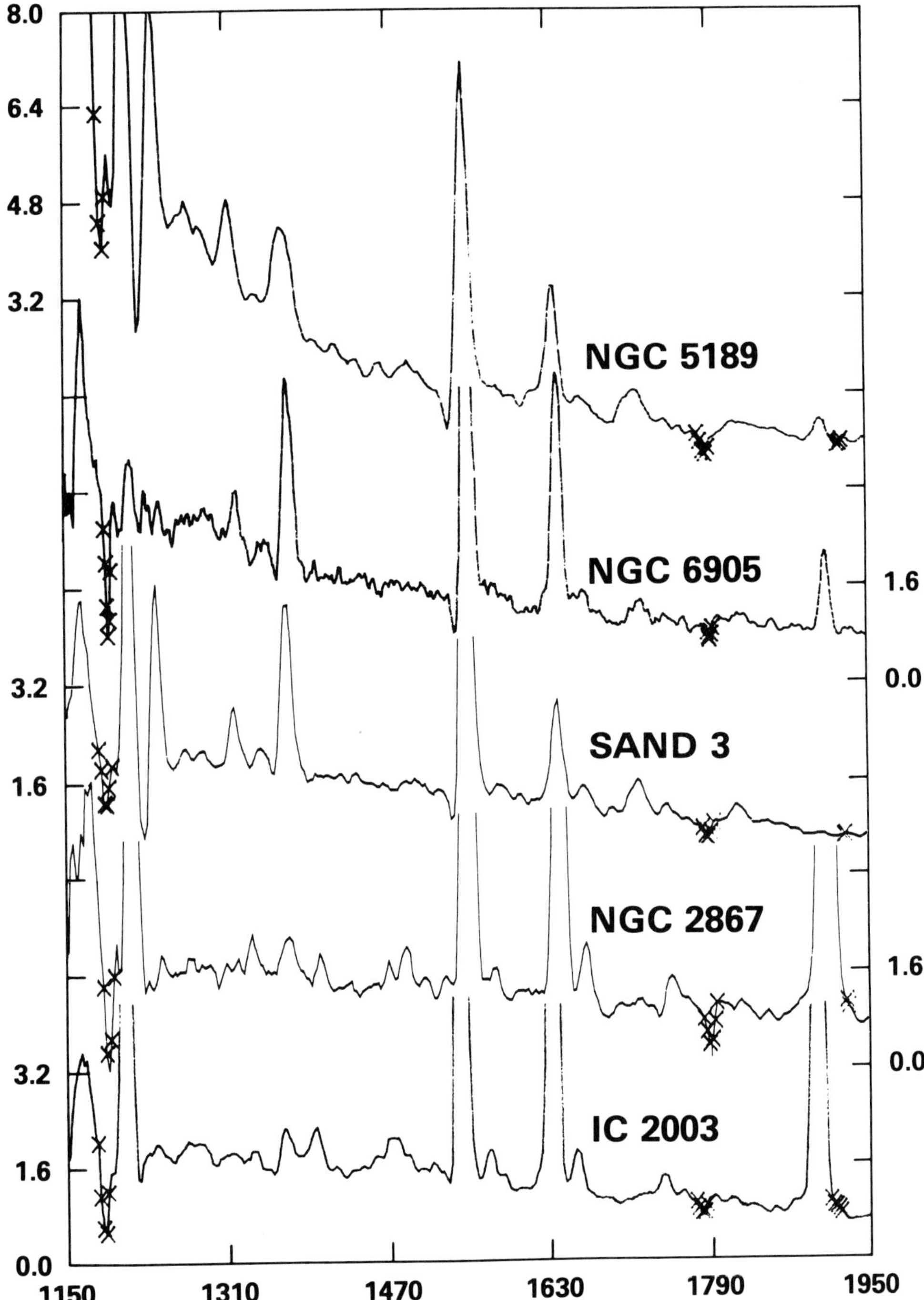

8.0
6.4
4.8
3.2
3.2
1.6
3.2
1.6
0.0
1.6
0.0
1.6
0.0
1150
1310
1470
1630
1790
1950
NGC 5189
NGC 6905
SAND 3
NGC 2867
IC 2003

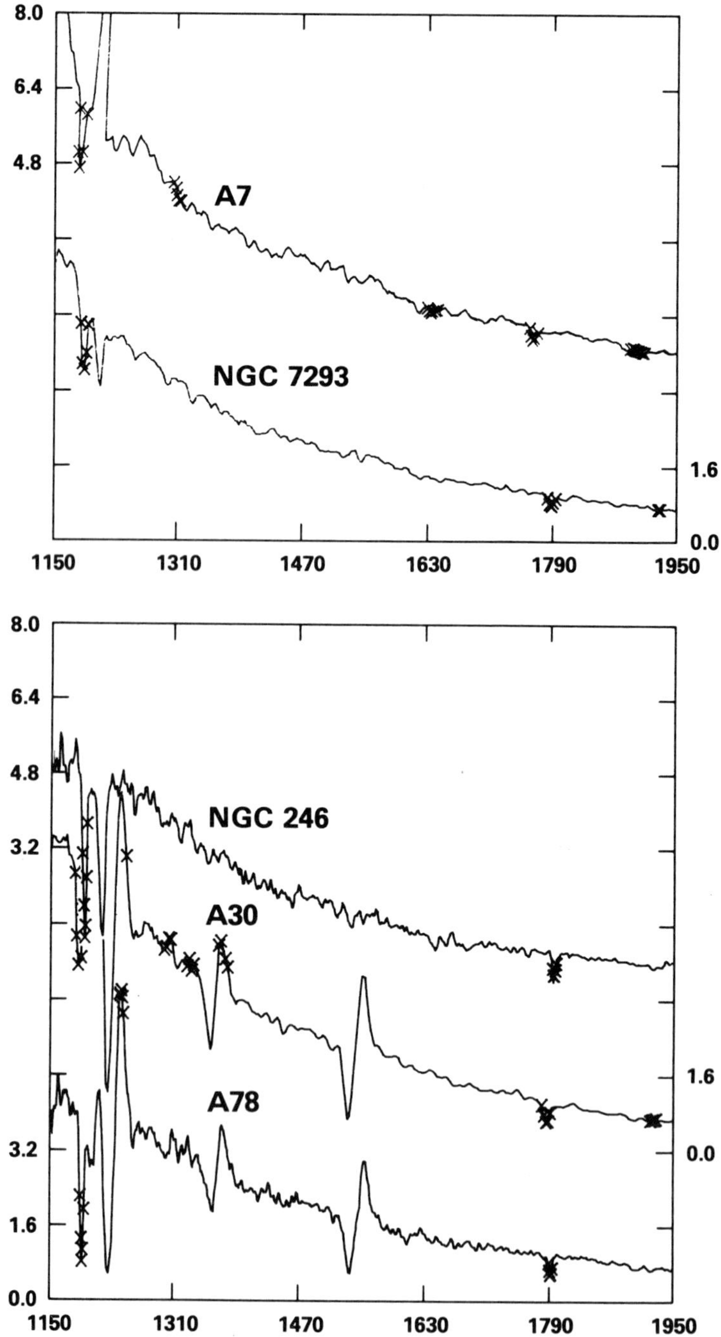
8.0
6.4
4.8
A7
NGC 7293
1.6
0.0
1150
1310
1470
1630
1790
1950
8.0
6.4
4.8
NGC 246
3.2
A30
1.6
0.0
A78
3.2
1.6
0.0
1150
1310
1470
1630
1790
1950

WEIDEMANN: In connection with your interpretation of the left hand points in your diagrams as a "high mass tail", I wish to point out that, in Schonberner's theoretical plots, the abscissa is actually the time, t. The assumption is that the nebular radius, $R_n \propto t$, corresponding to a constant expansion velocity; but objects with $R_n < 0.1$ pc are optically thick, and $R_n > R_n$ (ionized). It follows that the corresponding points should be shifted to the right. There is a further shift to the right if one considers the expansion velocity to be an increasing function of t in young nebulae (cf. Sabbadin et al., 1982, Astron. Astrophys. 110, 105), since a longer time is required to reach a given R.

KWOK: If the abscissa is, indeed, a time axis, would it not be better to plot against time by dividing each data point by the corresponding expansion velocity?

HEAP: Yes, it would be better - and practical if the expansion velocity of each nebula were known! Most nebulae have expansion velocities of around 20 km s^{-1}, but there are exceptions - in my sample, NGC 2392 has an expansion velocity in excess of 50 km s^{-1}.

SEATON: I hope that I have not given the impression that I thought there were no problems in reconciling optical and ultraviolet photometry with determinations of Zanstra temperatures. The point I wished to make was that I do not think that ultraviolet photometry indicates a need for systematic revision of temperatures obtained by the Zanstra method, as was suggested earlier by attempts to interpret the ANS ultraviolet observations. At the risk of oversimplifying the problem, I would say that optical and ultraviolet colour temperatures are broadly in agreement with Zanstra temperatures.

ALLER: The disagreement between Zanstra temperatures and those appropriate to the spectral class, for objects such as the nucleus of NGC 2392, has been recognised for many years. The binary hypothesis offers a possible solution and certainly holds in some cases (e.g. NGC 1501); but in others, we must consider more sophisticated models with winds, coronae etc. Help is needed from stellar atmosphere experts who can compute model atmospheres not in hydrostatic equilibrium.

MÉNDEZ: Concerning the "continuous" objects, we have recently found that two of the prototypes of this class (NGC 3242 and NGC 7009) are, in fact, absorption-line objects. Therefore, at the present time, it would seem preferable not to use the "continuous" classification in statistical discussions.

HEAP: In my talk, I applied the term "continuous" to those stars classified by Aller (1968) as having continuous spectra in the visual and ultraviolet regions. The central star of NGC 7009 is observed to have a strong wind in the ultraviolet.

RENZINI: If PN with massive nuclei ($M \gtrsim 1\ M_\odot$) do actually exist, we shall have to search for them among those nebulae with visually undetected central stars. Massive PN nuclei should be visually very faint ($M_v > 8$) and should be very hot ($2 \times 10^5 < T_* < 3 \times 10^5$ K). Is there any chance of detecting such stars with IUE?

HEAP: Certainly, the ultraviolet (particularly the region shortward of Lyman α, where the contrast of the star relative to the surrounding nebula should be a maximum) offers some improvement in the possibility of detecting massive stars, but the IUE data have not yet been studied systematically with this improvement in mind.

POTTASCH: We have looked at the ultraviolet spectra of several nebulae with faint central stars. Many do not show any continuum, nebular or stellar, in the far ultraviolet. Examples are NGC 6741 and NGC 6445 (although these objects are heavily reddened). For other nebulae, such as NGC 2440, the observed continuum is mainly nebular, which gives an upper limit, $m_v < 16$.

KALER: It is very important that those who observe with IUE and fail to detect a central star should establish an upper limit to the magnitude and publish it. In this way, we can at least place a limit on the Zanstra temperature.

HARRINGTON: Even when a strong continuum is present in IUE spectra, it is sometimes hard to separate the stellar and nebular contributions. In the case of IC 2165, it would appear that about a half of the observed SWP continuum is stellar.

ALLER: Under conditions of good seeing, with the Lick Observatory 3 m telescope I saw a condensation at the centre of IC 2165 that looked very much like a central star.

ISAACMAN: Is the lack of faint central stars in PN of small radius an evolutionary or an observational selection effect (i.e. faint stars would be difficult to detect in young, optically thick nebulae)?

HEAP: It is a selection effect. I chose only those nebulae for which I was sure that the star was detectable.

BOHANNAN: Is it significant that the Wolf-Rayet central stars are all of the WC sub-class? Why are there no WN's?

RENZINI: Post final flash PN nuclei can retain (for a while) a tiny skin containing both C and N - then mass loss removes this skin, exposing the underlying C-rich region, where N is virtually absent. So, stellar evolution theory predicts that there should be a few WCN and many WC nuclei, which, I believe, is observed.

MENDEZ: Concerning the determination of visual magnitudes of central stars, I would like to suggest the use of monochromatic fluxes from photoelectric spectrophotometry, particularly in the case of high surface brightness nebulae. Our observations of NGC 6790 and NGC 7009 show that the magnitudes of Shao and Liller are too bright by about one magnitude. A good visual magnitude is important for the determination of the colour temperatures of central stars.

CLEGG: Following on from this point, I wish to report that, in a survey of PN fluxes, we have found that quoted UBV magnitudes of central stars of small PN were systematically too bright. For NGC 3242, for example, a comparison of optical spectrophotometry (with nebular continuum subtracted), IUE spectra and UBV magnitudes shows the latter to be too bright. A reliable colour temperature can be derived from the merged IUE and optical spectra.

DISTANCES OF THE CENTRAL STARS AND THEIR POSITION IN THE HR DIAGRAM

S.R. POTTASCH

ABSTRACT:

Determination of the distances to individual planetary nebulae are discussed. Especially those methods which are independent of assumed nebular properties (mass, absolute flux, etc.) are assembled and discussed. In this way, reasonable approximations to the distance can be obtained for about 50 planetary nebulae. The accuracy of the distances is tested by comparing nebular properties derived from these distances with the properties of nebulae at the galactic center or in the Magellanic clouds. A comparison is also made with the statistical distance determinations; the conclusion is that the assumption of constant mass often leads to an overestimate of the distance, while the assumption of constant Hβ flux leads to distances having individual uncertainties of up to a factor of 3.

The central star temperature determination is summarized. Individual central stars are placed on the HR diagram and compared with theoretical predictions. Deductions concerning the evolution which can be made from the observations are discussed.

A. INTRODUCTION

In placing individual stars on the HR diagram the greatest uncertainties are the distance and the temperature of the star. Both of these problems will be discussed in turn, with emphasis on the distance because of the difficulties involved in its determination.

The usual methods of distance determination in astronomy, triginometric parallax and/or spectroscopic parallax, are seldom applicable to planetary nebula. Therefore methods have been devised for distance determination which are based on assumed nebular properties. The method in most common use is the so-called 'Shklovskii method', which assumes all nebulae have the same ionized mass. A second method assumes all nebulae have the same absolute Hβ flux. There is growing evidence that these assumptions are both incorrect. In fact they may have systematic errors affecting large and small nebulae in different ways.

D. R. Flower (ed.), Planetary Nebulae, 391–409.

At present there are several methods available to determine individual distances, independent of any assumptions concerning the physical properties of the nebula. These distances give independent information on the properties of the nebulae, especially mass and density. These resulting values are then compared with those determined for nebulae found in the galactic center, whose distance is well established. This data allows an interesting comparison with the results of the constant ionized mass and constant absolute Hβ flux methods.

B. DISTANCES TO INDIVIDUAL NEBULAE

Seven methods are considered as independent ways of determining the distance. Some of the methods are more reliable than others, most have a limited application. We shall discuss each of the methods, and give some results for the more reliable methods.

(1) Spectroscopic parallax. Some planetary nebulae are excited by stars which have binary companions. If the spectral type and luminosity class of the companion can be measured, a spectroscopic distance for the system can be determined. This method may have a wide application, since it is estimated that at least 10% of the exciting stars are binaries. The method may be applied both to visual binaries (where both stars are seen separately) and to spectroscopic binaries where the 'normal' star dominates the spectrum. At present, however, only a very limited number of such cases have been well studied. The resultant distances are shown in Table 1; the nebulae are all nearby. This is consistent with the fact that the nebulae all have large angular diameters.

TABLE 1 - SPECTROSCOPIC DISTANCES

NEBULA	SPECTRAL TYPE COMPANION	m_V	M_V	E_{B-V}	d
NGC 246	KO V	14.3	5.9	0.01	470 pc
1514	AO III	9.42	-0.2	0.45	400
2346	A2 V	11.12	1.4	0.22	640
3132	AO V	10.06	0.7	0.07	670
A35	G8 III-IV	9.63	1.9	0	360

The spectral types and magnitudes are taken from Lutz (1977), Mendez (1978), Jacoby (1981), and Lutz (1978).

(2) Expansion distances. Radial velocity measurements of the nebulae show a splitting of emission lines. This is interpreted as expansion of the nebulae, with velocities typically of 20 km s^{-1}. One may compare this expansion velocity with angular expansions derived from comparing the location of knots, filaments edges, and other features seen on both old and new photographs. If spherical symmetry

obtains, then the distance calculation is straight forward, and is given as follows

$$d = \frac{100\ v}{4.74\ \dot{\Theta}}\ \text{pc}$$

where the distance d is in pc, $\dot{\Theta}$ is the angular expansion in arc sec per 100 years, and v is the measured radial velocity (km s^{-1}).

Angular expansion rates have been determined by a number of investigators (see the references given under Table 2). The rates are difficult to measure, partly because of non-uniform shrinkage of the plate emulsion and partly because of the difficulty of finding sharply defined features to measure. The interpretation also has uncertainties. Firstly, the angular expansion may occur in a part of the nebula that is moving at a velocity different than the observed radial velocity. Secondly, expansions measured at the edge of optically thick nebulae may be influenced by motion of the ionization front, in addition to bulk motion of the gas. This second problem may be avoided if a well defined feature is measured.

Measurements of angular expansion rates and resultant distances are given in Table 2. Where two values are given for an individual nebula they are from different observers.

TABLE 2 - ANGULAR EXPANSION RATE AND RESULTANT DISTANCES

NEBULA	$\dot{\Theta}$ $\frac{\text{arc sec}}{\text{100 years}}$	Θ	v	d
NGC 246	1.4 ±0.5	100"	38 km s^{-1}	570 pc
3242	0.83 ±0.25	15"	30	760
3587	2.0 ±1.0	99""	41	430
6572	0.81 ±0.10	4.9	16	420
6720	0.9 ±0.1	40"	30	700
7009	0.75 ±0.3	14"	21	600
7662	1.0 ±0.6		26	550
	0.6 ±0.17	10"	26	900
2392	0.72 ±0.06	6"	18	530
			54	1600

Angular expansion rates from Latypov (1955), Chudovicheva (1964), Liller et al., (1966), Liller and Liller (1968).

(3) Distance determined on the basis of membership in a stellar group. Other than the planetary nebulae near the galactic center, or in an extragalactic system, there is only one well established case of a nebula which is a member of a stellar group. That is the planetary nebula discovered by Pease (Ps-1) in the globular cluster M15, in 1928. The distance of M15 is given as 10.1 kpc (Harris, 1976).

(4) Interstellar extinction distances. This method is simple: by measuring the spectral type and B and V magnitudes of field stars close to the nebula in the plane of the sky, one can obtain both their

distance and their color excess (E_{B-V}). These two quantities should define a monotonic curve on which the planetary nebula may also be placed. If E_{B-V} is known for the nebula, its distance follows directly.

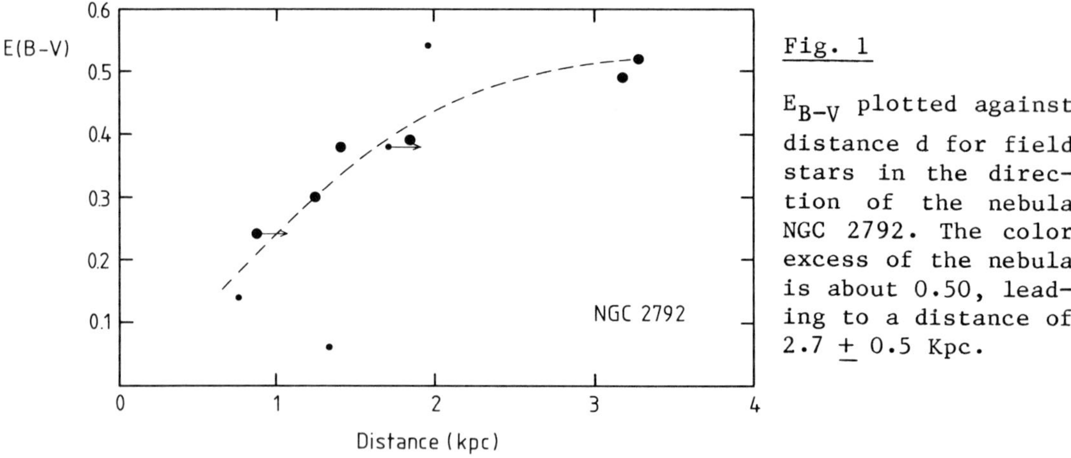

Fig. 1

E_{B-V} plotted against distance d for field stars in the direction of the nebula NGC 2792. The color excess of the nebula is about 0.50, leading to a distance of 2.7 $\pm$ 0.5 Kpc.

An example of one such curve is shown in Fig. 1. The method has been used, as well for objects other than nebulae, e.g. novae. It is generally considered reliable if carefully applied. A summary of the method and details of the results to 1976 is given by Acker (1978).

This method can be applied to all nebulae with measurable extinction that are reasonably close to the galactic plane. There is no further limitation in principle, although in practice several problems arise because:

A. An insufficient number of measurements of individual stars close to a given nebula have been made. This requires using measurements as far as 5° from the nebula, which often introduces errors because of the patchiness of the interstellar extinction.

B. Determining a correct distance to the field stars in the line-of-sight requires that their absolute magnitude be known. This can be determined from the spectrum, and sometimes, if one is careful, from the photometric colors as well. The limiting magnitudes for accurate spectroscopy and photometry consequently limit the method to moderately nearby nebulae. Only now are sufficiently accurate measurements becoming available (Acker, 1978; Gathier and Pottasch, 1983).

In a similar, but less accurate way, distances for a much larger number of nebulae can be determined. From the photometric and spectroscopic data in astronomical catalogues one can make distance vs. E_{B-V} plots in many areas of the sky. Since a sufficient number of stars is usually not available in a small region, average values of the extinction vs. distance over larger areas of the sky must be used.
Lucke (1978) has made such plots (see also Acker, 1976). It is not ideal for our purposes because the actual extinction shows variations on a scale considerably smaller than is shown in the diagram. These variations have been averaged out in the diagram. On the other hand, general trends in the extinction can be clearly seen and give a useful

first approximation to the distance. At distances greater than 300 to 500 pc above the plane, there is probably very little extinction and the method is of little use for nebulae this far the plane.

In Table 3, distances for more than 60 nebulae close to the galactic plane are given. One can judge the accuracy by comparing them with the more accurate (but preliminary) results of Gathier and Pottasch and the results of other methods. It appears that the accuracy is usually better than a factor of 2.

TABLE 3 - DISTANCES DETERMINED FROM AVERAGE 'EXTINCTION-DISTANCE' DIAGRAM NEAR THE GALACTIC PLANE

NEBULA	P.K. No.	E_{B-V}	d	NEBULA	P.K. No.	E_{B-V}	d	NEBULA	P.K. No.	E_{B-V}	d
N 40	120+ 9°1	0.50	0.8 kpc	N 5315	309- 4°2	0.42	1.3 kpc	N 6778	34- 6°1	0.23	1.0 kpc
IC 1747	130+ 1°1	0.67	3.0	N 6072	342+10°1	0.69	1.8	N 6790	37- 6°1	0.60	1.2
N 1501	144+ 6°1	0.74	1.4	N 5189	307- 3°1	0.40	0.8	N 6803	46- 4°1	0.48	1.7
N 2022	196-10°1	0.26	1.3	N 6153	341+ 5°1	0.71	1.8	N 6804	45- 4°1	0.62	2.0
IC 2149	166+10°1	0.31	1.2	N 6326	338- 8°1	0.25	0.8	BD+30	64+ 5°1	0.24	0.6
IC 2165	221-12°1	0.38	1.9	N 6439	11+ 5°1	0.53	1.3	He2-131	315+13°1	0.18	1.0
J 900	194- 2°1	0.56	2.0:	N 6369	2+ 5°1	1.43	1.5	N 6853	60- 3°1	0.05	0.25
N 2346	215+ 3°1	0.20	0.9	N 6445	8+ 3°1	0.83	2.5	N 6884	82+ 7°1	0.68	2.0
N 2438	231+ 4°2	0.29	2.0	N 6565	3- 4°5	0.30	1.3	N 6886	60- 7°2	0.58	2.0
N 2440	234+ 2°1	0.31	1.6	N 6563	358- 7°1	0.23	0.8	N 6894	69- 2°1	0.50	1.5
N 2452	243- 1°1	0.45	3.0	N 6567	11- 0°2	0.48	1.3	N 7008	93- 5°2	0.50	1.1
N 2792	265+ 4°1	0.57	2.5	N 6572	34+11°1	0.29	0.7	N 7026	89+ 0°1	0.65	2.3
N 2867	278- 5°1	0.28	2.0	N 6629	9- 5°1	0.66	1.6	Hu 1-2	86- 8°1	0.45	1.5
N 2818	261+ 8°1	0.20	1.8:	N 6720	63+13°1	0.07	0.35	IC 5217	100- 5°1	0.45	1.5
N 3211	286- 4°1	0.22	2.5	N 6751	29- 5°1	0.50	1.5	N 6578	10- 1°1	1.02	2.2
N 3918	294+ 4°1	0.28	1.3	N 6741	33- 2°1	0.83	1.4	N 6781	41- 2°1	0.85	1.5
N 5882	327+10°1	0.30	1.3	N 6772	33- 6°1	0.73	1.3	IC 5117	89- 5°1	0.87	2.5

(5) Comparison of 'forbidden line' n_e with recombination line flux measurement. The Hβ flux depends on density, nebular angular size, filling factor and distance. If the density is known from forbidden line ratio, the angular size is measured and the filling factor can be estimated from the observed geometry, the distance can be found. In practice the method gives unreliable results. There may be several reasons for this. First of all, the density is difficult to determine with sufficient accuracy from the forbidden lines, possibly because of variations within the nebula. Since the distance depends on the square of the density, an accurate value is required. Secondly, the geometry is often difficult to define with sufficient accuracy. The distance depends on the cube of the radius, which may be difficult to define precisely, either because of irregular structure or faint outer emission. Since the method is unreliable it is not of general interest. It may still be used as a check on the distances determined from other methods, as well as a means of gaining insight about the filling factor (see section C below).

(6) Stellar atmosphere analysis. Recently attempts have been made to explain the absorption line profiles observed in the spectra of some central stars. The models used predict the effective temperature and surface gravity of the star. The gravity, when coupled with an assumed stellar mass gives a value of the stellar radius. When these two values are combined with the measured visual magnitude of the star, the distance can be found.

The method can only be applied to stars which have an absorption line spectrum. Model atmospheres have been constructed for 5 central stars by Kudritzki et al. (1981). The method is in an early stage, and present accuracy is probably not better than a factor of 3 but it appears to be a promising approach for a limited number of stars.

(7) 21 cm hydrogen absorption line measurements. For nebulae located close to the galactic plane, with radio continuum fluxes greater than 100 mJy (at 21 cm), it is now possible to observe the 21 cm interstellar hydrogen absorption line. The strength of the absorption line is a measure of the distance. In this respect the method is similar to the extinction method. Another similarity is the necessity for calibrating the absorption - distance relation in different directions. Absorption line profiles give velocity information as well as line strength. Individual absorption profiles often correspond to a known spiral arm farther than the local arm. If the distance of the arm is known from other data, a lower limit for the distance of the planetary nebula can be obtained. If in addition the 21 cm absorption line is also measured in a nearby extragalactic source, it may be possible to pinpoint whether the nebula is located in or beyond the spiral arm.

As an example, Fig. 2 shows the 21 absorption profile of NGC 6537. The local arm and the Sagittarius arm, about 1.5 Kpc distant, are clearly evident. The Scutum arm 3.5 Kpc distant is completely absent. Hence the nebula must be at a distance of 2 to 3 Kpc. This method of distance determination has only recently been applied to planetary nebulae (Pottasch et al., 1981, 1983), although it has been used earlier for other difficult objects such as supernova remnants and pulsars.

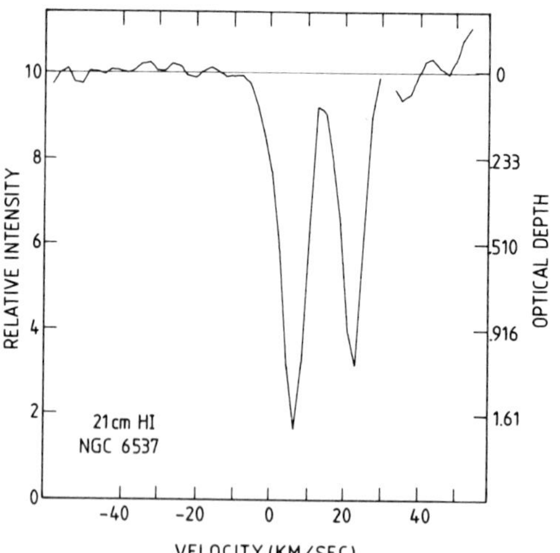

Fig. 2

The 21 cm absorption line profile in the direction of NGC 6537. Absorption due to the local arm and the Sagittarius arm are clearly seen, but no evidence for a further arm is present.

C. MASS AND DENSITY OF NEARBY NEBULAE.

When the distance is known it is possible to compare the various nebulae with each other, and especially to compare the nearby nebulae with those near the galactic center. To do this, the mass M and density n_{rms} are computed and plotted against each other as explained below.

This computation is done with a simple model: a sphere of uniform density n_e, with the material in clumps which fill a fraction ε of the total volume. The temperature T_e is also constant. The density and mass are then given by

$$n_e \, \varepsilon^{1/2} = 2.74 \times 10^4 \left(\frac{F_{H\beta} t^{0.88}}{\Theta^3 d}\right)^{1/2} \text{cm}^{-3} \quad (1)$$

$$M = 11.06 \, F_{H\beta} d^2 t^{0.88} n_e^{-1} \, M_\odot \quad (2)$$

where d is the distance in Kpc, $F_{H\beta}$ the Hβ flux corrected for extinction in units of 10^{-11} erg cm^{-2} s^{-1}. The angular radius θ is in arc sec and $t = 10^4 \, T_e$. M is quite strongly distance dependent (5/2 power) while n_e is much less dependent (1/2 power).

Once the distance has been determined, all the quantities in the above equations are specified, except the filling factor ε. To find an average value for ε, n_e has been determined from equation (1), assuming $\varepsilon = 1$. This density, called n_{rms}, may then be plotted against the density n_e determined for each nebula from the ratio of forbidden lines. Such a plot is shown as Fig. 3, where it can be seen that there is good agreement (close to the line $\varepsilon = 1$) between the two values. Thus $\varepsilon = 1$ is an acceptable value and there are no systematic departures from it. At first glance this is surprising, since examination of photographs of most of the (larger) nebulae show the presence of structure on different size scales. The answer may be: (1) the dissymmetry of the nebulae as a whole is at least partly taken into account in the determination of the angular radius, and (2) photographs

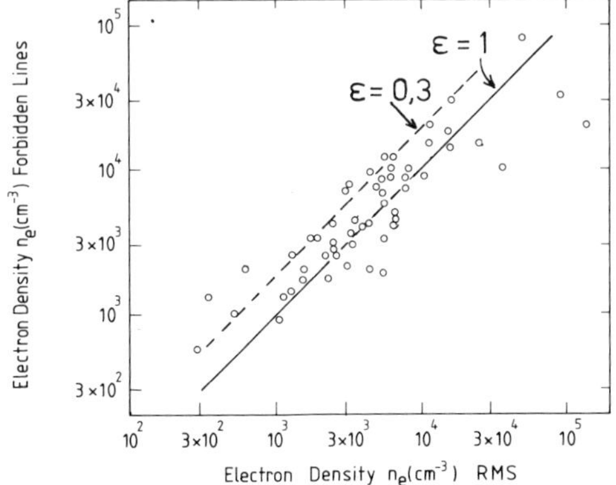

Fig. 3

The rms density obtained from eq.(1) is plotted against the electron density obtained from the ratio of forbidden lines for all nebulae for which distances could be found. A one to one correspondence is found, indicating that the filling does not systematically depart from unity.

exaggerate the importance of the smaller scale structure. On the other hand, nebulae with very low n_{rms} have not been included in Fig. 3, since no reliable forbidden line densities are available.

The ionized mass M and electron density n_e may now be determined, and are plotted against each other in Fig. 4. The names of the individual nebulae are given in the figure. From inspection of the figure, several conclusions can be drawn.

a) There is a wide range of ionized mass, from less than 10^{-4} $M_{\odot}$ to 1 $M_{\odot}$.
b) There is a clear relationship between the ionized mass and the electron density: the largest masses occur for nebulae with the lowest density.
c) A given density is correlated with only a small range of mass. Since the range of masses is considerably greater than the errors involved, the range is certainly real.

Comparable data is available for the 'galactic center nebulae'. Because the distance to these nebulae is well known this data will be presented before further discussion.

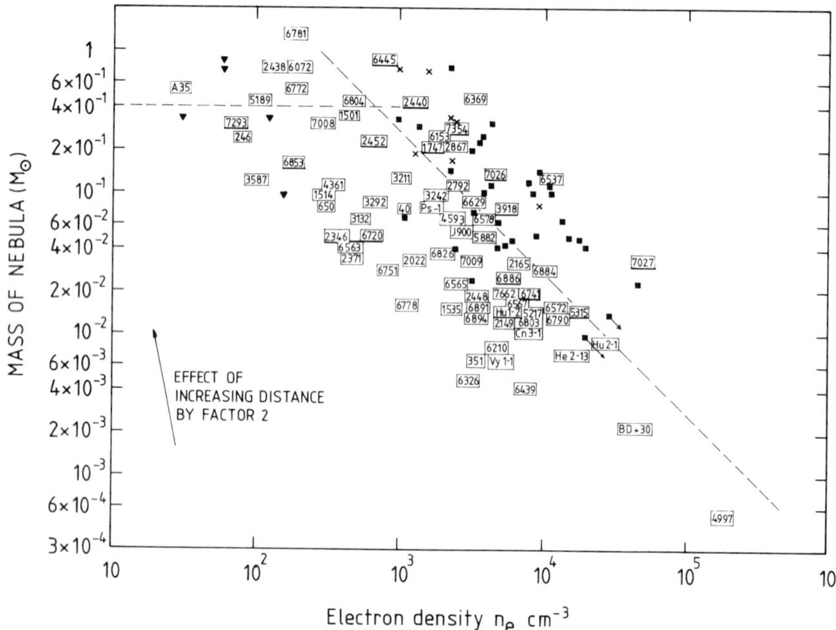

Fig. 4 The rms density is plotted against the mass of the nebula. The nearby nebulae are identified by NGC or other numbers. The filled squares are nebulae near the galactic center and observed with the VLA. The crosses are also galactic center nebula, but observed optically. The triangles are the few resolved nebulae in the Magellanic cloud. The lines are from Cudworth's proper motion study to calibrate distance scales for optically thin (horizontal line) and optically thick nebulae (sloping line).

D. GALACTIC CENTER NEBULAE

These nebulae are selected on the basis of two criteria. Firstly, they are within a few degrees of the galactic center on the plane of the sky. Using this criterion alone there is more than 90% certainty that the nebulae are actually near the center. Secondly, only nebulae with very high velocity ($V_{LSR} \geqslant 150$ km s^{-1}) are considered. It is very unlikely that nebulae selected by these criteria will be far from the center.

The distance to the galactic center used is 9 kpc. The error is probably less than 1 kpc. Because of the distance, many of the nebulae are quite small. The number of optically observed nebulae useful in mass determination is limited, because it is difficult to measure angular sizes of less than 2" accurately with optical telescopes. Radio continuum observations with the VLA have extended the number of nebulae which can be included, since this telescope can measure sizes an order of magnitude smaller. About 40 galactic center nebulae have now been measured with the VLA (Gathier et al., 1983).

The resultant masses and densities for the galactic center nebulae are shown in Fig. 4, as crosses (optical sizes) and filled squares (radio measurements). The masses have approximately the same variation with n_e as the nearby nebulae. A difference is that the galactic center nebuale occupy only the upper part of the range of Ne and M covered by the nearby objects. There are two possible reasons for this:

(a) The lower part of the range for the nearby nebulae is populated because the distance of these objects has been underestimated.

(b) There is a selection effect (discovery of the brightest nebulae) for the galactic center objects which causes one to pick out the more massive nebulae for observation.

While it is difficult to rule out possibility (a), discussion of selection effects in the next section makes it clear that (b) is the most likely explanation.

E. DISCUSSION OF THE NEBULAR MASS VARIATION

The fact that nebulae have a large range of ionized masses has two possible explanations:

a) The total mass ejected during formation of the nebula also varies over this range, and the ionized mass represents most of the total mass.

b) The total mass is considerably higher than the ionized mass and the ionization is limited because the nebulae are often optically thick in Lyman continuum radiation.

The first 'explanation' is likely to be wrong because it does not explain the variation of mass with density. On the other hand, the second 'explanation', predicts in a simple way just a mass-density relation. Consider a star of constant radiation surrounded by an expanding gas mass. If the gas is optically thick to the ionizing radiation, it absorbs all of the (constant) number of ionizing photons, K. In equilibrium, the number of ionizing photons is equal to the number of recombinations in the gas, $K = {n_e}^2\Delta v$, where Δv is the ionized volume. The mass ionized is

$$M_i \sim n_e \Delta v = K/n_e$$

The mass determined from observation has approximately this same variation with density for the higher density nebulae. We adopt this simple picture as a working hypothesis.

This picture has the important consequence that all nebulae with higher density are optically thick (ionization bounded) in the Lyman continuum. The dividing line is about $n_e \approx 3 \times 10^2$ cm^{-3} or somewhat higher and may be somewhat different for individual nebulae.

A further consequence is that the ionized masses found for the low density nebulae represent the total nebular mass, which appears to vary between 0.1 $M_\odot$ and 1 $M_\odot$.

F. RELATIONSHIP WITH STATISTICAL METHODS FOR DETERMINING DISTANCES

1. 'Shklovskii method'. This method assumes the ionized nebular mass is constant. It is usually calibrated using the mass of NGC 246, whose distance is determined from spectroscopic parallax. As can be seen from Fig. 4, the resulting mass for about 60% of the nebulae is within a factor 5 of this value. In other words, the distance of these 60% is determined only to within a factor 2. Larger errors occur for the high density nebulae. In some cases this method will lead to distance errors of an order of magnitude.

2. Proper motion studies. Cudworth (1974) has analysed measurements of the proper motions of the central stars of 51 nebulae. The measurements are not individually significant, so that they must be treated statistically. This was done by dividing the nebulae into two groups, those which are optically thin (density bounded) and optically thick (radiation bounded). The division was made on the basis of the nebular size, and is therefore related to the density.

For the optically thin nebulae Cudworth assumed that the nebular mass is constant (as above) and used the proper motion statistics to determine the value of the mass. The result, $M = 0.4$ $M_\odot$, is shown as a horizontal line in Fig. 4. It is a better fit to the masses of the low density nebulae. This should be so since the low mass, high density nebulae have been eliminated from this sample.

For the optically thick nebulae Cudworth assumed that the absolute Hβ flux is constant and determined the value of the constant from the proper motion statistics. The result is:

$$d = 5 \left[F_{H\beta} \right]^{-1/2}$$

where d is in Kpc and $F_{H\beta}$ is the measured Hβ flux in units of 10^{-11} erg cm^{-2} s^{-1}. This result is also plotted as a sloping line in Fig. 4. The agreement is considerably better than the constant mass assumption for nebulae having $n_e \geqslant 6 \times 10^2$ cm^{-3}. The precise position of the line in the diagram (or the constant in the above equation) is quite uncertain, since only 17 nebulae are involved in Cudworths study. Furthermore, the assumption of a unique absolute Hβ flux for all optically thick nebulae is a poor approximation (see below and Fig. 5).

G. ON THE DISPERSION OF THE POINTS IN FIG. 4.

The dispersion of the points in the mass-density diagram is larger than would be expected from the errors involved. Since the distance of individual objects is often uncertain, sometimes by as much as a factor 2, this must be a contributing factor. However there are reasons to think that it is not the overriding factor. The first is that for a given density in the optically thick region, the masses appear to vary by about a factor of 50. Such a large variation is well in excess of what the uncertainties in the distance would contribute. Secondly, the nebulae at the galactic center show a large dispersion, although it is only a factor of 20. This dispersion is clearly real because the nebulae are essentially all at the same distance.

The reason for the (smaller) dispersion in the optically thin region is clear: the (intrinsic) nebular masses vary by a factor of between 3 and 10. The reason for the dispersion in the optically thick

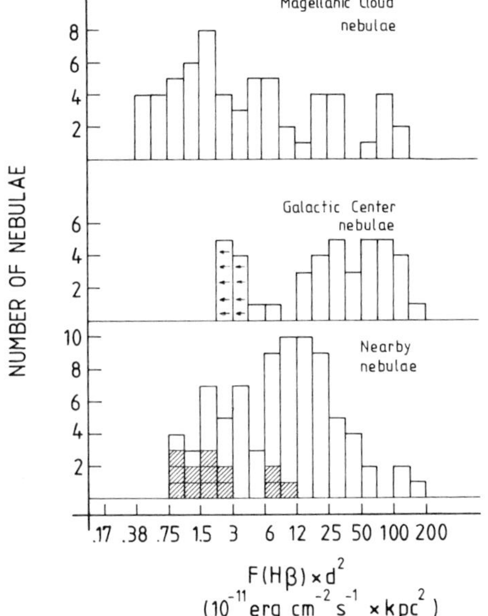

Fig. 5

Histograms of intrinsic Hβ flux.
(a) Magellanic cloud nebulae (Jacoby, 1980)
(b) Galactic center nebulae used in Fig. 4 (Gathier, et al., 1983).
(c) Nearby nebulae (shaded areas are optically thin nebulae).

region is different. It can best be understood by recalling that the sloping line in Fig. 4 is the locus of nebulae with a constant intrinsic value of Hβ flux. Other values of intrinsic Hβ flux will appear as parallel lines, small Hβ values to the left, larger values to the right. The dispersion in the optically thick nebulae can thus be explained by a large spread in the intrinsic Hβ flux.

There is other evidence for such a spread in the intrinsic flux. Recently, Jacoby (1980) made a very deep survey of selected regions in the Magellanic clouds and measured the integrated flux of the nebulae found. The results are shown in Fig. 5a, which is a histogram of the intrinsic Hβ flux (measured flux multiplied by d^2). It can be seen that the intrinsic flux varies by a factor of 400! For comparison,

Fig. 5c shows the intrinsic flux of the nearby nebulae discussed earlier. These can be divided into optically thin (shown as striped area) and optically thick nebulae. As can be seen, the optically thin nebulae all have small intrinsic Hβ flux. The low end of the distribution of nearby nebulae does not extend to such small values as is the case in the Magellanic clouds, and is probably not the result of a systematic overestimate of the distances of some of the nearby nebulae. This discrepancy is due to the fact that most of the nearby faint nebulae were not included in our sample, because they lie at high galactic latitudes and an independent value of the distance cannot be obtained for them. Fig. 5 includes only a few nebulae with diameters greater than 100", whereas more than 50 are known to exist!

Fig. 5b shows the distribution of intrinsic Hβ flux (really converted radio flux) of the galactic center nebulae in our sample. It does not extend to low intrinsic Hβ values, presumably because only the brighter objects have been selected for observation. This is evidence for the earlier statement that the lack of galactic center nebulae in the lower left side of Fig. 4 was a selection effect.

The mass-density dispersion of the optically thick nebulae is correlated with a very large variation in the intrinsic Hβ flux. The variation in flux is a direct consequence of intrinsic differences in the number of ionizing photons, which in turn follows from differences in radius and temperature of the central star.

H. THE EFFECTIVE TEMPERATURES OF THE CENTRAL STAR

There are a number of methods of determining the effective temperatures. The most important are:

1) Use of the spectral type for those stars with a normal O or Of type spectrum, which are generally the relatively cooler stars.
2) Zanstra temperatures
3) Comparison of the continuum emission between the visual and the ultraviolet (λ 1300 Å) with either a blackbody or a model atmosphere.
4) Energy balance or Stoy method.

Method 1) may only be applied to a limited number of central stars, whose spectra mimic quite closely the well studied normal O type stars. At present, model atmospheres are being constructed to reproduce the oberved spectra of higher temperature stars which have absorption line spectra (Mendez et al., 1981), which will increase the usefulness of this method.

The application of method 2) has been questioned for some time for the following reason. The method can be applied using either the hydrogen lines or the lines of ionized helium. The ratio of the flux shortward of λ 912 Å to the visual continuum is then computed. The ratio is then compared with that expected from a blackbody. The temperature is assumed to be that of the blackbody with the same ratio. The difficulty occurs because in a substantial number of stars the values of $T_Z(H)$ and $T_Z(HeII)$ are not the same; the latter value is consistently higher than the former.

There are two possible explanations for this difference.

Firstly the nebula may be optically thin in the radiation field which can ionize hydrogen. Since some photons which ionize hydrogen would then not contribute to the ratio, $T_Z(H)$ would only be a lower limit, and $T_Z(HeII)$ would be a better approximation. Secondly, it may not be correct to use a blackbody for comparison, but instead an atmosphere with emission in excess of that of a blackbody shortward of $\lambda = 228$ Å. Several of the models of Mendez et al.(1981) and Wesemael et al.(1980) have this characteristic. This would result in $T_Z(H)$ being the more nearly correct value and $T_Z(HeII)$ being excessively high.

Which of these two reasons is correct? In the last two sections it has been argued that Fig. 4 may be used to separate the optically thick nebulae from those that are optically thin. For the optically thick nebulae it is reasonable to assume that the second reason is correct and that $T_Z(H)$ is a good approximation to the effective temperature. The question still arises as to whether $T_Z(H)$ should be computed by comparing with a blackbody or with a model atmosphere. The remarkable fact is that for the three series of model atmospheres in the literature (Hummer and Mihalas, 1970; Wesemael et al., 1980; Mendez et al. 1981), the same temperature is obtained (within 10%), regardless of whether a blackbody or a model atmosphere is assumed. This is only true for $T_Z(H)$; it is definitely not true for $T_Z(HeII)$. The use of the atmosheric models of Hummer and Mihalas (1970) give a higher value of $T_Z(HeII)$ than a blackbody atmosphere (making the discrepancy worse). However the pure hydrogen models of Wesemael et al.(1980) and several of the models of Mendez et al.(1981) do just the opposite, lowering $T_Z(HeII)$, even to values lower than $T_Z(H)$. More accurate models are only a hope for the future, but it seems likely that $T_Z(H)$ will be relatively unaffected.

For the optically thin nebulae, method 3), comparison of continuum emission, must be applied. It can only be used when both visual and ultraviolet measurements are available because of the small change in slope with temperature of a blackbody curve for both objects. Several problems arise in this method. First of all, the extinction must be well determined because correction greatly affects the slope of the continuum. Results for heavily reddened nebulae are therefore less reliable. Secondly, current ultraviolet observation include the nebular continuum with the stellar continuum and the separation is often difficult. A counter-balance to these difficulties is that the interpretation is more straightforward. A blackbody and the model atmospheres mentioned above all predict approximately the same slope between $\lambda = 1300$ Å and $\lambda = 6000$ Å.

One of the difficulties mentioned above applies also to method 2). For the smaller nebulae it is sometimes very difficult to separate the stellar continuum from the nebular continuum. The problem is greatest for the hotter stars, because relatively more flux is used for nebular ionization and consequently produces more nebular continuum. Careful observational work is beginning (e.g. Kohoutek and Martin, 1981; Martin, 1981) but when 90% of the observed continuum emission is nebular, the resulting magnitude m_V or stellar continuum must be considered unreliable. This is particularly true of the very hot central stars of the nebulae Peimbert calls Type I, which have extremely high helium and nitrogen abundances and a strong concentration to the galactic plane, e.g. NGC 2440, 6302, 6445, 6741 and 7027.

For these hot, faint stars the most reliable temperatures can be obtained from method 4, the energy balance or Stoy method, which measures the excess energy of each ionizing photon. It is based on the measurement of the ratio of energy emitted in collisionally excited lines to that emitted in Hβ. The helium abundance and state of ionization must also be known, but the magnitude of the star does not have to be known. A recent discussion of this method, together with resultant temperatures, has been given by Preite-Martinez and Pottasch (1983). Some resultant stellar temperatures are given in Table 4. When the temperatures can be compared to those determined from other methods, the agreement is good.

TABLE 4 - STELLAR TEMPERATURES FROM ENERGY BALANCE METHOD

NEBULA	$\sum I(\text{coll.exc.}) / I(H\beta)$	T	NEBULA	$\sum I(\text{coll.exc.}) / I(H\beta)$	T
NGC 40	9.5	36.000 K	NGC 2165	41.7	115.000 K
IC 418	8.5	33.000	3211	47.9	135.000
NGC 6826	11.3	41.000	6886	55.4	165.000
6572	19.0	60.000	6741	66.9	230.000
1535	22.4	66.000	2440	70.8	250.000
3918	38.4	105.000	7027	73.6	270.000
7662	36.1	100.000	6302	92.7	350.000

I. POSITION OF THE STARS ON THE HR DIAGRAM.

When the distances and effective temperatures have been calculated only the luminosity must be found in order to place the stars on the HR diagram. The usual method for determining luminosity is to combine the temperature, visual magnitude m_V, and distance to obtain the stellar radius. The radius is only slightly dependent on the temperature for these hot objects. The radius is also little affected by the model atmosphere (or blackbody) used, to within about 20%. The radius and temperature then define the luminosity.

The greatest source of error for the hotter objects is the measurement of m_V as has been pointed out in the discussion of stellar continuum. This is probably why the hot objects listed above are not often plotted on the HR diagram. For optically thick nebulae the luminosity can be found without knowing the visual magnitude. This is because all the stellar radiation shortward of $\lambda = 912$ Å is absorbed in the nebula and transformed into (nebular) radiation longward of $\lambda = 912$ Å. All of this radiation can be measured, when ultraviolet and infrared observations above the atmosphere are available. The luminosity is thus a well determined quantity (its greatest uncertainty is the distance determination).

Even when far ultraviolet and infrared observations are not available, the luminosity can be reasonably well approximated as 100 times the luminosity in the Hβ line. This works empirically for such different objects as NGC 7027 and the low stellar temperature nebulae BD +30 3639. In NGC 7027 the ratio of forbidden emission lines to Hβ is much higher than in BD +30 3639, but this is compensated for in the latter object by the much higher stellar flux which is directly emitted longward of λ = 912 Å.

For optically thin nebulae this method will not work and one is dependent on calculating the radius using m_V. Since most of these nebulae are larger the influence of nebular emission in a small diaphragm is minimal.

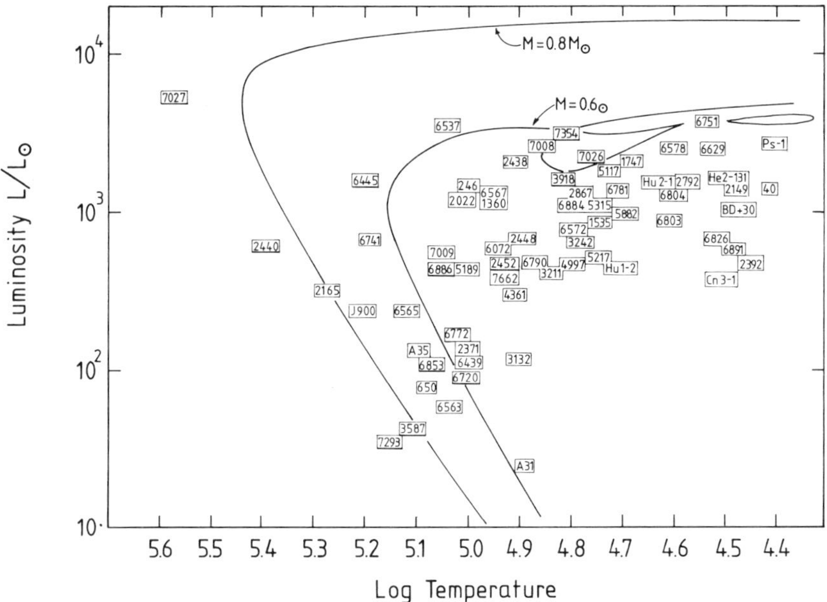

Fig. 6 Luminosity of the central star is plotted against temperature. Nebulae are identified by their NGC, etc., number. Those with high helium or nitrogen abundance are underlined.

The resultant HR diagram is shown as Fig. 6. In the diagram each star is identified by the NGC, etc., number of the nebula in which it is found. The stars underlined are those where the nebulae are helium and nitrogen or carbon rich. For orientation and comparison, the theoretical evolutionary tracks calculated by Paczynski (1971) are given for 0.6 $M_\odot$ and 0.8 $M_\odot$. The lower temperature stars appear to all have luminosities between $3 \times 10^2\ L_\odot$ and $5 \times 10^3\ L_\odot$. This range seems to remain fixed independent of stellar temperature. For stellar temperatures greater than 80.000 K much lower luminosities are observed as well, extending to almost 20 $L_\odot$.

As evolution progresses, the temperature and luminosity of the star change. Thus the time may be considered as an additional parameter, which may even be a measureable quantity: the nebular radius divided by the expansion velocity. Schonberner (1981) has plotted the absolute visual magnitude of the central star against this time t and compard this to theoretical expectation. He points out that the major uncertainty is distance. He has used the distances determined by assuming a constant nebular mass of M = 0.2 $M_\odot$. Fig. 7a is a plot of M_V against t using this assumption. It differs from the plot made by Schonberner, presumably because of a somewhat different selection of nebulae (those plotted in Fig. 6) and an observed value of the expansion velocity is used here instead of an assumed constant value of 20 km s^{-1}. Schonberner's plot shows less spread. He compares it to the theoretical curves, shown as solid lines in Fig. 7, and concludes that the central stars have a rather narrow range of mass.

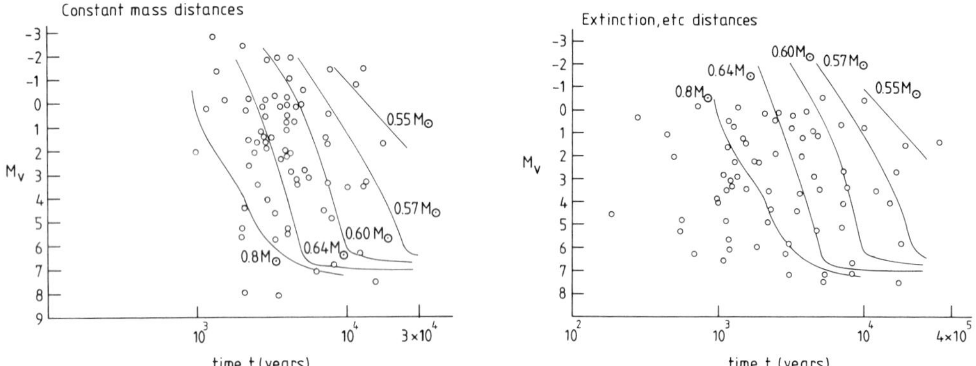

Fig. 7 Absolute visual magnitude is plotted against the nebular age t (in years), defined as the radius of the ionized region divided by the expansion velocity. Theoretical expectations for different masses are plotted as solid lines (Schonberner, 1981, except M = 0.8 $M_\odot$ taken from Paczynski, 1971).
(a) Distance determined from assumption of constant nebular mass.
(b) Distance determined by independent methods.

Fig. 7b is the same plot (M_V against t) but now using the distances we have described above. The greatest difference is the much larger spread in t, extending down to a few hundred years. One might conclude that this is inconsistent with Fig. 6 where it appears that most central stars have masses less than 0.6 $M_\odot$. The difficulty is that the age or time cannot be computed from the ratio of the ionized size to velocity for an optically thick nebula, because the ionized size is smaller than the actual size. All the nebulae known to be optically thin are at the right side of Fig. 7b with $t > 3 \times 10^3$ years.

Schonberner's conclusion appears correct, however. It is easier to see from a comparison to the observations and theoretical expectation in Fig. 6, where one would conclude that most of the stellar masses are in the range 0.5$M_\odot$ to 0.6 $M_\odot$. Some of the very hot stars may have masses exceeding this, possibly 0.8$M_\odot$ or somewhat higher.

CONCLUDING REMARKS

An important change has occurred in the last 15 years in the position of the central stars in the HR diagram. Earlier the luminosity was thought to extend to 10^4 to 10^5 $L_\odot$, and to increase from the low temperature end to a maximum value near 70.000 K (the Harman Seaton track). Now it appears that the luminosity is lower, within a factor 3 of 10^3 $L_\odot$, and it remains constant for a large part of the temperature range, until it begins to decrease at 10^5 K. What has caused this large change? While many factors are involved, it is primarily the growing realization that the large majority of nebulae are optically thick in Lyman continuum radiation. This has three consequences. First of all, the rather large correction for radiation not absorbed by the nebulae is no longer needed. Secondly, the temperature of some of the nebulae is lowered. Thirdly, the distance of some of the smaller nebulae is found to be considerably nearer.

The lower luminosity for the beginning phases of the central star, coupled with the realization that the mass probably lies within the range 0.5 $M_\odot$ to 0.7 $M_\odot$ provide a basis for further work on the pre-nebular phase of evolution.

REFERENCES

Acker, A., 1976, Publ. Obs. Astron. Strasbourg, Vol. 4, fasc. 1.
Acker, A., 1978, Astron. Astrophys. Supp. 33, 367.
Chudovicheva, O., 1964, Iz. Pulkova, Obs. 23, 154.
Cudworth, K.M., 1974, Astron., J., 79, 1384.
Gathier, R., Pottasch, S.R., 1983.
Gathier, R., Pottasch, S.R., Goss, W.M., v. Gorkom, J.H., 1983.
Harris, W.E., 1976, Astron. J. 81, 1095.
Hummer, O.G., Mihalas, D., 1970, Mon. Not. Roy. Astron. Soc., 147, 339.
Jacoby, G.H., 1980, Astrophys. J. Supp. 42,1.
Jacoby, G.H., 1981, Astrophys. J. 244, 903.
Kohoutek, L., Martin, W., 1981, Astron. Astrophys. 94, 365.
Latypov, A.A., 1955, Publ. Astron. Obs. Tashkent, 5,31.
Liller, M.H., Liller, W., 1968, Plan. Neb., IAU Symp. 34, ed. Osterbrock, D., O'Dell, C.R. (Reidel, Dordrecht).
Liller, M.H., Welther, B.L., Liller, W., 1966, Astrophys. J., 144, 280.
Lucke, P.B., 1978, Astron. Astrophys., 64, 367.
Lutz, J.H., 1977, Astron. Astrophys., 60, 93.
Lutz, J.H., 1978, Plan. Neb. IAU Symp. 76, Ed. Y. Terzian (Reidel, Dordrecht).
Martin, W., 1981, Astron. Astrophys., 98, 328.
Mendez, R., 1978, Mon. Not. Roy, Astron. Soc., 185, 647.
Mendez, R.H., Kudritzki, R.P., Gruschinski, J. Simon, K.P., 1981, Astron. Astrophys., 101, 323.
Pazynski, B., 1971, Acta Astron., 21, 417.
Pottasch, S.R., Gathier, R., Goss, W.M., 1983.
Pottasch, S.R., Goss, W.M., Arnal, E.M., Gathier, R., 1982, Astron. Astrophys., 106, 229.
Preite-Martinez, A., Pottasch, S.R., 1983.
Schoenberner, D., 1981, Astron. Astrophys., 103, 119.
Wesemael, F., Auer, L.H., Van Horn, H.M., Savedoff, M.P. 1980, Astrophys. J. Supp., 43, 159.

RODRIGUEZ: When using the extinction method, do you apply any correction for possible internal extinction?

POTTASCH: Present evidence suggests that, for most nebulae, a correction for internal extinction is not necessary. For many nebulae, the extinction may be determined by several methods: (a) comparison of radio and Hβ emission; (b) λ 2200 feature; (c) Balmer decrement; (d) He II λ 4686/λ 1640 ratio. All these methods give the same extinction, when interpreted with a "standard" interstellar extinction curve. If a substantial part of the extinction arose from internal nebular dust, one would not expect this good agreement because the internal dust is expected to have properties which are substantially different from the interstellar dust.

OSTERBROCK: The progress in getting distances of PN shown in this work is very gratifying. It is, therefore, important to make quantitative estimates of the accuracy of these distances. The ideal case for the extinction method would be when all the stars were exactly along the line of sight to the PN and their intrinsic colors were exactly known. In practice, there is some scatter in angle and the intrinsic colors are not precisely known. You could evaluate the errors in the extinction method by applying it to a few stars of accurately known distance, in fields of about the same richness as the PN fields. Have you done anything like this, or do you intend to do so? Can you give any numerical estimate of the probable errors in the extinction method?

POTTASCH: We are in the process of analysing the errors involved in this method. The errors depend on many factors and will vary from nebula to nebula. If extensive observations have been made in the region close to the nebula (within 1°) and stars are observed with extinctions 50% higher than the nebular extinction, the extinction distance is probably accurate to within 30%. At the other extreme, when it is necessary to use observations as far away as 5° or even 10° from the nebula, the distance derived may be within only a factor 2 of the correct value.

ALLER: The level of excitation of NGC 6741 resembles that of NGC 7662 and our theoretical models give $T_* \approx 95\,000$ K. We could not reproduce the observed spectrum with a stellar flux corresponding to $T_* \approx 200\,000$ K, as you suggest.

POTTASCH: The ratio of the energy in collisionally excited lines to Hβ is twice as high in NGC 6741 as in NGC 7662. This observational fact can most easily be explained if the exciting star of NGC 6741 has a considerably higher temperature than that of NGC 7662.

WEIDEMANN: In your HR diagram, the luminosities of the lower temperature objects cluster around 10^3 L_o. Although this behaviour is qualitatively what is expected from evolutionary calculations (low mass nuclei spend most of their life at high luminosity, high mass nuclei at lower luminosity), your result is quantitatively difficult to understand - nuclei of 0.55 $M_\odot$ already reside at $10^{3.2}$ $L_\odot$. I conclude that either the distances are still underestimated or the temperatures are too low. The smaller distances would present the additional problem of a higher PN birth rate and increase the existing discrepancy with the White Dwarf birth rate (cf. Weidemann, 1977, Astron. Astrophys. 61, L27).

POTTASCH: The distances of only the optically thick nebulae in my sample are smaller than previously adopted values, and then by only about 50%, on the average, as compared with Cudworth's values. The value of the local birthrate of PN, on the other hand, is mainly determined by the distances of the large, optically thin nebulae, and I presented evidence that their distances have probably been under-estimated in previous studies.

COHEN: I have the impression that, when the observed infrared contribution is included, a number of nuclei are as luminous at 10^4 $L_{\odot}$! Are we selecting the highest mass objects or using incorrect distances?

POTTASCH: I have checked that the luminosities plotted in the HR diagram exceed the total observed luminosity, including the infrared contribution. The high luminosity which you quote,especially for BD + 30° 3639, arises from using too large a value for the distance.

THE DISTANCES OF PN AND THE GALACTIC ROTATION CURVE

S.E. Schneider, Y. Terzian
Cornell University, Ithaca, N.Y., USA

A. Purgathofer
Institut fur Astronomie der Universitat Wien, Austria

M. Perinotto
Astrophysical Observatory of Arcetri, Largo E. Fermi 5, Firenze, Italy

The problem of determining the distance scale for planetary nebulae (PN) is approached through the kinematics of this subpopulation in the galaxy. To this end, we have compiled a catalogue of all known radial velocities for 457 galactic PN with standardized error statistics. External (calibration?) errors of the same magnitude as internal spectral line deviations are noted.

A subsample of 62 PN lying in the galactic longitude ranges 20° to 50° and 310° to 340° and having distances determined by Acker was chosen for examination since radial velocities vary rapidly with distance in these directions. We find that Acker's distances must be increased by a factor 1.5 (± 0.1) $R_{\odot}/9$ kpc for velocities and distances to correspond with the known rotation curve interior to the solar circle. This technique is relatively more sensitive to the distant PN, but locally we similarly find that distances must be increased to give the PN a scale height commensurate with their velocity dispersion.

The scatter of kinematical distances from the rescaled distances is small - less than that attributable to the expected velocity dispersion alone in most cases. Since most of these distances were based upon Schlovskii's method, this small deviation lends support to such distance determinations when properly scaled.

Conversely, we are able to estimate the galactic rotation exterior to the sun. We find that the rotation curve steadily rises out to 6 kpc (and possibly 11 kpc) beyond the solar circle, in agreement with previous determinations.

OSTERBROCK: Does the check on distances which you propose rely on the assumption that PN have circular (or nearly circular) orbits in the Galaxy, like the interstellar gas from which the rotation curve which you use was derived? Is there observational evidence to justify such an assumption?

SCHNEIDER: While the velocities of individual, old disk stars may be peculiar, on the average they take part in the general Galactic rotation. For example, in the solar vicinity, the asymmetric drift of the mean PN rotation is observed to be only about 15 km s^{-1}. The close agreement of our predicted velocities with the observed velocities suggests that this asymmetric drift does not increase much beyond this value, to within 4 kpc of the Galactic centre.

ALLER: Plots of PN radial velocities against longitude display a marked departure from a double sine wave, as Minkowski showed long ago. There are many orbits of high eccentricity. I do not understand how Galactic rotation theory can be applied fruitfully to such a mixture of orbits.

SCHNEIDER: It is true that plots of PN velocities along a given direction, regardless of distance, show a large scatter. Towards the Galactic centre, the scatter is intrinsic to the PN orbits - they have a large radial velocity dispersion. However, none of the PN we use is as close as 3 kpc to the Galactic centre, and the scatter of points about the predicted velocity/distance curves is small. This result, and preliminary computer simulations of the evolution of orbits in the presence of a ring of molecular clouds, indicate that the apparent velocity dispersions should be and are sufficiently small to obtain meaningful results.

ACKER: In 1976 and 1980, I studied the relationships between chemical, spatial and kinematic parameters of PN. Using a sample of 330 PN with known radial velocities, I found the usually accepted values for the rotation parameters. However, the velocity dispersion is high because the majority of PN belong to an old disk population. I believe that it is very difficult to calibrate a distance scale using kinematic criteria alone.

SECTION VI

PLANETARY NEBULAE IN A GALACTIC AND EXTRAGALACTIC CONTEXT

A RADIO SEARCH FOR GALACTIC CENTER PLANETARY NEBULAE

Richard Isaacman
Sterrewacht Leiden (The Netherlands) and
United Kingdom Infrared Telescope

SUMMARY

This review summarizes the techniques and results of a λ21cm and λ6cm search with the Westerbork telescope for planetary nebulae within 2° of the galactic center. After accounting for background sources and compact HII regions it appears that there are ∿ 300 planetaries within 300 pc of Sgr A. This is consistent with a galactic population of ∿ 21000 and agrees with the birthrate of white dwarfs. The surface density of galactic center planetaries falls off with galactic latitude approximately as b^{-1} and is best accounted for by a bulge with a mass of $1 \times 10^{10}\ M_\odot$.

INTRODUCTION

Searches for planetary nebulae (PN) near the galactic center aim at answering these questions:

(1) What is the distribution and total number of planetaries in the galactic bulge?

(2) Are their properties (lifetimes, masses, luminosities) similar to those in the solar neighborhood?

(3) What does the distribution imply about the structure of the bulge and the total number of planetaries in the Galaxy?

When studying the mass distribution of the bulge it is convenient to seek a set of relaxed Population II "tracers" which we assume to be characterized by a Maxwellian velocity distribution, with no systematic orbital motion. If this is true then the galactic gravitational field, and hence the mass distribution, can be read from the tracer distribution in a simple way provided that the velocity dispersion σ is known. Globular clusters have been used in this way to study the structure of M 31 (Tremaine et al. 1975) as well as our own Galaxy (Oort 1977).

D. R. Flower (ed.), Planetary Nebulae, 415–422.

Planetary nebulae can also be used. These were first shown by Minkowski (1948) to share Population II dynamics in the galactic bulge (if not in the disk). Their principal advantage over other tracers is that they are, at least above $|b| \gtrsim 2°$, discoverable via optical spectral lines as well as radio emission.

Within 2° of the center, where optical observations are no longer useful, planetaries suffer a single tremendous disadvantage: their lack of radio spectral emission makes them very difficult to distinguish from background sources and, in particular, compact HII regions. For that reason, much effort must be directed towards extracting the true distribution of planetaries from the observed distribution of all sources in a radio survey. This must be done mostly in a statistical way; identifying individual planetaries in a radio sample is much more difficult.

THE OBSERVATIONS: HOW MANY PLANETARIES?

To date, the only radio search specifically aimed at discovering planetary nebulae in the highly-obscured central bulge has been the Westerbork survey by Wouterloot and Dekker (1979) and by myself (Isaacman 1980, 1981ab). The Westerbork program concentrated on five 21-cm fields all centered within about 1.5° (i.e. about 225 pc) of the galactic center. Accounting for overlap in the fields, these covered some 3 square degrees at sensitivities ranging from about 1 mJy (1σ) at the field centers to about eight times worse at the edges. A total of 119 sources were found, of which 50 were later observed (though not all detected) at λ6cm with Westerbork and with the VLA.

Not all -- or even most -- of these 119 are planetaries, of course. Contaminating the sample are (a) nonthermal galactic sources such as supernova remnants and radio stars, (b) extragalactic background sources, and (c) compact HII regions. It is easy to show on statistical grounds that at most $\sim$1 object from the first category will be detected as a compact source in the survey. Hence supernovae, UV Ceti stars, radio binaries and the like are not an issue. Extragalactic background sources are also easily accounted for statistically because their spatial distribution is isotropic and because their flux and spectral index distributions are known (Willis et al, 1977; Willis and Miley 1979).

Compact HII regions are more difficult to extract from the data because of their similarity to planetaries at the distance of the galactic center. On an individual basis the two can be distinguished in high-resolution radio data since planetaries will often show characteristic shell structure. Several of the Westerbork objects have been observed this way at λ6cm with the VLA A-array (Isaacman 1980a and unpublished work). More generally, we know that HII regions occupy a broad range of ionized masses and emission measures (Israel 1976; Habing and Israel 1979) and thus can be either bright or dim in the radio continuum, whereas the ionized masses of planetaries seem limited to a few tenths of a solar mass (Pottasch 1980) and should consequently rarely be brighter than several tens of mJy at the galactic center. A strong tendency for the flat-spectrum (i.e. thermal) sources in the Westerbork sample to be weak should therefore imply that planetaries dominate the distribution. This is definitely the case: nearly 90% of the flat-spectrum sources have $S_{21} < 50$ mJy. On this basis we conclude that planetary nebulae greatly outnumber compact HII regions in the inner 300 pc of the galactic bulge.

Among the 85 weak (< 50 mJy) 21-cm sources we expect on the basis of earlier Westerbork source counts (Willis et al. 1977) that $\sim$ 59 will be extragalactic. From the spectral index arguments we therefore estimate that there are $\sim$ 25 planetaries in the data set. A similar argument using the Westerbork 6-cm fields and the background counts by Wall and Cooke (1975) gives the same answer.

NEBULA LIFETIMES AND RADIO FLUX DISTRIBUTION

In an aperture synthesis survey we must model the intrinsic flux density distribution of the sources in order to extract their spatial distribution. The radio flux density distribution of nearby planetaries can be reasonably well reproduced using a simple model characterized by an expanding spherical shell of $0.2M_\odot$ ionized by a central star that follows the evolutionary track proposed by Seaton (1966). In this model PN are density bounded; at the distance of the galactic center they would have peak flux densities of about 60 mJy at 21cm (Isaacman 1980).

The agreement with the fluxes of local PN is improved if the luminosity and temperature of the central star in the model are scaled down by factors of 10 and 2, respectively, following the results of Pottasch et al. (1978). In this case most nebulae are ionization bounded and will emit a maximum of $\sim$ 30 mJy at λ21cm at a distance of 9 kpc.

It is not, however, sufficient to translate a model for local PN to the distance of the galactic center for the purpose of modelling the flux density distribution. The random velocities of PN in the bulge are much higher than in the disk, so ram pressure effects from the interstellar medium (ISM) can be important. If the velocity through the ISM (V_{ISM}) is much greater than the shell expansion velocity V_e, then it is possible to show (Isaacman 1979) that the nebula will start to break up on a time scale

$$(1) \quad T \sim 6 \times 10^5 \left[\frac{M}{\rho V_e V_{ism}^2} \right]^{1/3} \quad \text{yr}$$

where M is the shell mass in $M_\odot$, velocities are in km/s, and ρ is the density of the ISM in $M_\odot$ pc^{-3}. The density in the inner bulge can be as high as $\sim$ 1 $M_\odot$ pc^{-3} due to molecular clouds (Bania, 1977); if $M = 0.2\ M_\odot$, $V_e = 25$ km/s, and $V_{ism} = 125$ km/s -- approximately planetaries' velocity dispersion -- then galactic center PN will live only ∿ 5000 yr. This is a factor ∿ 5 shorter than local PN, so the flux density distribution of galactic center planetaries will be affected accordingly.

In order to reconcile the number of PN in the Westerbork data with the number of optically-identified planetaries in the bulge -- after correcting the latter number for extinction -- we require that ∿ 75% of the galactic center PN be ionization bounded. The resultant model flux density distribution predicts that ∿ 1/4 of radio-identified PN should have flux densities between 30 mJy and 60 mJy, and that very few -- primarily local objects -- will be brighter than 60 mJy. This prediction appears to have been borne out in recent VLA observations by Gathier et al (1982), as shown in Figure 1. They measured the radio fluxes of a few dozen Perek and Kohoutek (PK) planetaries close to the galactic center.

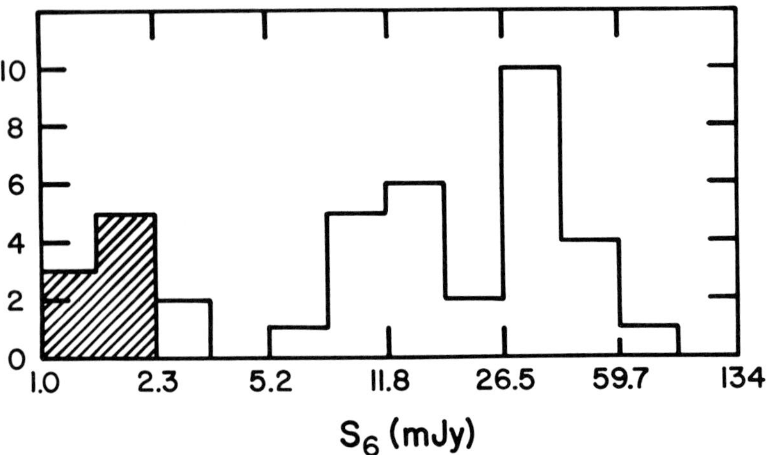

Figure 1. Radio fluxes of optically-identified galactic center planetary nebulae by Gathier et al. (1982). Hatched areas are upper limits.

THE SPATIAL DISTRIBUTION OF PN AND MASS MODELS OF THE BULGE

By using the model flux densities to correct the observed source distribution for the loss in sensitivity at the field edges we obtain the PN surface density shown by the filled circles in Figure 2. The open circles are counts of planetaries from the PK catalogue in a wedge about the galactic center that corresponds roughly to the area of the Westerbork survey. These optical counts are corrected for extinction based on an intrinsic Hβ brightness distribution derived from measurements of local nebulae.

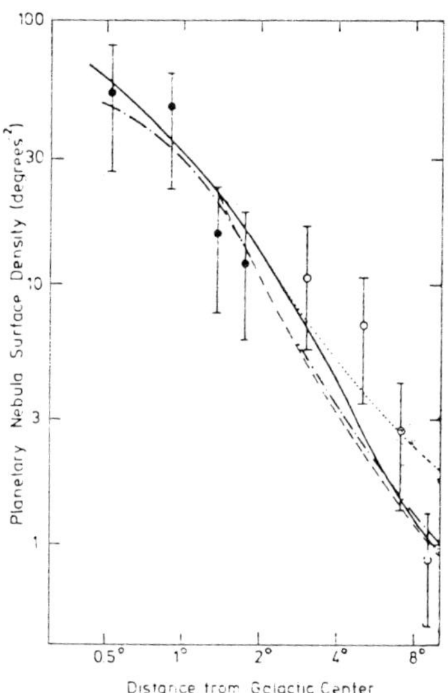

Figure 2. Corrected radio (filled circles) and optical (open circles) surface densities of galactic center planetary nebulae. Fitted curves are distributions predicted from mass models of the bulge (see text).

The average corrected source density in the radio survey (within 2° of the galactic center) leads to a PN surface density of 23.9 deg^{-2}. Within 2° (314 pc) of the center, we therefore expect about 300 nebulae. Mass models of the galactic center (e.g. Sanders and Lowinger 1972) predict masses of $\sim 2 \times 10^9$ $M_\odot$ in this region, giving a relative number of 1.5×10^{-7} $M_\odot^{-1}$. The corresponding birthrate is 3×10^{-11} $M_\odot^{-1}$ yr^{-1}. For a galactic mass of 1.4×10^{11} $M_\odot$ and a local mass density of 0.13 $M_\odot$ pc^{-3}, these figures imply a total of $\sim$ 21000 PN throughout the Galaxy, and a local birthrate of 2.9×10^{-3} kpc^{-3} yr^{-1}, comparable to white dwarf birth rates in the solar neighborhood (Weidemann 1968). A total of 21000 is in excellent agreement with most recent estimates (Alloin et al. 1976, Cahn and Wyatt 1976; Acker 1978).

The surface density shown in Figure 2 can be used to test mass models of the bulge in a straightforward way provided that PN motions are characterized solely by a single velocity dispersion σ. For axisymmetric bulge mass distributions the space density of planetaries $\rho(r,z)$ is simply:

$$(2) \qquad \rho(r,z) = \rho_c \exp \{[\phi(r,z) - \phi(0,0)]/\sigma^2\}$$

where $\rho_c = \rho(0,0)$ and ϕ is the gravitational potential associated with the mass distribution. Radial velocity data for galactic center optical PN give σ = 134 km/sec (Oort 1977).

Using this assumption I have investigated two classes of mass models (Isaacman 1981b):

(a) An ellipsoidal bulge with a power-law density profile, like the one applied by Sanders and Lowinger (1972) to infrared data. Such a distribution is characterized by its total mass and the exponent of the power law.

(b) A "thickened disk" model devised by Miyamoto and Nagai (1975). This is a three-dimensional generalization of a class of thin, disk-like models derived by Toomre (1963) and is characterized by a total mass, two length parameters (whose ratio determines the degree of flattening of the system), and an "order" parameter that determines the steepness of the density distribution.

The best-fit models from each class are shown as the curves in Figure 2. These are:

Solid line: An ellipsoidal bulge of mass 0.9 X 10^{10} $M_{\odot}$ surrounded by an exponential disk, and having a power-law density profile with exponent 1.8. The axial ratio of the bulge is 0.4 (Okuda et al. 1977).

Dotted line: Same as solid line, but without a disk. Note that only points at $|b| > 4°$ are affected.

Dashed line: Zero-order thickened disk with scale length 200 pc and a mass of 1.3 X 10^{10} $M_{\odot}$.

Dashed-dot line: Second-order thickened disk with scale length 300 pc and mass 1.0 X 10^{10} $M_{\odot}$.

It is extremely gratifying that both classes of models are consistent with bulge masses of $\sim$ 1 X 10^{10} $M_{\odot}$. (I refer here to the mass within a radius of 1 kpc.) Moreover, for the power-law models the technique converges on the same exponent derived by Sanders and Lowinger (1972).

FUTURE WORK: INFRARED OBSERVATIONS

The thrust of future observations should be towards identifying individual planetaries. The infrared seems to be the most suitable regime, and a program is now underway at the United Kingdom Infrared Telescope (UKIRT) in Hawaii to observe several of the best Westerbork candidates in the near-IR and at 10μm.

Several infrared spectral features show some promise for separating planetaries from HII regions. Cohen and Barlow (1980) have noted that PN show a much more pronounced 9.7μm silicate feature than lower-excitation objects. The 10.5μm [SIV] and 12.8μm [NeII] lines are also excitation indicators, the ratio [SIV]/[NeII] tending to be higher in PN than in compact HII regions. However, the proximity of the [SIV] line to the silicate feature makes this technique difficult to apply to the faint objects of the Westerbork survey.

The unidentified molecular lines at 3.3μm, 3.4μm, 6.2μm, 7.7μm, 8.6μm, and 11.3μm are also particularly strong in planetaries, though some are found as well in a variety of bright infrared sources. So far, only one of the Westerbork sources -- 19W32 -- has been observed spectrally in the infrared, and was easily detected in the Brackett γ and 3.3μm lines (see Figure 3) at UKIRT. 19W32 is the only optically-identified object in the Westerbork survey (Isaacman et al 1980). It shows strong [OIII] emission at λ5007 Å and so is definitely a planetary: the first planetary nebula ever discovered by radio observations! No doubt the UKIRT program will yield more such objects.

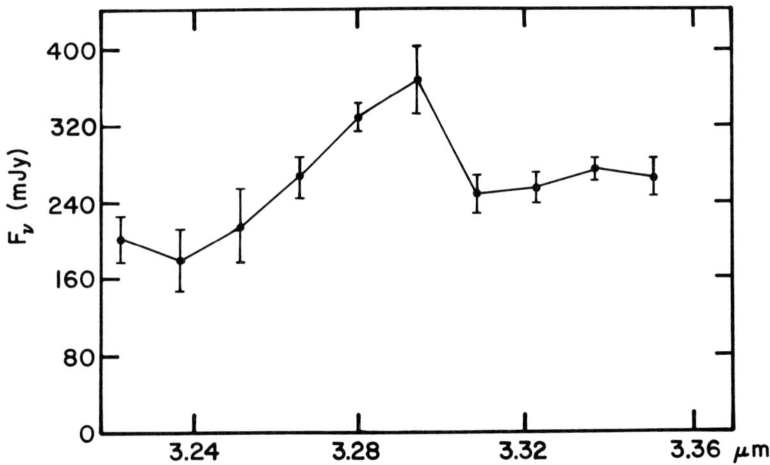

Figure 3. 3.3μm emission line in Westerbork planetary nebula 19W32.

REFERENCES

Acker, A.: 1978, Astron. Astrophys. Suppl. 33, 367.
Alloin, D., Cruz-Gonzales, C., Peimbert, M.: 1976, Astrophys.J. 205,74.
Bania, T.: 1977, Astrophys. J. 216, 381.
Cahn, J., Wyatt, S.: 1976, Astrophys. J. 210, 508.
Cohen, M., Barlow, M. J.: 1980, Astrophys. J. 238, 585.
Gathier, R., Pottasch, S., Goss, W., van Gorkum, J.: 1982 Astron. & Astrophys. (Submitted).
Habing, H., Israel, F.: 1979, Ann. Rev. Astron. Ap. 17, 345.
Isaacman, R.: 1979, Astron. & Astrophys. 77, 327.
Isaacman, R.: 1980, Astron. & Astrophys. 81, 359.
Isaacman, R.: 1981a, Astron. Astrophys. Suppl. 43, 405.
Isaacman, R.: 1981b, Astron. & Astrophys. 95, 46.
Isaacman, R., Wouterloot, J., Habing, H.: 1980, Astron. & Astrophys. 81, 359.
Israel, F. P.: 1976, Ph.D. thesis Leiden University.
Minkowski, R.: 1948, Publ. Astr. Soc. Pacific 60, 386.
Miyamoto, M., Nagai, R.: 1975, Publ. Astron. Soc. Japan 27, 533.
Okuda, H., Maihara, T., Oda, N., Sugiyama, T.: 1977, Nature 265, 515.
Oort, J.: 1976, Publ. Astr. Soc. Pacific 88, 596.
Oort, J.: 1977, Astrophys. J. Lett. 218, L97.
Pottasch, S.: 1980, Astron. & Astrophys., 89, 336.

Pottasch, S., Wesselius, P., Wu, C.-C., Fieten, H., van Duinen, R.: 1978, Astron. & Astrophys. 62, 95.
Sanders. R., Lowinger, T.: 1972, Astron. J. 77, 292.
Seaton, M.: 1966, Mon. Not. Roy. Astr. Soc. 132, 113.
Tremaine, S., Ostriker, J., Spitzer, L.: 1975, Astrophys. J. 196, 407.
Wall, J., Cooke, D.: 1975, Mon. Not. Roy. Astr. Soc. 171, 9.
Weidemann, V.: 1968, Ann. Rev. Astron. Astrophys. 6, 351.
Willis, A., Miley, G.: 1979, Astron. & Astrophys. 76, 65.
Willis, A., Oosterbaan, C., Le Poole, R., de Ruiter, H., Strom, R., Valentijn, E., Katgert, P., Katgert-Merkelijn, J: 1977, Proc. I.A.U. Symp. 74.
Wouterloot, J., Dekker, E.: 1979, Astron. & Astrophys. Suppl. 36, 323.

DINERSTEIN: How do you separate PN from H II regions? While it is true that the best observed H II regions are much brighter than PN, consideration of the luminosity function of OB stars suggests that there could be many lower luminosity H II regions which may be detected in this way.

ISAACMAN: This is the most difficult problem facing us. In practice, we eliminate all objects with S > 60 mJy.

DINERSTEIN: Regarding the structure of PN, many young H II regions also show shell structure (e.g. W3A, W3(OH)), so this might not be a good way of discriminating between PN and H II regions. Furthermore, the 3.3 μm feature, which you mentioned as being observed in one PN candidate, is also observed in the spectra of many H II regions and other types of objects.

ISAACMAN: Yes, the 3.3 μm feature is seen in a variety of objects. It would be better to look at some of the infrared lines, such as [S IV] 10.5 μm, in order to decide which objects are PN. The S IV/Ne II line ratio might provide a means of distinguishing PN.

ZUCKERMAN: Many years ago, Osterbrock pointed out that PN observed toward the Galactic center appeared to have smaller radii than those observed locally. He suggested a number of possible explanations. More recently, we suggested an additional possibility - that the masses of PN near the Galactic center are systematically smaller than those near the Sun. Since the stars in the Galactic bulge are, on the average, different from the stars near the Sun, it is not unreasonable to suppose that the resulting PN differ also. Therefore, your assumption that the Galactic center PN belong to basically the same population as nearby PN seems questionable. Furthermore, by making such an assumption, you relinquish the possibility of discovering any systematic differences which might exist.

ISAACMAN: Optical observations of Galactic centre PN are very likely to suffer from severe selection effects. My assumption regarding the properties of these objects has little to do with their dynamical or evolutionary population; I suppose only that their ionized masses be similar to those of local PN. Gathier's VLA observations give strong support to this assumption.

VLA OBSERVATIONS OF PLANETARY NEBULAE AT THE GALACTIC CENTRE

R. Gathier, S.R. Pottasch, W.M. Goss
Kapteyn Astronomical Institute, Groningen, The Netherlands

J.H. van Gorkom
Very Large Array, Socorro, New Mexico, USA

The Very Large Array (VLA) in Socorro, New Mexico, has been used to measure the 6 cm continuum flux densities and the angular sizes of 42 planetary nebulae (PN) in the direction of the galactic centre (GC). These were all optically confirmed PN for which the radial velocities (and positions on the sky) make it very likely that they are close to the GC. With a detection limit of about 1 mJy, 34 PN were detected. Their flux densities range from 2 to 100 mJy. Initially we used a configuration of the VLA with an instrumental resolution of 1". About 80% of the detected PN could be clearly resolved with this resolution. The unresolved PN were observed again with a configuration of the VLA that has a resolution of 0".4. For all but one of the 34 detected PN we could determine reliable angular sizes. The inferred total ionized masses range from < 0.01 to $\sim 0.5\ M_{\odot}$, assuming a distance to the GC of 9 kpc. The results argue strongly against the use of the Shklovsky method for distance determinations. Previous measurements of PN at the GC showed that their luminosities were substantially higher than those for nearby PN (Pottasch, 1980). The luminosity distribution of the PN in our sample is broader towards lower luminosities (up to the detection limit of the observations), but the luminosities are still high compared with nearby PN. We interpret this as a selection effect: by studying only optically confirmed PN, the intrinsically brightest PN are selected.

S.R. Pottasch, 1980, Astron. Astrophys. 89, 336.

TERZIAN: If I understand correctly, you have assumed that all the PN that you observed have a distance of 9 kpc. Since all these PN are observed optically, did you check the "extinction" distances of each one? It is perhaps surprising that, given the high extinction towards the Galactic centre, you can see all of them out to 9 kpc!

GATHIER: We did, indeed, assume a distance of 9 kpc for the whole sample. We have not yet tried to determine individual extinctions; this will be difficult, as the Hβ fluxes of most of these objects are not accurately determined.

SCHNEIDER: The radial velocity dispersion of PN both in our Galaxy and in M 31 appears to be greater than 70 km s^{-1} out to at least 3 kpc from their centres, so I would question your distance scale.
GATHIER: From the distribution of PN as a function of Galactic longitude, we estimate that 80% of the objects in our sample are close to the Galactic centre. Allowing for the fact that most of our objects have high radial velocities, in excess of 150 km s^{-1}, we estimate that at least 90% are close to the Galactic centre.

BIRTHRATE OF PN

D.C.V. Mallik
Indian Institute of Astrophysics, Bangalore, India

Recent observations of planetary nebulae have called into question the Shklovsky method of measuring distances. For those planetaries for which independent distance and electron density determinations are available, it is found that the ionized mass and the radius are linearly correlated (Maciel and Pottasch, 1980) and also that the ionized masses increase with decreasing electron density (Pottasch, 1981). These relations imply that the nebulae are optically thick in Ly continuum radiation and the distances based on the Shklovsky method are over-estimates. Using an empirically determined mass-radius relationship Maciel and Pottasch have obtained new distances for the nebulae in the catalogue of Milne and Aller (1975). We have used the more complete catalogue of Cahn and Kaler (1971) to obtain distances corrected for possible variations in the ionized mass and have compiled a new list of local planetaries. We obtain a surface density of 15 ± 3 kpc^{-2} and a planar number density of 44 ± 4 kpc^{-3}.

Using the galactic centre PN density derived by Isaacman (1981) and the local density determined here, a radial scale factor of 2.16 kpc is obtained which leads to a total number of 28 000 in the Galaxy. The lifetime of planetary nebulae has been calculated keeping in mind that initially they are optically thick. Assuming that the ionizing photon luminosity remains constant it is found that the lifetime is more or less independent of the ionizing luminosity and the shell mass. The derived lifetime yields a birthrate of $(2.4 \pm 0.2) \times 10^{-3}$ kpc^{-3} y^{-1} for planetaries in the solar neighbourhood. Theoretical estimates of the birthrate based on the Initial Mass Functions due to Lequeux (1979) and Miller and Scalo (1979) are consistent with the observational birthrate.

Isaacman, R.B.: 1981, Astron. Astrophys. Suppl. 43, 405.
Lequeux, J.: 1979, Astron. Astrophys. 80, 35.
Maciel, W.J. and Pottasch, S.R.: 1980, Astron. Astrophys. 88, 1.
Miller, G.E. and Scalo, J.M.: 1979, Ap. J. Suppl. 41, 513.
Milne, D.K. and Aller, L.H.: 1975, Astron. Astrophys. 38, 183.
Cahn, J.H. and Kaler, J.B.: 1971, Ap. J. Suppl. 22, 319.
Pottasch, S.R.: 1981, in Physical Processes in Red Giants, ed. I. Iben Jr. and A. Renzini, D. Reidel, p. 147.

SERRANO: What is the mass range you have taken when calculating the birthrate of PN from the IMF? If your rate is lower than that predicted by integrating the IMF, may we conclude that not all stars in that mass range (particularly low mass stars) give rise to PN.

MALLIK: The birthrate of PN is equated to the observed deathrate of main sequence stars between 0.95 $M_\odot$ (corresponding to an age of 9×10^9 or 10^{10} y) and M_u, with $1 \leqslant M_u \leqslant 5\ M_\odot$. Contributions of more massive progenitors do not change the rate much in view of the steepness of the IMF towards lower masses; but the possibility that a fraction of the lower mass stars does not pass through the PN phase cannot be ruled out.

WEIDEMANN: About 30% of the White Dwarfs with masses below 0.55 $M_\odot$ have not evolved through the PN stage - thus the lower mass limit will be above 1 $M_\odot$. As to the high mass limit, new results of Reimers and Koester (1982, Astron. Astrophys., in press) show the presence of White Dwarfs with progenitor masses up to 7 $M_\odot$ or 8 $M_\odot$, all the way to the beginning of non-degenerate carbon burning.

TERZIAN: Dr. Daub, who is not present, recently derived that there are 14 000 PN per 10^{11} $M_\odot$ of Galactic matter. Hence, for a Galactic mass of 2×10^{11} $M_\odot$, we expect 28 000 PN, in agreement with your value.

WEINBERGER: Why did you exclude PN with linear radii greater than 0.4 pc?

MALLIK: The mass-radius relation of Maciel and Pottasch was established within the range of radii 0.01 - 0.40 pc. Beyond 0.40 pc, there is large scatter and distances become uncertain.

PEIMBERT: How did you determine the mass density distribution of the Galaxy?

MALLIK: The surface density distribution of PN in the Galaxy was determined essentially from two points: the value give by Isaacman for Galactic centre PN and the local surface density derived here. Based on a distance to the Galactic centre of 9 kpc, this leads to a radial scale factor of 2.2 kpc.

PLANETARY NEBULAE IN THE MAGELLANIC CLOUDS

George H. Jacoby
Kitt Peak National Observatory
Tucson, Arizona

ABSTRACT

The identification and masses of Magellanic Cloud planetary nebulae are discussed. The masses are shown to be uncertain and should not be directly compared to values for galactic planetaries.

The kinematics suggest that the planetary nebulae belong to a younger rather than an older population. Abundance analyses show the Magellanic Cloud planetaries to be deficient in most elements, but the abundances of helium and carbon are comparable to values found for galactic planetaries.

1. INTRODUCTION

The Magellanic Clouds are the nearest galaxies in which a large number of planetary nebulae can be studied in detail. Only in the Clouds can all the planetary nebulae be identified and a significant number of planetaries at a known distance be spatially resolved. These aspects of the Magellanic Cloud planetary nebulae have been exploited to improve our estimates of the number of planetaries in our Galaxy and also to provide values for their shell masses.

The chemical composition of planetary nebulae in the Clouds provides a contrast to abundances derived from the more recently formed HII regions, thereby indicating the importance of planetary nebulae in the chemical evolution of these galaxies. Coupled with an understanding of the kinematics, a clearer view of the history of the Clouds may emerge. Recent efforts have been directed towards determining the chemical abundances using linear optical and UV detectors. These observations provide reliable results and a better understanding of the role of planetary nebulae in the Magellanic Clouds and the Galaxy.

Table 1 lists some of the properties of the Magellanic Clouds which are relevant to the study of their planetary nebulae.

D. R. Flower (ed.), Planetary Nebulae, 427–442.

Table 1.

Properties of the Magellanic Clouds

	SMC		LMC	
distance, kpc.	62-69		46-57	
Angular size, deg.	6		10	
Mass, $M_\odot$	1.6×10^9	(1,2)	6.0×10^9	(2)
M_{HI}, $M_\odot$	0.5×10^9	(2)	0.5×10^9	(3)
M_V	-16.7	(4)	-18.7	(4)
V_R, Km-s^{-1}	161	(5)	278	(5)
Dominant Age, yrs.	3×10^9	(6)	4×10^9	(7)

References:

(1) Tully et al (1978)

(2) Hindmann (1967)

(3) McGee and Milton (1966)

(4) Allen (1973)

(5) Lin and Lynden-Bell (1982)

(6) Hawkins and Brück (1982)

(7) Stryker (1981)

2. IDENTIFICATIONS OF PLANETARY NEBULAE

2.1. Surveys

At the distance of the Magellanic Clouds, even the largest planetary nebulae are nearly stellar, having diameters less than about 3 arcsec. The morphology technique used for identifying galactic planetaries is therefore not generally applicable to the Cloud planetaries, so secondary techniques must be used. These other methods rely on the basic property that planetary nebulae are bright emission-line sources but have very wea continua. To discriminate from other emission-line objects such as Be stars, it is important to consider the [OIII] λ5007 line which is generally strong in planetaries but not in emission-line stars. Exclusive use of the [OIII] line, however, can select against finding low excitation planetary nebulae and an alternative line, perhaps [OII] λ3727, can be employed to identify this class of planetary. Some estimate of the continuum intensity is also necessary to provide a means to reject compact HII regions and certain emission-line stars; large scale direct plates are particularly useful to determine diameters to further discriminate against HII regions.

Thus the criteria used to identify planetary nebulae in the Clouds require that candidates have: (1) stellar or nearly stellar diameters, (2) strong emission lines, including [OIII], and (3) little or no continuum.

The two instrumental methods which have been used to survey the Clouds are the objective prism and the on-line/off-line techniques. The objective prism method has the advantage of allowing identifications with a single plate while simultaneously providing spectral information on the excitation and the continuum intensity. It is, however, subject to crowding confusion in dense stellar fields such as the bar region of the LMC where spectra from the multitudes of stars overlap.

The on-line/off-line technique allows identification of fainter objects in very dense fields, while providing plate material suitable for determining accurate positions. It has the disadvantage of requiring two plates per field and a relatively tedious blinking procedure to locate candidates. Both methods have been used successfully, although the objective prism technique has been the most popular.

Planetary nebula surveys in the SMC have been reported by Lindsay (1955,1956,1961), Koelbloed (1956), Henize and Westerlund (1963), Sanduleak, MacConnell, and Phillip (1978), Jacoby (1980), and Sanduleak and Pesch (1981). Koelbloed's identifications are unfortunately irrecoverable because no positions or finding charts were published. Lindsay's objects have been sifted by Henize and Westerlund, and later by Sanduleak, MacConnell, and Phillip to exclude compact HII regions and variable emission-line stars. The list given by Sanduleak, MacConnell and Phillip is probably the most reliable set of planetary nebula identifications in the Clouds and includes finding charts for 28 SMC objects.

Jacoby's survey includes very faint planetaries with the goal of measuring the luminosity function for planetary nebulae in the bars of the SMC and LMC. He identified 19 new SMC planetaries, some of which are 3 magnitudes fainter than previous limits. Also included is the first spatially resolved planetary in an external galaxy, having a diameter of approximately 2.8 arcsec.

Sanduleak and Pesch obtained very deep objective prism plates in which the SMC Bar is nearly saturated due to crowding of overlapping spectra. They identify 6 planetary nebula candidates; two of these (numbers 30 and 31) fall in regions included on my plates. They are just visible at the plate limit, but number 31 is brighter on the off-line plate. Number 30 is extremely faint in [OIII] and probably would not normally be identified as a planetary on objective prism plates. These two objects may be late-type stars with strong absorption bands, or possibly variable stars. Although the possibility that they are very low excitation planetary nebulae cannot be ruled out from the [OIII] plates alone, the criterion used by Sanduleak and Pesch requiring their candidates to have strong [OIII] emission appears to be at odds with the direct plates. For now, these two objects should be considered uncertain.

Surveys of the LMC have been reported by Westerlund and Rodgers (1959), Lindsay and Mullan (1963), Westerlund and Smith (1964), Sanduleak, MacConnell and Phillip (1978), Fehrenbach, Duflot and Acker (1978), Webster (1978), and Jacoby (1980). Again the objects found by Westerlund and Rodgers and Lindsay and Mullan were later re-examined by Westerlund and Smith and Sanduleak, MacConnell and Phillip to reject compact HII regions. Fehrenbach, Duflot and Acker also compared previous identifications to their plate material and noted possible spectral inconsistencies plus one new identification.

Sanduleak, MacConnell and Phillip provide coordinates good to 1 arcmin for 102 LMC planetaries with reliable identifications. Of these, 41 finding charts are published by Westerlund and Smith, so locating the remaining 61 objects may be difficult due to the imprecise positions currently available. Gull (1982) reports improved positions for the bright SMC and LMC planetaries, but these are not yet generally available.

Jacoby's survey identified 34 faint planetary nebulae in the LMC bar which are used to determine the LMC planetary nebula luminosity function. Included in the list are 8 extended planetaries having seeing-corrected diameters between 0.56 and 3.75 arcsec. This survey is the only LMC survey using the on-line/off-line technique, which is especially appropriate in the crowded regions in the LMC bar. Webster (1978) reported examining deep objective prism plates of the LMC bar to test for crowding confusion, but no details are available, other than the conclusion that few bright planetaries would be masked by the stellar spectra. Planetaries much fainter than the limits of the surveys up to that time would probably be considerably more difficult to identify on objective prism plates.

To date, there are a total of 51 planetary nebulae known in the SMC and 137 in the LMC as a result of these surveys.

2.2. The Number of Planetaries in the Clouds

Although current technology in astronomical instrumentation is adequate to identify and, therefore, to count nearly all the planetary nebulae in the Magellanic Clouds, the required observations have not been undertaken. Until such data is obtained, we can only estimate the total number of nebulae based on the numbers already known.

Henize and Westerlund (1963) first estimated the number of planetaries in the SMC to be 300 based on the 30 objects known at that time. The factor of 10 is derived from an estimate of the fractional lifetime represented by the luminosity limits of the observed sample. Westerlund and Smith (1964) argued similarly for 450 planetaries in the LMC based on 45 objects. But not all the objects in those samples are true planetaries, nor are those samples spatially complete. The survey by Sanduleak, MacConnell and Phillip (1978) would suggest 280 and 1020 planetaries based on 28 and 102 objects in the SMC and LMC, respectively, again using the lifetime factor of 10.

An alternative approach was taken by Jacoby (1980) who sampled selected regions in the bars of the Clouds to determine the number ratio of faint (3 to 6 magnitudes below the brightest) to bright (0 to 3 magnitudes below the brightest) planetary nebulae in those regions. He then scaled that ratio to the entire galaxy. Because his sample did not include planetaries with luminosities comparable to the faintest identifiable galactic planetary nebulae, a correction was required to account for luminosity incompleteness. This factor was obtained by examining the number ratio of very faint planetaries (6 to 8 magnitudes below the most luminous) to all brighter planetaries in the solar neighborhood. The ratio was found to be about 2. Despite the usual cautions concerning the distances to galactic planetaries, the above ratio is identical to that obtained by extrapolating a theoretical luminosity function beyond the observational limit. Additional corrections were included to account for obscuration by dust, spatial incompleteness of the brighter surveys, and crowding in the LMC bar. Jacoby finds the total estimated number of planetary nebulae in the SMC and LMC to be 285 and 996, respectively, in surprisingly good agreement with the result obtained by scaling the numbers given by Sanduleak et al by the lifetime factor of 10. This is not completely coincidental because the observed luminosity function (see Section 2.4.) agrees rather well with the theoretical one used by Henize and Westerlund (1963) to derive the lifetime factor.

One expects that for galaxies of similar age, composition, and initial stellar mass function, the planetary nebula birthrates per unit star would also be similar. We can compare the birthrates in the two Clouds by dividing the number of nebulae by the mass of the stellar component of each galaxy. The masses listed in Table 1 must be reduced by the HI mass, since the non-stellar components (e.g., neutral hydrogen) do not directly produce planetary nebulae. We then find the mass specific numbers of planetary nebulae to be 2.6×10^{-7} PN/$M_{\odot}$ in the SMC and 1.8×10^{-7} PN/$M_{\odot}$ in the LMC. These numbers suggest either an overabundance of planetaries in the SMC or an underabundance in the LMC.

Jacoby suggests using the visual luminosity specific number rather than the mass specific number due to difficulties in measuring the masses of galaxies. For the Clouds, he finds 7×10^{-7} PN/$L_{\odot}$ in the SMC and 4×10^{-7} PN/$L_{\odot}$ in the LMC, again indicating a greater population of planetaries in the SMC. This argues that the basic assumptions are incorrect -- either the age, initial mass function, or composition is different in the SMC. The metallicity is known to be deficient compared to the LMC while the age is thought to be comparable (see Section 4). The initial mass function is yet to be determined.

We can, nevertheless, apply the above specific numbers to the mass and luminosity of our Galaxy to obtain 23,000 to 34,000 planetaries from the mass specific numbers. Using the preferred indicator, the luminosity specific numbers, we obtain 6,000 to 10,000 planetaries.

2.3. Shell Masses

Due to the uncertainties in the galactic distance scale, shell masses for planetary nebulae have been subject to some debate. The hope of exploiting the Magellanic Cloud nebulae, whose distances are reasonably well known, to determine shell masses, has proven to be no easier because the nebulae are, in general, spatially unresolved. Therefore, assumptions and indirect methods have been employed, resulting in a range of derived shell masses comparable to the range seen for shell masses derived for galactic planetary nebulae.

Seaton (1968) took advantage of the fact that the nebula shell attains maximum Hβ luminosity and maximum surface brightness simultaneously. By equating the maximum Hβ flux from the Magellanic Cloud planetaries to the maximum Hβ surface brightness of galactic planetaries and solving for the radii of the brightest galactic planetaries, the distances and shell masses can be readily computed. Seaton thus finds the average shell mass to be $0.17M_{\odot}$. His basic assumptions include: (1) T_e=8,000°K; (2) there are no relevant differences between Magellanic Cloud and galactic planetaries; and (3) the nebulae are optically thin at maximum Hβ flux; that is, the entire shell is ionized.

The above assumed temperature is systematically too low by about 4,000 degrees (Aller et al 1981), due to lower metal abundances than in galactic planetaries. This results in an Hβ emissivity which is systematically too high for the Cloud objects and leads to an underestimate of the shell mass by about 40 percent. This effect is partially offset by the expected higher masses of the average LMC population which is somewhat younger than that in the Galaxy (Butcher 1977, Stryker 1981).

The greatest source of uncertainty in Seaton's approach is the assumption that the nebula shells are brightest when fully ionized. This assumption derives from an evolutionary track for a brightening central star, but this effect has become difficult to support (Paczynski 1971, Pottasch et al 1978, Renzini 1979). In the situation where the central star does not increase in luminosity, the nebula will have its greatest surface brightness when it is very small, due to the very high electron density, and it will monotonically fade throughout its lifetime. Until the radius has expanded to about 0.1 to 0.2pc, the nebula is not yet fully ionized. This effect has been dramatically demonstrated by the mass-radius "relation" given by Pottasch (1980) in which the derived nebula mass (in solar masses) follows the approximate relation $7r_{pc}^{1.8}$. This suggests that for an unresolved nebula in the LMC (diameter less than 1 arcsec, r<0.13pc), the maximum derived mass will be less than $0.5M_{\odot}$, even considering the significant scatter in the mass-radius relation.

Webster (1969a) performed a similar analysis and the same reservations apply. Note that Webster gives hydrogen masses, not total shell masses, of about $0.11M_{\odot}$ for 4 galactic planetary nebulae. This corresponds to a total shell mass of about $0.16M_{\odot}$. Although this value is higher than $0.09M_{\odot}$ calculated from the mass-radius relation for these nebulae which have radii of about 0.09pc, the dispersion about the relation at this radius extends to $0.19M_{\odot}$ (see Pottasch 1980, Figure 2).

On the other hand, one of the four planetaries, NGC 7662, is a well-known triple shell planetary having a maximum radius of 67 arcsec (Kaler 1974). The existence of ionized material at this radius, corresponding to 0.8pc suggests that ionizing radiation escapes the central nebula. This implies an optically thin nebula, at least in certain directions.

Webster also derived upper limits to 3 Magellanic Cloud planetaries which span the range of the brightest to the faintest identified at that time. The total masses, derived by assuming a radius of 2 arcsec (0.27pc), range from 2.0 to $0.7M_{\odot}$. The lower value is more likely to be correct because the brighter objects would be much smaller than 1 arcsec in radius. The smaller value is also probably too high because such planetaries would be easily resolved on plates taken during good seeing conditions, yet this has not been reported.

Pottasch (1980) derives total masses for 3 SMC planetaries based on a density dependent method. Because the distance to the SMC is known, a priori, the method is not strongly dependent on density as is Seaton's (1966) method for galactic planetaries. Pottasch finds the masses to be about $0.35M_{\odot}$.

Webster (1976) uses a method similar to Pottasch to derive similar results, but cautions that inhomogeneities in the nebulae can lead to measured forbidden line densities which are inappropriate to the mass calculation.

Jacoby (1980) was able to spatially resolve 9 faint planetary nebulae in the Clouds to obtain diameters directly for the first time. The availability of physical diameters, distances, and fluxes allows a complete solution of the inverted Shklovskii method:

$$M(M_{\odot}) = 2.65 \times 10^{-6} \quad [d^5(\text{pc})r^3(\text{arcsec})\ F(H\alpha)\varepsilon]^{1/2}$$

where ε is the filling factor, usually about 0.6. Unfortunately, the uncertainties in the individual parameters are non-negligible. In particular, the measured diameters are, in some cases, unreliable due to the convolution of seeing effects and inadequate definition of the nebula edge. Furthermore, the so-called known distances to the Clouds are accurate to perhaps 15 percent (van den Bergh 1975). An additional distance uncertainty is the position of the nebulae within the Cloud, which may affect the distance by up to 4 percent. These factors result in an uncertainty in the derived masses of 50 percent due to the impreciseness of the distance alone! For the 9 objects, the masses are calculated to range from 0.02 to $0.8M_{\odot}$ thereby spanning the usual quoted values for planetary nebula shells. The conclusion to be drawn is that the Magellanic Cloud planetary nebulae can provide a set of objects at a nearly uniform distance, but the distance is subject to sufficient uncertainty that the derived masses may not be directly applicable to our Galaxy.

Other factors which must be considered when generalizing the properties of the Cloud planetaries to our Galaxy are: (1) there is a compositional difference, and hence a systematic difference in electron temperatures and possibly in central star properties; (2) the progenitor stars in the Clouds are drawn from a somewhat younger stellar population due to a recent (3-5 billion years ago) burst of star information; (3) the likelihood that the brightest and best observed nebulae are not yet fully ionized.

2.4. The Luminosity Function

The planetary nebula luminosity function serves several purposes: (1) it provides an observational constraint on models of the combined evolution of nebular shells plus central stars; (2) it is a check on the distances to the Magellanic Clouds and other Local Group galaxies by direct comparison; (3) it provides a means of extrapolating the observed number of planetary nebulae from a luminosity-limited sample to the total number of planetaries in a galaxy. Also, Svestka (1962) used the luminosity function to infer that the brightest of Lindsay's (1961) and Koelbloed's (1956) identifications in the SMC were not likely to be true planetary nebulae, as the luminosity function exhibited an excess of overly bright objects when compared to the planetary nebulae in our Galaxy.

There have been few attempts to measure the luminosity function for galactic planetary nebulae because of the uncertainties in the distance determination. Those for the SMC published by Lindsay (1961) and Koelbloed (1956) were contaminated by HII regions and difficult to interpret. Henize and Westerlund (1963) suggested a simple model of an expanding sphere and deduced the luminosity function based on the relative lifetimes of such a nebula as it expanded to fainter magnitudes.

By extending the identifications of Magellanic Cloud planetary nebulae over a 6 magnitude range, Jacoby (1980) was able to directly measure the luminosity function. The overall shape of the curve is adequately represented by the simple expanding sphere model, although the model underestimates the number of bright planetaries. Pottasch (1982) has compared the luminosity function obtained by Jacoby with that derived from a set of galactic planetary nebulae for which he feels the distances are reasonably accurate. The agreement is generally good, but there are proportionately more faint than bright planetaries in the Galactic sample. Tentatively, this disparity can be attributed to the small number of bright planetaries in both authors' samples.

Using the luminosity function from the Clouds, Jacoby extrapolates the available surveys in 8 additional Local Group galaxies to estimate their total numbers of planetary nebulae. He finds the average mass specific number for the Local Group to be $2.1 \pm 1.5 \times 10^{-7} \mathrm{PN}/M_{\odot}$ while the visual luminosity specific number is $6.1 \pm 2.2 \times 10^{-7} \mathrm{PN}/L_{\odot}$. The relatively smaller dispersion about the luminosity specific number suggests it is a better indicator of the planetary nebulae population in a galaxy. Applying this value to our galaxy, along with the absolute visual luminosity of -20.6 (de Vaucouleurs and Pence 1978), a total number of 9100 ± 3300 planetary nebulae is estimated.

3. KINEMATICS OF PLANETARY NEBULAE IN THE CLOUDS

Until very recently extensive velocity data for the older stellar components in the Clouds were available only from planetary nebulae. Freeman, Illingworth, and Oemler (1982) have combined their results for the LMC globular cluster system with those of Ford (1970) and Searle and Smith (1982) to obtain velocities for 59 clusters. The younger blue clusters are found to behave similarly to the HI and HII components while the older red clusters lie in a disklike system but have different rotational parameters. Furthermore, there was no kinematical evidence for an old halo population.

Planetary nebulae are usually associated with an old disk component. We would like to test this association in the LMC by comparing the velocity information for the planetary nebulae with the kinematic solutions for the young and old populations. Interpretation of such a comparison is very difficult because we expect planetaries to form from progenitors of varying ages, especially in the LMC where a major burst of star information occurred 3-5 billion years ago (Butcher 1977, Stryker 1981).

Feast (1968) found that the brightest planetaries exhibited a dispersion of 15 km-s^{-1} about the HII region kinematic solution, and Smith and Weedman (1972) found a similar result for a spatial subset of the planetaries. In contrast, Webster (1969b) found the dispersion of the planetaries to be only 8.2 km-s^{-1} about the rotation curve defined by the planetaries themselves.

Freeman, Illingworth, and Oemler compare their results for the blue and red clusters with the velocities for the 35 LMC planetaries taken from the table of Feitzinger and Weiss (1979) which is derived from the measurements by Feast, Webster, and Smith and Weedman. They find that the planetary nebulae generally follow the kinematic solution for the young population derived from HI and HII data, but with significant scatter. This is attributed to a wide range of ages for planetary nebula progenitors and to deviations in the velocity field which are evident in the HI data. A third consideration is inaccurate velocities for planetaries with asymmetric shells as reported by Smith and Weedman.

It would be best to be able to preselect the planetaries which exhibit properties of younger objects, perhaps excessive nitrogen and helium abundances indicative of dredge-up in the late evolution of higher mass progenitors (Kaler, Iben, and Becker 1978). A subset of young LMC planetaries would be expected to follow the HI solution very closely and have a velocity dispersion comparable to the HII region dispersion of 7.5 km-s^{-1} (Smith and Weedman 1971).

Unfortunately, abundances are available for too few objects to allow compositional sorting. Three planetaries have characteristics indicating high helium abundances -- WS 7 (Osmer 1976), WS 9 and WS 38 (Webster 1976). [Note: The prefix WS refers to the catalogue of Westerlund and Smith (1964)]. Of these, no velocity is available for WS 9, and WS 7 has a velocity affected by an asymmetric shell. Using Smith and Weedman's velocity for WS 7, it falls only 2 km-s^{-1} above the HI solution; but, WS 38 falls about 23 km-s^{-1} above the solution. Clearly more abundance data are needed to assess the population separation.

Interpretation of the situation will still be complicated even if a single population is considered. Feitzinger and Schmidt-Kaler (1982) find evidence for a double structure in the HI rotation curve which is especially evident in the southern part of the LMC. A suggestion of this effect was found by Smith and Weedman (1972) from the planetary nebula velocity dispersion which increases dramatically in the south. Thus, the measured velocity dispersion may not be indicative of the z-motions of the planetary nebulae in a simple circular motion velocity field, but rather of the deviations of that field from circularity.

One final point regarding LMC objects should be mentioned. Feitzinger, Isserstedt, and Schmidt-Kaler (1977) and Lin and Lynden-Bell (1982) report evidence for a transverse velocity of approximately 250 km-s^{-1} in the plane of the sky for the LMC. Due to its large angular size, an apparent radial velocity gradient on the order of 30 km-s^{-1} is introduced across the face of the LMC. None of the analyses discussed here include this effect.

Velocities for 12 SMC planetaries are currently available. Data on more SMC planetaries are of particular interest due to the anomalous grouping found by Feast and also by Webster. The SMC planetaries segregate into two velocity groups differing by about 40 km-s^{-1}, but with no apparent spatial preference. Recent HI observations by McGee and Newton (1982) reveal 4 radial velocity groups at 115, 134, 167, and 192 km-s^{-1}. The groups at 134 and 167 km-s^{-1} appear to be the most massive and those authors find that stellar radial velocities tend to clump near these values as well. The 12 planetaries prefer velocities of 112 and 150 km-s^{-1}, but no definitive statement can be made with so few objects.

On the other hand, Walker (1982) reports 85 additional velocities for Magellanic Cloud planetary nebulae. Although no details are yet available, the bimodal velocity distribution in the SMC is not seen when more data are considered. This is clearly an area where much more observational work is needed. The identification and velocity measurement of a significant number of planetary nebulae in the SMC should help clarify the kinematics of this galaxy.

4. CHEMICAL ABUNDANCES OF MAGELLANIC CLOUD PLANETARY NEBULAE

The first suggestion of low abundances in the Cloud planetary nebulae was based on objective prism spectra of the SMC by Sanduleak, MacConnell, and Hoover (1972). They noted that with the exception of only one planetary in the SMC (N67), none had detectable [NII] lines, implying a general nitrogen deficiency. Webster (1976) has shown that excitation differences between planetary nebulae in the SMC (low excitation) and other galaxies (moderate to high excitation) can invalidate such a comparison, but did find evidence for the low SMC nitrogen abundance when low excitation nebulae are compared.

Osmer (1976) performed the first detailed abundance analysis for Magellanic Cloud planetaries using a photoelectric scanner to observe 3 planetary nebulae in the SMC (N2, N54, N67) and LMC (WS7, WS8, WS40). [Note: The prefix N refers to Henize (1956).] He found general underabundances of oxygen and nitrogen by about a factor of 3 in the Cloud planetaries when compared to Galactic planetaries. He also found evidence for a 50 percent overabundance of helium, but this has since been traced to the uncertainties in his photometry as well as to the inclusion of the anomalous planetary N67. His analysis of the ISM nitrogen enrichment rate indicated that planetary nebulae in the Clouds are an inadequate source by a factor of 10.

Dufour and Killen (1977) reobserved N67 and WS7, and additionally observed WS33 plus 3 SMC HII regions. The results of their image-tube spectra indicate a helium abundance comparable to Galactic planetaries, but N67 does appear enriched. They found very high ratios of N/O in the Cloud planetaries relative to their respective HII regions, while N/H is comparable to Galactic planetaries and O/H is comparable to Cloud HII regions. They also conclude that nitrogen enrichment of the Cloud ISM is provided by sources other than planetary nebulae.

Webster (1976) observed 7 planetary nebulae in the SMC (N2, N5, N38, N43, N44, N70, N87) and the LMC (WS2, WS8, WS9, WS24, WS33, WS38) with a spectracon electronographic camera, an image-tube spectrograph, and a photoelectric photometer. Helium is again found to be comparable to Galactic planetaries while oxygen is a factor of 3 below Galactic bulge planetaries. The O/H ratio is about the same in both Clouds but a factor of 2 higher than in their respective HII regions.

Webster (1978) combined the previously available data with additional spectra from the Robinson-Wampler scanner for 19 planetary nebulae in the Clouds. Among these are 5 "HN" objects which exhibit high excitation, strong [NII] lines, and high helium abundance ($\frac{He}{H} \simeq 0.14$). Although planetary nebulae in this category are found in our Galaxy, they occur somewhat less frequently. Peimbert and Serrano (1980) estimate their frequency at 10 percent as compared with 25 percent in the Clouds, but this may be due to selection effects if they are derived from young progenitors which are more likely to be obscured by interstellar extinction in the plane of the Galaxy. Webster concludes that any derived overabundance of helium in Cloud planetary nebulae is probably due to observing HN objects rather than to an actual effect of the planetaries being in the Clouds.

Aller et al (1981) used the IPCS photon-counting spectrograph to observe 7 SMC planetaries (N2, N5, N43, N44, N54, N70, N87). They find the helium abundance to be comparable to Galactic planetaries but slightly higher than the SMC HII regions. Their results are adequately summarized by the values in Table 2, except for the O/H ratio which is nearly a factor of 2 lower than in the table. This can be directly attributed to the additional 5 planetary nebulae in the sample, as the O/H ratios found by Aller et al agree very well with those of Maran et al for the two objects in common. Aller et al conclude that the SMC planetaries contribute nearly enough nitrogen to enrich the ISM, but that massive stars may also be involved.

Maran et al (1982) used IUE to observe N2 and N5 in the SMC and WS40 in the LMC. The advantage offered by the UV data allowed those authors to measure the carbon abundance in an extragalactic planetary nebulae for the first time. Although their sample is small, they find results which generally agree very well with previous investigators. They conclude that planetary nebulae in the Magellanic Clouds are the dominant source of carbon enrichment of the ISM. Their results are summarized in Table 2. The abundances for the Cloud HII regions are from Dufour, Shields and Talbot (1982), while values for Galactic planetary nebulae are averages.

The general conclusions from Table 2 and the previous discussions are: (1) the abundances in the SMC and LMC planetary nebulae are similar; (2) the helium abundance is comparable for planetaries in the SMC, LMC, and the Galaxy, and perhaps 25 percent enhanced above galactically local HII regions; (3) a class of high helium abundance planetaries exists in all three galaxies; (4) oxygen is underabundant by a factor of 2 relative to galactic planetaries but comparable to local HII regions; (5) nitrogen is underabundant by a factor of 5 relative to Galactic planetaries but overabundant by a factor of 3 to 8 relative to local HII regions; (6) the neon abundance is a factor of 2 below Galactic planetaries but is comparable to local HII regions; (7) the carbon abundance is comparable to Galactic planetaries but is very much enhanced relative to the Cloud HII regions.

Thus the overall picture is one of metal deficiency in the Cloud ISM, with the SMC having slightly lower abundances than the LMC. Although the differences between the SMC and LMC can be attributed to the higher gas content of the SMC (Pagel et al 1978), the Clouds appear to have been chemically enriched only in the last few billion years (Butler, Demarque, and Smith 1982). This pattern is consistent with a burst of star formation having recently occurred in these galaxies (Butcher 1977, Stryker 1981, Hawkins and Brück 1982).

The question of responsibility for ISM enrichment has been considered by Williams (1982). He finds that classical novae are likely to be the dominant source of nitrogen in the Clouds as they return approximately 8 times as much nitrogen to the ISM as do the planetaries. Although non-negligible, the nova contribution to carbon enrichment is lower than that of the planetaries.

Table 2.

Chemical Composition of Magellanic Cloud Planetary Nebulae*

	SMC PN relative to:		LMC PN relative to:		Galactic PN
	HII regions	Galactic PN	HII regions	Galactic PN	$12 + \text{Log}\,\frac{X}{H}$
He	1.3	1.0	1.3	1.0	11.03
O	1.9	0.5	0.9	0.5	8.51
N	8.1	0.2	3.6	0.2	8.18
Ne	2.1	0.4	0.9	0.4	8.02
C	38	0.8	6.3	0.7	8.85

*Original data taken from Maran et al (1982).

The anomalous planetary N67 in the SMC appears to have a high helium and nitrogen abundance. It has an extraordinarily high [OIII] electron temperature of 25,000K (Osmer 1976, Dufour and Killen 1977) which leads to a low calculated oxygen abundance. Webster (1978) suggests that the derived temperature is overestimated due to collisional de-excitation of the λ5007 line. Aller et al (1981) further suggest that N67 is atypical. As such it may prove to be a very interesting object in itself but no generalizations should be based on its properties.

5. SUMMARY

We currently know of 51 planetary nebulae in the SMC and 137 in the LMC; estimates for the total numbers are 300 and 1000. The derived shell masses exhibit a range from below $0.1M_{\odot}$ to $0.8M_{\odot}$. One should be aware that the nebula shells, particularly for the brighter, denser planetaries, may not be fully ionized so that derived masses will underestimate the total shell mass. Furthermore, the derived masses are uncertain due to the uncertainty in the distances to the Magellanic Clouds.

Kinematically, the LMC planetary nebulae tend to associate with a younger rather than an older stellar population. However, the complicated details of the LMC velocity systems preclude a simple interpretation of the kinematics of the planetaries. Currently available data on the SMC planetaries are still too sparse to provide an understanding of the kinematic situation there.

Improvements in instrumentation, in particular the availability of UV data from IUE, has significantly advanced our knowledge of chemical abundances in the Clouds. Relative to galactic planetaries, the Cloud nebulae are a factor of 2 to 4 underabundant in O, N, and Ne, but have similar helium and carbon abundances. Carbon produced by planetary nebulae is probably the dominant source of ISM carbon enrichment in the Clouds. Nitrogen appears to originate primarily from other sources.

The overall view, based on kinematics and compositions of the various stellar components of the Clouds, is one in which most of the planetary nebulae are derived from progenitors having ages of 3-5 billion years.

REFERENCES

Allen,C.W.: 1973, in "Astrophysical Quantities", (3d ed.; London:Athlone), p.287.
Aller,L.H., Keyes,C.D., Ross, J.E., and O'Mara,B.J.: 1981, M.N.R.A.S. 194,613.
Butcher,H.: 1977, Ap. J. 216,372.
Butler,D., Demarque,P., and Smith,H.A.: 1982, Ap. J. 257,592.
de Vaucouleurs,G. and Pence,W.D.: 1978, Astron. J. 83,1163.
Dufour,R.J. and Killen,R.M.: 1977, Ap. J. 211,68.
Dufour,R.J., Shields, G.A., and Talbot,R.J. Jr.: 1982, Ap J. 252,461.
Feast,M.W.: 1968, M.N.R.A.S. 140,345.
Fehrenbach,Ch., Duflot,M., and Acker,A.: 1978, Astron. Astrophys. Suppl. 33,115.
Feitzinger,J.V. and Schmidt-Kaler,Th.: 1982, Ap. J. 257,587.
Feitzinger,J.V. and Weiss,G.: 1979, Astron. Astrophys. Suppl. 37.575.
Feitzinger,J.V., Isserstedt,J., and Schmidt-Kaler,Th.: 1977, Astron. Astrophys. 57,265.
Ford,H.C.: 1970, Ph.D. thesis, Univ. of Wisc.
Freeman,K.C., Illingworth,G., and Oemler,A. Jr.: 1982, preprint.
Gull,T.R.: 1982, priv. comm.
Hawkins,M.R.S. and Brück,M.T.: 1982, M.N.R.A.S. 198,935.
Henize,K.G.: 1956, Ap. J. Suppl. Ser. 2,315.
Henize,K.G. and Westerlund,B.E.: 1963, Ap. J. 137,747.
Hindmann,J.V.: 1967, Australian J. Phys. 20,147.
Jacoby,G.H.: 1980, Ap. J. Suppl. 42,1.
Kaler,J.B.: 1974, Astron. J. 79,574.
Kaler,J.B., Iben,I., and Becker,S.A.: 1978, Ap. J. Letters 224,L63.
Koelbloed,D.: 1956, Observatory 76,894.
Lin,D.N.C. and Lynden-Bell,D.: 1982, M.N.R.A.S. 198,707.
Lindsay,E.M.: 1955, M.N.R.A.S. 115,248.
Lindsay,E.M.: 1956, M.N.R.A.S. 116,649.
Lindsay,E.M.: 1961, Astron. J. 66,169.
Lindsay,E.M. and Mullan,D.J.: 1963, Irish Astron. J. 6,51.
Maran,S.P., Aller,L.H., Gull,T.R., and Stecher,T.P.: 1982, Ap. J. Letters 253,L43.
McGee,R.X. and Milton,J.A.: 1966, Australian J. Phys. 19,343.
McGee,R.X. and Newton,L.M.: 1982, Proc. Astron. Soc. Aust. 4,189.
Osmer,P.S.: 1976, Ap. J. 203,352.
Paczynski,B.: 1971, Acta Astronomica 21,417.
Pagel,B.E., Edmunds,M.G., Forsbury,R., and Webster,B.L.: 1978, M.N.R.A.S. 184,569.

Peimbert,M. and Serrano,A.: 1980, Rev. Mex. Astron. y Astrofis. 5,9.
Pottasch,S.R.: 1980, Astron. Astrophys. 89,336.
Pottasch,S.R.: 1982, in "Planetary Nebulae" (preprint).
Pottasch,S.R., Wesselius,P.R., Wu,C.C., Fieten,H., and van Duinen,R.J.: 1978, Astron. Astrophys. 62,95.
Renzini,A.: 1979, in "Stars and Star Systems", Proc. Fourth European Regional Meeting, ed. Westerlund,B.E. (Dordrecht:Reidel), p.155.
Sanduleak,N., MacConnell,D.J., and Hoover,P.S.: 1972, Nature 237,28.
Sanduleak,N., MacConnell,D.J., and Phillip,A.G.D.: 1978, Pub. Astron. Soc. Pac. 90,621.
Sanduleak,N. and Pesch,P.: 1981, Pub. Astron. Soc. Pac. 93,431.
Searle,L. and Smith, H.: 1982, Ap. J., in press.
Seaton,M.J.: 1966, M.N.R.A.S. 132,113.
Seaton,M.J.: 1968, Astrophys. Letters 2,55.
Smith,M.G. and Weedman,D.W.: 1971, Ap. J. 169,271.
Smith,M.G. and Weedman,D.W.: 1972, Ap. J. 177,595.
Stryker,L.L.: 1981, Ph.D. thesis, Yale Univ.
Svestka,Z.: 1962, Bull. Astro. Inst. Czech. 13,35.
Tully,B.M., Bottinelli,L., Fisher,J.R., Gouguenheim,L., Sancisi,R., and van Woerden,H.: 1978, Astron. Astrophys. 63,37.
Walker,A.R.: 1982, SAAO Report, p.15.
Webster,B.L.: 1969a, M.N.R.A.S. 143,79.
Webster,B.L.: 1969b, M.N.R.A.S. 143,97.
Webster,B.L.: 1976, M.N.R.A.S. 174,513.
Webster,B.L.: 1978, in "IAU Symposium No. 76, Planetary Nebulae Observations and Theory", ed. Y. Terzian (Dordrecht:Reidel), p.11.
Westerlund,B.E. and Rodgers,A.W.: 1959, Observatory 79,132.
Westerlund,B.E. and Smith,L.F.: 1964, M.N.R.A.S. 127,449.
Williams,R.E.: 1982, preprint.
van den Bergh,S.: 1975, in "Galaxies and the Universe", ed. A. Sandage, M. Sandage and J. Kristian (Chicago : Univ. of Chicago Press), p.509.

PEIMBERT: Is there a selection effect, towards higher masses and higher luminosities, in the sample studied?

JACOBY: Certainly, the abundance studies concentrate on the most luminous PN, which may also be the most massive. The derived range of masses of the resolved PN indicates that at least a few of those observed are not of high mass origin.

ZUCKERMAN: If the brighter PN in the Clouds have $m_V \approx 16$, as you have indicated, then it might be possible to measure the size of these objects in the [O III] lines using optical interferometric techniques.

BARLOW: With reference to the high mass (0.8 $M_\odot$) obtained for the 3 arcsec diameter PN which you discovered in the SMC, spectra which I took of it at the AAT showed [N II] λ 6584 to be twice as strong as Hα; this result will lower the mass estimate.

SERRANO: The masses that you have derived suggest that many of the PN are of type I and may be expected to have strong [N II] lines. This would mean, in turn, that the masses have been overestimated and cast doubt on the suggestion that there has been a recent burst of star formation in the Clouds. In this context, it is worth noting that Cohen has found an age-metallicity relation which does not support such a burst.

JACOBY: Only half of the masses are determined from Hα fluxes - the rest are based on [O III] fluxes and an estimate of the Hα/[O III] flux ratio. Furthermore, only two of the objects have very high masses. The others may not be type I objects and the masses are likely to be correct.

HENIZE: It was exciting to see a resolved image of a Magellanic Cloud PN. How many have you resolved?

JACOBY: Nine have been resolved, but only one in the SMC (1980, Ap. J. Suppl. 42, 1).

PEIMBERT: The very large carbon abundances of some of these objects probably imply that some O and Ne have been produced by the stars themselves. Therefore it is very important to obtain S/H and Ar/H ratios to compare with the results derived from H II regions.

ALLER: In both the Clouds, we have found that S and Ar have essentially the same abundances as in the ambient interstellar medium.

RENZINI: As both Clouds are full of AGB stars which are also carbon stars, it is not surprising that many PN in the Clouds are carbon-rich.

PLANETARY NEBULAE IN LOCAL GROUP GALAXIES

Holland C. Ford
Space Telescope Science Institute, Homewood Campus
Baltimore, MD 21218

ABSTRACT

Recent surveys for planetary nebulae have given the first identifications in Fornax, NGC 6822, M33, IC 10, Leo A, Sextans A, Pegasus, WLM, NGC 404, and M81, and extended the identifications in the SMC, the LMC, and M31. Observations of planetaries have established chemical compositions in old or intermediate age populations in 8 Local Group galaxies. The chemical compositions show that i) the helium abundance is higher in planetary nebulae than in H II regions in the same galaxy, and ii) nitrogen is overabundant relative to H II regions by factors of 4 to 100. Planetary nebulae are not a major source of helium in star-forming galaxies, and are a major source of nitrogen. The planetary in Fornax has a relatively high O abundance, and, together with Fornax's carbon stars, establishes the presence of at least 2 stellar populations. The abundance gradient derived from 3 planetaries in M31 is very shallow, and gives high abundances at ~ 20 kpc. By using planetary nebulae as standard candles, upper and lower distance limits have been set for 10 Local Group candidates, and a new distance estimated for M81.

1. RECENT SURVEYS FOR PLANETARY NEBULAE

New surveys for planetary nebulae in Local Group Galaxies have significantly extended the identifications reviewed by Ford (1978). These searches are important because they pinpoint stars that can be used to study the chemical composition and kinematics of otherwise mostly inaccessible populations. The surveys and their pertinent characteristics are summarized in Table 1. The galaxies in Table 1 are grouped according to certain, possible, or non-membership in the Local Group. The latter two groups will be discussed in Section 3.

The survey by Danziger et al. (1978) is noteworthy because it establishes the only identification of a planetary nebula in a spheroidal galaxy. It is not surprising that Fornax, the most massive of the spheroidal galaxies ($M \sim 2\times10^7\ M_\odot$), has a planetary nebula. Using Jacoby's (1980) mass specific number, 2.1×10^{-7} planetary nebulae $M_\odot^{-1}$, we expect roughly 4 nebulae in Fornax. From another point of

D. R. Flower (ed.), Planetary Nebulae, 443–460.

view, Fornax's mass is comparable to the mass of the galaxy's globular cluster system, wherein there is one luminous planetary in M15, and a faint planetary possibly associated with NGC 6401 (Peterson, 1977). Danziger et al. surveyed Fornax in Hα, and sampled approximately 2.3 magnitudes of the luminosity function. Because [OIII] λ5007 is typically 2 to 5 times brighter than Hα (c.f. Table 2), a deep survey in [OIII] λ5007 might reveal one or two more faint nebulae.

Killen and Dufour (1982) searched NGC 6822 for planetary nebulae and H II regions. They used Hα, [OIII] λ5007, and continuum plates taken with the CTIO 4-m telescope to find emission line sources and estimate their excitation. They found 31 diffuse nebulae and 36 stellar nebulae, and singled out 8 of the latter as possible planetary nebulae. Dufour and Talent (1980) used spectrophotometry to confirm that one of the nebulae (S 33) is indeed a planetary (cf. Section 2). The author used an unpublished survey by Sedwick and Ford (1976) to confirm that five of Killen and Dufour's candidates (S 10, S 14, S 16, S 29, and S 33) are probably planetary nebulae. One of their candidates, S 26, most likely is a star, and another, S 23, isn't on either Lick Hα or [OIII] λ5007 image tube plates, even though they appear to reach a fainter limiting magnitude than Killen and Dufour's plates. The last Killen and Dufour candidate, S 30, which is not in the area surveyed by Sedwick and Ford, probably is a planetary nebula. Sedwick and Ford found two additional faint candidates, which brings the total number of identifications back to 8. Killen and Dufour noted that the planetary nebulae are mostly in or near NGC 6822's bar, which presumably contains old stars.

Lawrie and Ford (1982) used a Velocity Modulating Camera (VMC) to isolate planetary nebulae in the central 250 parsecs of M31 and M32. By combining the slow f/17 Lick 3-m Cassegrain focus with a very narrow band (2.1Å FWHM) [OIII] λ5007 filter, they were able to suppress the galaxy's light and detect planetary nebulae to within a few arc seconds (a few tens of parsecs) of their nuclei. The VMC photographs revealed 42 nebulae in the center of M31, including 19 new identifications, and 5 new nebulae in the central 8" (r = 25 pc) of M32. Ford and Jenner confirmed the 5 new nebulae (and found still another, #28) with a video camera picture which is reproduced in Figure 1. Nebula number 27 projects 3.75" (12 pc) from M32's nucleus. The number of new nebulae (6) in M32's center is equal to the number predicted from the number per unit light in the outer regions. We expect that there are one to two more nebulae in the saturated core shown in Figure 1.

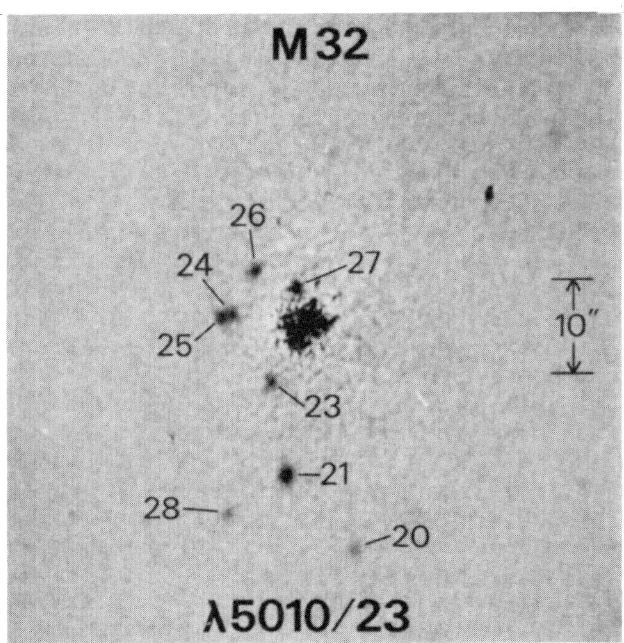

Figure 1. Planetary nebulae in the center of M32. The picture is a 12.5-min exposure with the video camera on the KPNO 4-m telescope. Galaxy light was removed by subracting a smoothed and normalized off-band picture. The dark area in the center is due to camera saturation rather than [OIII] emission.

The VMC survey swept velocity space in the centers of M31 and M32 with a sequence of photographs taken through a temperature tuned 2.1 Å interference filter. Lawrie (1978, 1983) used the photographs to measure the radial velocities of 30 planetary nebulae within 250 parsecs of M31's nucleus, and to place velocity limits on 10 more. He used a maximum likelihood method to derive a radial velocity dispersion of 155 ± 22 km s^{-1}, which he combined with the virial theorem and light distribution to calculate a mass-to-light ratio for the inner nuclear bulge, $M/L_V = 9.6 \pm 3.6$. His M/L_V agrees with previous results based on dynamical studies of galaxy nuclei (Faber and Jackson, 1976; Williams, 1977), but is inconsistent with the large M/L_V predicted by dwarf-enriched stellar population models (Spinrad and Taylor, 1971; Williams, 1976).

Ford, Jacoby, and Jenner (1983) extended previous surveys of M31 (cf. Ford, 1978) to large distances from the center. Using the KPNO 0.9-m and 4-m telescopes, they found planetaries out to projected distances of 34 kpc. They have measured radial velocities of 49 of the most distant nebulae, and Jacoby and Ford (1983) have determined chemical abundances in planetary nebulae at 10 kpc, 17 kpc, and 34 kpc (cf. Section 2).

Eason and Ford (1983) used a long exposure (165-min) [OIII] λ5015/32 KPNO 4-m plate of M33 to improve the limited coverage of an earlier unpublished Lick survey by Ford and Jenner which identified 20 planetaries. The distribution of 58 planetary nebulae in M33 is shown schematically in Figure 2. The schematic representation of the spiral structure was taken from Sandage and Humphreys (1980). The H II regions in M33 are not very bright in [OIII] λ5007. Consequently, the detection limit was relatively constant across the galaxy. The nebulae do not show a strong concentration toward the center, reflecting the absence of a nuclear bulge which in the Milky Way Galaxy and M31 gives rise to a strong central concentration.

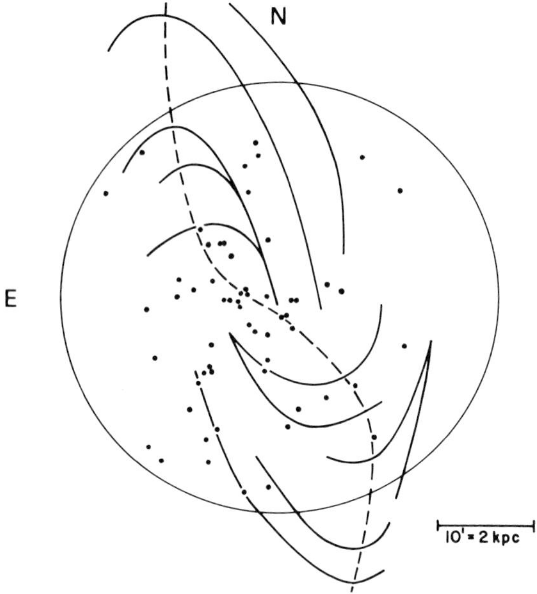

Figure 2. A schematic representation of the distribution of planetary nebulae in M33. The schematic spiral arms were taken from Sandage and Humphreys (1980). The distribution is noticeably asymmetrical about the major axis. The circle delimits the field of the 4-m telescope.

The planetary nebula distribution in Figure 2 is strikingly asymmetrical about M33's major axis. The ratio of the number on the SE side to the number on the NW side is between 1.7 and 1.8, depending on the value of the major-axis position angle. The binomial probability of observing a chance ratio greater than 1.5 is 0.1, which suggests that the distribution may have a physical origin. The asymmetry can be understood if we idealize the distribution of nebulae as a uniform density ellipsoid that is bisected by a thin sheet of dust which obscures the far side. Let q be the number of far side nebulae which can be seen through the disk divided by the number which can be seen on the near side. The ratio of the number of nebulae which project across the major axis onto the far side to the number which project onto the near side is then given by

$$R = \frac{\pi - (1-q)\cos^{-1}\theta(b/a,i)}{q\pi + (1-q)\cos^{-1}\theta(b/a,i)} \quad (1)$$

where b/a is the ratio of the ellipsoid's minor axis to major axis, i is the inclination of the minor axis to the line of sight and θ is given by

$$\theta = \frac{(b/a)\tan i}{\sqrt{1 + (b/a)^2\tan^2 i}} \quad (2)$$

The optical depth through M33's disk can be estimated from the average H I column density ($\sim 9.5 \times 10^{20}$ atoms cm^{-2}; Rogstad and Shostak, 1972), using $A(V) = 3 \times n(H\ I) \times 1.69 \times 10^{-22}$ mag cm^2 (Spitzer, 1978), A(V) = 0.5 mag. If we assume that the planetary luminosity function n(m) is constant, the value of q is 0.28. We can now solve equation (1) for the b/a which gives the observed R for a specified position angle and inclination i.

In reality, M33's geometry is rather more complicated than our simple geometrical model. Sandage and Humphreys (1980) have shown that the position angle and inclination respectively change from 49°/40° in the inner 2 kpc to 15°/65° in the outer parts of the galaxy. The respective solutions are b/a = 1.0 and b/a = 0.5. Because our symmetrical model does not represent the bent galaxy, we cannot derive a unique b/a. We can, however, conclude that the planetaries must have an inflated distribution. A thin bent disk cannot reproduce the observed distribution other than by chance.

Jacoby's (1978, 1980) Magellanic Clouds survey, which is reviewed elsewhere in this volume, established the [OIII] planetary nebula luminosity function over a six magnitude range. Jacoby noted that in the absence of strong central star evolution over the 20,000-year nebula lifetime, the [OIII] luminosity evolution will be controlled by the nebula's expansion velocity (i.e. the rate of change of the density). There is relatively strong observational evidence that both the form (approximately uniform magnitude density) and the scaling of luminosity functions are similar from one galaxy to another. Jacoby and Lesser (1981) and Ford and Jenner (1979) found that the dispersion in the magnitudes of the brightest planetaries in M31, M32, NGC 185, NGC 205, the LMC, and the SMC is less than 25%. Jacoby found that the LMC and SMC luminosity functions are statistically indistinguishable, and Lawrie and Ford (1982) showed that the first three magnitudes of the luminosity function in the center of M31 matches that of the Clouds.

Jacoby (1980) estimated the total number of planetaries (within an 8-mag interval) in Local Group galaxies by using his luminosity function to extrapolate from the observed bright end to the unobserved faint end. He showed that the derived luminosity-specific number of planetary nebulae per unit luminosity is approximately consant in 8 galaxies, with a value of 6.1×10^{-7} planetaries $L_{\odot}^{-1}$.

2. CHEMICAL COMPOSITIONS

Chemical abundances have been measured in planetary nebulae in 8 Local Group galaxies. Because the nebulae in the more distant galaxies are very faint, there is considerable variation in the observational uncertainties and the number of atomic species measured. In spite of these difficulties, some interesting trends appear to be emerging.

The data are summarized in Tables 2, 3, and 4. Table 2 is a compilation of reddening corrected line intensities for Fornax (Danziger et al., 1978), NGC 185 (Jenner and Ford, 1976), M32 (Jenner and Ford, 1978), NGC 6822 (Dufour and Talent, 1980), and M31 (Jacoby and Ford, 1983). The line intensities in NGC 185 and M31 are published here for the first time. The Hβ fluxes and luminosities of the nebulae are given in the last row of Table 2. Many of the fluxes were measured through small spectroscopic apertures, and thus are lower limits. It is evident that the Hβ luminosities of the brighter planeteries in galaxies of widely differing masses are relatively constant.

The ionic abundances of the nebulae are given in Table 3. Both Danziger et al. (1978) and Dufour and Talent (1980) calculated ionic abundances for the case of a uniform temperature and the case of temperature fluctuations represented by Piembert and Costero's (1969) fluctuating temperature with $t_2 = 0.035$. All of the ionic abundances in Table 3 are based on a uniform temperature throughout the nebula. Table 3 shows that the O^{++} abundances of oxygen are relatively high in the Fornax planetary and in the M31 10 kpc and 17 kpc planetaries. Consequently, independent of the ionization correction scheme which is used, the abundance of oxygen must be relatively high in those planetaries.

Atomic abundances are given in Table 4. I have included the abundances of H II regions in those galaxies where data is available. In addition to the average abundances of planetary nebulae in the SMC and LMC, I have included N67 and N97, helium and nitrogen-rich nebulae in the respective galaxies which Dufour and Talent considered most similar to the planetary observed in NGC 6822.

Before discussing abundances in Table 4, I will first briefly discuss the ages, or equivalently the masses, of the progenitor stars. Planetary nebulae near the sun have an average distance from the plane $\langle |z| \rangle = 150$ pc (Osterbrock, 1974), which corresponds to precursor masses $M \cong 1.5 M_{\odot}$ (Peimbert, 1978) and ages of $\sim 2 \times 10^{9}$

years. Peimbert (1978) used chemcial composition and kinematics to classify galactic planetary nebulae into the following 4 types: Type I, nebulae which are helium and nitrogen rich and that appear to originate in young stars; Type II, an intermediate population with a characteristic mass of $M \sim 1.5\ M_{\odot}$; Type III, which have $|\Delta V| > 60$ km s^{-1} and $|z| > 0.8$ kpc; and Type IV, a metal-deficient halo population. Tinsley (1978) used models to show that a steady star formation rate in the galaxy will yield approximately equal numbers of 0.9 to 1.2 $M_{\odot}$, 1.2 to 2 $M_{\odot}$, and 2 $M_{\odot}$ to 5 $M_{\odot}$ progenitors. That distribution apparently does not represent galactic planetary precursors, because the majority of planetaries are Type II. If there is a declining star formation rate the respective intervals will provide 65%, 20%, and 15% of the precursors. From an observational point of view, several arguments suggest that planetary nebulae in Local Group galaxies primarily derive from old and intermediate age populations. The stars in M32 and in M31's and the galaxy's nuclear bulges produce planetary nebulae copiously. There is no evidence for recent star formation in either the galaxy's or M31's nuclear bulge, nor in M32, which has no detectable gas (cf. Ford and Jenner, 1975) and has not had star formation more recently than 5×10^{9} years (O'Connell, 1980). Feast (1968) has shown that planetaries in the LMC have approximately twice the velocity dispersion of extreme Population I stars. The planetary nebulae in M33, which has vigorous star formation, may be distributed in an ellipsoidal distribution (cf. Section 1) rather than in a thin disk. There is no star formation in Fornax, although the carbon star population (Frogel et al., 1981) shows that there has been star formation within the last 2 to 8 billion years. In view of these considerations, we conclude that planetary progenitors are typically 2 or more billion years old $M \leq 1.5\ M_{\odot}$).

Turning now to Table 4, I first note, as have many previous authors, that the helium abundance in planetary nebulae is typically higher than in H II regions in the same galaxy. There can be little doubt that planetary shells are helium enriched. However, if we use Tinsley's (1978) criterion that X(PN)/X(ISM) be greater than 7 for planetaries to produce most of an element in our galaxy, it does not appear that planetary nebulae are a major source of helium in star-forming galaxies. Table 4 also shows that the nitrogen abundance in planetaries is consistently higher than in H II regions, by factors of 4.0, 4.5, 5.5, 4.2, and 123 in the galaxy, M31, the SMC, the LMC, and NGC 6822. The average abundances in Table 4 are based on "typical" planetaries. In addition to these there are populations of helium-rich and nitrogen-rich planetaries in the galaxy (Peimbert's Type I; Peimbert, 1978), the LMC (N97, P07, and P09; Osmer, 1976; Dufour and Killen, 1977; and Aller, 1983), the SMC (N67; Osmer, 1976; Dufour and Killen, 1977) and NGC 6822 (Dufour and Talent, 1980). Between the "typical" planetaries, which have a nitrogen enrichment of 4 to 5 relative to the ISM, and the smaller populations of helium-rich planetaries, which have nitrogen enrichments of 20 to 100, there is ample data to substantiate earlier conclusions (e.g. Tinsley, 1978) that planetary nebulae are most likely a major source of nitrogen enrichment in galaxies. The only qualification of this conclusion is William's (1982) demonstration that novae may return even more nitrogen to the interstellar medium than planetaries.

There are no obvious trends in the oxygen abundance. Oxygen is underabundant in the planetaries studied in NGC 185 and NGC 6822, and in some planetaries such as the helium-rich planetaries in the Clouds. There are weak correlations between N/O and He/H and between N/O and O/H, which suggest that nitrogen is at least partially a secondary product of nucleosynthesis (cf. Danziger et al., 1978 ; Talbot and Arnett, 1974).

I will now consider the galaxies in Table 4 individually. The analysis by Danziger et al. (1978) of the Fornax planetary shows that there are at least two stellar populations in Fornax. The first population is characterized by Fornax's 4 globular clusters, which are clearly old, metal-deficient clusters that are similar to M15 and M92 (van den Bergh, 1969; Danziger, 1973; Zinn and Persson, 1981). By analogy with the galaxy (cf. Sandage, Freeman, and Stokes, 1970), this old population presumably had its genesis during Fornax's earliest star-formation epoch. In contrast to the metal-poor globular clusters, the Fornax planetary nebula has a nearly normal oxygen abundance, which is unlike the oxygen-deficient planetary K648 in M15 and the galactic halo planetaries PK 49+88°1 and 108-76°1 (Hawley and Miller, 1978). The discovery of carbon stars in Fornax (Frogel et al., 1982) with probable ages between 2 and 8 billion years (inferred from comparison with carbon stars in the Magellanic Clouds; Mould and Aaronson, 1980) establishes star formation and continuing metal enrichment long after the globular clusters formed. Although the detailed evolution from red giants to planetary nebulae is poorly understood, it seems likely that carbon stars and the carbon-rich protoplanetary nebulae discussed by Zuckerman (1978) are two links in the evolutionary chain. Consequently, it is plausible to suppose that the Fornax planetary and the carbon stars are members of the same intermediate age population.

The considerable metal enrichment revealed by the presence of the oxygen-enriched planetary and the carbon-enriched carbon stars points to an origin in nucleosynthesis-enriched gas, rather than in primordial gas that was retained or captured by Fornax. The mass loss rate from stellar evolution in Fornax, estimated from comparison with NGC 147 (Ford, Jacoby, and Jenner 1977), is $\sim 3 \times 10^{-5}\ M_{\odot}\ yr^{-1}$. Consequently, even if Fornax did not retain any gas after the first star formation epoch, it produces enough gas to fuel star formation. However, planetary nebula shells, which have expansion velocities larger than the escape velocity ($v_{escap} \lesssim 10$ km s^{-1}), may be able to power a cool thermal wind (cf. Ford, et al., 1977) that would keep the galaxy relatively gas free. If the mass loss rate is too high for planetary shells to power a warm wind, Fornax may be an example of an elliptical galaxy which accumulates gas until star formation begins and is subsequently quenched by a supernova-powered, hot-pulsed wind (Sanders, 1981).

Dufour and Talent (1980) have shown that there is an extreme overabundance of nitrogen in the NGC 6822 planetary. Although considerable uncertainty in the nitrogen abundance is introduced by correction for unseen N^{++}, Dufour and Talent note that N^+, O^+, and S^+ have similar ionization potentials, and thus originate in the same regions of the nebulae. Consequently, the great strength of the [NII] lines in the NGC 6822 planetary must point to a nitrogen overabundance relative to galactic planetaries. A similar argument applies to the planetaries in M32, which have stronger nitrogen lines that galactic planetaries. Although Jenner and Ford (1979) could not derive O and N abundances because of the difficulty of measuring the planetary's temperature (which suggests a high O abundance), the similarity of the line intensities in M32-1 and the NGC 6822 planetary suggests a considerable overabundance of nitrogen. In spite of the possibly similar overabundances of nitrogen, the ages of the progenitor stars in the two galaxies may be very different. Because of the NGC 6822 planetary's similarity to galactic helium and nitrogen-rich stars, Dufour and Talent conclude that it derives from a young star. By contrast, O'Connell's (1980) spectral synthesis of M32 suggests that there has not been any star formation in M32 more recently than 5 x 10^9 years ago.

The planetary in NGC 185 has a high helium and nitrogen abundance relative to the nebulae in Fornax and typical nebulae in the clouds. Although the nitrogen abundance is very uncertain because of a large ionization correction, the nebulae may be a helium and nitrogen-rich planetary that had its origin in one of NGC 185's small population of OB stars (Baade, 1951; Hodge, 1963) rather than the old, metal poor populations which produce most of NGC 185's light.

The M31 abundances reveal several interesting facts. First, the oxygen abundance in an old star(i.e a planetary nebula) and an H II region (No. 685; Baade and Arp, 1964) are surprisingly high at 17 kpc from the center. The oxygen abundances are very nearly equal to those in the Orion nebulae. Blair et al. (1982) used empirically calibrated abundances in H II regions and also found high abundances at large radii. Jacoby and Ford's (1983) oxygen abundance in the H II region is a direct determination based on a well-measured temperature, and falls near the mean fit of O/H versus R (Blair et al., 1982), thereby supporting their empirical calibration at 17 kpc.

Oxygen and neon, and probably nitrogen, are relatively underabundant in the planetary at 34 Kpc. The oxygen abundance is a little less than in PK 49+88° 1, a galactic halo planetary (Hawley and Miller, 1978). We believe the remote 34 kpc nebula is a member of a metal-poor halo population in M31.

Although an abundance gradient based only on the three planetary nebulae in Table 4 cannot be considered well determined, there is as yet no other data available. Abundance gradients in M31 and the galaxy are summarized in Table 5. The reader should note that different investigators derive widely differing galactic abundance gradients. Barker (1978) found weak evidence for a nitrogen abundance gradient, and gave a formal value of approximately half that in Table 5. Aller and Czyzak (1982) find little evidence for a galactic nitrogen abundance gradient. Table 5 shows that the gradient in both the gas and stars is very shallow in M31, and much shallower than in the galaxy.

3. PLANETARY NEBULAE AS STANDARD CANDLES

Ford and Jenner (1979) used unpublished KPNO 4-m telescope video camera photographs to show that the dispersion in the reddening corrected [OIII] λ5007 magnitudes of the brightest planetary nebulae in M31, M32, NGC 205, and NGC 185 is ~ 25%. This relative luminosity constancy in galaxies with a wide range in chemcial composition and number of nebulae suggests that planetaries may be useful as standard candles. In particular, it is reasonable to suppose that a comparison of two well-populated homogeneous populations, such as those found in the nuclear bulges of two nearly identical spirals, will lead to an even smaller dispersion between the first ranked nebulae, or some combination of brightest nebulae. Following this line of reasoning, Ford and Jenner (1979) used video camera [OIII] λ5007 on-band/off-band pictures to find and measure the magnitudes of planetary nebulae in the nuclear bulges of M81 (Sb, inclination = 58°) and M31 (Sb, inclination = 77°).

Figure 3 shows the sum of twelve 12.5-min on-band pictures and an equivalent sum of off-band pictures. Both pictures have been flattened by subtracting a smoothed off-band picture. Using the criterion that a nebula must appear in each of three on-band pictures taken on three separate nights, there are eight very faint, but definite identifications.

The galactic-extinction corrected differences between the 1st brightest nebulae, the average of the 2nd through 4th nebulae, and the cumulative luminosity distributions lead to preliminary M81 to M31 distance ratios, which do not include corrections for internal extinction, of 4.0, 3.6, and 3.5. These ratios are significantly lower than those determined by Sandage and Tammann (4.9; 1974) and de Vaucouleurs (5.3; 1979).

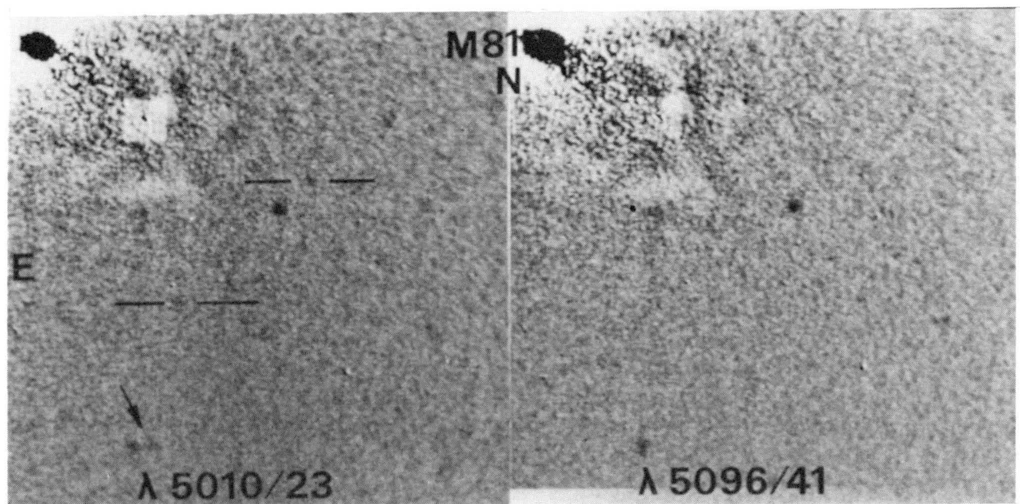

Figure 3. Three planetary nebulae in M81's nuclear bulge. The pictures are a superposition of 12 12.5-min KPNO telescope video camera exposures. The galaxy light has been removed by subtracting a smoothed off-band summation.

Jacoby and Lesser (1980) looked for planetary nebulae in 10 galaxies whose Local Group membership is uncertain. Using the video camera on the KPNO 2.1-m telescope, they found 7 planetaries in 5 of the galaxies (cf. Table 1). Because planetaries are faint during most of their lifetime (Jacoby 1980), they could not derive reliable distances from the observed fluxes of one or two nebulae in a sparsely populated galaxy. However, they were able to place upper limits on the distances by assuming that the nebulae are as intrinsically luminous as the planetaries in well populated galaxies such as M31 and the Magellanic Clouds. These upper limits suggest that IC 10 is a definite Local Group member and that Sextans A and WLM could be Local Group members. In those galaxies where they did not detect planetaries, they were able to set rough lower limits to the distances by finding the distance at which at least one planetary should have been found. Their lower limits allow them to exclude Sextans B, NGC 3109, and NGC 6946 from Local Group membership. The membership designations in Table 1 are based on Jacoby and Lesser's survey and the kinematical considerations which they cited. Finally, we note that Ford and Jenner used KPNO video camera pictures to find 2 faint planetary nebulae and extended high excitation nuclear emission in NGC 404, a galaxy excluded from Local Group membership by kinematical considerations (Yahil et al., 1977). The 2 nebulae and the asymmetrical, extended nuclear emission are shown in Figure 4.

Table 1

Recent Surveys For Planetary Nebulae

Galaxy	Telescope	Detector or Plates	Filters: On-Line ion, λ_c/FWHM	Filters: Off-Line λ_c/FWHM	Nebulae Found		References
			Local Group Members				
Sculptor Fornax	UK 1.2-m	098	Hα + [NII] interference filter	RG 630	0 1		Danziger, Dopita, Hawarden, and Webster (1978)
NGC 6822	CTIO 4-m	IIIa-J 127 04	[OIII] λ5000/70 Hα + [NII] λ6560/120	GG 385, GG 495 GG 385, GG 495	8		Killen and Dufour (1982)
M32 Nucleus M33	Lick 3-m KPNO 4-m	Velocity Modulating Camera IIIa-F IIa-D	[OIII] $\lambda_c$tuneable/2.1 [OIII] λ 5015/32	$\lambda_c$tunable/2.1 OG 570	6 (new) 58		This paper. Eason and Ford (1983)
M31 Nucleus M31 Halo	Lick 3-m KPNO 4-m	Velocity Modulating Camera IIIa-J, IIa-D IIIa-F, IIa-D	[OIII] $\lambda_c$tuneable/2.1 [OIII] GG 475 [OIII] λ5015/23	$\lambda_c$tuneable/2.1 OG 570 OG 570	19 (new) 47		Lawrie and Ford (1982) Ford, Jacoby, and Jenner (1983)
M31 Disk and Halo	KPNO 0.9-m	KPNO Image Intensifier	[OIII] λ5007/21	λ5300/130	40		Ford and Jacoby (1983)
			Possible Local Group Members			Membership	
IC 10	KPNO 2.1-m	Video Camera	[OIII] λ5007/21	λ5300/130	1	yes(?)	Jacoby and Lesser (1981)
Leo A	"	"	"	"	2	yes(?)	"
Sextans A	"	"	"	"	1	(?)	"
Pegasus	"	"	"	"	1	(?)	"
WLM	"	"	"	"	2	yes(?)	"
			Galaxies Beyond the Local Group				
IC 342	KPNO 2.1-m	Video Camera	[OIII] λ5007/21	λ5300/130	0	no	Jacoby and Lesser (1981)
Sextans B	"	"	"	"	0	no(?)	"
NGC 3109	"	"	"	"	0	no	"
GR 8	"	"	"	"	0	no(?)	"
NGC 6946	"	"	"	"	0	no	"
NGC 404	KPNO 4-m	Video Camera	[OIII]5007/20	λ5300/200	2	no	This paper
M 81	"	"	"	"	8	no	Ford and Jenner (1979)

Table 2

Reddening Corrected Line Intensities in Local Group Planetary Nebulae

Ion	λ	Fornax	NGC 185	M32	NGC 6822	M31		
						10 kpc	17 kpc	34 kpc
[OII]	3727	25.2	< 7	63	43.8	117	144	34:
[NeIII]	3869	35.7	45	-	30.5	74	143	128
HeI	3889	14.4	21	96	-	-	16:	64:
Hε + [NeIII]	3970	21.0	-	-	31.0	31	-	88
Hδ	4102	26.1	-	30:	25.4	27	-	80:
Hγ	4341	47.3	52	36	47.1	57	57	48
[OIII]	4363	3.9	30	-	39.1	9.4:	15	26
HeI	4471	5.5	-	-	-	-	-	-
HeII	4686	-	<10	<15	71.9	-	-	-
Hβ	4861	100.0	100	100	100.0	100	100	100
[OIII]	4959	200.8	327	292	410.0	355	500	375
[OIII]	5007	579.5	967	897	1160.0	1079	1471	1122
HeI	5876	16.3	25	12	14.6	22:	26:	-
[NII]	6548	0.2	-	94	134.0	-	31:	-
Hα	6563	290.1	280	264	272.0	281	290	302
[NII]	6584	7.1	23	250	399.0	57	84:	<35
HeI	6678	6.0	-	-	-	-	-	-
[SII]	6724	<2.2	-	94:	<10	-	-	-
[ArIII]	7136	4.5	31	-	-	-	-	-
[OII]	7324	<2.2	< 7	-	-	-	-	-
Distance (kpc)		190	~600	<652	557		652	
log F(Hβ) ergs $cm^{-2}s^{-1}$		-13.8	-14.24	-14.75	-14.83	-15.09	-15.20	-15.44
log L(Hβ) ergs s^{-1}		34.8	>34.3	<35.0	34.7	>34.6	>34.51	>34.27

TABLE 3

Local Group Planetary Nebulae Ionic Abundances
Relative to H by Number

Ion	Fornax	NGC 185	M32	NGC 6822	M31		
					10 Kpc	17 Kpc	34 Kpc
He^{+}	0.13	0.21	0.089	0.12	0.16:	0.21:	-
He^{++}	0.00	< 0.009	$\leq$ 0.0012	0.066	-	-	-
O^{+}	1.4×10^{-5}	$< 7.7 \times 10^{-7}$	-	0.51×10^{-5}	7.6×10^{-5}	2.8×10^{-5}	0.62×10^{-5}
O^{++}	1.99×10^{-4}	7.3×10^{-5}	-	0.81×10^{-4}	3.05×10^{-4}	3.1×10^{-4}	1.16×10^{-4}
N^{+}	0.15×10^{-5}	1.4×10^{-6}	-	0.24×10^{-5}	3.3×10^{-5}	1.2×10^{-5}	$<2.2 \times 10^{-6}$
Ne^{++}	3.8×10^{-5}	6.4×10^{-6}	-	0.50×10^{-5}	6.05×10^{-5}	7.0×10^{-5}	0.28×10^{-5}
Ar^{++}	0.54×10^{-6}	9.5×10^{-6}	-	-	-	-	-
S^{+}	$<0.12 \times 10^{-6}$	-	-	$<0.12 \times 10^{-6}$	-	-	-
S^{++}	-	-	-	$<1.86 \times 10^{-6}$	-	-	-

Table 4

Chemical Abundances in Local Group Planetary Nebulae and H II Regions

Element	Fornax	NGC 185	M32	NGC 6822		SMC			LMC			Galaxy		M31			
				PN	HII	N67	Avg	HII	N97	Avg	HII	PN	Orion	10 kpc PN	17 kpc PN	17 kpc HII	34 kpc PN
He	11.08	11.32	11.00:	11.27	10.92	11.27	11.02	10.92	11.26	11.07	10.97	11.04	11.00	11.20	11.32:	10.93	-
O	8.51	7.88	-	8.12	8.4	7.7	8.16	8.10	7.8	8.34	8.43	8.9	8.7	8.58	8.66	8.52	8.09
N	7.41	>8.14	-	8.79	6.7	7.6	7.19	6.45	8.3	7.64	7.02	8.3	7.7	8.22	7.90	7.25	<7.6
Ne	-	6.81:	-	6.91	7.7	-	7.60	7.58	8.0	7.64	7.77	8.3	7.9	7.88	8.09	7.63	6.47
A	5.91	7.02	-	-	-	-	-	-	-	6.41	6.35	-	-	-	-	-	-
Abundance Sources	1	2	3	4	5	6	7	8	6	9	8	10	10	11	11	11	11
Galaxy Mass	~2×10^7 (1)	~2×10^8 (1)	5×10^8 (3)	1.4×10^9 (2)		1.5×10^9 (2)			6×10^9 (2)			1.8×10^{11} (4)		1.85×10^{11} (5)			
Galaxy Type	Spheroidal	dE0	E2	Ir IV-V		Ir IV			SBc III-IV			Sbc		Sb			

Abundances with considerable uncertainty are marked with a colon.

Mass References

1. Derived from the luminosity by assuming that it has the same mass-to-light ratio as M32.
2. van den Bergh (1968)
3. Ford (1978)
4. Schmidt (1965)
5. Rubin and Ford (1970)

Abundance References

1. Danziger et al., 1978
2. Jenner and Ford, 1978
3. Jenner, Ford, and Jacoby, 1979
4. Dufour and Talent, 1980
5. Lequeux et al., 1979
6. Osmer 1976, Dufour and Killen, 1977
7. Aller et al., 1981
8. Aller, Keyes, and Czyzak, 1979
9. Aller, 1983
10. Peimbert, 1978
11. Jacoby and Ford, 1983

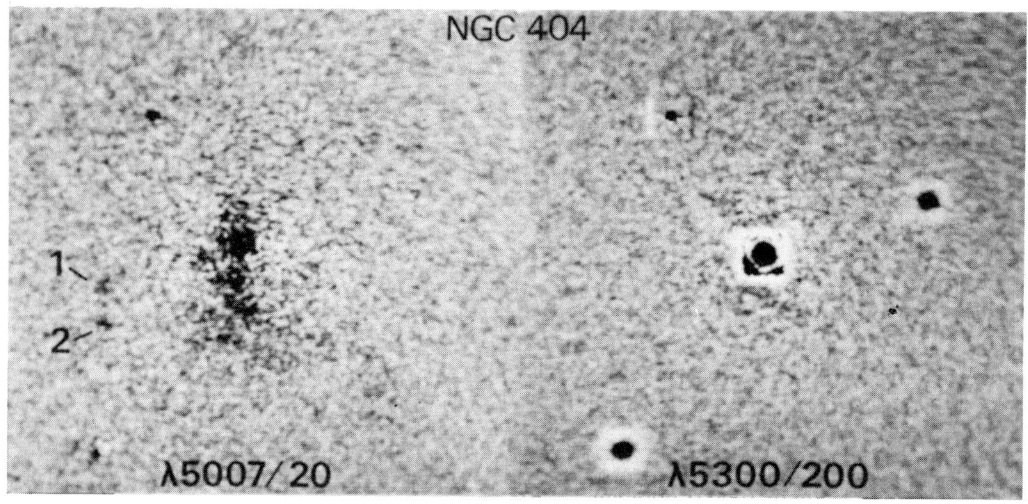

Figure 4. An [OIII] λ5007 pair of on-band off-band pictures of NGC 404. Two candidate planetary nebulae are marked. The on-band picture shows [OIII] emission extending asymmetrically out of the nucleus.

Table 5

Abundance Gradients in M31 and the Milky Way Galaxy

	M31 PN	M31 HII	MWG PN	MWG HII
$\frac{d\ \log\ (O/H)}{dR}$	-0.023	-0.029	-0.06	-0.13
$\frac{d\ \log\ (N/H)}{dR}$	$\leq$-0.024	-0.027	-0.18	-0.23
Source	1	2	3	3

1. Jacoby and Ford, 1983
2. Blair, Kirshner, and Chevalier, 1982
3. Peimbert, 1978

REFERENCES

Aller, L. H. 1983, This volume.

Aller, L. H. and Czyzak, S. J. 1982, Ap. J. Suppl., in press

Aller, L. H., Keyes, C. D., and Czyzak, S. J. 1979, Proc. Natl. Acad. Sci. USA, 76, 1525

Aller, L. H., Keyes, C. D., Ross, J. E., and O'Mara, B. J. 1981, Mon. Not. R. Astr. Soc., 194, 613

Barker, T. 1978, Ap. J., 220, 193

Baade, W. 1951, Pub. University of Michigan 10, 7

Baade, W. and Arp, H. 1964, Ap. J. 139, 1027

Blair, W. P., Kirshner, R. P., and Chevalier, R. A. 1982, Ap. J., 254, 50

Danziger, I. J. 1973, Ap. J., 181, 641

Danziger, I. J., Dopita, M. A., Harwarden, T. G., and Webster, B. L. 1978, Ap. J., 220, 458

Dufour, R.J. and Killen, R. M. 1977, Ap. J., 211 68

Dufour, R. J. and Talent, D. L. 1980, Ap. J., 235, 22

Eason, E. L. E. and Ford, H. C. 1983, in preparation

Faber, S. M. and Jackson, R. E. 1976, Ap. J., 204, 668

Feast, M. W. 1968, Mon. Not. R. Astr. Soc., 140, 345

Ford, H. C. 1978, in Planetary Nebulae (I.A.U. Symp. No. 76), p. 19, ed. Y. Terzian

Ford, H. C. and Jacoby, G. H. 1978, Ap. J. 219, 437

Ford, H. C., Jacoby, G. H., and Jenner, D. C. 1977, Ap. J. 213, 18

Ford, H. C., Jacoby, G. H., and Jenner, D. C. 1979, Ap. J. 227, 391

Ford, H. C, Jacoby, G. H., and Jenner, D. C. 1983, in preparation

Ford, H. C. and Jenner, D. C. 1975, Ap. J., 202, 365

Ford, H. C. and Jenner, D. C. 1979, B.A.A.S, 10, No. 4, 665

Frogel, J. A., Blanco, V. M., McCarthy, M. F., and Cohen, J. G. 1982, Ap. J., 252, 133

Hawley, S. A. and Miller, J. S. 1978, Ap. J., 220, 609

Hodge, P. W. 1963, A. J., 68, 691

Jacoby, G. H. 1978, Ap. J. 226, 540

Jacoby, G. H. 1980, Ap. J. Suppl., 42, 1

Jacoby, G. H. and Ford, H. C. 1983, in preparation

Jacoby, G. H. and Lesser, M. P. 1981, A. J., 86, 185

Jenner, D. C. and Ford, H. C. 1978, in Planetary Nebulae (I.A.U. Symp. No. 76), p. 246, ed. Y. Terzian

Jenner, D.C., Ford, H. C. and Jacoby, G. H. 1979, Ap. J., 227, 391

Killen, R. M., and Dufour, R. J. 1982, Pub. Astr. Soc. Pac. (June)

Lawrie, D. G. 1978, Ph.D. Thesis, Univ. of Calif., Los Angeles

Lawrie, D. G. 1983, in preparation

Lawrie, D. G. and Ford, H. C. 1982, Ap. J., 256, 120

Lequeux, J., Peimbert, M., Rayo, J. F., Serrano, A., and Torres-Peimbert, S. 1979, Astr. Ap.,

Mould, J. and Aaronson, M. 1980, Ap. J., 240, 464

O'Connell, R. W. 1980, Ap. J., 236, 430

Osterbrock, D. E. 1974, Astrophysics of Gaseous Nebulae, W. H. Freeman and Co., San Francisco

Peimbert, M. 1978, in Planetary Nebulae (I.A.U. Symp. No. 76), p. 215, ed, Y. Terzian
Peimbert, M. and Costero, R. 1969, Bol. Obs. Tonantzintla y Tacubaya, 5, 3
Peterson, A. W. 1977, Pub. Astr. Soc. Pac., 89, 129
Rogstad, D. H. and Shostak, G. S. 1972, Ap. J., 176, 315
Sandage, A., Freeman, K. C., and Stokes, N. R. 1970, Ap. J., 160, 831
Sandage, A., and Humphreys, R. M. 1980, Ap. J. (Letters), 236, L1
Sandage, A. and Tammann, G. A. 1974, Ap. J., 190, 525
Sanders, R. H. 1981, Ap. J., 244, 820
Sedwick, K. E. and Ford, H. C. 1976, Bull. Am. Astr. Soc., 8, 568
Spinrad, H. and Taylor, B. J. 1971, Ap. J. Suppl., 22, 445
Spitzer, L. 1978, Physical Processes in the Interstellar Medium, Wiley, New York
Talbot, R. J., Jr., and Arnett, W. D. 1974, Ap. J., 190, 605
Tinsley, B. 1978, in Planetary Nebulae (I.A.U. Symp. No. 76), p. 341, ed. Y. Terzian
van den Bergh, S. 1969, Ap. J. Suppl., 14, 145
Williams, T. B. 1976, Ap. J., 209, 716
Williams, T. B. 1977, Ap. J. 214, 685
Williams, R. E. 1982 (preprint)
Yahil, A., Tammann, G. A., and Sandage, A. 1977, Ap. J., 217, 903
Zinn, R. and Persson, S. E. 1981, Ap. J., 247, 849

FLOWER: Which lines are available for abundance determinations?

FORD: The ionization corrections are based on [O II] λ 3727 and [O III] λ 5007 and are applied to [N II] λ 6584 and [Ne III] λ 3868.

PAPP: I would like to point out that the distribution of PN you have found in the outer disk of M 31 follows very well the observed H I and optical warp.

REAY: From the relatively high signal/noise of the spectrum which you showed, it appears that it would be possible, at higher dispersion, to use velocity measurements to discriminate against background sources.

FORD: The photon rate is about 2 s^{-1} in [O III] λ 5007 for the brighter PN when using a 4 m telescope. Although the shells could be resolved, we believe that there is little or no confusion with H II regions. The [O III] magnitudes and absence of a stellar continuum in our spectrophotometric scans support this conclusion.

DOPITA: The metallicity gradient of M 31 has recently been derived (Dopita, Binette, d'Odorico and Benvenuti, Ap. J., in press) from the data of Blair, Kirshner and Chevalier. Our result is a gradient of -0.05 ± 0.02 dex kpc^{-1}.

FORD: Although your result is somewhat larger than the values given in my Table 5, it is still relatively shallow.

ZUCKERMAN: I hope that observations of PN will be extended to more distant PN, perhaps using the Space Telescope.

LAWRIE: Observations of PN are already being extended outwards using ground-based instruments. John Graham and I have identified 9 strong PN candidates in the Sculptor group, NGC 300. We hope to obtain [O III] λ 5007 magnitudes for these objects in the autumn.

THE EFFECTS OF MASS AND METALLICITY UPON PN FORMATION

K.A. Papp
University of Waterloo

C.R. Purton
Dominion Radio Astrophysical Observatory

S. Kwok
Herzberg Institute for Astrophysics

We construct a parameterized function which describes the possible dependence of planetary nebulae formation upon metal abundance and stellar mass. Data on galaxies in the local group compared with predictions made from the parameterized function indicate that heavy element abundance is the principal agent influencing the formation of planetary nebulae; stars which are rich in heavy elements are the progenitors of planetary nebulae. Our analysis, when compared with the observations, argues for a modest degree of pre-enrichment in a few of the sample galaxies. The heavy element dependence of planetary nebulae formation also accounts for the deficit of planetary nebulae in the nuclei of NGC 221 and NGC 224, and in the bulge of our galaxy.

WEHRSE: Model calculations for luminous M-star atmospheres show that the geometrical extensions and temperature distributions depend strongly on the metallicity. Do your computations show such correlations?

PAPP: No. Our model shows only a correlation of PN formation with metallicity - high metallicity stars are more likely to form PN.

SERRANO: You are probably overestimating the number of low mass stars, first, because the IMF flattens out at low masses, and, second, because a constant BRF would be more appropriate.

PAPP: Our IMF is well within the limits set by Miller and Scalo, but, even if one adopts an IMF which levels off at 1.5 $M_\odot$ or 2 $M_\odot$, the results will not change very much, as they are most sensitive to the form of the IMF between the turn-off mass and 1.2 $M_\odot$.

A constant BRF for the irregulars and spirals would reduce our problems. In particular, a constant BRF for the SMC would help to explain the very large number of PN without having to invoke an extended burst of star formation. There will be little or no change for spiral galaxies.

KWOK: A small value for the low-mass cut-off implies that the transition time from the AGB to PN is short, even for low mass stars.

PAPP: Too high a value for the low-mass cut-off would mean that we could not explain the large numbers of PN in elliptical galaxies without invoking extreme conditions of star formation at the birth of the galaxy. However, a small value does imply that the transition time must be short, certainly very short compared with the AGB-lifetime, even for low-mass stars.

RENZINI: When comparing theoretical PN birthrates with PN counts, one has to make an assumption regarding the PN lifetimes. You have implicitly assumed that the PN lifetime is independent of the initial mass of the parent star - an assumption which is very probably incorrect. If this assumption is relaxed, the results of your analysis will change dramatically.

PLANETARY NEBULAE AND THE CHEMICAL EVOLUTION OF GALAXIES

Alfonso Serrano
Instituto de Astronomía, Universidad Nacional Autónoma de México

1. INTRODUCTION

Tinsley (1978) has done an excellent review that illustrates the methods and concepts that can be developed to assess the effects of planetary nebulae (PN) on the long-term history of the galaxy. Tinsley concluded that research in PN could put constraints on the past rate of star formation and provide information on chemical enrichment by low mass stars.

In fact, the relationship between PN and chemical evolution of galaxies is twofold; one aspect of this relationship is the constraints produced by studies of PN on the fundamental parameters of chemical evolution: accretion of gas in galaxies, yields of primary and secondary elements, the initial mass function (IMF) and the variations in time and location of the stellar formation rate (SFR). On the other hand, chemical evolutionary models of galaxies, based on independent evidence, can clarify and put constraints on our theoretical understanding of the mixing and ejection processes in the stars precursors of PN. In this review some recent examples of this twofold relationship are illustrated.

2. MASSES OF THE PROGENITORS

PN are of interest to study the chemical evolution of galaxies and to test theories of nucleosynthesis. Firstly, because the initial mass of the stars that produce them is relatively low (1 to 5 $M_\odot$) and this means high numbers, long stellar lifetimes and, hence, information on the conditions of star formation (abundances, IMF, SFR) long ago. Secondly, because PN are produced in the final stages of the active stellar lifetime and their abundances reflect not only the composition at the time and place of birth, but also show the effect of newly synthesized material . In this sense, they contribute to the enrichment of chemical abundances in the interstellar medium. For both reasons, the fundamental parameters that affect models of chemical evolution are the nebular abundances and masses as a function of the initial mass of the star (see e.g. Iben and Truran 1978; Renzini and Voli 1981; Iben and Renzini 1982,

D. R. Flower (ed.), Planetary Nebulae, 463–472.

and references therein).

The number of objects for which a crude estimate of their masses at the zero age main sequence can be made is very small (see Méndez and Niemela 1981; Peimbert and Serrano 1980; Calvet and Peimbert 1982); binary central stars, as in NGC 3132 and 2346, or probable members of a cluster, as NGC 2818, are rare to find. Moreover, the known masses happen to be very similar (about 2.4 $M_{\odot}$). Hence, in order to estimate the dependance on initial stellar mass, statistical arguments must be used.

1.1 TYPE I PN: THOSE WITH MASSIVE PROGENITORS

Greig (1971) has classified PN into two main groups according to their morphology. Cudworth (1974) finds that the kinematics of these groups correspond to stars of 1 and 1.5 $M_{\odot}$, respectively. On the other hand, Peimbert (1978) has divided PN into four groups, according to their He and N abundances and to their kinematics (see also Peimbert's review in this Symposium).

From a review of the best observed PN, Peimbert and Serrano (1980) suggest that Peimbert's Type I PN are more massive than Types II and III and that the dividing line between them corresponds to objects of $\sim 2.4\ M_{\odot}$. In Figure 1, a comparison is shown between He/H and N/O

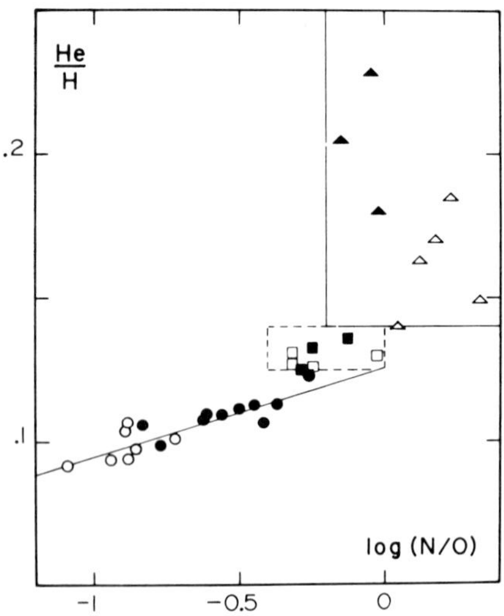

Fig. 1. Comparison of N(He)/N(H) and N(N)/N(O) ratios. Full and open symbols represent PN in the direction of the galactic center and galactic anticenter respectively. Triangles are Type I, squares are Type I-II and circles are Types II or III. The solid line is the least squares solution for PN of Types II and III.

abundance ratios. There is a clear separation between Type I PN and the rest. The three objects of known initial stellar mass mentioned above are all of Type I-II. This suggests that stars more massive than $\sim 2.4\ M_\odot$ give rise to Type I PN with large overabundances of He and N, while stars with mass smaller than $2.4\ M_\odot$ would give rise to PN Types II and III. According to Peimbert and Serrano, the fraction of Type I PN is the 10 to 30% range.

A knowledge of the fraction of Type I PN and of the lower mass limit for their formation can be combined with the stellar birthrate to estimate stars of the various types of PN. Adopting Serrano's (1978) IMF and a fraction of 20%, Peimbert and Serrano find that stars in the 2.4 to $4.6\ M_\odot$ range produce PN of Type I, while stars in the 1 to $2.4\ M_\odot$ range produce PN of Types II and III. The upper limit of $4.6\ M_\odot$ should correspond to the minimum mass, M_W, required to produce a degenerate carbon-oxygen core of $1.4\ M_\odot$. Renzini and Voli (1981) obtained similar values of M_W if the parameter η in the Reimer's (1975) formula for the mass loss rate is 1/3. It is also similar to the value obtained by van den Heuvel (1975). However, Romanishin and Angel (1980) have obtained, from counts of faint blue objects in open clusters, that M_W is probably $\sim 7\ M_\odot$. Such a value would imply a higher η and either a fraction of Type I PN higher than 20%, or that stars of $\sim 1\ M_\odot$ very seldom become PN (see, however, Renzini 1981).

1.2. THE ENVELOPE MASSES

It is well known that the electron densities derived from forbidden lines are usually higher than those derived from the flux in Hβ. Thus, lower and upper bounds for the envelope mass can be derived by assuming either extreme density fluctuations or the density indicated by the Hβ flux, respectively. Taking the filling factor found by Torres-Peimbert and Peimbert (1977) and interpolating between the two limits to the mass, Peimbert (1981) has obtained a mean envelope mass for PN of $0.09\ M_\odot$ using the Cahn and Kaler (1971) distance scale, and of $0.25\ M_\odot$ using the Cudworth (1974) distance scale. On the other hand, using the IMF mentioned above, Serrano and Peimbert (1981a) obtained average nebular masses of $0.6\ M_\odot$ and $0.33\ M_\odot$ for Renzini and Voli's cases A and B, respectively. Agreement between theory and observations would favor Renzini and Voli's case B and Cudworth's distance scale. Thus, it appears that Wood and Cahn (1977) have overestimated the luminosity at which a star of a given mass give rise to a PN.

2. ABUNDANCE GRADIENTS

Several authors have obtained abundance gradients from PN (D'Odorico *et al.* 1976, Aller 1976, Torres-Peimbert and Peimbert 1977, Barker 1978). Peimbert and Serrano (1980) used a large sample of the best observed PN and obtained He, N and O abundances as a function of galactocentric distance, as shown in Figure 2. For Type II and III PN there is a correlation between abundance and position for the three elements. Moreover, PN

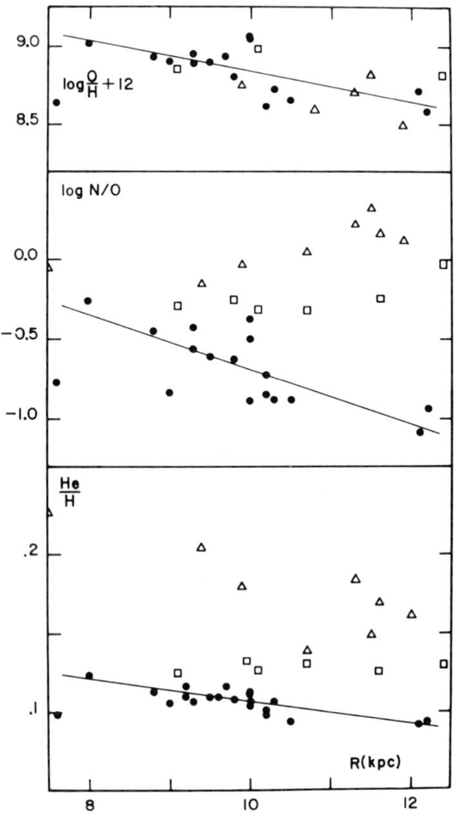

Fig. 2. Abundance ratios of O/H, N/O and He/H compared to the galactocentric distance. A value $R_{\odot}$= 10 kpc was adopted. Symbols as in Figure 1. Lines are least square solutions to PN of Types II and III, without NGC 5307.

of Type I also show a good fit to the gradient of O/H defined by those of Types II and III. On the contrary, N/H and He/H values in PN of Type I do not show any correlation with position.

This difference arises, as Renzini and Voli pointed out, because the first dredge-up dominates in less massive stars, giving rise to relatively small He and N surface enhancements, while the third dredge-up dominates for stars with M> 2.5 $M_{\odot}$, resulting in much higher He and N enhancements at the surface.

Notice, however, that the observed slope Δ(He/H)/Δ(N/O) in Figure 1 for Types II and III PN is ∿ 0.03, while models of stellar evolution predict a very high slope in this mass range (see e.g. Renzini and Voli's Figure 10). This indicates that mixing of N into the envelope has been underestimated for low mass stars (see Iben and Renzini 1982).

Pagel (1978) has shown that in an exponential disk the yield of primary elements, p, is proportional to the gradient of heavy elements. Assuming that O is a constant fraction of the heavy elements abundance, Z, Peimbert and Serrano (1980) obtained, from the gradient of O/H apparent in Figure 2, a yield p= 0.008. This value of p is intermediate between that of metal poor galaxies, p= 0.003, and those of galactic H II regions, p= 0.01, or of the metal rich galaxy M83 (p= 0.014). This fact led Peimbert and Serrano (1982) to suggest that p increases with metallicity and that this is due to a decrease of the amount of low mass stars ($M< 1\ M_{\odot}$) with Z.

It must be mentioned, however, that Kaler (1980) does not find evidence for a galactic gradient of O/H in PN. He divides PN into age groups and finds that the mean O/H for each group increases with decreasing age, revealing the oxygen enrichment of the galaxy from the formation of the early halo to the present.

3. HELIUM PRODUCTION

A long standing problem in the chemical evolution of galaxies has been the observed helium to heavy elements enrichment ratio by mass, $\Delta Y/\Delta Z=3\pm1$ (Serrano and Peimbert 1981a, and references therein). Hacyan *et al.* (1976) produce chemical evolution models of the solar neighborhood without being able to fit $\Delta Y/\Delta Z$ and p at the same time.

Helium production is considerably increased when PN and their progenitor stars are taken into account. In Figure 3, He production of a

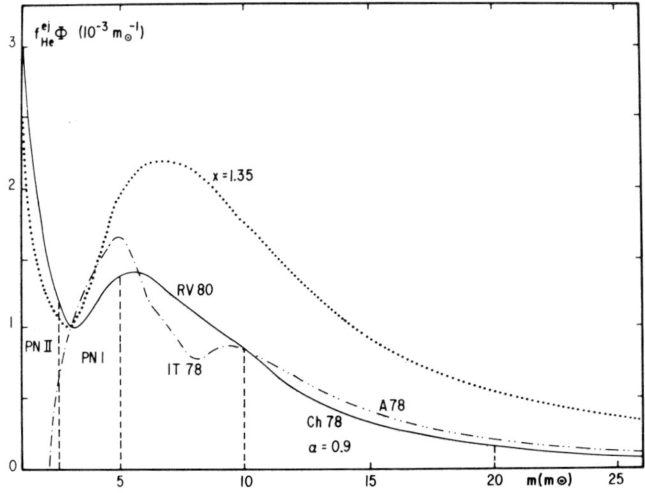

Fig. 3. Weighted fraction of the mass of the star contributing to the He production, per unit stellar mass.

TABLE 1

$\Delta Y/\Delta Z$ FOR DIFFERENT ASSUMPTIONS ABOUT MASS LOSS

Models adopted for mass loss		Properties at G= 0.1		Maximum $\Delta Y/\Delta Z$
Intermediate mass stars	Massive stars (α)	ΔY	t(10^9 y)	
RV80 (Std.)	0.9	0.032	12.19	3.06
"	0.8	0.032	"	2.28
"	0.0	0.031	"	1.01
IT78	0.9	0.026	12.22	2.47
no	0.9	0.013	"	1.23

RV80= Renzini and Voli 1980; IT78= Iben and Truran 1978; α of Chiosi *et al.* 1978.

generation of stars is shown. In is clear from this figure (taken from Serrano and Peimbert 1981a) that intermediate mass stars contribute with 2/3 of the newly formed He ejected per generation of stars. PN of Type I contribute with 1/6 of the total He production, the same as those of Type II. In Table 1, $\Delta Y/\Delta Z$ is shown for different assumptions about mass loss in intermediate and in massive stars. Chiosi and Matteucci (1982) have also constructed detailed models of the chemical evolution of the solar neighborhood in which $\Delta Y/\Delta Z$ is consistent with the observed value. It must be stressed that not only the third dredge-up but also envelope burning (as in Renzini and Voli 1981) is necessary to account for the observed $\Delta Y/\Delta Z$.

4. CARBON PRODUCTION

Observed C abundance in PN seem to agree with the predictions of evolution models of intermediate mass stars (Figure 4). Exceptions are the halo PN, and two Type I PN. The halo PN seem to have larger C/O values than those predicted by stellar models of metal poor stars (Peimbert 1981; Torres-Peimbert et al. 1981). The opposite effect is shown by the type I PN NGC 6853 and, particularly, NGC 6302 (not shown in Fig. 4); these planetaries have relatively low C abundance but high He/H.

From Figure 4 it would seem that models with envelope burning ($\alpha \neq 0$) produce abundances nearer to the observed ones than models without it. If case B is preferred, as other arguments suggest, then models with higher mass loss rates reproduce better the observed abundances. PN of Types II and III have lower He abundances than the models, but this is probably due to the model initial abundances.

A further evidence in this direction comes from a comparison of the C/O values produced by models with varying α and the observed C/O $\sim$ 0.58 in the solar neighborhood. As it is shown in Figure 5 and Table 2, there

TABLE 2

MODEL C/O FOR DIFFERENT VALUES OF α

α = ℓ/Hp	C/O*
0	0.95
1	0.95
1.5	0.71
2	0.53

* Total C/O ratio by number. The contribution of C(>8)/O is 0.25 in these units.

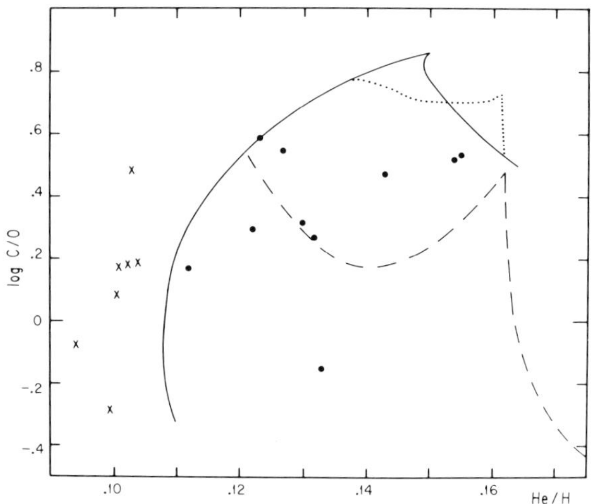

Fig. 4. Model surface abundances at the time of PN ejection (from Renzini and Voli 1981) are represented for a model with α= 0 and case A by a continuous line. Also shown models with α= 2 represented by dashed and dotted lines for cases A and B respectively. Case B corresponds to lower mass planetaries and α measures the importance of envelope burning. Observed points for PN of Types II and III (X) are taken from Peimbert 1981, and those of Type I PN(.) from Peimbert and Torres-Peimbert (1982).

is galactic overproduction of C unless envelope burning is effective,i.e. α >2 (Serrano and Peimbert 1981b). Moreover, as shown in Table 3, the lower mass limit for PN I formation is consistent with the discussion in §1, only if α >2.

5. NITROGEN PRODUCTION

To study the galactic enrichment of N it is necessary to know which stars are responsible for it and if the production is due to primary or secondary mechanisms.

TABLE 3

MINIMUM MASS REQUIRED TO PRODUCE PLANETARY NEBULAE OF TYPE I*

$a = \ell/Hp$	m_{min} ($M_\odot$)
0	...
1	7.4
1.5	4.3
2	3.4

*From models by Renzini and Voli (1981)

PN of Types II and III show moderate production while PN of Type I show much larger overabundances of N (Peimbert 1978). Primary production of N, if present, should be easiest to achieve in PN of Type I. However, Serrano and Peimbert (1982) have shown that the N/O vs O/H diagram for external galaxies can only be explained if N is mainly a secondary product of the low mass stars (nevertheless, for a different interpretation, see Edmunds and Pagel, 1978).

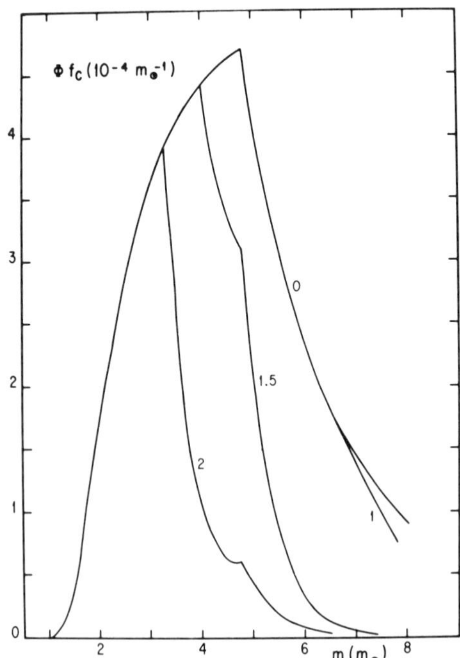

Fig. 5. Weighted fraction of the newly formed carbon, $f_c\Phi$, as a function of initial stellar mass. The area under the curve gives the fraction of an average stellar mass, in a generation of stars, which is ejected as newly formed carbon, in the $1 \leq m/M_\odot \leq 8$ range, $<f_c(1-8)>$.

It is a pleasure to acknowledge a critical reading of the manuscript of this paper by R. Costero.

REFERENCES

Aller, L.J.: 1976, Publ. Astron. Soc. Pacific 88, 574.
Barker, T.: 1978, Astrophys. J. 220, 193.
Cahn, J.H. and Kaler, J.B.: 1971, Astrophys. J. Suppl. 22, 319.
Calvet, N. and Peimbert, M.: 1982, Rev. Mexicana Astron. Astrof., submitted.
Chiosi, C., Nasi, E., and Sreenivasan, S.R.: 1978, Astron. Astrophys. 63, 103.
Chiosi, C., and Matteucci, F.M: 1982, Astron. Astrophys. 105, 140.
Cudworth, K.M.: 1974, Astrophys. J. 79, 1384.
D'Odorico, S., Peimbert, M., and Sabbadin, F.: 1976, Astron. Astrophys. 47, 341.
Edmunds, M.G. and Pagel, B.E.J.: 1978, Monthly Notices Roy. Astron. Soc. 185, 77p.
Greig, W.E.: 1971, Astron. Astrophys. 10, 161.
Hacyan, S.,Dultzin-Hacyan, D., Torres-Peimbert, S., and Peimbert, M.: 1976, Rev. Mexicana Astron. Astrof. 1, 355.
Iben Jr. I. and Renzini, A.: 1982, preprint (to appear in Ann. Rev. Astron. Astrophys.).
Iben Jr. I. and Truran, J.W.: 1978, Astrophys. J. 220, 980.
Kaler, J.B.: 1980, Astrophys. J. 239, 78.
Méndez, R.H. and Niemela, V.S.: 1981, Astrophys. J. 250, 240.
Pagel, B.E.J.: 1978, paper presented at the Fourth European Regional Meeting in Astron., Uppsala.
Peimbert, M.: 1978, in Y. Terzian (ed.), "Planetary Nebula: Observations and Theory, IAU Symp. No. 76", Dordrecht: Reidel, p. 215.
Peimbert, M.: 1981, in I. Iben Jr. and A. Renzini (eds.), "Physical Processes in Red Giants", Dordrecht: Reidel, p. 409.
Peimbert, M. and Serrano, A.: 1980, Rev. Mexicana Astron. Astrof. 5, 9.
Peimbert, M. and Serrano, A.: 1982, Monthly Notices Roy. Astron. Soc. 198, 563.
Reimers, D.: 1975, Mém. Soc. Roy. Sci. Liege, 6o serie 8, 369.
Renzini, A.: 1981, in I. Iben Jr. and A. Renzini (eds.), "Physical Processes in Red Giants", Dordrecht: Reidel.
Renzini, A. and Voli, M.: 1981, Astron. Astrophys. 94, 175.
Romanishin, W. and Angel, J.R.P.: 1980, Astrophys. J. 235, 992.
Serrano, A.: 1978, Ph. D. Thesis, University of Sussex.
Serrano, A. and Peimbert, M.: 1981a, Rev. Mexicana Astron. Astrof. 5,109.
Serrano, A. and Peimbert, M.: 1981b, Rev. Mexicana Astron. Astrof. 6, 41.
Tinsley, B.M.: 1978, in Y. Terzian (ed.), "Planetary Nebulae: Observations and Theory, IAU Symp. No. 76", Dordrecht: Reidel, p. 341.
Torres-Peimbert, S. and Peimbert, M.: 1977, Rev. Mexicana Astron. Astrof. 2, 181.
Torres-Peimbert, S., Rayo, J.F.,and Peimbert, M.: 1981, Rev. Mexicana Astron. Astrof. 6, 315.
Van den Heuvel, E.P.J.: 1975, Astrophys. J. (Letters) 196, L121.
Wood, P.R. and Cahn, J.H.: 1977, Astrophys. J. 211, 499.

KWOK: A more realistic value for η is 2.5, rather than 0.3.
The fact that mass-loss from AGB stars is the major contributor to the ISM was recognized by Woolf as early as 1971.

SERRANO: I would agree that η should be larger than 0.3, but smaller than 2.5. Regarding mass-loss from AGB stars, the current (unsolved) problem is to determine the composition of the mass which is lost in order to compare with the observed abundances in the ISM.

PLANETARY NEBULAE AND SEYFERT GALAXIES - SIMILARITIES AND DIFFERENCES

Donald E. Osterbrock
Lick Observatory, Board of Studies in Astronomy and Astrophysics
University of California, Santa Cruz

ABSTRACT. Knowledge gained in the study of planetary nebulae has been, and can be further, transferred to understanding active galactic nuclei. Photoionization is the main energy-input mechanism in the narrow-line regions, and probably although by no means certainly in the broad-line regions as well. There are many detailed differences because of the much "harder" input spectrum in active galactic nuclei, compared with planetary nebulae. A tentative model of the structure of the gas distribution in a Seyfert-galaxy nucleus is presented.

The concept of Seyfert galaxies dates back to the paper in which Seyfert (1943) stated that a very small proportion of galaxies have spectra showing many high-ionization emission lines localized in their nuclei. These emission features are similar to those found in planetary nebulae, as Seyfert reported. Invariably, these galaxies contain very luminous nuclei. Of the workers before Seyfert, Hubble (1926) had particularly remarked on the planetary-nebula-like emission-line spectra of three of them, NGC 1068, 4051, and 4151. Thus the earliest spectroscopic observers already noted the close similarity between the emission-line spectra of Seyfert galaxies and planetary nebulae. Many more galaxies, indeed at some level or other, nearly *all* spiral and irregular galaxies, have in their spectra the generally lower ionization-level emission lines of H II regions.

Since planetary nebulae, and the mechanisms by which their emission-line spectra are formed are fairly well understood (although not completely so), it is clearly advantageous to use the knowledge gained from these objects in analyzing and interpreting Seyfert galaxies. In particular, there are many planetary nebulae in our Galaxy, some of them quite close to us, which can easily be resolved, while very few Seyfert galaxies are close enough for us to get any resolved optical information on their emission-line emitting nuclei. Thus planetary nebulae may be regarded not only as interesting objects in themselves, but as guides and test objects to understanding the structure of Seyfert galaxy nuclei.

Clearly not only the similarities must be studied, but the differences as well. One obvious dissimilarity between the spectra of

D. R. Flower (ed.), Planetary Nebulae, 473–486.

Seyfert galaxies and planetary nebulae is in the widths of their emission lines. Part of the standard definition of a Seyfert galaxy is that the emission lines be strong and broad (Weedman 1977); actually this means in practice noticeably broader than in typical galaxies, whose line widths are often approximately 300 km/sec full width at half maximum (FWHM). Planetary nebulae of course have much smaller line widths, typically 50 km/sec FWHM, resulting from expansion (Wilson 1948).

Seyfert galaxies may be classified into two types from their spectra. Seyfert 1 galaxies are those in which the H I emission lines are noticeably broader than the forbidden lines, such as [O III] $\lambda\lambda$4959, 5007, while Seyfert 2 galaxies are those in which the H I and forbidden emission lines have similar widths (Khachikian and Weedman 1974). Radio galaxies, the optical counterparts of strong radio sources, have bright nuclei and emission-line spectra similar in many ways to Seyfert galaxies (Baade and Minkowski 1975), although they are much rarer per unit volume of space. They also can be divided into two groups, broad-line radio galaxies, (BLRG), with H I emission lines noticeably broader than the forbidden lines, as in Seyfert 1 galaxies, and narrow-line radio galaxies (NLRG), with H I and forbidden emission lines with similar widths, as in Seyfert 2 galaxies (Osterbrock, Koski and Phillips 1976). Seyfert galaxies and radio galaxies together are usually referred to as active galaxies. QSO's (quasistellar objects), and quasars (quasistellar radio sources) are clearly similar to them in many ways and are often included within the class of active galactic nuclei.

The FWHM of the emission lines in Seyfert 2 and NLRG cover a broad range, approximately 300 to 1200 km/sec, with a typical value of about 500 or 600 km/sec. There is no systematic difference between the widths in the two groups. Note that NGC 1068, which has a FWHM 1200 ± 150 km/sec, is often cited as a typical Seyfert 2, but actually is no more so than NGC 7027 is a typical planetary nebula (Koski 1978; Cohen and Osterbrock 1981; Shuder and Osterbrock 1981).

In Seyfert 2 and NLRG the measured H I Balmer-line ratios are generally steeper than predicted by recombination theory. However, just as in planetary nebulae, the differences between the observed and recombination gradients can generally be ascribed to interstellar reddening. In well observed objects several different H I line ratios generally give approximately the same amount of extinction (Osterbrock and Miller 1975; Koski 1978; Cohen and Osterbrock 1981). However, very recent measurements of [O II] and [S II] line ratios, comparing lines in the violet and red spectral regions, yield on the average somewhat lower amounts of extinction (Malkan 1982). If the interpretation is correct, they suggest that $H\alpha/H\beta = 3.6 \pm 0.3$, somewhat larger than the recombination value 2.9. The method of deriving the amount of extinction is based on the assumption that the effective electron densities and temperatures in the [O II] and [S II] are the same, which is questionable, but a range of model calculations, of the type described below, could be substituted for this assumption. Another assumption in all the reddening calculations is that the wavelength dependence of the extinction in active galactic nuclei is the same as in our Galaxy. This is by no means obvious, but probably is not too serious when the results are used only to interpolate the extinction as a function of wavelength.

It seems well established that the ionization mechanism in the Seyfert 2 and NLRG is photoionization by a "hard" spectrum, extending to high energies (Collin-Souffrin 1978). The most convincing evidence for the conclusion that photoionization is important is that the electron temperature T calculated from [O III] and [N II] emission-line ratios is in many cases of order 1-2 x 10^4 K. Particularly for [O III], this is much smaller than the temperature expected under conditions of "collisional ionization", that is, direct input of mechanical energy which is converted into heat, and thermal ionization under relatively low-density conditions. On the other hand, temperatures of this order are expected to result from photoionization over a wide range of input spectra, because of the strong thermostatic effect of cooling by collisionally excited line radiation. No star or mixture of stars could give the wide range of ionization, from [O I] and [S II] to [Ne V] and [Fe VII] observed in many Seyfert 2 and NLRG. On the other hand, a power-law, featureless-continuum spectrum, extrapolated from observational data in the optical region to the extreme ultraviolet, will give just this type of spectrum. Often the featureless continuum, which must be determined by decomposing the observed continuous spectrum into normal-galaxy and power law components, can be represented as $F_\nu \propto \nu^{-\alpha}$, with $\alpha \approx 1$. Qualitatively, the high-energy photons of such a power-law continuum, beyond the exponential tails of the continuous spectra of even the hottest known stars, will produce highly ionized species close to the source, as well as a long partially ionized zone in which [O I] and [S II] will be the strongest optical lines emitted (Mitton 1972). Furthermore, the number of ionizing photons in such an extrapolated power-law spectrum is sufficient, in all well observed cases, to balance the total number of recombinations indicated by the H I emission lines (Osterbrock and Miller 1975; Costero and Osterbrock 1977; Koski 1878).

Calculated photoionization models, with input spectra of the form $L_\nu \propto \nu^{-\alpha}$, with $\alpha \approx 1$, do approximately agree with the observed emission-line spectrum of the Crab Nebula, known from the work of Woltjer (1958) to be photoionized by a synchrotron power-law continuum, agrees well with the emission-line spectra of Seyfert 2 and NLRG, except for the He I and He II lines. They are stronger in the Crab Nebula because of the high He abundance in this object. An even better match to the active-galactic-nuclei spectra is provided by a linear combination of the spectra of the Crab Nebula and of the planetary nebula NGC 7027 (Koski 1978). Since NGC 7027 has a very hot central star, this indicates that the photoionization continua in Seyfert 2 and NLRG probably turn down somewhat at high photon energies.

Seyfert 1 and BLRG have broad H I, He I and He II emission lines with FWHM in the range 1 - 7 x 10^3 km/sec, and with full widths at zero intensity (FWOI) ranging up to 2.8 x 10^4 km/sec. Here it is not so clear what the energy input to the ionized gas is. There are essentially no diagnostics that give information on T, or on electron density except that it is so high that all forbidden lines are collisionally deexcited, requiring $N_e \gtrsim 10^8$ cm^{-3}. A few Seyfert 1 galaxies observed with the IUE (Wu, Boggess and Gull 1982), and many high redshift quasars and QSOs show broad C III] λ1909 emission. This requires $N_e \lesssim 10^{10}$ cm^{-3} in the broad-line regions of these objects, and presumably of all Seyfert 1 and BLRG.

The best working hypothesis is that the main ionization source in the broad-line regions (BLR) is also photoionization. This is based on the linear relationship over several powers of ten between the luminosity in the Hα or Hβ emission lines, summed over the total profile, broad plus narrow components (Yee 1980; Shuder 1981; see also Osterbrock 1978). This is just the relationship expected from photoionization by an ultraviolet spectrum that is the extension, with more or less the same power law, of the observed optical featureless continuum (Searle and Sargent 1968). The photoionization interpretation is not required by this observation; any source of ionization that is directly proportional to the optical featureless continuum would also satisfy the data. Photoionization is simply the most straightforward explanation. It requires that the ratio of the number of ionizing photons emitted by the source to the number of Hβ or Hα photons emitted by the ionizing gas be essentially constant.

For pure recombination, this means that the "covering factor" or fraction of ionizing photons observed by the gas be constant from the Seyfert 2 and NLRG at lower luminosities, through Seyfert 1 and BLRG at intermediate luminosity, to low redshift QSOs and quasars that are the highest luminosity objects in which Hβ or Hα has been measured and compared quantitatively with the optical featureless continuum. Furthermore, for a covering factor $\Omega/4\pi = 1$, the exponent in the power law $L_\nu \propto \nu^{-\alpha}$ must be $\alpha \gtrsim 1$; this can be accepted for the Seyfert and radio galaxies but for high-luminosity QSOs and quasars, in which the Lyman limit at λ912 is directly observable, the data suggest $\Omega/4\pi \approx 0.3$ is more likely (Yee 1980). This would require an index $\alpha \gtrsim 0.6$. If this value of the exponent were used for all the objects, it would suggest that the luminosities in the featureless continuum of the Seyfert galaxies have been undercorrected for reddening by dust. An alternate interpretation would be that $\Omega/4\pi$ varies smoothly from ~ 1 at low luminosities to ~ 0.3 at high luminosities, and that the number of Hα or Hβ photons emitted per photoionization varies smoothly in just such a way as to compensate the change in covering factor. The third interpretation of course is that processes other than photoionization also transfer energy from the source to the gas, and lead ultimately to emission of H I line photons, in just such a way as to mimic photoionization. Understanding this puzzle is a necessity to understanding active galactic nuclei, QSOs, and quasars.

The Balmer decrements of the broad H I lines in BLRG and Seyfert 1 galaxies cannot be matched by recombination plus reddening alone. At the high densities of the BLR, Lα excapes only slowly and both the 2^2S and 2^2P terms of H I are populated; both collisional excitation and radiative-transfer effects involving these levels are therefore expected to occur and thus to modify the relative intensities of the lower Balmer lines (Netzer 1975; Osterbrock, Koski and Phillips 1975, 1976). Yet, although there is a scatter about the reddening line considerably larger than the observational errors, the broad-line

Balmer decrements on the average approximately follow it, indicating that dust also does play an important role (Osterbrock 1977).

In QSOs and quasars (Baldwin 1977) and in several Seyfert 1 galaxies observed with the IUE (Wu, Boggess and Gull 1980), the measured Lα/Hβ ratios are much different from the recombination values. These measurements provide extremely important additional information for sorting out the effects of optical depths in the lines, collisional excitation, and dust on the emission processes in the BLR. Infrared measurements of H I provide still further information (Soifer *et al.* 1981). Theoretical treatments taking all these effects into account are inevitably very complicated and involve massive calculations, even though the physical situation assumed is necessarily highly simplified (Davidson and Netzer 1979; Collin-Souffrin, Dumont and Tully 1982). Yet surely, in the end, they will be necessary for understanding the BLR. Dust clearly may be very important, not only in modifying the emergent radiation from the BLR, but also by its effects in scattering and absolving ionizing and line radiation within that radiation. It should not be omitted from the calculations for supposed "simplicity" (Osterbrock 1979).

Many Seyfert galaxies have intermediate-type spectra, with H I line profiles combining strong broad and narrow components. These objects are often referred to as Seyfert 1.5 galaxies, to indicate they combine with characteristics of Seyfert 1's and 2's. Examples can be found with almost any relative strengths of broad and narrow components (Osterbrock and Koski 1976; Osterbrock 1977). They are included in the very good proportionality relationship between featureless continuum and Hα or Hβ luminosities if the total, broad plus narrow, H I profiles are measured. Most of the BLRG have Balmer-line profiles with strong narrow-line components, more nearly similar to Seyfert 1.5 galaxies than to Seyfert 1's (Grandi and Osterbrock 1978).

The narrow emission-line spectra of Seyfert 1 galaxies are very similar to those of Seyfert 2 and NLRG. The equivalent widths of the narrow lines in the Seyfert 1's are relatively smaller, however, indicating that not as large a fraction of the ionizing radiation reaches the narrow-line region (NLR) (Osterbrock 1978). Also, detailed study shows that in many cases the [O III] (λ4959 + λ5007)/λ4363 ratio is relatively small, indicating either a relatively high T in the NLR, or alternatively if $T \approx 1\text{-}2 \times 10^4$, a relatively high N_e ($N_e \approx 10^6 - 10^7$ cm^{-3}). Because of the continuity of the narrow-line spectra, and the fact that many of these objects and many more of the Seyfert 2 and NLRG have [O III] ratios that do indicate $T \approx 1\text{-}2 \times 10^4$ and $N_e \lesssim 10^4$, it has been argued that photoionization is the main energy-input mechanism to the NLR in all these objects, and that therefore $T \approx 1\text{-}2 \times 10^4$ in all of them. On this interpretation the objects with relatively small (λ4959 + λ5007)/λ4363 are objects with relatively high N_e in their NLR. However, from comparisions of [O III] and [Ne III] emission-line strengths, Heckman and Balick (1979) have suggested that T instead may be higher in the objects with relatively strong λ4363. Some observational evidence to support this conclusion has been found by Cohen (1981) from additional spectrophotometric measurements of intermediate-type Seyfert galaxies.

The indicated higher temperatures in the NLR of many Seyfert 1, 1.5 and BLRG suggests that an additional heating mechanism is effective in them. It may be the radio plasma which, in the BLRG, is streaming out from the nuclei to the radio-emitting lobes, generally symmetrically placed far outside the optical galaxy, at opposite ends of the axis of rotation (see e.g. Miley 1980). There are many indications that Seyfert 1 and 1.5 galaxies are objects in which similar plasma is generated at the nucleus but does not get outside the galaxy to the lobes, perhaps because it is ejected in directions along which it encounters more ambient gas and is slowed down or stopped (Ulvestad, Wilson and Sramek 1981). If the ultraviolet radiation field is sufficiently intense, the main ionization process will be photoionization, but additional heat will be delivered to the ionized gas by the frictional slowing down or stopping of the ionized gas. This will tend to raise its temperature without significantly raising the level of ionization. Theoretical work in this direction would be of great interest in exploring this possible difference between Seyfert galaxies and planetary nebulae.

As to the level of ionization, [Ne V] and [Fe VII] are observed in many Seyfert 1 and 1.5 galaxies, but few Seyfert 2's. [Ne V] is observed in several planetary nebulae, and [Fe VII] only weakly in a few. The energies required to produce these ions are 97 and 99 ev respectively. This suggests there may be a cut-off or turndown of the ionizing continuum in Seyfert 2 galaxies around 100 ev (Cohen 1981). [Fe X] and [Fe XI] are observed in a significant number of Seyfert 1 and 1.5 galaxies and [Fe XIV] in the unusual high-ionization Seyfert 1 object III Zw 77. They are not observed in any planetary nebulae. The ionization energies to produce them are 234, 262, and 361 ev respectively. Simplified calculations show that all these observed lines can be produced in photoionization models with power-law sources, and the observed correlations suggest, but do not prove, that they are formed in this way (Grandi 1978; Osterbrock 1981a). The line widths are larger for the high-ionization lines, and since they must be formed close to the source of ionizing photons, this suggests that gravitational accelerations are important (Wilson 1979, Osterbrock 1981a).

The [Fe X] and [Fe XI] emission lines indicate ionizing photons at least up to 300 ev in many Seyfert galaxies, and [Fe XIV] up to at least 500 ev in III Zw 77. These are X-ray energies. Within the past decade measurements made in space have shown that every Seyfert 1 or 1.5 galaxy is an X-ray source, and that nearly every galaxy observed with X-ray luminosity above some threshhold is a Seyfert 1 or 1.5 (see e. g. Culhane 1978). There is a very good correlation between the luminosity in the <u>broad</u> component of Hα and the X-ray luminosity. "Pure" Seyfert 2 galaxies are weak or non-existent X-ray emitters; a few Seyfert 2 or narrow emission-line galaxies identified as X-ray sources turned out on close inspection with high-quality data to have very weak but definitely present broad components of Hα emission (Shuder 1980, Veron <u>et al</u>. 1980). There is not a good correlation between the X-ray luminosity of a galaxy and the strength of the high-ionization optical lines [Fe X], [Fe XI], and [Fe XIV]. Perhaps this is because when photoionization occurs it destroys the ionizing photon.

Very high quality X-ray measurements are now becoming available for many Seyfert 1 and 1.5 galaxies, and their interpretation may be expected to provide good information on the geometrical structure, orientation and covering factor in these objects, particularly in the BLR (Lawrence and Elvis 1972; Mushotzky 1982; Maccacaro, Perola and Elvis 1982).

Abundances of the elements cannot be measured in Seyfert galaxies with the precision possible in planetary nebulae, because we do not yet know the detailed structure of the galactic nuclei, and therefore cannot calculate accurate models for them. Standard diagnostic methods, as used in the early planetary-nebula work, show the abundances are approximately "normal" in Seyfert and radio galaxies. The most readily apparent difference between the spectra of Seyfert galaxies and planetary nebulae is that in some of the former the [Fe VII] lines are relatively strong, while in the latter they are invariably weak, as mentioned above. This can be traced to the fact that the Fe abundance is approximately "normal" in Seyfert galaxies, while it is considerably subabundant in the gas in planetaries. The straightforward interpretation is that the missing Fe is locked up in solid components in dust particles in planetaries, but not in Seyferts (Shields 1975). Quite recently Gaskell, Shields and Wampler (1981) have found evidence that Fe, Si, Al, Mg and C all have normal abundances in quasars, and have argued that since none of these elements are depleted as in planetary nebulae and the interstellar medium, there is no dust in quasars or in the BLR of active galactic nuclei, which they take to be physically similar. Since the same reasoning would seem to show that there is no doubt in the NLR either, from the normal abundance of Fe, while there is strong evidence that dust is there and weak suggestions of dust in the BLR as well, the question must be regarded as still open.

All the active galaxies have detectable featureless continua, which can be found by decomposing the observed spectrum into emission lines plus a normal-galaxy spectrum with absorption lines plus a power-law continuum (Koski 1978; Yee and Oke 1978; Shuder 1981). In Seyfert 1 and BLRG the featureless continuum is usually very strong. Typically the fraction of the continuous spectrum near Hβ in the featureless continuum is $f_{FC} \approx 0.9$, or in many cases undetectably different from 1. For typical Seyfert 2 and NLRG on the other hand, $f_{FC} \approx 0.1$ to 0.4. Thus there is a very strong correlation between the presence of broad emission-line components, and the presence of a strong featureless continuum. This correlation suggests that the physical mechanism that produces the high velocities observed in the BLR is intimately connected with the mechanism that produces the featureless continuum. Cyg A is the outstanding exception to this correlation; recent high-quality spectral scans show it has $f_{FC} \approx 0.6$, above the upper limit of previously observed NLRG (Osterbrock 1982). In its radio properties also, Cyg A is more similar to the N galaxies that are BLRG, than to the cD and E galaxies that are NLRG like itself (Grandi and Osterbrock 1978).

An interesting group of objects, recognized only relatively recently, are Seyfert galaxies with weak broad Hα emission components, and in some cases very weak broad Hβ emission, combined with fairly strong narrow emission-like spectra. All of them have fairly small f_{FC}. Since these properties are close to but not identical with those of Seyfert 2's, I have called them Seyfert 1.8 or 1.9 galaxies (Osterbrock 1981b). The broad-line component Hα/Hβ ratios are very large in all these objects, suggesting that extinction by dust is important in the BLR. However the galaxies in which they are found are not, as a group, seen nearly edge-on, as is, for instance NGC 4235, a Seyfert 1 galaxy that shows the effects of strong extinction by dust in its plane (Abell, Eastmond and Jenner 1978). In addition to the five published examples of Seyfert 1.8 or 1.9 galaxies (Osterbrock 1981b), more recently classified objects of this type are Mrk 728, 744, 766, 1179, and 1218.

A very striking difference between the spectra of Seyfert 1 and BLRG is that the former usually show strong Fe II emission, while the latter almost invariably do not (Osterbrock 1977, Grandi and Osterbrock 1978). The Fe II profiles are either the same as the H I broad-line profiles or slightly narrower than them; the Fe II emission lines are clearly associated with the BLR but perhaps are weighted slightly more strongly toward the lower-velocity parts of it (Phillips 1977, 1978a). Analysis of the optical Fe II emission multiplets alone seemed to favor slightly resonance-fluorescence as their excitation mechanism (Phillips 1978b). However, more recent comparisons of the ultraviolet, higher-excitation multiplets with the optical ones have greatly strengthened the idea that the Fe II emission arises by collisional excitation in heated, mostly neutral regions (Collin-Souffrin _et al_. 1979, 1980; Wills, Netzer and Wills 1980; Grandi 1981).

The great majority of galaxies that show emission lines in their spectra are objects in which the photoionization source is early-type stars. At a low enough level, essentially every late-type spiral galaxy and irregular has emission lines of this type. Many of these objects have O-type stars even in their nuclei, which are thus giant H II regions. Some of them have so many stars that the nebular emission lines are very strong, and radio and X-ray emission can be observed from the integrated effect of the supernova remnants in them. These are the "star-burst" nuclei of Weedman _et al_. (1981). Most early-type spiral galaxies also have nuclear emission lines; in general they have larger [N II]/Hα and [S II]/Hα ratios than the H II region galaxies. It has been suggested that the interstellar gas in these early-type spirals is collisionally ("shock") heated (Heckman 1980). However, further observational and theoretical study makes it seem much more likely that in fact the gas is photoionized by a featureless, power-law continuum, with a much lower luminosity than in typical Seyfert galaxies, and thus essentially undetectable in the integrated-galaxy optical-continuous spectrum (Keel 1982).

In the active galaxies, the line profiles are broader than in nearly all such narrow emission-line galaxies. Still, the narrow-line profiles of active galactic nuclei are barely resolvable with most scanner data obtained to date, and only the line widths have been determined (e.g. Koski 1978; Shuder and Osterbrock 1981; Feldman _et al_.

1982). Recently, considerably higher-resolution measurements have revealed that the narrow lines are characteristically asymmetric, with a sharper fall-off to the red than to the blue. An interpretation in terms of extinction by dust in the NLR, with the gas flowing radially outward, is suggested by these profiles (Heckman _et al_. 1981). Several other groups are actively making similar profile measurements, and it is clear that they will add greatly to our understanding of active galactic nuclei.

Likewise an atlas of broad-line profiles in Seyfert 1 galaxies has recently been published by Osterbrock and Shuder (1982). They exhibit a very great range of widths, and are symmetric in some cases, asymmetric to the red in others, and to the blue in still others. Although it is difficult to draw any but the most general information from the profiles alone, they will clearly be extremely useful in testing any models that are calculated from definite physical pictures, as for instance Capriotti, Foltz and Byard (1979) or Raine and Smith (1981)..

Perhaps the most widely adopted basic idea is that the observed high velocities result from radial, radiation-pressure driven flows (e.g. Blumenthal and Mathews 1975). An alternate picture is that the emitting regions are flattened, perhaps by rotation (Shields 1977; Osterbrock 1978). Recent very high angular-resolution measurements with the VLA of some of the nearest Seyfert galaxies show that the regions of weakly radio-emitting plasma in these objects, which appear to be closely associated with the NLR, have disk-like distributions associated with the planes of the galaxies. In some cases, there appear on the VLA maps two small, oppositely directed "jets", which lie along a line that is not the same as the rotation axis, but rather is inclined to it. This suggests that there may be an inner, small disk-like structure, with size of order of 1 pc, whose rotation axis is tipped with respect to that of the rest of the galaxy (Wilson and Willis 1980; Ulvestad, Wilson and Sramek 1981). It is natural to associate this small region with the BLR, and Tohline and Osterbrock (1982) have given theoretical and observational evidence that supports such a tipped structure.

A pictorial representation of such a model is shown in Figure 1. The tipped BLR is shown crosshatched; at its center is the photoionization source, perhaps an accretion disk around a black hole. The BLR is supposed to be optically thick to ionizing radiation along nearly all rays in its equatorial plane, but not at its poles, so that ionizing photons mostly escape only within a cone about this axis, which is also the axis along which radio plasma can escape. In a nearly "pure" Seyfert 1 galaxy the optical depth of the BLR is large and the cone shrinks to an angle of nearly 0°; in the opposite extreme the disk is optically thin in all directions, the object is a nearly pure Seyfert 2, and the cone opens out to 90°. The NLR is composed of density condensations, each of which is mostly highly ionized on the face toward the photoionization source. Each condensation is optically thick, and is only slightly ionized on the side facing away from the source. The most distant condensations are not as highly ionized at their front faces as those nearer the source. The BLR has a similar structure, consisting of condensations also.

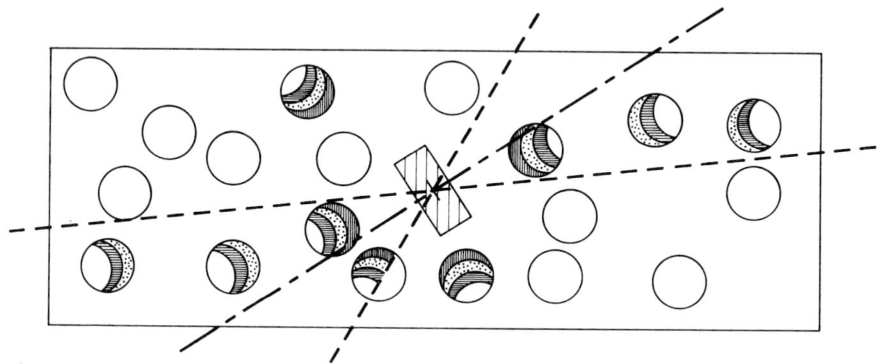

Fig. 1. Schematic drawing of model for gas distribution in active galactic nuclei.

The drawing is an extreme simplification; there are many more condensations than shown, and the fraction of them that are struck by ionizing radiation varies smoothly with angle. The only slightly ionized regions of the dense condensations in the BLR are the Fe II emitting regions; presumably they are heated by plasma. In the BLRG, on the other hand, the observations suggest that the flow is along the rotation axis out into the radio lobes, and perhaps better collimation in this situation leads to less of the neutral parts of the condensations being heated in this way. On this picture the Seyfert 1.8 and 1.9 nuclei might be understood as cases in which the tipped BLR is seen nearly edge on, so that there is strong extinction in its outer parts, but the main galaxy is not, so the NLR has on the average no more extinction than in typical Seyferts.

The entire model is highly schematic. Much further work will be necessary to make it physically definite, and to test it. There are clearly additional complications - for instance, spectropolarimetric measurements show that although in NGC 4151 the forbidden lines and the narrow components of the H I lines have the same polarization (Schmidt and Miller 1980), in NGC 1068 the forbidden lines and permitted lines have different polarizations, indicating that they cannot arise in identical regions (Angel _et al_. 1976). NGC 4151 is a fairly typical Seyfert 1.5; NGC 1068 is an abnormal Seyfert 2. Clearly many problems remain, but one may hope by combining data from all spectral regions either to reject this type of model, and pass on to a better one, or to find that it meets, at least for a time, all the observational tests. The ultimate aim is to understand the structure and physical nature of all the emitting regions, and of the mechanisms by which energy is "generated" (assembled and released) almost certainly by gravitational processes. The ionizing radiation emitted by the central source, perhaps as a function of position and angle, must be the input spectrum that leads to the observed emission-line spectrum.

I am very grateful to my collaborators and colleagues at Lick Observatory - G. R. Blumenthal, W. G. Mathews, J. S. Miller, J. M. Shuder, and O. Dahari, and to long-term visitors such as G. A. Shields and G. J. Ferland for many stimulating and helpful discussions on the subjects treated in this paper. I am also grateful to the National Science Foundation for continued support of my research, most recently under grant AST 79-19227.

REFERENCES

Abell, G. O., Eastmond, T. S. and Jenner, D. C.: 1978, Ap. J. 221, L1.
Angel, J. R. P., Stockman, H. S., Woolf, N. J., Beaver, E. A. and Martin, P. G.: 1976, Ap. J. 206, L5.
Baade, W. and Minkowski, R.: 1954, Ap. J. 119, 206.
Baldwin, J. A.: 1977, M.N.R.A.S. 178, 67P.
Blumenthal, G. R. and Mathews, W. G.: 1975, Ap. J. 198, 517.
Capriotti, E., Foltz, C. and Byard, P.: 1979, Ap. J., 230, 681.
Cohen, R. D.: 1981, UCSC Ph.D. Thesis.
Cohen, R. D. and Osterbrock, D. E.: 1981, Ap. J. 243, 81.
Collin-Souffrin, S.: 1978, Physica Scripta 17, 293.
Collin-Souffrin, S., Dumont, S., Heidmann, N and Joly, M.: 1980, A.A 83, 190.
Collin-Souffrin, S., Dumont, S. and Tully, J.: 1982, A.A. 106, 362.
Collin-Souffrin, S., Joly, M., Heidmann, N. and Dumont, S.: 1979, A.A. 72, 293.
Costero, R. and Osterbrock, D. E.: 1977, Ap. J. 211, 675.
Culhane, J. L.: 1978, Q.J.R.A.S. 19, 1.
Davidson, K. and Netzer, H.: 1979, Rev. Mod. Phys. 51, 715.
Feldman, F. R., Weedman, D. W., Balzano, V. A. and Ramsey, L. W.: 1982, Ap. J. 256, 429.
Gaskell, C. M., Shields, G. A. and Wampler, E. J.: 1981, Ap. J. 249,443.
Grandi, S. A.: 1978, Ap. J. 221, 501.
_____. 1981, Ap. J. 251, 451.
Grandi, S. A. and Osterbrock, D. E.: 1978, Ap. J. 220, 783.
Heckman, T. M.: 1980, A.A. 87, 152.
Heckman, T. M. and Balick, B.: 1979, A.A. 79, 350.
Heckman, T. M., Miley, G. K., van Bruegel, W. J. M. and Butcher, H. R.: 1981, Ap. J. 247, 403.
Hubble, E. P.: 1926, Ap. J. 64, 321.
Keel, W. C.: 1982, UCSC Ph.D. Thesis.
Khachikian, E. Ye and Weedman, D. W.: 1974, Ap. J. 192, 581.
Koski, A. T.: 1978, Ap. J. 223, 56.
Lawrence, A. and Elvis, M.: 1982, Ap. J. 256, 410.
Malkan, M. A.: 1982, Ap. J. in press.
Miley, G.: 1980, Ann. Rev. Ast. Ap. 18, 165.
Maccacaro, T., Perola, G. C. and Elvis, M.: 1982, Ap. J. 257, 47.
Mitton, S. and Mitton, J.: 1972, M.N.R.A.S. 158, 245.
Mushotzky, R. F.: 1982, Ap. J. 256, 92.

Netzer, H.: 1975, M.N.R.A.S. 171, 395.
Osterbrock, D. E.: 1977, Ap. J. 215, 733.
_____. 1978, Proc. Nat. Acad. Sci. 75, 540.
_____. 1979, A. J. 84, 901.
_____. 1981a, Ap. J. 246, 696.
_____. 1981b, Ap. J. 249, 42.
_____. 1982, P.A.S.P., submitted.
Osterbrock, D. E. and Koski, A. T.: 1976, M.N.R.A.S. 1976, 61P.
Osterbrock, D. E., Koski, A. T. and Phillips, M. M.: 1975, Ap. J. 197, L41.
_____. 1976, Ap. J. 206, 898.
Osterbrock, D. E. and Miller, J. S.: 1975, Ap. J. 197, 535.
Osterbrock, D. E. and Shuder, J. M.: 1982, Ap. J. Supp. 49, 149.
Phillips, M. M.: 1977, Ap. J. 215, 746.
_____. 1978a, Ap. J. Supp. 38, 187.
_____. 1978b, Ap. J. 226, 736.
Raine, D. J. and Smith, A.: 1981, M.N.R.A.S. 197, 339.
Schmidt, G. D. and Miller, J. S.: 1980, Ap. J. 240, 750.
Searle, L. and Sargent, W. L. W.: 1968, Ap. J. 153, 1003.
Seyfert, C. K.: 1943, Ap. J. 97, 23.
Shields, G. A.: 1975, Ap. J. 195, 475.
_____. 1977, Ap. Letters, 18, 119.
Shuder, J. M.: 1980, Ap. J. 240, 32.
_____. 1981, Ap. J. 244, 12.
Shuder, J. M. and Osterbrock, D. E.: 1981, Ap. J. 250, 55.
Soifer, B. T., Neugebauer, G., Oke, J. B. and Matthews, K.: 1981, Ap. J. 243, 369.
Tohline, J.E. and Osterbrock, D. E.: 1982, Ap. J. 252, L49.
Ulvestad, J. J., Wilson, A. S. and Sramek, R. A.: 1981, Ap. J. 247, 419.
Veron, P., Lindblad, P. O., Zuiderwijk, E. J., Veron, M. P. and Adam, G.: 1980, A.A. 87, 245.
Weedman, D. W., Feldman, R. F., Balzano, V. A., Ramsey, L. W., Sramek, R. A. and Wu, C. C.: 1981, Ap. J. 248, 105.
Weedman, D. W.: 1977, Ann. Rev. Astron. Astrophys. 15, 69.
Wilson, A. S.: 1979, Q.J.R.A.S. 19, 1.
Wilson, A. S. and Willis, A. G.: 1980, Ap. J. 240, 429.
Wilson, O. C.: 1948, Ap. J. 108, 201.
Wills, B. J., Netzer, H. and Wills, D.: 1980, Ap. J. 242, L1.
Woltjer, L.: 1958, B.A.N. 14, 39.
Wu, C. C., Boggess, A. and Gull, T. R.: 1980, Ap. J. 242, 14.
_____. 1982, Ap. J. in press.
Yee, H. K. C.: 1980, Ap. J. 241, 894.
Yee, H. K. C. and Oke, J. B.: 1978, Ap. J. 226, 753.

ALLER: The presence of strong [Fe VII] and [Ca V] lines in the spectra of Seyfert galaxies suggests that there is little loss of refractory elements from the gas phase. This differs from the situation in PN, where Fe and Ca are depleted. However, observations show that dust is present in Seyferts! How might these facts be explained?

OSTERBROCK: There is no doubt that the [Fe VII] lines are stronger in Seyfert 1 galaxies than in PN. The [Fe VII] line profiles are basically the same as those of other lines formed in the narrow-line emitting regions. Yet the Balmer decrements show that dust is present in or very near the narrow-line regions. Perhaps the dust in Seyfert 1 galaxies differs from that in PN, or perhaps it is just outside the ionized regions.

PEIMBERT: What are your views on the evolution of galaxies with broad-line emitting regions?

OSTERBROCK: Essentially all Seyfert galaxies are spirals, and nearly every Seyfert is a SB spiral, or is distorted, or has a close companion galaxy. Evidently, the deviation from a circularly symmetric gravitational field is necessary for the Seyfert phenomenon to occur. Thus, it seems possible, and perhaps likely, that every spiral galaxy with these properties will, given enough time, ultimately become a Seyfert.

Recent spectral studies of "normal" spiral galaxies, particularly by William C. Keel at Lick Observatory, have shown that many of them have emission line spectra that can be understood in terms of photoionization by a power law spectrum ($L_\nu \propto \nu^{-\alpha}$), but with a luminosity much smaller than in typical Seyferts. These objects can thus be regarded as "weak" examples of the Seyfert phenomenon. Perhaps, in some of them, the energy source will grow and they will become fully developed Seyferts. In others, the energy source may remain weak or become exhausted.

SEATON: To what extent can the anomalous Balmer decrements be explained by reddening? The basic theory of recombination spectra has been pursued further for Seyferts than for PN, particularly by including effects of line transfer. Perhaps studies of PN have something to gain from the work on Seyferts in this field.

OSTERBROCK: In the narrow-line regions, the Balmer decrements can be fitted pretty well by recombination plus reddening. In the broad-line regions, deviations from the recombination decrements are large and seem to indicate that reddening, collisional excitation (from n = 1 and n = 2), and radiative transfer effects all play a role. Many theoretical computations, from one side or another of this problem, have been made, but the real physical situation is undoubtedly much more complicated than that assumed in the models.

NUSSBAUMER: Have the electron temperatures in the narrow and broad-line regions been reliably determined?

OSTERBROCK: In many cases, yes, for the [O III] and [N II] narrow-line emitting regions. Also, in one case (III Zw 77), from [Fe VII], although the result is uncertain because the reddening correction is not known. In the broad-line regions, there are no good diagnostics, although the Fe II lines indirectly indicate $T \approx 10^4$ K, and the broad-line spectra of the brighter Seyferts that have been observed with

IUE (and in QSO's with large red-shift, in which the ultraviolet spectrum can be observed) indicate that $1 \times 10^4 \lesssim T \lesssim 2 \times 10^4$ K is consistent with the observed C III] and C IV line strengths.

SURDEJ: With your proposed model, how can you escape the conclusion that Seyfert 1 (Seyfert 2) galaxies are galaxies seen edge-on (face-on)?

OSTERBROCK: In the proposed model, the main velocity field in the broad-line region is rotation, but there is a smaller "turbulent" velocity field. This produces the finite height of the broad-line region shown in the Figure. Thus, viewed even from along the axis, the broad-line profiles are significantly wider than the profiles of the forbidden lines produced in the narrow-line regions. There is a large range of line widths in Seyfert 1 galaxies, from about 0.01 c to about 0.10 c. According to the model, part of this spread arises from differences in rotational and turbulent velocities, part from projection effects. I believe that the observational data show the disks to be tipped, in many cases, with respect to the planes of the spiral galaxies in which they occur. The spectra of the Seyfert 1 galaxies with even the narrowest broad-lines typically also show Fe II emission, high stages of ionization in the narrow lines, and have small [O III] λ 5007/Hβ ratios - all of which differentiate them from Seyfert 2 galaxies.

COSTERO: It seems that the spectra of type I PN and Seyfert 2 galaxies are very similar, both showing very low and very high excitation lines. Would you comment on this?

OSTERBROCK: I agree with you. However, in Seyfert 2 galaxies, the [S II], [O I], and [N I] lines are typically stronger (relative to Hβ) than in type I PN. I believe that the photoionizing spectra are roughly the same in both types of objects but that the approximate power-law spectra, probably with a cut-off around 100 eV in the Seyfert 2 galaxies, contain more high energy photons than even the hottest central stars of type I PN, thus producing larger "transition zones" in which O^o, N^o, S^+, H^o, H^+, and electrons coexist.

ROCHE: There are clear differences in the infrared (10 μm) spectra of narrow-line galaxies and Seyferts. The former show strong emission in the narrow dust features, while the Seyferts generally have featureless continua with no direct evidence of dust emission.

FINAL REVIEW

Yervant Terzian
Cornell University, NAIC, Ithaca, N.Y.

What splendid laboratories have planetary nebulae been! We have heard of important observations in the ultraviolet, optical, infrared, and radio spectral ranges, all rich in astrophysical information. We are now certain that planetary nebulae represent a crucial phase of stellar evolution. We have also begun to study the fine details of the evolution of planetary nebulae and our endless aim is to understand these objects completely.

The observational advances of planetary nebulae during the last few years have been dominated by the very successful 'International Ultraviolet Explorer'. Numerous important studies have been made at UV wavelengths, some examples showing the excellent UV spectra are shown in Figure 1 (compiled from Boggess et al. 1981). A classical example of an in depth study of a planetary nebula, NGC 7662, by Harrington et al. (1982) shows the vast and accurate information which can be obtained by combining the UV observations with optical, IR, and radio results. Figure 2 shows the UV spectrum of NGC 7662 and its decomposition into the stellar and nebular spectra, indicating a star with $T_s = 10^5$ K. The IUE has also produced some excellent high resolution spectra. The particular study of the CIII] λ1907/1909 ratio has proved useful for determining electron densities in planetary nebulae. Figure 3 shows several examples of the CIII] observations (Feibelman et al. 1980, and Feibelman et al. 1981). These high resolution spectra also indicate clearly the expansion of the nebular envelopes (NGC 3242, Hu 1-2). The theoretical computations for the line ratio λ1907/1909 as a function of density have been made by Loulergue and Nussbaumer (1976) and are shown in Figure 4.

Observational advances have also made possible the panoramic imaging of planetary nebulae in selected optical emission lines. Such studies have resulted in the determination of the electron temperature and density gradients in many planetary nebulae (Reay and Worswick 1982).

The remarkable optical spectrum of A30, shown in Figure 5, (Hazard et al. 1980, Jacoby 1979) has all the characteristics of a planetary

D. R. Flower (ed.), Planetary Nebulae, 487–499.

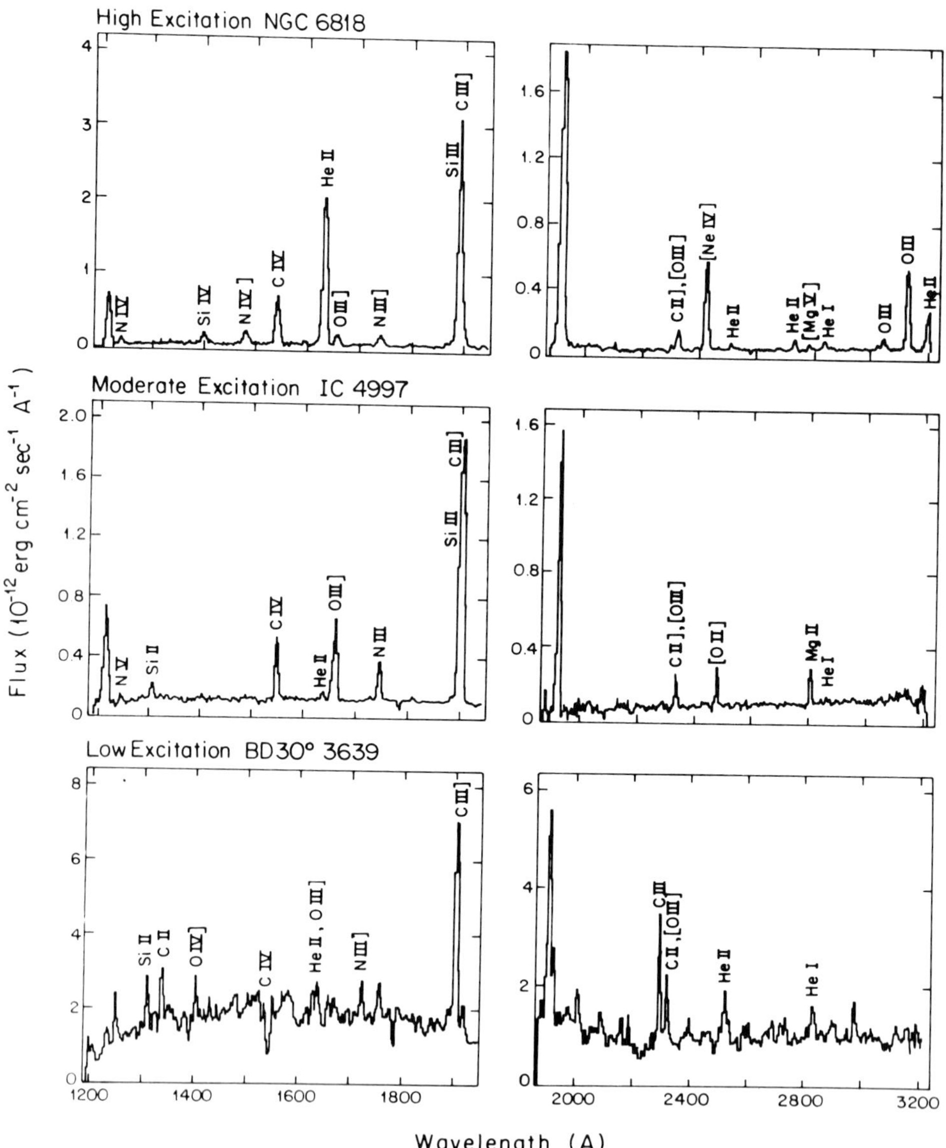

Figure 1. Three IUE representative spectra of planetary nebulae.

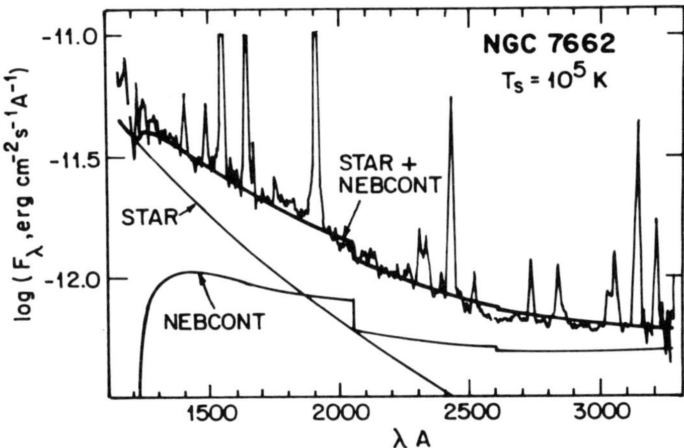

Figure 2. The ultraviolet spectrum of NGC 7662, where the stellar and nebular continua are also shown separately.

nebula except that it shows no Balmer lines. This may be due to material recently ejected from the central star, which may have been depleted of hydrogen due to previous explosions. Such new observational results are of primary importance for the study of the post red giant evolution, and in general for the evolution of the chemical abundances in the galaxy.

In the infrared spectral region the Kuiper Airborne Observatory has been used to study the infrared emission lines in many planetary nebulae. Recently accurate measurements of SIII at 18.7μ, NeV at 24.3μ, and OIV at 25.8μ, have been reported (Shure et al. 1982). Such information is essential for realistic determinations of the chemical abundances. High resolution observations in the infrared continuum have also been performed indicating the dust distribution in these objects. Most observations tend to indicate that the dust surrounds the ionized nebular gas, but there is some evidence that small amounts of dust may co-exist with the gas at the outer parts of the nebulae. It also appears that younger nebulae contain more dust and molecules compared to the older extended objects. There has also been some evidence of a size and temperature distribution of the grains.

In the radio spectral range the Very Large Array has begun to provide very high resolution radio images of planetary nebulae. Future observations with the VLA (and with the Space Telescope) will permit angular resolutions of $\sim 0\overset{''}{.}1$, which corresponds to 5×10^{-4} pc at a distance of 1 kpc. In many cases this resolution represents $\sim$ 1/200 of the size of the nebulae. Recent radio interferometric observations (Turner and Terzian 1982) with an effective resolution of ∿ 1" have already begun to study the compact 'stellar' planetary nebulae.

The technological advances which have permitted us to study planetary nebulae in the ultraviolet, optical, infrared and radio spectral regions have made possible a more complete analysis of the

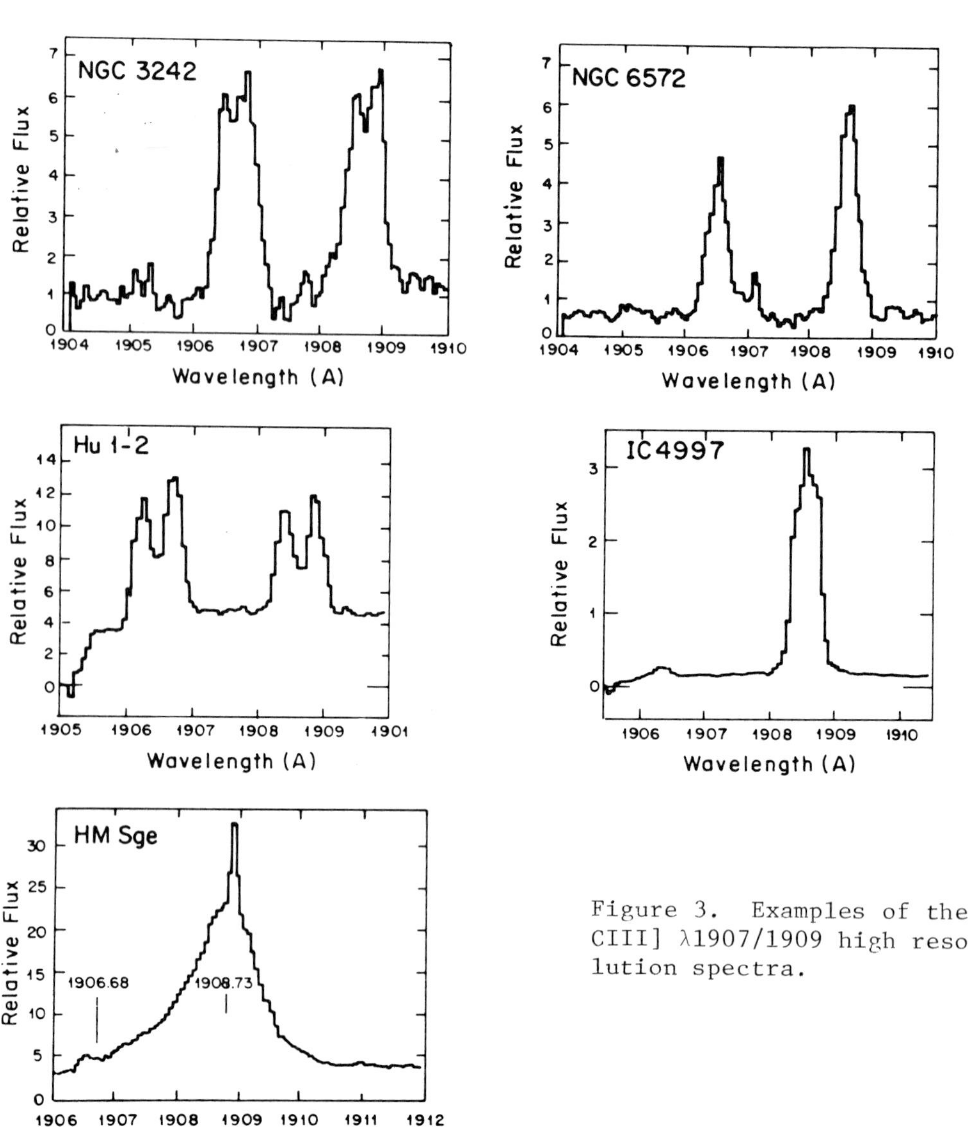

Figure 3. Examples of the CIII] λ1907/1909 high resolution spectra.

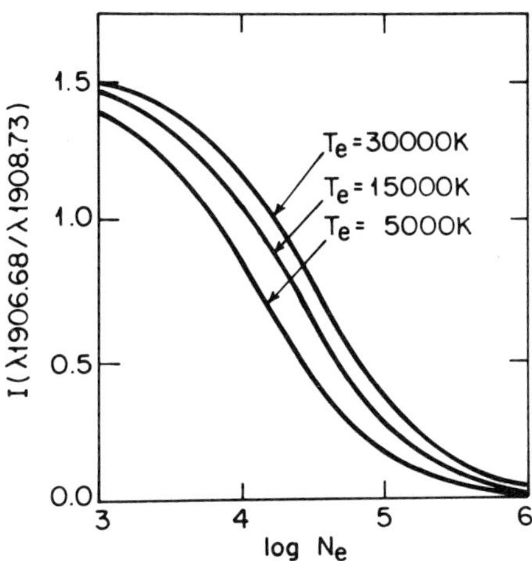

Figure 4. The theoretical line ratio for CIII] λ1907/1909 as a function of electron density for various electron temperatures.

excitation and ionization states of planetary nebulae, and in particular have allowed us to derive more realistic chemical abundances. Very recently Aller and Czyzak (1982) have completed a study of the chemical abundances of planetary nebulae and have concluded that pronounced chemical composition differences do exist in planetary nebulae, and that some are nitrogen rich and others are carbon rich. Nineteen different chemical elements have been detected in planetary nebulae, and abundance determinations exist for about 80 objects. There are however several outstanding problems including the marginal indications of abundance gradients in the galaxy, the very high He/H ratios (>0.18) determined for several nebulae, and the lack of a

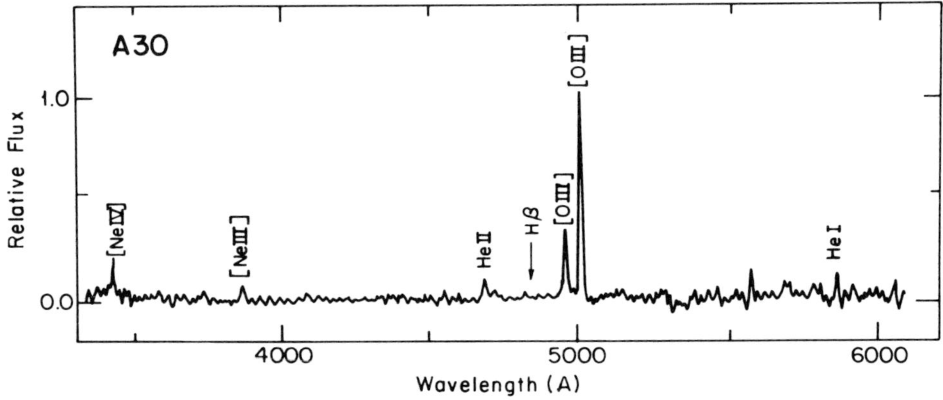

Figure 5. The optical spectrum of A30, note the absence of Hβ.

systematic study of abundances in halo objects. It is also remarkable that almost no attempts have been made to estimate the errors involved in determining chemical abundances, it is strongly suggested that such efforts be made in the future.

An important realization in the studies of planetary nebulae has been the possible large mass range of the progenitor stars from ~ 0.8 up to 6 or even 8 $M_\odot$. The mass distribution of the nuclei of planetary nebulae shows a strong peak at ~ 0.6 $M_\odot$, but extends slightly over 1 $M_\odot$. Therefore an important realization has been that the amount of ejected material back into the interstellar medium may be very substantial, and in some cases may be a few solar masses. It also appears that a large fraction of the ejected material is in a neutral state.

Mass loss from evolved stars and proto-planetary nebulae like IRC+10216, and CRL 2688 derived from CO observations (Knapp et al. 1982) show that $\dot{M} \sim 10^{-4}$ to 10^{-7} $M_\odot$/yr. Other nebulae show similar mass loses like NGC 7027, $\dot{M} = 4.0 \times 10^{-4}$; IC 418, $\dot{M} = 5.0 \times 10^{-6}$; and NGC 6543, $\dot{M} = 5.1 \times 10^{-6}$ $M_\odot$/yr derived from CO observations. The estimated mass of the molecular envelope around NGC 7027 is ~ 5 $M_\odot$, compared to 0.5 $M_\odot$ for the ionized mass. The mass of the central star of NGC 7027 is estimated to be ~ 1 $M_\odot$, resulting into a total progenitor stellar mass for NGC 7027 of 6 to 7 $M_\odot$. Recent UV studies of planetary nebulae nuclei (Heap 1980, Perinotto et al. 1981, Pottasch 1981) indicate wind driven mass loss inferred from the presence of P Cygni profiles in their spectra. This mass loss rate ranges from $\sim 10^{-6}$ to $\sim 10^{-10}$ $M_\odot$/yr.

The successful UV observations of planetary nebulae and their central stars in the Magellanic Clouds (Maran et al. 1982, Stecher et al. 1982) have provided additional information on the planetary nebulae stellar masses which are found to be ~ 1 $M_\odot$ from ~ 4 $M_\odot$ progenitor stars, hence indicating a mass loss of ~ 3 $M_\odot$ per star in various winds and explosions. And very recently Rodriguez and Moran (1982) may have detected a neutral hydrogen envelope around NGC 6302, indicating further that significant mass loss takes place from the progenitors of planetary nebulae (Figure 6).

The realization that the progenitor stars of planetary nebulae return a large mass fraction of processed matter back into the interstellar medium represents a fundamental conclusion. The chemical evolution of the galactic interstellar medium from light elements like H, and He into heavier ones like C, O, Mg, and Si is probably dominated by mass loss from the stellar progenitors of planetary nebulae. Since the rate of formation of planetary nebulae in the Galaxy is of the order of ~ 1 per year, then to a first approximation 10^{10} planetary nebulae were formed during the age of our galaxy, and perhaps 10^{10} $M_\odot$ of processed and perhaps reprocessed material has been provided to the interstellar medium.

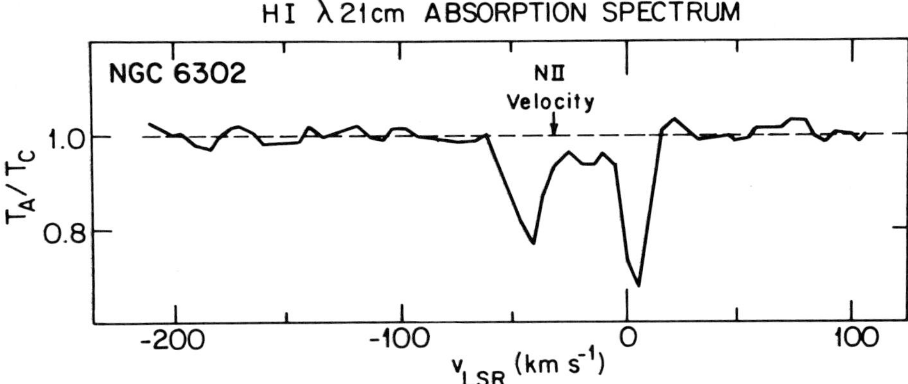

Figure 6. The λ21 cm HI absorption spectrum in the direction of NGC 6302. The arrow marks the nebular radial velocity from optical [N II] observations. The absorption at -40 km/sec may be due to the expanding HI shell of the nebula.

Careful and more sensitive observations of planetary nebulae have shown that many have double shells. More than half a dozen objects are now known with 'giant' halos (Terzian 1980). The mass involved in these outer halos must be significant - and probably is of the order of a solar mass or more.

There are now at least three examples of planetary nebulae with triple shells NGC 7009, NGC 7662, and NGC 6826. These objects have two inner shells and an outer halo. NGC 6826 recently discussed by Feibelman (1981) has an inner shell with a size of ∿ 10", and outer shell with a size of 25", and a halo 130" in size. Louise (1981) has also shown that on longer exposure photographs of planetary nebulae outer filamentary structure is normally seen. The length of photographic exposure seems to make great difference on what is observed. It seems likely that outer fainter halos around classical planetary nebulae may be common. This result has implications on the total mass ejected from the planetary nebulae progenitors, and possibly on the origin of the nebulae. Figure 7 shows the very symmetric bubble planetary nebula 268 + 11°1 (Longmore 1977), and the filamentary nebula Abell 43 (Jacoby 1982) for a morphological comparison. Figure 8 shows two nebulae with outer giant halos, NGC 6542 and NGC 6826 (Millikan 1974), note that the central images are overexposed in order to detect the outer fainter envelopes. Many other nebulae show predominantly a bipolarity in their morphological structure. This striking appearance is most obvious in the radio maps of these objects (Terzian et al. 1974) where the observed brightness distribution is not affected by

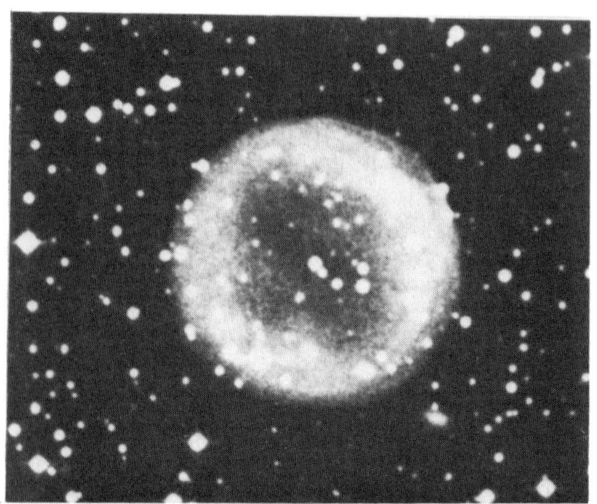

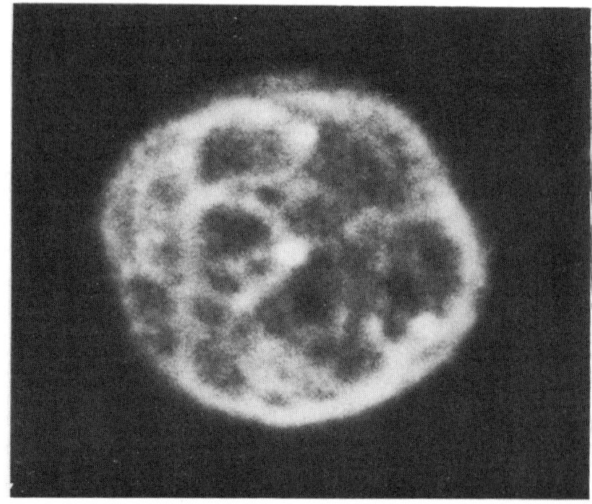

Figure 7. A symmetric nebula (286 + 11°1) and a filamentary (A 43 courtesy of Kitt Peak National Observatory) one.

dust. Figure 9 shows four bipolar nebulae with a striking resemblance (The Red Rectangle, M2-9, NGC 6302 and NGC 2346 courtesy of M. Cohen, L. Kohoutek, and R. Minkowski) strongly suggesting that some common physical processes govern the nebular ejection from their central stars.

The origin of planetary nebulae remains uncertain. Although sudden multiple ejections from a central star are suggestive due to the observed double and multiple shells, continuous mass loss and stellar winds have been given some attention (Kwok et al. 1978). Normal stellar winds however will not produce planetary nebulae, and a 'super-wind' is necessary with a mass loss $\dot{M} \sim 10^{-5}$ to 10^{-4} $M_{\odot}$/yr. Figure 10 shows two examples of nebulae with double shells, NGC 2392 where the two shells seem to be detached, and NGC 3242 where no discontinuity is seen between the shells.

Five years ago during the IAU Symposium on Planetary Nebulae at Cornell University not a word was said on the importance of magnetic fields in planetary nebulae! Is it that we understand that magnetic fields are totally unimportant, or is it that we do not know how to deal with magnetic fields? Some have discussed that the bipolarity in the morphology of planetary nebulae has a natural explanation in the existence of a bipolar magnetic field, however no suggestions have been made as to how the field arises, and the required fields of $\sim 10^{-3}$ to 10^{-4} gauss are difficult to produce. Others, feel that the static magnetic fields do not contribute directly to the form and structure of planetary nebulae. It is perhaps possible that the stellar magnetic fields can influence the structure of the nebula at the very early

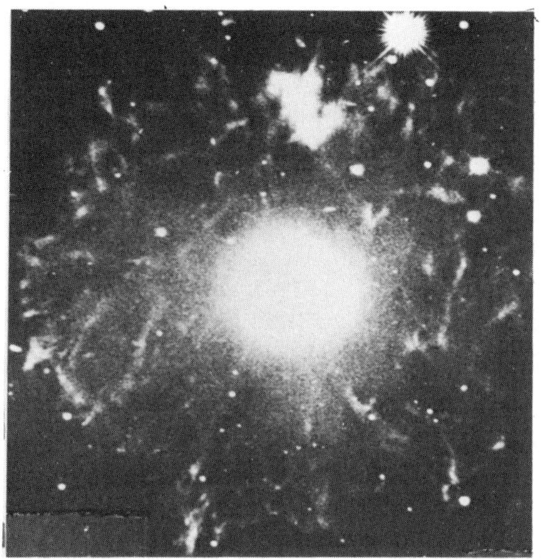

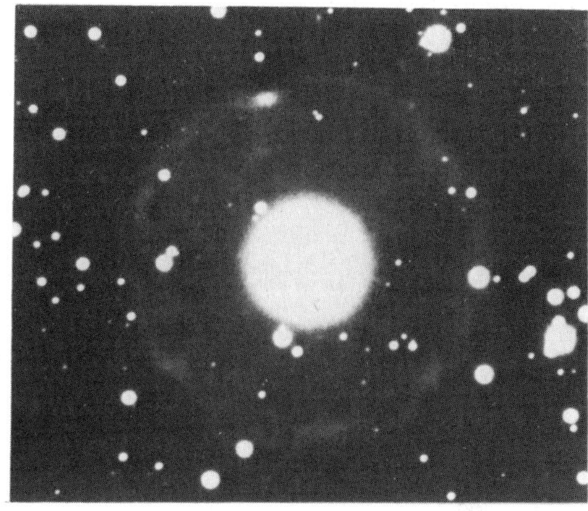

Figure 8. Giant halos around NGC 6543 and NGC 6826.

stages when the nebula is still part of the stellar expanded atmosphere. After ejection of the envelope the slowly expanding shell may retain part of its original structure. The problem of internal magnetic fields merits additional study. More recently Nussbaumer (1982) has suggested that magnetic flux, randomly emerging from stellar surfaces, may be responsible for certain types of stellar winds which may produce planetary nebulae. It is fair to conclude that we still do not have a fundamental understanding of the physical processes which produce planetary nebulae, and the explanation of the morphological types of planetary nebulae remains unknown.

Concerning the central stars of planetary nebulae there now exists a sample of 60 well observed stars which have a reported surface temperature range from $\sim$ 30000 to 350000K. There are however differences expressed by various authors in the physical parameters of the central stars and these should be resolved with further accurate observations. The existing theoretical model atmospheres of the central stars of planetary nebulae seem oversimplified and future work should concentrate in relaxing the assumptions, in particular a small fraction of metals should be introduced in the atmospheric composition.

It is encouraging to see that serious theoretical research is being performed on the gas dynamics of the nebulae. The prediction that the inner shocked stellar wind gas may be very hot with temperatures of the order of 3×10^6 K is an interesting new development and should be explored further.

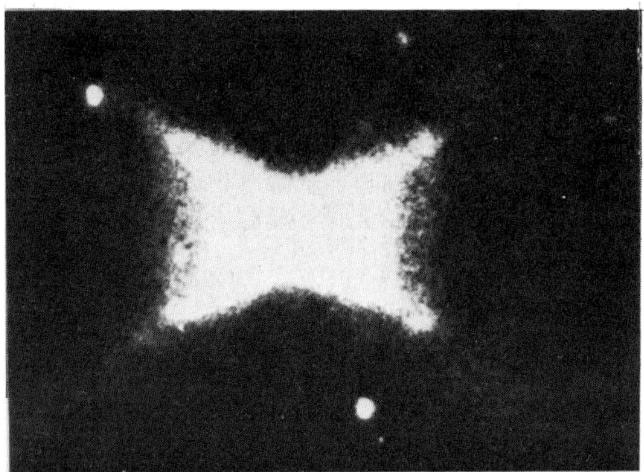

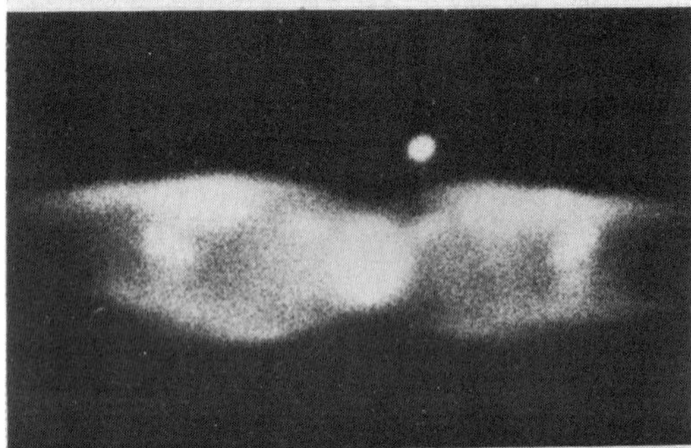

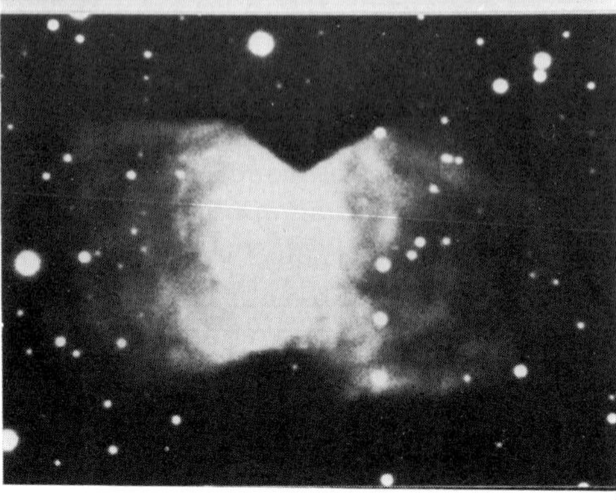

It is also important to note that there now exist far better theoretical calculations on atomic parameters, although more work remains to be done. The construction of models has become standard practice for the interpretation of the nebular spectra and we have to ask if the physics is complete and accurate, and if the procedures are sound. We have seen, for example, that for realistic models one needs a density distribution in a gaseous nebula, rather than adopting a constant density.

Perhaps the most disturbing problem in the study of planetary nebulae has been the determination of the distance scale to these objects. Although some progress has been made by using a variety of distance determination methods, progress in this area has been slow, and individual distances may have errors of at least a factor of two. This situation makes it difficult to derive accurate luminosities, and also introduces a large error in the statistics and distribution of planetary nebulae in the galaxy. Nevertheless, the total number of these objects in the galaxy has been estimated to be about 28,000 with a local density of about

Figure 9. The bipolar nebulae The Red Rectangle, M2-9, NGC 6302 and NGC 2346, suggesting common physical processes for their ejection.

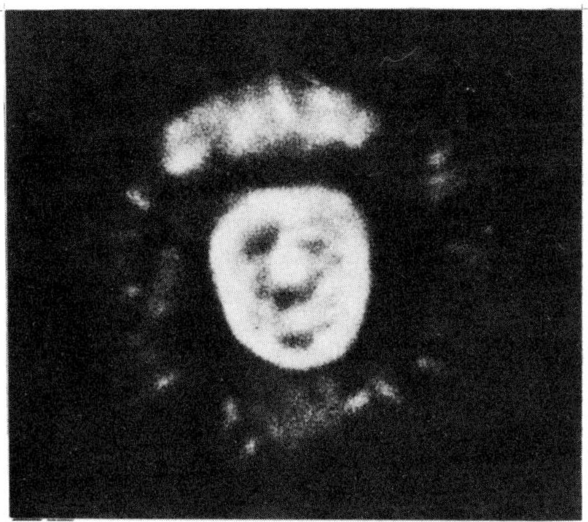
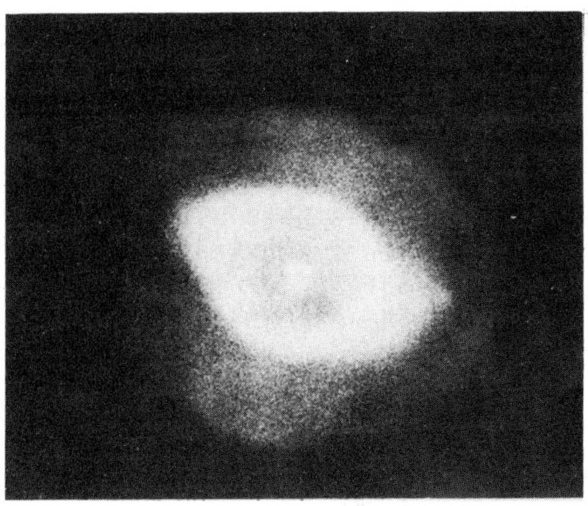

Figure 10. The detached double shell nebula NGC 2392, and the double shell nebula NGC 3242 where no discontinuity is seen between the shells (Palomar Observatory, California Institute of Technology).

50 nebulae within one kpc from the sun. The birthrate of these objects is $\sim$ 1 per year in the galaxy, however somewhat higher estimates have also been reported. The actual number of detected planetary nebulae in the galaxy is now about 1400.

Remarkable progress has been evident in the studies of planetary nebulae in nearby galaxies despite the very difficult observational task. More than 100 nebulae have been detected in each of the LMC and M31, and more than 50 in each of the SMC and M33. Preliminary results indicate an overabundance of C and N in these objects compared to our galaxy, and these important results should be examined more extensively.

It is clear that the classical picture of a planetary nebula as a hot bright and ionized envelope represents an oversimplified description. Today we have begun to understand the various stages of planetary nebulae from the red giant progenitor stars to the contracting white dwarfs. This evolution is rich in variations both for the central stars and their envelopes.

The next few years promise to be very fruitful in the study and understanding of planetary nebulae. The excellent recent theoretical work on the post red giant evolution, the formulation of model atmospheres of the nuclei of planetary nebulae, the stellar winds, the gas dynamics of the nebulae, and on the atomic data promise to yield a deeper understanding of the mildly explosive late evolutionary stellar stages of stars between 0.6 and a few solar masses.

On the observational scene the IUE work will be supplemented by optical observations with the Space Telescope, by infrared observations with the Infrared Astronomical Satellite and by radio observations with the Very Large Array.

This work was supported in part by NAIC which is operated by Cornell University for the National Science Foundation.

REFERENCES

Aller, L.H., and Czyzak, S.J., 1982, Preprint.
Boggess, A., Feibelman, W.A., and McCracken, C.W., 1981, The Universe at Ultraviolet Wavelengths, NASA-CP2171, ed. R. Chapman.
Feibelman, W.A., Boggess, A., Hobbs, R.W., and McCracken, C.W., 1980, Ap. J. 241, 725.
Feibelman, W.A., 1981, PASP, 93, 719.
Feibelman, W.A., Boggess, A., McCracken, C.W., and Hobbs, R.W., 1981, Ap. J. 246, 807.
Harrington, J.P., Seaton, M.J., Adams, S., and Lutz, J.H., 1982, MNRAS, 199, 517.
Hazard, C., Terlevich, R., Morton, D.C., Sargent, W.L.W., and Ferland, G., 1980, Nature, 463.
Heap, S.R., 1980, Bull. Am. Astr. Soc., 12, 540.
Jacoby, G.H., 1979, PASP, 91, 754.
Jacoby, G.H., 1982, private communication.
Knapp, G.R., Phillips, T.G., Leighton, R.B., Lo, K.Y., Wannier, P.G., and Wootten, H.A., 1982, Ap. J., 252, 616.
Kwok, S., Purton, C.R., and FitzGerald, P.M., 1978, Ap. J., 219, L125.
Longmore, A.J., 1977, private communication.
Louise, R., 1981, Astron. Astroph., 102, 303.
Loulergue, M., and Nussbaumer, H., 1976, Astron. Astroph., 51, 163.
Maran, S.P., Aller, L.H., Gull, T.R., and Stecher, T.P., 1982, Ap. J. 253, L47.
Millikan, A.G., 1974, A.J., 79, 1259.
Nussbaumer, J., 1982, Astron. Astroph., 110, L1.
Perinotto, M., Benvenuti, P., and Cacciari, G., 1981, Effects of Mass Loss on Stellar Evolution, ed. C. Choisi, and R. Stallio (Reidel, Dordrecht), 45.
Pottasch, S.R., 1981, Physical Processes in Red Giants, ed. I. Iben and A. Renzini (Reidel, Dordrecht), 447.

Reay, N.K., and Worswick, S.P., 1982, MNRAS, 199, 581.
Rodriguez, L.F., and Moran, J.M., 1982, Nature, 299, 323.
Shure, M.A., Herter, T., Houck, J.R., Briotta, D.A., Forrest, W.J., Gull, G.E., and McCarthy, J.F., 1982, Preprint.
Stecher, T.P., Maran, S.P., Gull, T.R., Aller, L.H. and Savedoff, M.P., 1982, Preprint.
Terzian, Y., Balick, B., and Bignell, C., 1974, Ap. J., 188, 257.
Terzian, Y., 1980, Quarterly Journal Roy. Astr. Soc., 21, 81.
Turner, K., and Terzian, Y., 1982, Preprint.

EDITOR'S NOTE: More recent computations of the C III line intensity ratio (Fig. 4 above) are to be found in H. Nussbaumer and H. Schild, 1979, Astron. Astrophys., 75, L17.

GENERAL DISCUSSION

Statistics ...

WEIDEMANN: A comment on the distance scales for PN. If one introduces a scale factor, K, with the normalisation K = 1 for Seaton (1968), then one finds, for other distance scales: K = 0.8 (Acker), K = 1.5 (Cudworth), K = 1.04 (Daub), K = 1.3 (Weidemann), K = 1.3 (Jacoby), and K = 1.2 (Schneider et al., contributed paper at this meeting), giving PN birth rates, $1.3 \times 10^{-12} \lesssim \chi_{PN} \lesssim 14 \times 10^{-12}\ pc^{-3}\ y^{-1}$ (since $\chi_{PN} \propto K^{-4}$). As White Dwarfs probably do not all go through the PN stage, there remains a serious discrepancy with the White Dwarf birth rate ($\chi_{WD} \approx 1.5 \times 10^{-12}\ pc^{-3}\ y^{-1}$) for $K \lesssim 1$.

JACOBY: The (1978) value of 10 000 PN in M 31 was derived before the luminosity function was observed. The newer data suggest that there are about 20 000 PN in M 31, corresponding to 10 000 - 15 000 in our Galaxy. This translates into K = 1.3 (compared with K = 1 for Cahn and Kaler), as suggested by Dr. Weidemann.

SERRANO: Do the scale heights of the low and high mass parts of the White Dwarf distribution differ? If so, could this explain the discrepancy with the PN scale height?

WEIDEMANN: The scale height of the White Dwarfs is larger than that of the PN, according to Green's Palomar study; but the mass distribution of the White Dwarfs is so narrow about the average ($0.58\ M_{\odot}$) that it is not possible to determine the scale height as a function of the mass.

RENZINI: These discrepancies (in scale heights, statistics etc.) may arise as a consequence of the naive assumption that all PN have the same lifetime, irrespective of the initial mass of the parent star.

Formation of the nebulae ...

KAHN: We now have the picture in which the total mass in a PN is much larger than that seen in the form of ionized gas. This mass has been ejected by a slow superwind. Later, a fast (2 000 km s^{-1}) wind pushes this gas away from the central star and starts to compress it into a shell.

If the primary ejection was not spherically symmetrical, then the swept-up shell will show the same sort of angular dependence in density, but amplified. As far as I know, there is not yet a decent calculation of this effect, but it seems worth doing because there is a good chance that it will explain the origin of bipolar structures.

It is important, in this connection, that the ambient gas distribution (derived from the superwind) should extend to a large distance from the central star in all directions. There must be no chance of hot shocked gas escaping from the bubble through a hole in the jacket.

WANNIER: With regard to an asymmetric superwind, I wish to point out that, for the four circumstellar molecular clouds which we have mapped, there is no indication yet of a departure from circular symmetry. The symmetry is especially striking in IRC + 10 216, which has been

D. R. Flower (ed.), Planetary Nebulae, 501–504.

extensively mapped, despite the indications of elongation in the inner infrared source.

HEAP: Before coming to this Symposium, I looked, with Allan Sweigart, at the question of rotation of Red Giant stars. If angular momentum is conserved, so that the core can spin up, then the outer edge of the core (i.e. the central star) of a Red Giant will rotate at 10 km s^{-1} if its initial mass was 1.5 $M_\odot$ and at 800 km s^{-1} if its initial mass was 3.5 $M_\odot$. So, perhaps rotationally enhanced ejection plays a role in the superwind.

PEIMBERT: If the core spins up, we should expect three phases in the mass loss process, the first being more or less spherically symmetric, the second with mass loss preferentially occurring close to the equatorial plane, and the third with confinement near the equator but not near the poles, owing to the presence of material ejected during the second phase. An object which shows evidence of these three phases is NGC 2440, which has an outer spherical halo, very faint and smooth, a bright ring or disk near the centre, and "lobes" or "ansae", presumably produced by the confinement of the disk.

RENZINI: Some White Dwarfs have strong magnetic fields, of the order of 10^3 gauss. When they were PN nuclei, their high speed (say, 3000 km s^{-1}) wind was probably very asymmetric, which may help in giving funny shapes to previously ejected, almost spherical shells.

KAHN: I disagree - it should be remembered that the fast wind gas is trapped in a bubble which is much larger than the linear dimensions of the inward facing shock. It does not really matter, in such a case, how the fast wind leaves the central star. All that happens to it is that it contributes thermal energy to the hot shocked wind bubble, which will expand quite subsonically and will not be affected particularly by the way in which the gas is injected.

NUSSBAUMER: In my suggestion about the origin of PN (1982, Astron. Astrophys. 110, L1), I gave a mass loss formula $\dot{M} \propto B^2$. About 5 - 10% of White Dwarfs have $10^4 \lesssim B \lesssim 10^5$ gauss. Fields lower than 10^3 gauss could not, at present, be detected, but, in analogy to the Sun, the mean magnetic field may be the result of magnetic flux tubes of much higher local field strengths. Under such conditions, magnetic fields would dominate all other pressure terms and could easily produce mass loss. The same remark applies to Red Giants, where convection could produce locally strong emerging magnetic flux. Thus, I think that magnetic fields could, at certain stages in the evolution towards and during the PN stage, be of fundamental importance.

ZUCKERMAN: Regarding the comparison between symbiotic stars and PN, it is interesting to note that, in the former case, we (very probably) have a close binary star but relatively little mass ejection (as judged from infrared and radio observations), whereas PN contain much more ejected mass but show little evidence for close binary central stars. This seems difficult to reconcile with a picture in which mass ejection in PN is supposed to be enhanced by the presence of a close companion.

TERZIAN: A very small percentage (perhaps about 1%) of known PN have binary stars. There should be many more, since more than one half of the field stars are in binary stellar systems. Perhaps we should intensify our searches for binary PN nuclei.

KEYES: It should be remembered that there are two broad classes of symbiotic stars: those with infrared (dust) emission (D-type), and those without (S-type). The D-type symbiotics are radio sources, and it has been suggested that they are proto-PN. The S-type symbiotics include all known symbiotic binaries but only one radio source, AG Peg; this is the only S-type symbiotic that we can definitely say contains a subdwarf (see my contributed paper, this volume). It has been suggested that super-critical wind accretion on a disk about a main sequence star may supply the ionizing photons that produce the observed emission spectrum. This might also account for the lack of radio emission from S-types, since there would be no mass lost from the system to produce it.

A further, interesting aspect of symbiotics is that the observed emission line spectra of some of them resemble those of Seyfert galaxies. Several symbiotics display a wide range of ionization (up to [Fe VII]) and composite emission line profiles (both narrow and broad components). Electron temperatures and densities in these symbiotics are similar to those in the Seyferts. Therefore, the symbiotics provide a nearby and much brighter laboratory in which to study plasmas that are similar in some respects to those found in Seyferts and QSO's.

Characteristics of the central stars ...

HEAP: This morning, Pottasch showed us two displays of the mass distribution of the central stars of PN. The distribution in the L/T_{eff} (HR) diagram implied a median mass well under 0.6 $M_{\odot}$, whereas the distribution in the M_V/R_{neb} ("Schönberner") diagram implies a median mass of 0.6 $M_{\odot}$. The same distance scale was used in both diagrams, so what is the reason for the discrepancy?

POTTASCH: In the diagram of M_V against the age of the nebulae, as deduced from the ionized size and expansion velocity, the points extend very far to the left of the diagram. If the ages of these nebulae were correct, they would indicate that the central stars must have very high mass indeed. The difficulty in interpreting this diagram is that optically thick nebulae should not be placed on it. This is because the ionized mass may be much smaller than the total mass and, consequently, the derived age is much too low. I think that most of the nebulae whose derived ages are less than 3-4 x 10^3 y are optically thick and should be removed from the diagram. This is why it seems preferable to use the HR diagram, where all nebulae can be represented.

KALER: When interpreting results (temperatures, luminosities etc.) obtained for the central stars, it should be remembered that we still have poor data on magnitudes and expansion velocities for many objects.

ALLER: The question of the temperatures of PN nuclei is fundamental, and we must explain the discrepancies that occur. The energy distribution of the central star that is required to model the spectrum of the nebula often corresponds to a much lower temperature than proposed by Pottasch, at least for a number of the hottest objects. The stellar temperatures which we derive from nebular models tend to be in harmony with those suggested by Seaton.

HUMMER: If the effect that Abbott and I have obtained in the one model (contribution to second poster session, this volume) is anything like universal, then the He II continuum becomes closer to a blackbody at the effective temperature of the star. This would have the consequence that the He II Zanstra temperatures have a more nearly universal meaning as blackbody temperatures.

SEATON: Two different approaches to the determination of central star temperatures and luminosities have been presented, by Schönberner and Weidemann, on the one hand, and by Pottasch, on the other. The former obtain distances assuming the nebulae to be optically thin and plot absolute visual magnitudes against nebular radii. The latter uses individual distances (the accuracy of which might, in many cases, be questioned) and plots nebular masses against electron densities. From this work, it is concluded that most nebulae are optically thick.

In neither case is much of an attempt made to estimate the optical thickness of individual nebulae, although this can be done (by comparing H I and He II Zanstra temperatures, or the presence of outer halos, for example).

Dust in PN ...

HOUCK: I think that there is strong evidence for a considerable amount of dust inside the ionized gas of many PN. For example, the 12 μm "photo" of NGC 7027, shown by Terzian, follows the "free-free" radio contours (the nearest strong fine-structure line is [Ne II] 12.8μm).

BARLOW: The depletion of the C IV resonance lines, observed in several PN, definitely implies the presence of dust inside the ionized region. However, no more than $\tau_D \approx 0.1$ is required to explain the observations. Shields has shown that gas-phase iron and some other heavy elements are heavily depleted in PN. If these elements were locked up in grains which had a reasonable size distribution, they could give rise to $\tau_D \approx 0.1$.

If a significant amount of carbon was tied up in grains, the dust optical depth could be embarrassingly large. I believe that the C/O ratio derived from observations of gas-phase carbon correctly reflects the total carbon abundance.

SEATON: The total thermal infra-red emission is a significant fraction (say, 10 - 20%) of the total energy emitted by many PN. One can understand how the dust is heated if it is inside the ionized region. I think that the problem is made more difficult if the dust is assumed to be outside the ionized region.

KWOK: There is no doubt that there is dust outside - the question is, what is the temperature of this dust?

PEIMBERT: It should be added that the observations of Aitken and Roche definitely show that the unidentified features arise outside the ionized zone in NGC 7027.

ABSTRACTS OF CONTRIBUTED PAPERS

POSTER SESSION I

DETECTION AND STUDY OF SECONDARY STRUCTURES IN SOME PLANETARY NEBULAE

R. Louise
Observatoire de Marseille (LA N°237-CNRS), 2 Place Le Verrier, F-13248 Marseille Cedex 4, France

In order to detect faint nebulosities associated with planetary nebulae, long exposure plates are made on nine selected nebulae, using a large bandwidth ($\Delta\lambda$ = 50 Å) interference filter coupled with an ITT image tube. Some peculiar features are observed, but they do not all account for "secondary structures" following Louise's terminology. We discuss the difficulties encountered by the photographic method.

Spectrophotometric observations are made for one nebula, IC 418. Contour map of [NII]/Hα ratio is derived. It is shown that this ratio increases towards the outer extended envelope of the nebula. These observations are made with the IDS system of the ESO in Chile. We obtained 65 spectra covering the outer parts of IC 418.

One of the typical features of secondary structures is the enhancement of [NII] line with respect to Hα. In addition, filamentary structures appear sharper in [NII] than in Hα. This is fairly illustrated by NGC 650-1.

HIGH-SPATIAL RESOLUTION OBSERVATIONS OF PLANETARY NEBULAE

C.T. Hua
Laboratoire d'Astronomie Spatiale de Marseille

R. Louise
Observatoire de Marseille

Monochromatic images in Hα, Hβ, [NII] λ 6584, [SII] λ 6717 and [OIII] λ 5007 lines are presented for morphological study of planetary nebulae. Narrow bandpass ($\Delta\lambda$ = 5 to 10 Å) interference filters are generally used in order to discriminate peculiar structures existing in different emission lines. However, large bandwidths ($\Delta\lambda$ = 50 Å) along with long exposures, are also necessary in searching for faint nebulosities associated with planetaries.

Three faint objects of the Abell's list of old planetary nebulae have been observed through narrow band filters, by means of an image tube (A33, A36 in Chile) or the image photon counting device (A79 at the Haute Provence Observatory). Following the Hα/[NII] intensity ratio, a discussion is given about the distance previously derived with some assumptions concerning the measured red fluxes.

KINEMATICS OF ABELL 30

N.K. Reay, P.D. Atherton
Astronomy Group, Blackett Laboratory, Imperial College, London SW7 2BZ, U.K.

K. Taylor
P.O. Box 296, Anglo-Australian Observatory, Epping, NSW 2121, Australia

We have obtained seeing limited spectra of the [OIII] λ 5007 Å line at a velocity resolution of 18 km s^{-1} over the envelope of Abell 30.

The compact system of ansae near to the central star has been studied in some detail, and individual ansae are shown to be moving, relative to the star, with a line of sight velocity of 22 to 25 km s^{-1}.

Two of the four ansae previously identified (Jacoby, 1979; Hazard et al., 1980) are shown to form a pair, expanding symmetrically with respect to the star with a radial velocity of ± 25 km s^{-1}. The brighter of the other two ansae has a radial velocity of + 22 km s^{-1}, and the symmetric disposition of these ansae about the star (Jacoby, 1979) is suggestive that these also form a symmetrically expanding pair.

The outer envelope is shown to be expanding at 40 km s^{-1} with an age of 10^4 to 10^5 years. The age of the ansae is $\sim 1.5 \times 10^3$ years, consistent with them being ejected at a late epoch from an evolved star.

A DYNAMICAL AND CHEMICAL STUDY OF NGC 6302

I.J. Danziger, D. Baade
European Southern Observatory, Karl-Schwarzschild-Str. 2,
D-8046 Garching, F.R. Germany

P.D. Atherton
Imperial College of Science & Technology, The Blackett
Laboratory, Prince Consort Road, London SW7 2BZ, U.K.

K. Taylor
Anglo-Australian Observatory, P.O. Box 296, Epping, NSW 2121,
Australia

A. Boksenberg
Royal Greenwich Observatory, Hailsham, East Sussex BN27 1RP,
U.K.

From five spectrograms obtained at five different positions in the nebula, relative ionic concentrations have been derived with respect to the nucleus. They show that the degree of excitation generally decreases with distance from the nucleus. But there are also areas with locally enhanced or attenuated excitation. Taurus data, a series of two-dimensional monochromatic images centered on [OIII] λ 5007, have been used to construct a two-dimensional velocity map. It shows a large-scale structure similar to the one of direct images with the biconical pattern being at least partly present. Areas of locally lower radial velocity which seem to be inversion symmetrically distributed with respect to the centre, are also distinguished. They do not have pronounced counterparts on direct images. The cavity model suggested by Barral et al. (1982, MNRAS 199, 95) for NGC 6302 and Icke's biconical flow model (1981, Ap. J. 247, 152) are discussed.

NEUTRAL HYDROGEN ASSOCIATED WITH THE PLANETARY NEBULA NGC 6302

L.F. Rodríguez
Instituto de Astronomía, UNAM, Apdo. Postal 70-264,
04510 México, D.F., Mexico

J.M. Moran
Harvard-Smithsonian Center for Astrophysics, 60 Garden St.,
Cambridge, MA 02138, USA

Observations of HI in absorption made with the Very Large Array towards the thermal radio emission of the planetary nebula NGC 6302 show two velocity components at 6 and -40 km s^{-1} (radial velocity with respect to the local standard of rest). The 6 km s^{-1} component is almost certainly due to a line-of-sight cloud, but the -40 km s^{-1} component is most probably associated with NGC 6302. We interpret this absorption component as coming from the neutral, outer part of an expanding ($\sim$ 10 km s^{-1}) ring whose inner part is ionized and produces the thermal continuum. The mass in atomic hydrogen of the outer (neutral) part of the ring is $\sim$ 0.06 $M_\odot$. NGC 6302 is in an evolutionary stage intermediate to those of protoplanetary nebulae such as GL 2688 and evolved planetary nebula such as NCG 7293. This is the first detection of neutral hydrogen associated with a planetary nebula.

THE KINEMATICAL STRUCTURE OF THE BIPOLAR NEBULAE M2-9 and M1-91

U. Carsenty, J. Solf
Max-Planck-Institut für Astronomie, D-6900 Heidelberg, FRG;
and Centro Astronómico Hispano-Alemán, Almeria, Spain

Both nebulae are of similar appearance consisting of a central core and two highly symmetrical, elongated lobes. Using the large vertical Coude spectrograph of the 2.2-m telescope on Calar Alto, Spain, we obtained long-slit spectra in the red of high spectral (12 km s^{-1} FWHM) and spatial (2") resolution from various positions within the nebulae. Our data indicate high similarity in the kinematical structure of both nebulae. The central cores are dominated by very broad emission lines. The Hα profiles (width ≈ 1600 km s^{-1} at 5% level in M2-9) exhibit an absorption feature blue-shifted by ≈ 20 km s^{-1} relative to the emission maximum, similar to profiles observed in some Herbig-Be-stars. The [NII] profiles (width ≈ 150 km s^{-1} at 5%) show some structure, probably due to kinematical effects in the core region. The lobes are dominated by strong, rather symmetrical, narrow (≈ 20 km s^{-1} FWHM) emission lines. However, Hα presents the superposition of a weaker second component, which is considerably broader (total width ≈ 170 km s^{-1}) and red-shifted relative to the narrow component. We propose that this broad component represents emission originating in the central core and being scattered by dust particles in the ionized gas flowing radially outwards through the lobes. Thus, the observed velocity difference of ≈ 20 km s^{-1} between both line components directly measures the outflow velocity of the gas-dust mixture. Using this result and the observed difference in the radial velocity of the lobes, we derive an inclination angle of the polar axis with respect to the line-of-sight of ≈ 60° (M2-9) and ≈ 85° (M1-91). Assuming that the mass flow fills up the volume of a cone, we deduce an aperture angle ≈ 40° for both objects from the width of the narrow lines, comparable with the geometrical appearance of the lobe structure. Our data do not support a rotation of M2-9 around its polar axis.

HIGH-RESOLUTION SPECTROSCOPY OF NGC 7026

J. Solf
Max-Planck-Institut für Astronomie, D-6900 Heidelberg, FRG.

R. Weinberger
Max-Planck-Institut für Astronomie, D-6900 Heidelberg, FRG; and Institut für Astronomie der Universität Innsbruck, A-6020 Innsbruck, Austria

Up to now, the geometrical and kinematical structure of the well-known bizarre nebula NGC 7026 has not been discussed in the literature. Using the large vertical Coude spectrograph of the 2.2 m telescope on Calar Alto, Spain, we obtained long-slit spectra covering the nebula at 5 different position angles, in the ranges from 4730 to 5050 Å and from 6470 to 6770 Å. The high spectral (up to 6 km s^{-1} FWHM) and spatial resolution (seeing-limited ≲ 2") reveals a rather complex structure in the lines of Hα, Hβ, [OIII] λλ 4959, 5007, HeI 6678, HeII 6560, [NII] λλ 6548, 6583, and [SII] λλ 6716, 6731. Generally, the lines exhibit a double "bowed" appearance; both components consist of several condensations of small angular extent. The velocity field suggests a non-spherical expansion of an elongated thin shell structure. The observations can be explained by an ovoidal or "bipolar" configuration of the nebula consisting of an expanding equatorial toroid (V_{exp} = 54 km s^{-1} in [SII]) and two blobs moving at higher velocities outwards along the polar axis (inclination angle with respect to the line-of-sight:

75°). The geometrical and kinematical structure observed in the lines of various excitation degrees indicates a pronounced ionization stratification and allows to derive the dependence of the expansion velocities on the radial distance inside the nebulae. No noticeable extinction within the nebula has been found. The bipolar structure of NGC 7026 resembles that of some other planetary nebulae and might be caused by an equatorial concentration of the circumstellar material lost during the late phase by the progenitor asymptotic giant-branch star. Using distances and interstellar extinctions of 48 stars within 1° of the planetary, we determined a distance of 2180 (≈ ± 700) pc for the nebula.

IONIC ABUNDANCES OF S III, O IV AND NeV FROM INFRARED OBSERVATIONS OF FINE STRUCTURE LINES IN EIGHT PLANETARY NEBULAE

M.A. Shure, T.L. Herter, J.R. Houck, D.A. Briotta, Jr., W.J. Forrest, G.E. Gull and J.F. McCarthy
Center for Radiophysics and Space Research, Cornell University, Ithaca, NY 14853, USA

The Kuiper Airborne Observatory has been used to make measurements of the infrared forbidden lines of [SIII] 18.72μm, [NeV] 24.28μm and [OIV] 25.87μm in eight planetary nebulae. In all cases the beam was larger than the emitting region. The observed line fluxes are used to determine ionic abundances under the assumption of constant density throughout the relevant volume as determined by optical observations. In some cases the NeV near UV lines are used in conjunction with the infrared measurements to determine the electron temperature in the NeV emission regions. The [SIII] 33.47μm line can be used with the [SIII] 18.72μm line flux to characterize the clumping within the nebulae.

OBSERVATIONS OF THE 3.3μm EMISSION FEATURE IN PLANETARY NEBULAE

W.P.J. Martin
Max-Planck-Institut für Astronomie, Königstuhl, D-6900 Heidelberg 1, FRG

Absolute fluxes of the 3.3μm features were measured for 12 planetary nebulae. Both narrow-band photometry and low resolution spectrophotometry were used. Photometry was performed with narrow bandpass filters centered at 3.28μm and 3.72μm($\lambda/\Delta\lambda$ = 14). Using a CVF spectrophotometer, the 3.1μm - 3.8μm spectra of four nebulae (IC 418, IC 2149, NGC 6543, NGC 6572) were obtained. The values of the fluxes measured with the two different methods agree well.

The 3.3μm feature appears in the spectrum of each nebula. In addition, the spectral scans show the 3.4μm feature to be present in IC 2149, to be weak or probably absent in NGC 6572 and to be absent in IC 418 and NGC 6543.

The intensities of the 3.3μm feature of the planetary nebulae in our sample are correlated with the total infrared emission (taken from Mosley, H. 1980, Ap. J. 238, 892 and Cohen, M. and Barlow, M.J. 1980, Ap. J. 238, 585). This implies that the 3.3μm emission is associated with the major dust component. No correlations between this feature and other parameters of either the nebula or the central star were found.

LOW TEMPERATURE DIELECTRONIC RECOMBINATION COEFFICIENTS FOR IONS OF C, N AND O

H. Nussbaumer
Institute of Astronomy, ETH Zentrum, CH 8092 Zurich, Switzerland

P.J. Storey
Department of Physics & Astronomy, University College London, Gower Street, London WC1E 6BT, UK

Dielectronic recombination coefficients have been calculated for some ions of C, N and O by Storey (1981, Mon. Not. R. astr. Soc., 195, 27P). Using the same approach, we have extended those calculations to all other ions of C, N and O for which a dielectronic contribution to the total recombination coefficient might be expected at nebular temperatures. Recombination coefficients have been calculated in the temperature range from 5000 K up to the temperature at which the Burgess general formula becomes valid. The total dielectronic recombination coefficients are fitted to a simple function of the electron temperature.

RECOMBINATION SPECTRA OF PLANETARY NEBULAE

P.O. Bogdanovich, Z.B. Rudzikas
Institute of Physics of the Academy of Sciences of the Lithuanian SSR, Vilnius, USSR

T.H. Feklistova
W. Struve Astrophysical Observatory, Tartu, USSR

A.F. Kholtygin, A.A. Nikitin, A.A. Sapar
Leningrad State University, Leningrad, USSR

The lines of the transitions between the subordinate levels of the CIII, NIII etc. ions are observed in the spectra of planetary nebulae (PN) [1]. Their theoretical intensities may be found by solving the stationarity equations and accounting for both the recombination and cascade radiative transitions. It is possible to calculate the recombination spectra in various approaches: the single- or multi-configuration approximations (SCA and MCA) making use of both the superposition of configurations (SC) or the multiconfigurational Hartree-Fock-Jucys equations [2], taking into consideration the contribution of the dielectronic recombination to the intensities of the recombination lines. The energy spectra, the transition probabilities etc., as a rule ought to be calculated in the intermediate coupling scheme [2]. Both analytical or numerical (e.g. Hartree-Fock) wave functions may be adopted.

In the framework of the above-mentioned approximations we have calculated the probabilities of many transitions in the ions CIII, NIII, OIII etc. In Table 1 some transition probabilities (in 10^8 s^{-1}) are presented as examples.

The transition probabilities found were used for calculations of the intensities of the recombination lines of the ions under consideration. It turned out that the accounting for the correlation effects improves essentially the coincidence of the theoretical lines and those observed in the spectra of PN. Taking into account the two-electron transitions we can explain the appearance and calculate the intensities of the lines caused by the transitions from doubly excited configurations in CIII and NIII.

Ion	Transition	$\lambda_{aver.}$	A_{SCA}	A_{SC}
CIII	$2p3p\ ^3P - 2s5f\ ^3F$	4156	0	1.3
CIII	$2s3s\ ^3S - 2s3p\ ^3P$	4650	0.88	0.74
NIII	$2p3s\ ^2P - 2p3p\ ^2D$	4192	1.3	-
NIII	$2s^2 3p ^2P - 2s^2 3d ^2D$	4640	1.1	-
OIII	$2p3p\ ^1P \quad 2p3d\ ^3F$	2984	0.75	-

Table 1.

Table 2 contains the relative intensities (calculated $I_{theor.}$ and observed $I_{obs.}$) of some recombination lines, taking I(4656) = 1 for CIII, I(4640) = 1 for NIII and T_e = 10 000°K, observed in the spectra of PN NGC 7027 [3].

CIII	I(5696)	I(8500)	I(4070)	I(4156)	I(4187)
$I_{theor.}$	0.01	0.001	0.85	0.17	0.33
$I_{obs.}$	traces	-	0.75	0.18	0.41
NIII	**I(4097)**	**I(4379)**	**I(3999)**	**I(4192)**	
$I_{theor.}$	0.5	0.12	0.1	0.02	
$I_{obs.}$	0.6-1.7	0.12-0.22	0.2	0.01-0.03	

Table 2.

The relative intensities of the recombination lines were used to determine the relative abundance ($(X/H)\cdot 10^4$) of the ions considered in more than 20 PN [4]. Table 3 illustrates the results obtained.

PN	T_e	CIV	NIV	NV	C
7027	11500	6.4	17	0.65	17.8
7009	10600	5.8	27	-	18.1
2440	13500	7.8	38	3.0	-

Table 3.

It turned out that for some PN the abundance of C and, probably, N is higher than in the Sun.

1. Seaton, M.J.: 1980, Q.J.R.A.S., vol.21, 229.
2. Nikitin, A.A., Rudzikas, Z.B.: 1983, Foundations of the theory of the spectra of atoms and ions, Nauka, Moscow (in press).
3. Kaler, J.B.: 1976, Ap. J. Suppl., vol.31, 517.
4. Nikitin, A.A., Sapar, A.A., Feklistova, T.H., Kholtygin, A.F.: 1981, Astron. Journ. (USSR), vol.58, 101.

RADIATIVE TRANSFER EFFECTS DUE TO CURVATURE AND EXPANSION IN A DUSTY PLANETARY NEBULA

A. Peraiah
Indian Institute of Astrophysics, Bangalore, India

We have investigated the effects due to curvature and radial expansion in a planetary nebula, with hydrogen and helium. We have solved the radiative transfer equation with spherically symmetric approximation in the rest frame. We have included dust in static as well as in expanding media. The effects on the internal sources and the mean intensities at the internal points have been calculated. It is found that the effect due to the presence of dust is to reduce the mean intensities, and curvature effects on the internal sources are more pronounced than the effects due to radial expansion of the gas.

PROFILES AND INTENSITY RATIOS OF THE C IV λ1548, 1550 EMISSION LINES IN PLANETARY NEBULAE

W.A. Feibelman
Laboratory for Astronomy & Solar Physics, NASA-Goddard Space Flight Center, Greenbelt, Maryland 20771, USA

The C IV resonance doublet at λ1548, 1550 is an important diagnostic tool in the study of planetary nebulae. The predicted theoretical intensity ratio of 2 : 1 is, however, rarely observed in high dispersion

IUE spectrograms. The observed values for a sampling of 15 objects of differing excitation class range from a low of 0.74 to a high of 1.99. Variations in optical thickness, extinctions due to nebular dust, and interstellar absorption have been proposed as the cause for the deviation. Line profiles for the C IV doublet vary for different nebulae, encompassing a wide range of shapes which include: 1) narrow symmetric (Gaussian) profiles, 2) wide symmetric, 3) asymmetric with steep blue edge and extended red wing, 4) asymmetric with steep red edge and extended blue wing, 5) P Cygni profiles, 6) split peaks due to expansion velocities, and 7) multiple peaked or chaotic line structure. In addition to the main features, occasionally weak (subsidiary or ghost?) features occur that do not agree in intensity with the main line ratios but are not attributable to lines of other ions. Ultraviolet radial velocity displacements are in good agreement with optical data. Similar diversity in profiles and intensity ratios have been observed in proto-planetary nebulae and symbiotic stars. (Paper to appear in Astron. Astrophys.)

HIGH DISPERSION IUE OBSERVATIONS OF NGC 3918

M. Peña and S. Torres-Peimbert
Instituto de Astronomía, Universidad Nacional Autónoma de Mexico

Our results are based on SWP 3215, a 120 min large aperture exposure. The electron density derived from N IV], Si III] and C III] ratios are log N_e < 4.3, 5.0 and 3.5. The measured resonance line ratio 1239/1243 of N V is 1.2 ± 0.1 and 1549/1551 of C IV is 1.9 ± 0.2. Both ratios are expected to be 2.0. Furthermore, it had already been reported that the C IV and N V resonance lines are fainter than predicted from models by factors from ∿ 4 to ∿ 6 (Torres-Peimbert et al., 1980, The First Two Years of IUE, NASA CP-2171, 641). Internal dust cannot account for any significant deviation from 2 for the resonance doublet ratio of N V (Hummer and Kunasz,1980, Ap. J., 236, 609). However, the anomalous ratio of N V could be explained by a generally distributed hot ionized medium compatible with the C IV absorption found in approximately the same direction by Cowie et al. (1981, Ap. J., 248, 528).

More observations with different exposure times should be obtained; in our material the bright lines were overexposed and the ratios of C III] and C IV were derived from the wings.

ON THE STRENGTH OF THE CIV 155 nm RESONANCE LINES IN PLANETARY NEBULAE

J. Köppen, R. Wehrse
Institut für Theoretische Astrophysik, Heidelberg, FRG

Ionization models for NGC 6210, 7009, 3242 and II 2003 have been constructed from optical and IUE spectroscopic data. The CIV 155 nm resonance line is predicted about ten times stronger than observed. Radiative transfer calculations of the CIV lines in a spherical nebula, assuming partial frequency redistribution, were made to investigate the effects of dust absorption and an additional depopulation of the upper level.

The results of our calculations can be approximated by analytical formulae; the total flux F emitted in the line is decreased by dust absorption (optical depth τ_D) as $F(\tau_D)/F(\tau_D=0) = 1/(1 + \tau_D\sqrt{10})$, independent of the line optical depth ($\tau_\ell = 10^2 ... 10^4$). If the upper level is depopulated at a rate of δ times the spontaneous emission rate A of the CIV line, the emergent flux is roughly $F(\delta)/F(\delta=0) = 1/(1 + \delta\tau_\ell)$.

To weaken the CIV line by a factor of 10 in our models, we either need a dust optical depth of the C^{+3} zone of $\tau_D \sim 1$ or a de-excitation rate of $\delta A \approx 10^6\ s^{-1}$. We cannot find a process as fast as this rate. With respect to dust, we estimate from the 10 μm IR excess dust optical depths of less than 10^{-2} (cf. Köppen, 1977, Astron. Astrophys. 56, 189), which seem too small to explain the discrepancy in the CIV lines in these nebulae.

Shifts in the ionization balance due to e.g. dielectronic recombination do not appreciably alter CIV, as this line is a dominant coolant.

In these four objects, the CIV lines can be sufficiently weakened by assuming a thin shell model for the nebula with a relative thickness $\Delta R/R(\text{nebula}) \approx 0.3$. This is in accord with monochromatic images, and also brings NeIV 242 nm/NeIII 387 nm and NIV 149 nm/NIII 175 nm in agreement with observed line ratios.

THE FORMATION OF RESONANCE LINES IN GASEOUS NEBULAE PARTIALLY FILLED WITH DUST

R. Wehrse
Institut für Theoretische Astrophysik der Universität
Heidelberg, Im Neuenheimer Feld 294, D-6900 Heidelberg, FRG

The radiation field for a resonance line is calculated for a gaseous nebula which consists of a dust-free and a dust filled layer (1/2 or 1/4 of the total volume). For comparison we compute also the line profiles for corresponding homogeneous configurations, with and without dust.

The radiative transfer equation in plane parallel approximation is solved with a simplified version of the analytical method developed by Kalkofen and Wehrse. The results for various combinations of gas/dust ratios, ratio of collisional to radiative deexcitation, dust albedo and optical depth show that the dust is most efficient in reducing the total flux and the halfwidth when it is concentrated in a small volume.

PHYSICAL CONDITIONS IN THE PLANETARY NEBULA Hb 12

D.R. Flower, C.J. Penn
Department of Physics, University of Durham, UK

The planetary nebula Hb 12 has recently been observed at infrared (Aitken et al., 1979) and radio (Purton et al., 1982) wavelengths. The detection of a silicate emission feature in the 8-13 μm region suggests that the nebula is oxygen rich. A high emission measure is derived from the radio spectrum, implying a high intrinsic density.

We have observed Hb 12 with the IUE satellite and have combined these observations with optical measurements by Barker (1978) in order to determine physical conditions in the nebula. Our analysis of these observations suggests that there are two main regions in Hb 12: one emitting the [OIII] lines, with an electron density of almost 10^6 cm^{-3}, the other emitting the [OII] and [NII] lines, where the electron density is distinctly lower (although still in excess of 10^5 cm^{-3}). We find that the C/O abundance ratio is lower than the solar value by a factor of at least 2.

Aitken, D.K., Roche, P.F., Spenser, P.M. and Jones, B.: 1979, Astrophys. J. 233, 925.
Barker, T.: 1978, Astrophys. J. 219, 914.
Purton, C.R., Feldman, P.A., Marsh, K.A., Wright, A.E. and Allen, D.A.: 1982, Mon. Not. R. astr. Soc. 198, 321.

PHYSICAL CONDITIONS IN THE COMPACT PLANETARY NEBULA Sw St 1

M. Cohen
NASA Ames Research Center

D.R. Flower, A. Goharji
Department of Physics, University of Durham, UK

Sw St 1 is a compact and possibly young planetary nebula which has been recently observed at infra red (Aitken et al., 1979) and radio (Kwok et al., 1981) wavelengths. In the 8-13 μm region, a silicate emission feature is observed, suggesting that the nebular envelope is oxygen rich. The high emission measure determined from the radio observations implies a large value for the electron density.

We report optical ($4260 < \lambda < 6710$ Å) and IUE ultraviolet measurements of Sw St 1 which enable the interstellar reddening and physical conditions in the nebula to be determined. The IUE observations, taken at both low and high dispersion, serve to confirm that the envelope is oxygen rich (C/O = 0.5) and that the electron density is high (10^5 cm^{-3}).

Aitken, D.K., Roche, P.F., Spenser, P.M. and Jones, B.: 1979, Astrophys. J. 233, 925.
Kwok, S., Purton, C.R. and Keenan, D.W.: 1981, Astrophys. J. 250, 232.

OPTICAL AND UV NEBULAR SPECTRA OF NGC 40

M. Peimbert, S. Torres-Peimbert
Instituto de Astronomia, Universidad Nacional Autónoma de Mexico

R.E.S. Clegg, M.J. Seaton
Department of Physics & Astronomy, University College London, UK

The planetary nebula NGC 40 has a WC8 central star. From analysis of IUE spectra of the star, Benvenuti, Perinotto and Willis (1981, IAU Symposium 99) obtain the remarkable abundance ratio C/He $\sim$ 0.2 and probably no hydrogen. The question arises as to whether the nebular shell has abnormal abundances.

We have made observations of a bright region of nebulosity at about 14 arc sec NW of the star, optical observations at the Kitt Peak Observatory and UV observations with IUE. In the UV the fluxes in spectral lines are measured relative to those in the nebular continuum and hence put on the same scale as those measured at optical wavelengths relative to Hβ.

The abundances deduced - log N(C) = 9.1, log N(N) = 8.4, log N(O) = 9.0 - are typical of planetary nebulae and do not share the anomalies of the central star.

The C IV λ 1549 line is unexpectedly strong in the nebular spectrum. It may be produced by resonant scattering of λ 1549 from the star.

Newly-identified C II recombination lines have been detected in NGC 40 and IC 418. Their relative strengths are discussed.

THE IONIZATION STRUCTURE OF NGC 6720 AND NGC 7009

T. Barker
Wheaton College, Norton, Massachusetts, USA

Measurements of line intensities over a spectral range generally as great as 1300 Å to 11,000 Å have been made in four positions in the Ring Nebula and eight positions in NGC 7009. Ionic abudances determined from optical and UV lines are in good agreement, except that the C^{2+} abundance inferred from the optical 4267 Å recombination line is as much as 10 times higher than that measured from the 1906, 1909 Å CIII] lines. In both nebulae, this discrepancy is greatest nearest the central star. At the present time the most attractive explanation seems to be that the 4267 Å line is affected by resonance fluorescence due to light from central stars of planetaries.

The standard ionization correction formulae applied to O^+, O^{2+}, and N^+ abundances give consistent total O and N abundances in all positions in both nebulae. This result is particularly gratifying for NGC 7009 because previous studies have found that the standard formula for N gives erroneous results. These O and N abundances agree well with those found by including O^{3+}, N^{2+}, and N^{3+} abundances measured using the UV lines. The standard formula does not work for Ne, probably as a result of the different charge exchange rates for O and Ne. The average logarithmic abundances for NGC 6720 are: He = 11.04, O = 8.79, N = 8.34, Ne = 8.23:, C = 8.59, S = 6.99, Ar = 6.57. Preliminary estimates for NGC 7009 are: He = 11.06, O = 8.70, N = 8.07, Ne = 8.17:, C = 8.18, S = 7.19:, Ar = 6.40. The slightly higher N/O and C/O ratios in NGC 6720 suggest the possibility of enrichment due to mixing of processed material in the progenitor star.

The work on NGC 6720 has been published (1980, Ap. J. 240, 99; 1982, Ap. J. 253, 167); the results for NGC 7009 will be submitted to Ap. J.

CHEMICAL COMPOSITION OF THE PECULIAR PLANETARY NEBULA YM29

K.B. Kwitter
Hopkins Observatory, Williams College

G.H. Jacoby
Kitt Peak National Observatory

D.G. Lawrie
Ohio State University

YM29 is a nearby, filamentary, low surface brightness nebula whose spectrum exhibits lines of both high and low excitation. Johnson and Rubin (1971, Ap. J. 163, 151) observed YM29 as a weak radio source; despite uncertainties as to the thermal nature of the radio spectrum and their lack of a good optical spectrum, they classified YM29 as a peculiar planetary nebula. Based on photographic spectra, Liebowitz (1975, Ap. J. 196, 191) has reported that in YM29 the [N II] lines at λλ 6548, 6584 are several times the strength of Hα.

We have obtained spectrophotometry at several positions in YM29 in the region between λλ 3600-7000, using the KPNO # 1-36 with the IRS and the Perkins 72" at Lowell Observatory with the Ohio State IDS. We report here our analysis of one position in the bright northeastern portion of YM29. The following line intensities were found, after correction for interstellar extinction with c(Hβ) = 0.31:

[O II]	λ 3727 = 561		He I	λ 5876 =	11.9:
[Ne III]	λ 3869 =	55.2	[N II]	λ 6548 =	224
He II	λ 4686 =	8.7:	Hα	λ 6563 =	285
Hβ	λ 4861 =	100	[N II]	λ 6584 =	725
[O III]	λ 5007 =	266	[S II]	λ 6717 =	79.0
[N II]	λ 5755 =	11.2:	[S II]	λ 6731 =	60.0

The [N II] temperature is 10,000 ± 600 K; the [S II] density is 170 ± 50 cm^{-3}. These values lead to following the ionic concentrations (12 + log X^i/H^+):

O^+=8.34 O^{++}=7.97 N^+=8.08 S^+=6.47 Ne^{++}=7.75 He^+=10.94 He^{++}=9.85:

Correcting for unseen stages of ionization as usual according to the oxygen ionization, we derive these total abundances (12 + log X/H):

O = 8.53 N = 8.34 S = 6.66 Ne = 8.31 He = 11.12:

Despite uncertainty in the He I and He II line intensities, the total helium abundance is likely to be high due to the presence of neutral helium, particularly in this low excitation position. These enhanced helium and nitrogen abundances relative to solar or Orion values, along with the observed morphology, imply that YM29 belongs to Peimbert's Type I class of planetary nebulae, which presumably originate from more massive progenitors. Our neon abundance determination seems to suffer from the inflation commonly found for low excitation regions in nebulae.

OPTICALLY DERIVED CARBON ABUNDANCES IN PLANETARY NEBULAE

H.B. French
Department of Physics & Astronomy, University of Oklahoma, USA

Gas phase carbon abundances have been determined for a number of bright planetary nebulae from new photoelectric measurements of optical recombination lines (C II λ 4267, C III λ 4650 and C IV λ 4659). Because of blending problems for the latter two features, the abundance for any object may have substantial errors, but the average abundance should be reliable. For the twelve best observed planetaries, this average is $C/H = (8.4 \pm 2.9$ s.d.$) \times 10^{-4}$ by number. If it is assumed that the planetary progenitors had essentially solar abundances, then, based on Cameron's most recent results ($C/H = 4.2 \times 10^{-4}$, $O/H = 6.9 \times 10^{-4}$), it appears that the planetary ejecta have been enriched in carbon, presumably because of dredging of newly synthesized triple α carbon by helium shell flashes during the late evolution of the progenitor. The helium abundance is also slightly high, as would be expected in this interpretation. Since the mean planetary carbon abundance exceeds the solar oxygen abundance, it is possible that the progenitor became a carbon star prior to the ejection of the planetary; that may even have caused the ejection. Because the planetaries in this study were drawn from Peimbert's samples with relatively low mass progenitors ($1 - 2.5\ M_{\odot}$; these are not significantly helium- and nitrogen-rich objects), such a process might be a general feature of late double shell source evolution.

From simple estimates it appears that the contribution to Galactic carbon enrichment from planetaries is comparable to that from supernovae. Because the former represent long-lived stars, and the latter short-lived, we might expect that carbon and oxygen abundances in the Galaxy would not vary in lockstep in the way that, for example, neon and oxygen abundances appear to.

Further observations are underway both to expand the sample of planetaries, and, through higher spectral resolution, to improve the abundances in individual objects.

CHEMICAL ABUNDANCE DETERMINATIONS IN GASEOUS NEBULAE

Sueli M.V. Aldrovandi
Instituto Astronomico e Geofisico U.S.P., Av. Miguel Stefano, 4200 04301 - Sao Paulo, Brazil

The determination of the chemical abundances in planetary nebulae and HII regions from the observed intensities of emission lines usually needs a correction for the unseen ionization stages. An empirical correction was first introduced by Peimbert and Costero (1969) as a factor ICF (ionization correction factor) which was obtained assuming that ions with similar ionization potentials are equally populated. Recently, Natta et al. (1980) discussed the problem of sulphur abundance in nebulae and proposed an ICF for sulphur which takes into account the radiation field as well as the ionization equilibrium of O^+, O^{++}, S^{++} and S^{+3}. However they did not consider the effect of the charge exchange reactions which can be very important in determining the ionization equilibrium of the gas (Péquignot et al., 1978).

In this paper calculations for the ICFs of N, O, Ne, S and A are presented following the method proposed by Natta et al. but introducing the charge transfer reactions (Butler et al., 1980; Butler and Dalgarno, 1980). For each object the ICFs depend on the effective temperature of the central star and on the ratio n_{HI}/n_e. Taking this ratio equal to 10^{-2} or 10^{-3} and the effective temperature indicated by the He lines, the method for determining abundances was applied for several planetary nebulae and HII regions. Our results are usually less than or equal to those previously obtained (French, 1981; Natta et al., 1980; Torres-Peimbert and Peimbert, 1977). Good agreement was obtained for O and A abundances.

Butler, S.E. and Dalgarno, A.: 1980, Ap. J. 241, 838.
Butler, S.E., Heil, T.G. and Dalgarno, A.: 1980, Ap. J. 241, 442.
French, H.B.: 1981, Ap. J. 246, 434.
Natta, A., Panagia, N. and Preite-Martinez, A.: 1980, Ap. J. 242, 596.
Peimbert, M. and Costero, R.: 1969, Bol. Obs. Tonantzintla y Tacubaya 5, 3.
Pequignot, D., Aldrovandi, S.M.V. and Stasinska, G.: 1978, Astron. Astrophys. 63, 313.
Torres-Peimbert, S. and Peimbert, M.: 1977, Rev. Mex. Astron. Astrofis. 2, 181.

ABSTRACTS OF CONTRIBUTED PAPERS

POSTER SESSION II

OBSERVATIONS OF THE 30 μm FEATURE IN IRC + 10216

T. Herter, D.A. Briotta, Jr., G.E. Gull, J.R. Houck
Center for Radiophysics & Space Research, Cornell University, Ithaca, NY 14853, USA

New observations of the unidentified 30 μm spectral feature found in two planetary nebulae and four carbon stars are reported. The carbon star IRC + 10216 was observed from the Kuiper Airborne Observatory over the range 30 to 37 μm. These new data confirm the existence of the feature and determine its shape beyond 30 μm. The determination of the long wavelength end of the spectral excess depends critically on the assumed underlying continuum. A combined spectrum of the excess is presented and shows the feature to extend at least to 37 μm.

This work was supported by NASA grant 33-010-081.

OH/IR STARS: DARK PLANETARY NEBULAE?

J. Herman
Sterrewacht, Huygens Laboratorium, Leiden

The OH maser emission from OH/IR stars (type IIb) originates from the thick circumstellar dust shell. As the envelope is expanding and the strongest maser emission comes from the front- and the backside, where the gain pathlength is longest, we can measure a definite phaselag (and hence a diameter) between front- and backside of the shell for all sources that are variable.

A monitor program for 60 OH masers with the Dwingeloo Radio Telescope shows that most OH/IR stars are variable, with periods up to 2000 days. They appear to be extreme members of the Mira variables, the precursors of planetary nebulae. The monitor program reveals phase delays in the right sense, giving diameters of the envelopes typically of 10^{16} - 10^{17} cm. This is comparable with the sizes of the more massive planetary nebulae.

V.L.A. maps show pointlike structure for the strongest peaks and ring- or extended structure for the emission coming from halfway between the outer peaks, consistent with an expanding shell model. Combination of the two kinds of observations yields a direct measure of the distance and the three dimensional structure of the envelope.

THE MASER STRENGTH OF OH/IR STARS, THE EVOLUTION OF MASS LOSS AND THE FORMATION OF A PLANETARY NEBULA

B. Baud
Laboratorium voor Ruimteonderzoek, Groningen

H.J. Habing
Sterrewacht, Leiden

From observations we find that the OH luminosity L_{OH} of an OH/IR star increases with R^2, where R is the size of the masing region. From this correlation we deduce that the mass loss rate M, the expansion velocity v_e and L_{OH} are related by $L_{OH} \sim (M/v_e)^2$. Next we consider the large range that is observed in L_{OH} and the steep OH luminosity distribution for OH/IR stars. Both facts can be explained by the postulate that these objects undergo accelerated mass loss, and thus steadily increase their OH luminosity. We propose that OH/IR stars are at the extreme end of the Asymptotic Giant Branch and that many of them are in the process of blowing off their entire envelope in a superwind phase. Their mass loss rate during this superwind, as deduced from OH observations of the circumstellar shell, is given by a simple modification of the Reimers equation. This modification connects the superwind continuously to the Reimers wind and it provides observational evidence for the formation of a planetary nebula.

CATALOGUE OF CENTRAL STARS OF PLANETARY NEBULAE

A. Acker
Observatoire de Strasbourg, France

The catalogue contains 460 nuclei of 393 true and 67 possible planetary nebulae; 87 of these were discovered after the publication in 1967 of the catalogue of planetary nebulae of Perek and Kohoutek.

Produced by A. Acker (Strasbourg), M. Chopinet (Bordeaux), F. Gleizes (Montpellier), J. Marcout (Strasbourg), F. Ochsenbein (Centre de Données Stellaires de Strasbourg) and J.M. Roques (Montpellier), with the collaboration of the Observatoire de Haute Provence.

1. The data of observation, as well as bibliographical sources, are presented:

(a)-Denominations: the names HD, BD, CPD, ... of 90 stars are given; for all 460 objects the PK designation and the "usual name" of the nebula are indicated.

-Coordinates: the values α and δ are given for 1950, 1985, and 2000; annual precession refers to 1950. The XY coordinates measured on the Palomar or ESO charts are indicated, with the mean diameter of the nebula.

-Magnitude: the available photometric values or estimates are indicated for about 400 stars.

-Spectrum: for about 160 stars, a spectral study has been made.

-Velocity: the radial velocity is given for 249 objects, as well as the velocities of expansion and of the stellar wind if available. The proper motion has been given for 75 objects.

(b)-Distances have been listed for 265 objects; bibliographic references are given for the calculation of temperatures, luminosities, radii,... of the nuclei.

(c)-Notes: the possible nature of the object, the data concerning the binarity, variability, and other peculiarities are indicated.

2. Bibliographic references are listed. We give authors, year, name of the journal, volume and page numbers, and title of the article (682 papers).

3. Finding charts are provided for every object:
-BD or Cordoba map - field: $2^{o} \times 2^{o}$
-map from Palomar or ESO - field: 19' x 25'

This catalogue determines the direction of future observations, and brings out certain general properties of the planetary nebulae and their nuclei.

APPARENT MAGNITUDES OF PLANETARY NEBULAE NUCLEI

R.A. Shaw, J.B. Kaler
Astronomy Department, University of Illinois at Urbana-Champaign, USA

B and V magnitudes for the central stars of a number of planetary nebulae are presented. The observations were obtained between 1971 and 1981 with the University of Illinois one-meter telescope at Prairie Observatory. The average magnitudes presented are accurate extractions of the stellar continuum flux from the total (stellar plus nebular) measured flux (see Kaler, 1976, Astrophys. J., 210, 113).

The nebular continuum flux was calculated upon the best available values (and associated uncertainties) of the measured Hβ flux, the electron temperature, electron density (from which we obtained the contribution from 2-quantum emission), the He^+/H^+ and He^{++}/H^+ ratios, and the logarithmic extinction at Hβ. The uncertainties in the above quantities were propagated through the entire calculation to provide a correct evaluation of the resulting uncertainty in the quoted magnitude. Finally, when the central star contributed only a minimal fraction of the continuum, we were able to set realistic upper limits to the magnitudes.

The method used here is the best available for the determination of B and V central star magnitudes, and is probably the only reliable method for compact planetaries. As a test case, the B and V magnitudes for the nucleus of NGC 7662, which contributes only ≈ 20% of the total nebular continuum, agree well with those derived from the IUE data by Harrington et al. (1982, M.N.R.A.S., 199, 517).

UBV-OBSERVATIONS OF VARIABLE PLANETARY NEBULAE

E.B. Kostyakova
Sternberg State Astronomical Institute, Moscow, USSR

The UBV- and spectral observations of several variable planetary nebulae were continued at the Crimean Station of Sternberg Astronomical Institute.

During 1968-1981 the planetaries NGC 6572, IC 4997, Hu 2-1, and NGC 6891 showed systematic changes of total brightness within the ranges of $0\overset{m}{.}2 - 0\overset{m}{.}4$. Moreover, the nebula NGC 6572 became progressively brighter in filter V, but IC 4997 fainter in each of three filters. At the same time, the nebulae NGC 6720 and IC 3568 showed no variations of brightness exceeding $0\overset{m}{.}1$. (see Figure).

The preliminary results of spectral observations are in agreement with the photoelectric measurements, namely, the nebulae with larger changes of the UBV-brightness show larger variations of the emission line intensities.

The method of our photoelectric study was described in Mém. Soc. Roy. Sci. Liège, 6-e sér., tome V, p.473, 1973; the summary results of the UBV-observations in 1968-1980 are published in the Astron. Circular Acad. Sci. USSR, No.1166, p.4, 1981.

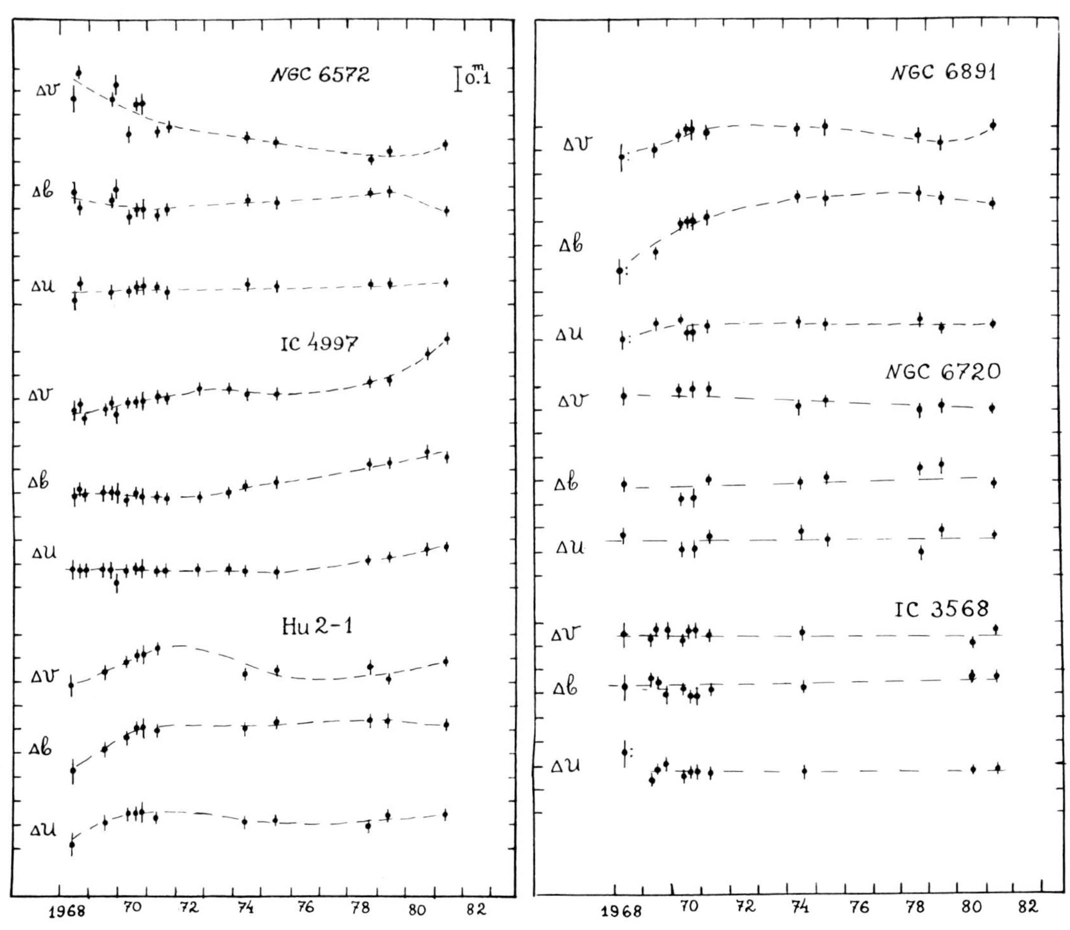

CONCERNING THE TEMPERATURES OF CENTRAL STARS OF PLANETARY NEBULAE

L. Kohoutek
Hamburg Observatory, Gojenbergsweg 112, D-2050 Hamburg 80, FRG

W. Martin
Max-Planck-Institut für Astronomie, Königstuhl, D-6900 Heidelberg 1, FRG

Recently Pottasch (1981, Astron. Astrophys. 94, L13) published extremely high effective temperatures of some central stars of planetary nebulae (> 200 000 K). Our study of planetary nebulae based on photoelectric photometry does not confirm his results. A histogram of T_z(HI) and T_z(HeII) shows smooth distribution of T_z with the maximum of about 48 000°K (HI) and 90 000°K (HeII), respectively; the effective temperature of none of the 62 planetary nuclei exceeds 120 000°K. We believe that the stellar temperatures reported by Pottasch are strongly overestimated due to the unreliable stellar magnitudes used; this conclusion follows from the investigation of the seven objects being common in Pottasch's and our sample:

Name	Design.	log T_z He II	log T_z H I	log $L/L_\odot$	$R/R_\odot$
IC 2165	221-12°1	5.01	4.78	3.86	0.27
J 900	194+ 2°1	4.97	4.74	3.84	0.32
NGC 2440	234+ 2°1	5.05	4.84	3.05	0.089
NGC 6565	3- 4°5	4.94	4.84	3.14	0.16
NGC 6741	33- 2°1	4.94	4.66	3.46	0.24
NGC 6884	82+ 7°1	4.92	4.75	3.71	0.34
NGC 6886	60- 7°2	4.97	4.74	4.30	0.54

In the H-R diagram these central stars lie in the range of the stellar mass 0.55 - 0.8 $M_\odot$ (Fig. 1).

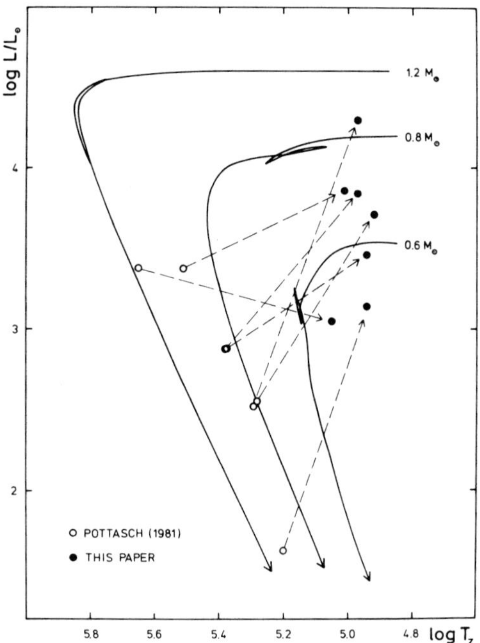

Figure 1. H-R diagram for seven planetary nuclei of our sample: open circles are results of Pottasch, filled circles are results of this paper. Solid curves are the predictions of Paczynski (1971).

UV RADIATION FROM CENTRAL STARS OF PLANETARY NEBULAE

M. Cerruti-Sola, M. Perinotto
Osservatorio Astrofisico di Arcetri, Firenze, Italy

C. Cacciari, P. Patriarchi
ESTEC Astronomy Division, Villafranca, Spain

The flux originating from the central stars of 27 planetary nebulae in the spectral range 1200-2000 Å has been deduced from the analysis of a large number of released IUE low resolution spectra.

The stellar UV continuum has been compared with black-body energy distributions. Preliminary colour temperatures have been derived for some central stars.

The main conclusions are:

1. Central stars of planetary nebulae with O-Of, OVI and continuous spectra appear to be reproduced in the interval 1200-2000 Å by black-body of very high to infinite temperature.

2. Considering only the spectral range 1500-2000 Å, a better representation is obtained with temperatures of 30-50,000 K for O-Of stars, or with substantially higher temperatures for objects with OVI or continuous spectra.

From these results it is evident that the spectral range 1200-1500 Å is particularly important for the evaluation of the colour temperature of central stars of planetary nebulae and that black-body temperatures deduced from the spectral range $\lambda > 1500$ Å cannot represent the 1200-1500 Å interval.

ULTRA-VIOLET SPECTROPHOTOMETRY OF SOME HOTTER CENTRAL STARS

R.E.S. Clegg, M.J. Seaton
University College London, UK

A spectrophotometric survey has been made for about 20 central stars of planetary nebulae, with emphasis mainly on hot stars. We use low-resolution IUE spectra, observed by ourselves or obtained from Data Center, together with, in some cases, results from optical observations. Data have been extracted and merged, regions of saturation eliminated, ITF errors corrected and nebular continua subtracted. Careful assessments have been made of reddening constants, and of data used to calculate Zanstra temperatures.

The stellar energy distributions have been compared with those for black-bodies, LTE line-blanketed models, and NLTE models. We find no evidence for conflict between colour temperatures and He II Zanstra temperatures (allowing in some cases, for the possibility of incomplete absorption by He II).

Our study includes the seven stars studied spectroscopically by Heap (1977) and the six stars for which Mendez et al. (1981) have made NLTE models. For all but one of the objects considered by Mendez et al., black-body colour temperatures and He II Zanstra temperatures are higher than those obtained from analysis of spectral features using static plane parallel NLTE models.

Our results do not provide confirmation for evidence of absorption by H_2^+, discussed by Heap and Stecher (1980).

AN OPTICAL AND ULTRAVIOLET STUDY OF NINE LOW-EXCITATION PLANETARY NEBULAE

S. Adams, M.J. Barlow
University College London, UK

Ultraviolet and optical observations of seven very low excitation nebulae (He 2-131, He 2-138, TC 1, M 1-11, M 1-12, M 1-26, H 2-1) and of two low-excitation nebulae (IC 418 and He 2-108) are discussed.

The very low excitation (VLE) objects are classed as having I([OIII] λ 5007)/I(Hβ) < 1, while the two low-excitation objects have [OIII]/Hβ < 2. No HeII emission lines are seen. Electron temperatures and densities are determined from forbidden line ratios. A nebular density gradient is inferred for He 2-108 and H 2-1. The C/O ratio is determined for IC 418, He 2-131 and TC 1; upper limits on C/O are obtained for He 2-138 and He 2-108.

IC 418, which has C/O > 1 (Harrington et al., 1980), is known to show SiC in emission between 10 and 12 μm. We have no UV data on M 1-11 which also shows SiC in emission. However He 2-108, He 2-131, He 2-138 and TC 1 all have C/O < 1, consistent with the infrared data which indicate silicate emission from the three that have been observed.

The reddening constant c(Hβ) is deduced from the λ 2200 interstellar absorption feature for six of the objects. In all cases the resultant value is less than that obtained from Radio/Hβ measurements. The discrepancy is particularly large for the case of M 1-26.

The stellar continuum of each object has been examined. Significant line blanketing occurs between 1200 Å and 1900 Å in all of the objects for which we have data. The HI and HeI Zanstra temperatures are found to be grossly inconsistent with the optical and UV colour temperatures when using black-body models, LTE line-blanketed plane-parallel models (Kurucz, 1979), or NLTE plane-parallel models (Mihalas, 1972). By contrast, the optical and UV continuum of IC 418 is well matched by a spherically extended NLTE model atmosphere with $T(\tau = 2/3) = 29\ 700$ K and log $g(\tau = 2/3) = 2.96$ (Model 6 of Kunasz, Hummer and Mihalas, 1975). This model also predicts the observed nebular fluxes in Hβ and HeI λ 5876 to within 15%. It is suggested that cooler extended model atmospheres will be required to adequately represent the other objects in our sample.

Harrington, J.P., Lutz, J.H., Seaton, M.J. and Stickland, D.J.: 1980, Mon. Not. R. astr. Soc., 191, 13.
Kunasz, P.B., Hummer, D.G. and Mihalas, D.: 1975, Astrophys. J., 202, 92.
Kurucz, R.L.: 1979, Astrophys. J. Suppl. 40, 1.
Mihalas, D.: 1972, NCAR-TN/STR-76.

NEBULAR ABUNDANCES AND CENTRAL STAR PARAMETERS FOR EIGHT PN IN THE MAGELLANIC CLOUDS

M.J. Barlow, S. Adams, M.J. Seaton, A.J. Willis
University College London, UK

A.R. Walker
South African Astronomical Observatory

We have obtained ultraviolet and optical spectrophotometry for two PN in the SMC (N87, L302) and for six in the LMC (N28, N66, N97, N141, N201, N203). The data were obtained with the IUE (eight nebulae), the 3.9 m AAT (six nebulae) and the 1.9 m SAAO reflector (two nebulae). Nebular temperatures, densities and abundances are presented. The nebular continua were calculated and subtracted from the observed continua, allowing the central star energy distributions and hydrogen and helium Zanstra temperatures to be derived. Our results for the effective temperatures and absolute magnitudes and luminosities of the central stars are presented, including data on two Wolf-Rayet central stars (WC5 and WC8) and one Of central star.

The 1150-2000 Å spectrum of LMC N97 is extremely rich and includes lines of SiII, SiIII] and FeIII. The overall spectrum appears to be a counterpart to that of the bipolar nebula NGC 6302 with, for example, significant amounts of nitrogen present in all stages from NI to NV. We find T_e(OIII) = 20,000 K, T_e(NII, SII, OII) = 13,000 K and N_e(SII, OII) = 2.6×10^3 cm^{-3}. The helium abundance, N(He)/N(H) = 0.18, agrees with that found by Osmer (1976) and is the same as that found by Aller et al. (1981) for NGC 6302. The heavy elements in N97 have about half the abundance relative to H compared to NGC 6302. We find for N97: N/H = 4(-4), O/H = 2.5(-4), C/H = 4(-5), Ne/H = 5.7(-5).

Aller, L.H., Ross, J.E., O'Mara, B.J. and Keyes, C.D.: 1981, MNRAS, 197,95
Osmer, P.S.: 1976, Astrophys. J., 203, 352.

WHY IS IC 4642 OF SUCH HIGH-EXCITATION CLASS?

C.J. Penn, D.R. Flower
Department of Physics, University of Durham

M.J. Barlow, M.J. Seaton
Department of Physics & Astronomy, University College London

L.H. Aller
Department of Astronomy, University of California, Los Angeles

We have observed IC 4642 with the AAT and the IUE. It is of exceptionally high-excitation class, as judged by ratios such as He II/Hβ and [Ne V]/He II.

We compare IC 4642 with the high-excitation planetary NGC 7662. They are found to have similar chemical compositions and both have central stars with T_Z (He II) ≃ 113 000 K. Values of T_Z (H I) are much smaller indicating that they are optically thin for H I. We assume similar nebular masses. The two stars are then found to have similar luminosities and the ratio of nebular radii is found to be R(IC 4642)/ R(NGC 7662) = 1.35.

That IC 4642 is of higher excitation class than NGC 7662 can be explained as a consequence of the difference in radii. The optical depths $\tau(\nu)$ for photo-ionization of He II are estimated using observed strengths of He I and He II lines and the following results obtained:

	NGC 7662	IC 4642
Threshold for He II photo-ionization	23	2.5
" " Ne IV " "	4	0.4 .

In both nebulae most quanta beyond the He II threshold are absorbed; it follows that the He II/Hβ ratios are proportional to R^3 which explains the difference in observed ratios. Beyond the Ne IV threshold NGC 7662 remains optically thick in the He II continuum but IC 4642 becomes optically thin; in consequence more quanta are available for the ionization of Ne IV in IC 4642 and the Ne V/Ne IV ratio is much larger.

EXTINCTION - DISTANCES TO PLANETARY NEBULAE

R. Gathier, S.R. Pottasch
Kapteyn Astronomical Institute, Groningen, The Netherlands

Individual distances to planetary nebulae (PN) which are independent of any assumption of average nebular characteristics, can be found if one knows the relation between interstellar extinction (E(B-V)) and distance along the line-of-sight to the PN, together with the E(B-V) towards the PN itself (Lutz, 1973 and Acker, 1978). We used VBLUW-photometry (Lub and Pel, 1977) to derive accurate E(B-V)'s and distances of stars up to V-magnitude + 14, within $0^{o}.3$ from the PN. Table 1 lists the PN we studied. The E(B-V)'s of the PN are derived from:

1) a comparison between radio-flux and H_{β}-flux
2) He II line intensities (Seaton, 1978)
3) 2200 Å feature (Pottasch et al., 1977)

NGC 2346	NGC 3918
NGC 2440	NGC 5189
NGC 2452	NGC 5315
NGC 2792	He2 -131
NGC 2867	NGC 6565
NGC 3132	NGC 6567
NGC 3211	

Table 1.

Acker: 1978, Astron. Astrophys. Suppl. 33, 367
Lub and Pel: 1977, Astron. Astrophys. 54, 137
Lutz: 1973, Ap. J. 181, 135
Pottasch, Wesselius, Wu and van Duinen: 1977, Astron. Astrophys. 54, 435.
Seaton: 1978, M.N.R.A.S. 185, 5P.

KINEMATIC DISTANCES OF PLANETARY NEBULAE

W.J. Maciel
Instituto Astronomico e Geofísico da USP, Sao Paulo, Brazil

S.R. Pottasch
Kapteyn Astronomical Institute, Groningen, The Netherlands

Most distances of planetary nebulae are known roughly within a factor of 2 or larger, except for some special objects for which the uncertainty can be as low as 50%. In the present work both IUE interstellar Lyman alpha profiles and 21 cm HI line surveys are used to infer the distances of four planetary nebulae.

From the measured Lyman alpha equivalent width the column density of neutral H can be determined. On the other hand, for a given 21 cm profile a LSR velocity can be obtained which corresponds to the same amount of gas as that producing the Lyman alpha absorption. Through the use of a galactic rotation curve the distance of the nebula can be estimated.

This procedure has been applied to the planetary nebulae NGC 7009, BD + 30°3639, NGC 7662 and IC 418, for which both Lyman alpha profiles and 21 cm observations were available. For these nebulae kinematic distances were obtained as well as plots of the H column density against distance. The results have been compared with distances obtained by different authors, in particular with extinction distances, which are generally considered as accurate. Finally, a discussion is included on the main sources of error on the derived distances.

This work was partly supported by CNPq and FAPESP - Brazil.

DISTANCE DETERMINATIONS FROM 21 cm INTERSTELLAR ABSORPTION-LINE MEASUREMENTS

S.R. Pottasch, R. Gathier and W.M. Goss
Kapteyn Astronomical Institute, Groningen, The Netherlands

HI observations at 21 cm have been made with the Westerbork Synthesis Radio Telescope in the direction of six planetary nebulae located in or near the galactic plane (N 7027, 2440, 6537, 6572, 7026, 7354).

Measurements of 15 other nebulae are in progress. The spectral resolution is 4 km s^{-1} with a total velocity range of about 125 km s^{-1}.

In all nebulae at least one absorption line was present, due to absorption in the interstellar medium in the local arm. For several of the nebulae a second absorption line was found, which from its velocity can be identified with the Perseus-arm or the Sagittarius-arm, indicating that these nebulae are either in or beyond the respective arm. Upper limits to the distance can be found by noting the absence of absorption from an arm further away. For example: the 21 cm line profile of NGC 2440 shows, besides local absorption, an absorption at a LSR-velocity ≈ +25 km/s. Assuming that this absorption is caused by galactic HI, this leads to a kinematic distance ≈ 2 kpc. The resultant distances are probably accurate to within 50%.

As a by-product of this work a limit can be placed on the amount of neutral hydrogen associated with the nebula as indicated by absorption at the nebular velocity. The value is often quite low. The results for NGC 7027 have recently been published (Pottasch et al., 1982).

Pottasch, S.R., Goss, W.M., Arnal, E.M., Gathier, R.: 1982, Astron. Astrophys. 106, 229.

OH/IR STARS NEAR THE GALACTIC CENTRE

H.J. Habing, F.M. Olnon
Sterrewacht, Huygens Laboratorium, Leiden

A. Winnberg, H.E. Matthews
Max-Planck-Institut für Radioastronomie, Bonn

B. Baud
Radio Astronomy, University of California, Berkeley

We have detected 34 OH/IR stars within 1 degree of the galactic centre by their OH emission line at 1612 MHz (18 cm) using the Effelsberg 100 m telescope and the Very Large Array. The spatial distribution and the distribution of the radial velocities show that practically all stars are within 150 pc from the Galactic centre, and that the number of foreground objects is very small. The projected distribution of the stars is similar to that of the surface brightness at 2.4 μm. Since the 2.4 μm radiation is supposed to be due to red giants, the OH/IR stars are probably members of the same population. The stars have considerable random velocities (velocity dispersion in one coordinate of 150 ± 50 km s^{-1}), but show general Galactic rotation. The high velocity dispersion is remarkable for objects of this population.

VELOCITY DISPERSION AND LUMINOSITY FUNCTION OF PLANETARY NEBULAE IN THE NUCLEAR BULGE OF M31

D.G. Lawrie
Ohio State University, USA

H.C. Ford
Space Telescope Science Institute

We used a sequence of velocity-modulated photographs to find and measure the radial velocities of faint planetary nebulae in the center of M31. The photographs were made with a Velocity Modulating Camera (VMC) which consists of a temperature-tuned 2.1 Å (FWHM) [O III] λ 5007 interference filter, a cooled, two-stage image intensifier, and a calibrating photomultiplier. The camera was mounted at the Cassegrain focus of the Shane 3 m telescope at Lick Observatory. We identified 19 new planetary nebulae, bringing the total number of known planetaries within 250 pc of M31's nucleus to 45. From the plate series, we derived radial velocities and relative brightnesses from 32 of the nebulae and placed radial velocity limits on the remaining nebulae in the field. By applying the method of maximum likelihood to the observed radial velocity distribution, we derive a mean heliocentric velocity of -309 (±25) km s^{-1} and a velocity dispersion of 155 (±22) km s^{-1} for the planetary nebulae.

The first three magnitudes of the planetary nebulae luminosity function, after correction for interstellar extinction in M31, is given by n(mag) = constant. We derive a planetary-to-luminosity ratio (PLR) of 69 ± 9 for the luminosity which corresponds to an integrated blue magnitude, m_B = 8.37. We combined the PLR with M31's integrated magnitude to estimate that there are 2800 ± 350 nebulae in the first three magnitudes of M31's luminosity function. By combining our observed luminosity distribution with Jacoby's (1980) Magellanic Cloud distribution, we estimate that M31 has 21,000 ± 2600 planetary nebulae within 8 magnitudes of the brightest nebulae.

SPECTROSCOPY OF THE PLANETARY NEBULA IN THE FORNAX GALAXY WITH THE INTERNATIONAL ULTRAVIOLET EXPLORER

L.H. Aller
Department of Astronomy, University of California, Los Angeles, California 90024, USA

S.P. Maran, T.R. Gull, T.P. Stecher
Laboratory for Astronomy and Solar Physics, NASA-Goddard Space Flight Center, Greenbelt, Maryland 20771, USA

The planetary nebula in the Fornax dwarf galaxy has been detected with the short wavelength spectrograph of the International Ultraviolet Explorer (IUE). At an estimated distance of 150 kpc, this nebula is probably the furthest planetary that can be observed with IUE. The object is the only known planetary nebula in the Fornax galaxy, which is a system of lower metallicity than the Small Magellanic Cloud. Individual exposures of 5 to 7 hours reveal continuum in the wavelength range from Lyman alpha to 1900 Å and C III] λ 1909 Å, the strongest emission line in this spectral region. Four such exposures have been coadded with the PDP 11/44 computer at the IUE Regional Data Analysis Facility at NASA-GSFC. The results, compared with the ground-based spectral measurements of Danziger et al. (1978), indicate that locally synthesized carbon was dredged to the envelope of the progenitor star of this nebula.

A NEW PLANETARY NEBULA WITH INDEPENDENTLY DETERMINED DISTANCE AND MASS

K.G. Henize
NASA Johnson Space Center

A.P. Fairall
University of Cape Town

A low surface brightness planetary nebula with weak ring structure and a diameter 22 arcsec has been discovered in the southwest edge of the open cluster NGC 6067. A calibrated spectrum shows strong He II 4686 (80% Hβ) and an Hα/Hβ ratio of 5.8. The Hα/Hβ ratio yields c = 0.88 which corresponds to $A_V = 1.88$. This is significantly greater than A_V for the cluster (1.17) and leads to a distance for the nebula of 3370 pc assuming extinction to be uniform with distance. This leads to a radius of 0.18 pc and a mass of 0.07 solar masses (assuming ε = 0.5).

SPECTROSCOPY OF EXTRAGALACTIC PLANETARY NEBULAE IN THE ULTRAVIOLET

T.R. Gull, S.P. Maran, T.P. Stecher
Laboratory for Astronomy and Solar Physics, NASA-Goddard Space Flight Center, Greenbelt, Maryland 20771, USA

L.H. Aller
Department of Astronomy, University of California, Los Angeles, California 90024, USA

Three high-excitation planetary nebulae in the Magellanic Clouds were successfully observed with the International Ultraviolet Explorer. Emission lines as well as nebular and stellar continua were detected. Fluxes in the lines 1550 C IV, 1640 He II, 1663 O III, and 1909 C III were measured in spectra of LMC P40, SMC N2, and SMC N5 obtained with the IUE short wavelength spectrograph; 2422 Ne IV was measured in P40 with the long wavelength spectrograph. The data were analyzed together with groundbased observations by Aller in order to derive ionization models and the nebular abundances of He, C, N, O, S, Ar. The C abundances are as large as those typically found in galactic planetaries, although the interstellar media of the Clouds are notably deficient in C. Thus, the C was synthesized in the progenitor stars and presumably was lifted to the stellar envelopes by convection prior to the ejection of the nebulae. Other planetary nebulae in the Clouds, as well as the planetary nebula in the Fornax galaxy, may be observable with IUE.

DISCOVERY OF A LARGE HIGH-EXCITATION PLANETARY NEBULA

J.N. Heckathorn
Computer Sciences Corporation

R.A. Fesen, T.R. Gull
NASA Goddard Space Flight Center

The discovery of a new planetary nebula from [O III] interference filter imagery is reported. The nebula, PN 136 + 5°1, is 15' in diameter and asymmetric in appearance. Spectrophotometry of the central regions indicate it is of excitation class between 8 and 10. A single bright condensation within the nebula exhibits emission lines of much lower ionization with [S II] λλ 6717, 6731 line intensities indicating an electron density of about 350 ± 150 cm^{-3}. Seeing limited [O III] imagery of this planetary nebula is also presented. (This article is appearing in Astron. Astrophys. in the near future).

BIPOLAR NEBULAE AND TYPE I PLANETARY NEBULAE

N. Calvet
Centro de Investigación de Astronomía, Mérida, Venezuela

M. Peimbert
Instituto de Astronomía, Universidad Nacional Autónoma de México

It is suggested that the bipolar nature of PN of Type I can be explained in terms of their relatively massive progenitors ($M_i \geqslant 2.4\ M_\odot$), that had to lose an appreciable fraction of their mass and angular momentum during their planetary nebula stage. The following objects are discussed in relation with this suggestion: NGC 6302, NGC 2346, NGC 2440, CRL 618, Mz-3 and M2-9. It is found that CRL 618 is overabundant in N/O by a factor of 5-10 compared with the Orion Nebula.

WIND-BLANKETED STELLAR ATMOSPHERES

D.C. Abbott, D.G. Hummer
Joint Institute for Laboratory Astrophysics

Radiation scattered by a stellar wind back into the photosphere alters the temperature-depth relation and thus the stellar flux distribution. The fraction of the radiation returned to the star at every wavelength has been calculated using stellar wind models accounting for approximately 10 000 lines. Model stellar atmospheres containing hydrogen and helium, both with and without the assumption of LTE, have been computed allowing for the reflected radiation. For realistic wind and stellar parameters relevant to central stars of planetary nebulae, we obtain a 25% increase in the surface temperature and in the optical brightness temperature, and a 2 order-of-magnitude increase in the flux in the He II continuum.

THE TEMPERATURES OF CENTRAL STARS OF PLANETARY NEBULAE: THE ENERGY-BALANCE METHOD

A. Preite-Martinez
Istituto di Astrofisica Spaziale, Frascati, Italy

S.R. Pottasch
Kapteyn Astronomical Institute, Groningen, The Netherlands

We present a method for determining the colour temperature of the ionizing continuum of the central star photoionizing a surrounding nebula. The method is based on the assumption that energy-balance holds in the photoionized nebula, and it is a generalization of Stoy's first derivation (Stoy, 1934) to a variety of possible situations in actual nebulae, namely to optically thin (Case I), partially thick (Case II), and completely optically thick nebulae (Case III).

FABRY-PÉROT RADIAL VELOCITIES OF S274: A PLANETARY NEBULA

E. Recillas-Cruz, P. Pismis
Instituto de Astronomia, UNAM, Mexico

The bright nebula S274 (YM29), 8' across has been classified as a planetary by Abell (1966) although it has been considered a SNR by other authors. We have determined radial velocities at 173 points on this nebula from four Fabry-Pérot interferograms. The velocity field exhibits a great deal of structure. The average expansion velocity is + 31.5 ± 8 km s^{-1}. The mean radial velocity of S274 is not well determined due to the nature of the velocity field, while the overall velocity (173 points) is + 33 ± 21 km s^{-1}. Points at the outer boundary yield an average of 22 ± 14 km s^{-1} while the average of the double points is 25 km s^{-1}. The age of expansion of the nebula is estimated at 6.8 x 10^3yr. The physical parameters of this object are consistent with those of a planetary nebula.

THREE SYMBIOTIC STARS CATALOGUED AS PLANETARY NEBULAE

L. Carrasco, R. Costero, A. Serrano
Instituto de Astronomía, Universidad Nacional Autónoma de México

Three stellar-like objects, previously identified as planetary nebulae, were observed with the Image Intensifier-SIT multichannel spectrograph at the 2.1 m telescope in San Pedro Mártir. The misidentified planetaries are Bl 3-11, He 2-417 and He 2-468. All of them show strong permitted H and He lines, detectable stellar continuum with absorption late-type bands and several other emission lines, including the unidentified λ 6830 Å feature. Clearly, the three objects are symbiotic stars, and not planetary nebulae. We present low-dispersion spectra for the three objects in the 3800-7000 Å spectral range.

OBJECT INDEX

This index lists the pages of references to individual planetary nebulae and other objects in the text of these Proceedings, or identified in the tables or diagrams therein. The index is arranged alphabetically according to the first letter occurring in each object name. Part A of the index lists Galactic (Milky Way) objects. The preferred names of Galactic PN are taken from the Catalogue of Perek and Kohoutek (1967), or from the supplementary lists of Kohoutek (Proc. IAU Symp. No. 76, 1978; these Proceedings, p. 17). The contribution on new and misclassified PN in this volume (Kohoutek, p. 17-30) is not included in the index. Part B of the index lists the page references to extragalactic PN or to galaxies which have been surveyed for PN. Part C lists the page references to Seyfert galaxies and related objects.

(A) GALACTIC

Name

Name

Name

Name

Name

Name

(B) EXTRAGALACTIC

Name

(C) SEYFERT GALAXIES AND RELATED OBJECTS

Name